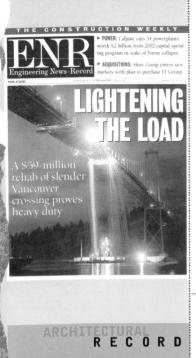

Student

Please enter my subscription for **Engineering News-Record**.

6 months ☐ $29.50 (Domestic)

Name

Address

City   State   Zip

☐ Payment enclosed   ☐ Bill me later

McGraw_Hill CONSTRUCTION ENR

5EN2DMHE

Friendly

Please enter my subscription for **Architectural Record**.

6 months ☐ $19.50 (Domestic)

Name

Address

City   State   Zip

☐ Payment enclosed   ☐ Bill me later

McGraw_Hill CONSTRUCTION Architectural Record

5AR2DMHE

Savings

Please enter my subscription for **Aviation Week & Space Technology**.

6 months ☐ $29.95 (Domestic)

Name

Address

City   State   Zip

☐ Payment enclosed   ☐ Bill me later

CAW34EDU

Construction Planning, Equipment, and Methods

Construction Planning, Equipment, and Methods

Sixth Edition

Robert L. Peurifoy, P.E.
Late Consulting Engineer
Austin, Texas

Clifford J. Schexnayder, P.E., Ph.D.
Eminent Scholar
Del E. Webb School of Construction
Arizona State University
Tempe, Arizona

Boston   Burr Ridge, IL   Dubuque, IA   Madison, WI   New York   San Francisco   St. Louis
Bangkok   Bogotá   Caracas   Kuala Lumpur   Lisbon   London   Madrid   Mexico City
Milan   Montreal   New Delhi   Santiago   Seoul   Singapore   Sydney   Taipei   Toronto

McGraw-Hill Higher Education

A Division of The McGraw-Hill Companies

CONSTRUCTION PLANNING, EQUIPMENT, AND METHODS
SIXTH EDITION

Published by McGraw-Hill, a business unit of The McGraw-Hill Companies, Inc., 1221 Avenue of the Americas, New York, NY 10020. Copyright © 2002, 1996, 1985, 1979, 1970, 1956 by The McGraw-Hill Companies, Inc. All rights reserved. No part of this publication may be reproduced or distributed in any form or by any means, or stored in a database or retrieval system, without the prior written consent of The McGraw-Hill Companies, Inc., including, but not limited to, in any network or other electronic storage or transmission, or broadcast for distance learning.

Some ancillaries, including electronic and print components, may not be available to customers outside the United States.

This book is printed on acid-free paper.

International   1 2 3 4 5 6 7 8 9 0 DOC/DOC 0 9 8 7 6 5 4 3 2 1
Domestic   4 5 6 7 8 9 0 DOC/DOC 0 9 8 7 6 5 4

ISBN 0-07-232176-8
ISBN 0-07-112257-5 (ISE)

General manager: *Thomas E. Casson*
Publisher: *Elizabeth A. Jones*
Executive editor: *Eric M. Munson*
Developmental editor: *Kate Scheinman*
Marketing manager: *Ann Caven*
Project manager: *Vicki Krug*
Production supervisor: *Sherry L. Kane*
Coordinator of freelance design: *Michelle D. Whitaker*
Freelance cover designer: *John Rokusek/Rokusek Design*
Cover image: *Boston Central Artery/Tunnel Project in front of South Station;*
photo by Clifford J. Schexnayder
Senior supplement producer: *David A. Welsh*
Media technology senior producer: *Phillip Meek*
Compositor: *Lachina Publishing Services*
Typeface: *10.5/12 Times Roman*
Printer: *R. R. Donnelley & Sons Company/Crawfordsville, IN*

Library of Congress Cataloging-in-Publication Data

Peurifoy, R. L. (Robert Leroy), 1902–
    Construction planning, equipment, and methods / Robert L. Peurifoy, Clifford J. Schexnayder.—6th ed.
        p.   cm.—(McGraw-Hill series in construction engineering and project management)
    Includes index.
    ISBN 0-07-232176-8
    1. Building. I. Schexnayder, Clifford J. II. Title. III. Series.

TH145 .P45   2002
624—dc21

2001030051
CIP

INTERNATIONAL EDITION ISBN 0-07-112257-5
Copyright © 2002. Exclusive rights by The McGraw-Hill Companies, Inc., for manufacture and export. This book cannot be re-exported from the country to which it is sold by McGraw-Hill. The International Edition is not available in North America.

www.mhhe.com

ABOUT THE AUTHORS

R. L. Peurifoy (1902–1995) was a retired Civil Engineering Professor and Consultant in the project management aspects of design and construction of engineered facilities. He taught construction at Texas A&M University and Oklahoma State University. In 1984 the Peurifoy Construction Research Award was instituted by the American Society of Civil Engineers upon recommendation of the Construction Research Council. This award was instituted to honor R. L. Peurifoy's exceptional leadership in construction education and research. The award recipients since the last edition of the book are:

1997	Don E. Hancher, University of Kentucky
1998	Keith C. Crandall, University of California Berkeley
1999	Paul M. Teicholz, Stanford University
2000	H. Randolph Thomas, Jr., Penn State University

Clifford J. Schexnayder holds the Eminent Scholar position at the Del E. Webb School of Construction, Arizona State University. He received his Ph.D. in Civil Engineering (Construction Management) from Purdue University, and a Masters and a Bachelors in Civil Engineering from Georgia Tech. A construction engineer with over 30 years of practical experience, Dr. Schexnayder has worked on major heavy/highway projects as a field engineer, estimator, and corporate Chief Engineer both for civilian contractors and as a member of the U.S. Army Corps of Engineers. After handling the asphalt paving of roads in northeast Thailand with the Army engineers, he joined a civilian contractor with responsibility for field layout of several major interstate highway projects. Later he moved to the home office handling field investigations and estimation of earthwork projects including strategic petroleum reserve sites, locks and dams, and interstate highway projects. As Chief Engineer he was responsible for a major earthwork containment project at the Savannah River Nuclear plant. He has taught construction equipment courses at Purdue, Louisiana Tech, Virginia Tech, U.S. Air Force Academy, Arizona State, and the U.S. Army Engineer School. Dr. Schexnayder is a registered professional engineer in eight states, as well as a member of the American Society of Civil Engineers. He served as chairman of the ASCE's Construction Division and on the task committee which formed the ASCE Construction Institute. He currently serves as chairman of the Transportation Research Board's Construction Section.

CONTENTS

CHAPTER **8**
Excavators   229

CHAPTER **9**
Finishing Equipment   270

PREFACE

"During the earlier years of construction, success frequently depended on one's ability to drive men, mules, and equipment to complete a project at the lowest possible cost."* The construction industry has moved a long way from mules pulling Fresno scrapers. The age of the laptop computer and the Internet is changing how we do business. With our computer systems, we now download data directly from machines. This abundance of data must, however, be turned into information that helps us better manage the machines. These technology improvements greatly enhance the constructor's ability to make equipment, planning, and construction method decisions.

This sixth edition follows in the tradition of the first five by providing the reader with fundamentals of machine selection and production estimating in a logical, simple, and concise format. With a grounding in these fundamentals, the constructor is prepared to evaluate those reams of computer-generated data and to develop programs that speed the decision process or that allow easy analysis of multiple options.

Significant changes have been made to this edition. The chapter on "Belt-Conveyor Systems" has been dropped, as it is a specialized subject. One new chapter has been added, "Finishing Equipment." Draglines and clamshells, which were covered in previous editions as part of "Excavation Equipment," are now a separate chapter. The "Excavator" chapter has been expanded to include material on specialty excavators. Each of the remaining chapters has undergone revision, ranging from simple clarification to major modifications, depending on the need to improve organization and presentation of concepts.

This book enjoys wide use as a practical reference by the profession and as a college textbook. I have followed the practice of including updated photographs from actual projects to illustrate equipment and methods. Additionally, the use of examples to reinforce the concepts through application has been continued. Based on professional practice, I have tried to present standard formats for analyzing production. Many companies use such formats to avoid errors when estimating production during the fast-paced efforts required for bid preparation.

To enhance the value of the book as a college textbook, I have updated and expanded the problems at the close of each chapter. The solutions to some problems are now included in the text at the end of the problem statements. Together with the examples, they facilitate learning and give students the confidence that they can master the subjects presented.

---

*R. L. Peurifoy, preface to the first edition 1956.

At the close of most chapters there are names, addresses, and in many cases the web addresses for manufactures of the construction equipment illustrated and described in the book. I am deeply grateful to the many individuals and firms who have supplied information and illustrations. Four individuals are owed a particular dept of gratitude for their support and efforts. Prof. John Zaniewski, Director, Harley O. Staggers National Transportation Center, West Virginia University, who drafted the revised chapter on "Asphalt Mix Production and Placement." Dr. Aviad Shapira, of the Technion-Israel Institute of Technology, Haifa, Israel, who drafted the revised section on Tower Cranes in the "Crane" chapter. Mr. Pat Gleuso of Neil F. Lampson, Inc., contributed many ideas and a critical review of the chapter on "Cranes" during preparation of the fifth edition. Mr. R. R. Walker of Tidewater Construction Corporation who, for the fifth edition, drafted the revised chapter on "Piles and Pile-Driving Equipment" and assembled many of the figures still included in that chapter. I would like to express my thanks for many useful comments and suggestions provided by the following reviewers:

L. Travis Chapin, *Bowling Green State University*

Larry G. Crowley, *Auburn University*

Jesus M. de la Garza, *Virginia Polytechnic Institute and State University*

F. H. (Bud) Griffis, *Brooklyn Institute of Technology*

Paul E. Harmon, *University of Nebraska–Lincoln*

Zohar Herbsman, *University of Florida*

C. William Ibbs, *University of California–Berkeley*

James Rowings, *Iowa State University*

Jeffrey Russell, *University of Wisconsin–Madison*

Richard Ryan, *University of Oklahoma*

Raymond F. Werkmeister, *University of Kentucky*

However, I take full responsibility for the material. Finally I wish to acknowledge the comments and suggestions for improvement received from persons using the book. We are all aware of how much our students help us to sharpen the subject presentation. Their questions and comments in the classroom have formed this revised book. For that and much more, I want to thank my students at Purdue, Louisiana Tech, Virginia Tech, Arizona State University, and the Air Force Academy who have over the years witnessed my classroom escapades explaining construction equipment and who have contributed so much helpful advice for clarifying the subject matter. Most importantly I thank my wife, Judy, who has typed chapters, proofread too many manuscripts, and was dragged to construction sites around the world. Without her support, this revision would not be a reality. I solicit comments on this edition.

Cliff Schexnayder
Chandler, Arizona

List of Abbreviations and Symbols

AAI	average annual investment
AASHTO	American Association of State Highway and Transportation Officials
ABC	Associated Builders and Contractors
AC and AR	asphalt grade designations
ACI	American Concrete Institute
ADT	articulated dump truck
AGC	Associated General Contractors of America
ANFO	an ammonium nitrate and fuel oil mixture
ANSI	American National Standards Institute
ASTM	American Society for Testing & Materials
AWPA	American Wood Preservers' Association
bcy	bank cubic yards
bhp	belt or brake horsepower
ccy	compacted cubic yards
CECE	Committee on European Construction Equipment
cf	cubic feet
cfm	cubic feet per minute
CII	Construction Industry Institute
cy	cubic yards
DOTs	departments of transportation
EBC	electric blasting cap
EVW	empty vehicle weight
FAR	Federal Acquisition Regulations
FHWA	Federal Highway Administration
FOB	free on board
FOG	fuel, oil, grease, and minor maintenance
fpm	feet per minute
fps	feet per second
ft	feet
ft-lb	foot-pound
fwhp	flywheel horsepower
fwhp-hr	flywheel horsepower hour

g/cc	grams per cubic centimeter
gph	gallons per hour
gpm	gallons per minute
GPS	global positioning system
GR	grade resistance
GVW	gross vehicle weight
hr	hours
ICAR	International Center for Aggregates Research
ID	inside diameter
IH	International Harvester
in. Hg	inchesof mercury
ISO	International Organization for Standardization
kip	1,000 1b
kPa	kilopascals
ksi	kips per square inch
LCD	liquid crystal device
lcy	loose cubic yard
LGP	low ground pressure
LL	liquid limit
MARR	minimum attractive rate of return
msec	millisecond
NAA	National Aggregates Association
NAHB	National Association of Home Builders
NIST	National Institute of Standards and Technology
NPW	net present worth
NRC	Nuclear Regulatory Commission
NSA	National Stone Association
NSPE	National Society of Professional Engineers
NVW	net vehicle weight
O&O	ownership and operation cost
OMC	optimum moisture content
OSHA	Occupational Safety and Health Act (Administration)
pcf	pounds per cubic foot
PCSA	Power Crane and Shovel Association
pen	penetration grade measurement unit
PETN	pentaerythritol tetranitrate
PI	plasticity index
PL	plastic limit
PPV	peak particle velocity

psf	pounds per square foot of pressure
psi	pounds per square inch of pressure
PWCAF	present worth compound factor
RAP	reclaimed asphalt pavement
RCC	roller-compacted concrete
ROPS	rollover protective structures
RR	rolling resistance
SAE	Society of Automotive Engineers
sec	second
SG	specific gravity
SPCAF	single compound amount factor
SR	stiffness ratio
sta.-yd	station-yards
sy	square yards
TMPH	ton-miles per hour
TNT	trinitrotoluene or trinitrotoluol
tph	tons per hour
TR	total resistance
TRB	Transportation Research Board
USCAF	uniform series, compound amount factor
USCRF	uniform series capital recovery factor
USPWF	uniform series present worth factor
USSFF	uniform series sinking fund factor
vpm	vibrations per minute
XL	extralong
yr	year

1

Introduction

Construction is the ultimate objective of a design, and machines make accomplishment of that objective possible. This book describes the fundamental concepts of machine utilization, which economically match machine capability to specific project construction requirements. Equipment is an economic investment and contractors must be able to apply the appropriate time value analytical formula to the decision process of machine purchase and utilization. The proof of how well the planner understands the work and coordinates the use of the company's equipment is in the bottom line when the contract is completed—at a profit or loss!

MACHINES MAKE IT POSSIBLE

The efforts of the engineer, who designs a project, and the constructor, who builds the project, are directed toward the same goal—creation of something that will improve the quality of life for mankind and serve the purpose for which it is built in a satisfactory manner. Construction is the ultimate objective of a design and machines make accomplishment of that objective possible. The most important aspect of estimating and building a construction project is determining production and cost, and controlling both during the progress of the work. This book introduces the constructor to the engineering fundamentals for planning, selection, and utilization of construction equipment. It enables one to analyze operational problems and to arrive at practical solutions for completing a task. It is about the application of engineering fundamentals and analysis to construction activities, and the economic comparison of machine choices.

THE CONSTRUCTION INDUSTRY

Over 1,145,000 businesses utilize heavy equipment while engaged in contract construction. The ability to win contracts and to perform them at a profit is determined for the construction contractor by two vital assets: people and equipment.

To be economically competitive, a contractor's equipment spread must be competitive, both mechanically and technologically. Old machines, which require costly repairs, cannot compete successfully with new equipment having lower repair costs and higher production rates.

In most cases, a piece of equipment does not work as a stand-alone unit. Pieces of equipment work in groups. An excavator loads trucks that haul material to a location on the project where it is required. At that point, the material is dumped and a dozer spreads the material and, after spreading, a roller compacts the material to the required density. Therefore a group of machines, in this example an excavator, haul trucks, a dozer, and a roller, constitute what is commonly referred to as an equipment spread. Because to perform the work the entire spread is required, a large equipment investment may be required even for small jobs. In some cases, the value of the equipment will be greater than the contract value. Therefore, not only must the contractor maintain a mechanically competitive equipment spread but individual projects must be planned and executed so that each machine complements the production capabilities of all the other pieces of equipment. The successful contractor is the one who accomplishes a continuing series of jobs and completes each individual job at the least possible cost.

Optimization in the management of an equipment spread is critical for a contractor, both in achieving a competitive pricing position and in accumulating the corporate operating capital required to finance the expansion of project performance capability. This book describes the basic operational characteristics of the major heavy construction equipment types. More important, however, it explains the fundamental concepts of machine utilization, which economically match machine capability to specific project construction requirements.

There are no unique solutions to the problem of selecting a machine to work on a particular construction project. All machine selection problems are influenced by external environmental conditions. To appreciate the conditions of the environment that influence the utilization of heavy construction equipment, one must understand the mechanics of how the construction industry operates.

EQUIPMENT-INTENSIVE OPERATIONS AND RISKS

By the nature of the product, the construction contractor works under a unique set of production conditions that directly affect equipment management. Whereas most manufacturing companies have a permanent factory where raw materials flow in and finished products flow out in a repetitive, assembly line process, a construction company carries its factory with it from job to job. At each new site, the company proceeds to set up and produce a one-of-a-kind project. If the construction work goes as planned, the job will be completed on time and with a profit.

A manufacturer spreads the cost of production mistakes across the volume of individual pieces produced and can adjust and "fine-tune" the production line manufacturing process when the product does not meet specifications. If the rate at which the product comes off the "production line" is less than expected, the manufacturer can easily analyze the impacting factors, within the closed environment of the plant. The typical construction contractor, because of the dollar magnitude and high resource requirements of a project, will have only a limited number of projects under contract during any one period of time. This limits the company's ability to spread the risk by volume and, thereby, protect itself from that one bad project that can send the company into bankruptcy.

Equipment-intensive projects are usually the very ones that present the greatest financial hazard. Earth and rock dam construction and canal work demand large concentrations of equipment. In addition, such work is usually bid on a unit-price basis and is subject to large variations between estimated and actual quantities. Highway work, the least profitable of all heavy construction activity, can require an equipment commitment that is greater than the amount that a contractor will be paid for completing a single project. Such a situation forces a contractor into a continuing sequence of jobs in order to support the long-term equipment payments. In such circumstances, a business condition of desperation can easily arise, pushing profit margins still lower in a business already fraught with risk. Since highway work is usually spread over several miles, its control and management can be very difficult.

Additional risk factors facing contractors in equipment-intensive work include financing structure, construction activity levels (the amount of work being put out for bid), labor legislation and agreements, and safety regulations. Project size and weather dependence both contribute to long project durations. Projects requiring two or more years to complete are not uncommon in the industry. As a means to ensure final completion of the entire project, owners typically retain 10% of the amount due a contractor for work complete. Although retainage is a receivable and shows as an asset on a contractor's company balance sheet, it cannot be utilized for operation and growth, and can cause serious cash flow problems for the firm. Many contractors consider cash flow to be the critical factor in any equipment decision.

Construction activity level is another risk factor. The level of construction activity effects the work volume of a construction company and work volume (backlog) seriously affects equipment decisions. Since a large portion of the construction activity in equipment-intensive contracting areas is attributable to government-sponsored projects, control of the funds for such work provides the bureaucracy with a direct method of regulating the economy. During slow or stagnant economic times, funds can be pumped into projects to stimulate the economy. In a like manner, funds are withheld and project starts delayed during boom and inflation periods. By being in the direct chain of action used by government to control *economic* cycles, a contractor is affected no matter which way the cycle is moving.

During such swings, a company with long-term equipment notes may be forced to seek work at severely reduced margins or in some cases with no margin for the short term. However, if such a bidding policy is pursued past the short term, it will quickly make equipment decisions moot, since the company will be broke!

The two other government-initiated actions that seriously affect the operating environment of the construction contractor are labor legislation and regulation and safety directives. In each of these areas, many regulations impact on a contractor's operations. These actions can directly influence equipment decisions. Legislative acts that exert direct pressure on equipment questions include the Davis-Bacon Act, which is concerned with wage rates, and the Occupational Safety and Health Act (OSHA), which specifies workplace safety requirements. Over one-half of the dollar volume of work in the equipment-intensive fields of construction is subject to wage determinations under the Davis-Bacon Act, and this strongly influences the labor costs incurred by contractors. OSHA, by its rollover protective structures (ROPS) mandate, substantially increased the cost of those pieces of construction equipment that had to have these structural elements included as part of the basic machine. That particular regulation had a single-point-in-time effect on equipment decisions, much like that resulting from the introduction of new equipment technology. Similarly, there remains the possibility of additional safety requirements. Sound and emissions are current areas of concern that would trigger other "step-jumps" in equipment cost if new regulations are imposed.

CONSTRUCTION CONTRACTS

Most heavy construction contractors work within a unique market situation. The job plans and specifications that are supplied by the client will dictate the sales conditions and product, but not the price. Almost all work in the heavy-equipment-intensive fields of construction is awarded on a bid basis, through either open or selective tender procedures. Under the design-bid-build method of contracting, the contractor states a price after estimating the cost based on a completed design supplied by the owner. The offered price includes overhead, project risk contingency, and the desired profit.

There is some movement to design-build contracts, where the contractor also has control of the project design. With a design-build project the contractor must state a guaranteed price before design is complete. This adds an additional element of risk, because estimating the quantities of materials required to complete the project becomes very subjective. But the advantage to the contractor is that the design can be matched in the most advantageous way to the contractor's construction skills. In either case, it is tacitly assumed that the winning contractor has been able to underbid the competition because of a more efficient work plan, lower overhead costs, or a willingness to accept a lower profit.

Not infrequently, however, the range between the high and low bids is much greater than these factors would justify. A primary cause of variance in bids is a contractor's inability to estimate costs accurately. Each construction project represents a custom-build situation that is subject to a new set of governing cost conditions, an atypical estimating situation. However, the largest portion of estimating variance is probably not caused by the differences between past and future projects but by a lack of accurate cost records. Most contractors have cost-reporting systems, but in numerous cases, the systems fail to allocate expenses to the proper sources, and therefore cause false conclusions when used as the historical database for estimating future work.

A construction company owner will frequently use both contract volume and contract turnover to measure the strength of the firm. Contract volume refers to the total dollar value of awarded contracts that a firm has on its books (under contract) *at any given time*. Contract turnover measures the dollar value of work that a firm completes during a specific *time interval*. Contract volume is a guide to the magnitude of resources a firm has committed at any one time, as well as to possible profit if the work is completed as estimated. But contract volume fails to answer any timing questions. A contractor who, with the same contract volume as the competition, is able to achieve a more rapid project completion, and therefore a higher capital turnover rate while maintaining the revenue-to-expense ratio, will be able to increase the firm's profits. Construction management's most effective procedure to maximize profit is to improve production and increase contract turnover. Contractors who finish work ahead of schedule usually make money.

THE RENT PAID FOR THE USE OF MONEY

What is commonly referred to as the time value of money is the difference—rent—that must be paid if one borrows some money for use today and returns the money at some future date. Many take this charge for granted, as the proliferation of credit cards testifies. This rent or added charge is termed *interest*. It is the profit and risk that the lender applies to the base amount of money that is borrowed. Interest, usually expressed as a percentage of the amount borrowed (owed), becomes due and payable at the close of each billing time period. It is typically stated as a yearly rate. As an example, if $1,000 is borrowed at 8% interest, then $0.08 \times \$1,000$, or $80, in interest plus the original $1,000 is owed after 1 year (yr). Therefore, the borrower would have to repay $1,080 at the end of a 1-yr time period. If this new total amount is not repaid at the end of the 1-yr period, the interest for the second year would be calculated based on the new total amount, $1,080, and thus the interest is *compounded*. Then after a 2-yr period, the amount owed would be $1,080 + (0.08 \times \$1,080)$, or $1,166.40. If the company's credit is good and it has borrowed the $1,000 from a bank, the banker normally does not care whether repayment is made after 1 yr at $1,080 or after 2 yr at $1,166.40. To the bank the three amounts, $1,000, $1,080, and $1,166.40, are equivalent. In other words, $1,000 today is equivalent to $1,080 1 yr in the future, which is also equivalent to $1,166.40 2 yr in the future. The three amounts are obviously not equal, they are *equivalent. Note that the concept of equivalence involves time and a specific rate of interest.* The three amounts are equivalent only for the case of an interest rate of 8%, and then only at the specified time intervals. Equivalence means that one sum or series differs from another only by the accrued, accumulated interest at rate i for n periods of time.

Note that in the example the principal amount was multiplied by an interest rate to obtain the amount of interest due. To generalize this concept, the following symbols are used:

P = a present single amount of money
F = a future single amount of money, after n periods of time
i = the rate of interest per period of time (usually 1 yr)
n = the number of time periods

Different situations involving an interest rate and time are presented next, and the appropriate analytical formulas are developed.

Equation for Single Payments

To calculate the future value F of a single payment P after n periods at an interest rate i, this calculation is used:

At the end of the first period: $F_1 = P + Pi$
At the end of the second period: $F_2 = P + Pi + (P + Pi)i = P(1 + i)^2$
At the end of the nth period: $F = P(1 + i)^n$

Or the future single amount of a present single amount is

$$F = P(1 + i)^n \qquad [1.1]$$

Note that F is related to P by a factor that depends only on i and n. This factor is termed the *single payment compound amount factor (SPCAF)*; it makes F *equivalent to P.*
    If a future amount F is given, the present amount P can be calculated by transposing the equation to

$$P = \frac{F}{(1 + i)^n} \qquad [1.2]$$

The factor $1/(1 + i)^n$ is known as the *present worth compound amount factor (PWCAF).*

EXAMPLE 1.1

A constructor wishes to set up a revolving line of credit at the bank to handle cash flow during a project. The constructor is required to borrow $12,000 to set up the account. The interest rate is 5% per year. If the borrowed amount and the interest are paid back after 3 yr, what will be the total amount of the repayment?
    To solve, use Eq. [1.1]

$$F = \$12{,}000(1 + 0.05)^3 = \$12{,}000(1.157625)$$

$$= \$13{,}891.50$$

The amount of interest is $1,891.50.

EXAMPLE 1.2

A constructor wants to set aside enough money today in an interest-bearing account to have $100,000 5 yr from now for the purchase of a replacement piece of equipment. If the company can receive 8% per year on its investment, how much should be set aside now to accrue the $100,000 5 yr from now?

To solve, use Eq. [1.2]

$$P = \frac{\$100,000}{(1 + 0.08)^5} = \frac{\$100,000}{(1.469328)}$$

$$= \$68,058.32$$

In Examples 1.1 and 1.2 the explanations of equivalent, single payments now and in the future were equated. Four parameters were involved: P, F, i, and n. *Given any three parameters, the fourth can easily be calculated.*

Formulas for a Uniform Series of Payments

Often payments or receipts occur at regular intervals, and such uniform values can be handled using additional functions. First, let us define another symbol:

A = uniform *end-of-period* payments or receipts

continuing for a duration of n periods.

If this uniform amount A is invested at the end of each period for n periods at a rate of interest i per period, then the total equivalent amount F at the end of the n periods will be

$$F = A[(1 + i)^{n-1} + (1 + i)^{n-2} + \cdots + (1 + i) + 1]$$

By multiplying both sides of the equation by $(1 + i)$ and subtracting the result from the original equation, we obtain

$$Fi = A(1 + i)^n - 1$$

which can be rearranged to

$$F = A\left[\frac{(1 + i)^n - 1}{i}\right] \tag{1.3}$$

The relationship $[(1 + i)^n - 1]/i$ is known as the *uniform series, compound amount factor (USCAF)*.

The relationship can be rearranged to yield

$$A = F\left[\frac{i}{(1 + i)^n - 1}\right] \tag{1.4}$$

The relationship $i/[(1 + i)^n - 1]$ is known as the *uniform series sinking fund factor (USSFF)* because it determines the uniform end-of-period investment A that must be made to provide an amount F at the end of n periods.

To determine the equivalent uniform period series required to replace a present value of P, simply substitute Eq. [1.1] for F into Eq. [1.4] and rearrange. The resulting equation is

$$P = A\left[\frac{(1 + i)^n - 1}{i(1 + i)^n}\right] \qquad [1.5]$$

This relationship is known as the *uniform series present worth factor (USPWF)*.

By inverting Eq. [1.5] the equivalent uniform series end-of-period value A can be obtained from a present value P. The equation is

$$A = P\left[\frac{i(1 + i)^n}{(1 + i)^n - 1}\right] \qquad [1.6]$$

This relationship is known as the *uniform series capital recovery factor (USCRF)*.

As an aid to understanding the six preceding equivalence relationships, appropriate cash flow diagrams can be drawn. *Cash flow diagrams* are drawings where the horizontal line represents time and the vertical arrows represent cash flows at specific times (up positive, down negative). The cash flow diagrams for each relationship are shown in Fig. 1.1. These relationships form the basis for

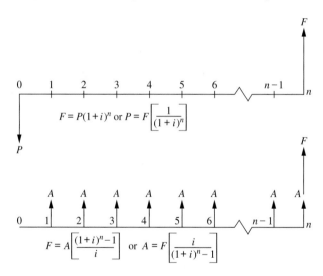

FIGURE 1.1 | Cash flow diagrams.

many complicated engineering economy studies involving the time value of money, and many texts specifically address this subject.

Most engineering economy problems are more complicated than the examples we have considered and must be broken down into parts. Example 1.3 illustrates how this is done and shows the use of the six equivalency relationships.

EXAMPLE 1.3

A machine cost $45,000 to purchase. Fuel, oil, grease, (FOG) and minor maintenance are estimated to cost $12.34 per operating hour. A set of tires cost $3,200 to replace and their estimated life is 2,800 use hours. A $6,000 major repair will probably be required after 4,200 hr of use. The machine is expected to last for 8,400 hr, after which it will be sold at a price (salvage value) equal to 10% of the original purchase price. A final set of new tires will not be purchased before the sale. How much should the owner of the machine charge per hour of use, if it is expected that the machine will operate 1,400 hr per year? The company's cost of capital rate is 7%.

First solve for n, the life

$$n = \frac{8,400 \text{ hr}}{1,400 \text{ hr per yr}} = 6 \text{ yr}$$

Usually ownership and operating cost components are calculated separately. Tires are considered an operating cost element because they wear out much faster than the basic machine. Therefore, before calculating the ownership cost of a machine having pneumatic tires, the cost of the tires should be subtracted from the purchase price.

$$\$45,000 - \$3,200 = \$41,800$$

Now the annualized purchase expense can be calculated using the *uniform series capital recovery factor*

$$A_{\text{ownership}} = \$41,800 \left[\frac{0.07(1 + 0.07)^6}{(1 + 0.07)^6 - 1} \right] = \$41,800 \times 0.209796 = \$8,769.46$$

$$\text{Salvage} = \$45,000 \times 0.1 = \$4,500$$

The annualized *value* of the salvage amount 6 yr in the future can be calculated using the *uniform series sinking fund factor*

$$A_{\text{salvage}} = \$4,500 \left[\frac{0.07}{(1 + 0.07)^6 - 1} \right] = \$4,500 \times 0.139796 = \$629.08$$

The annualized *value* of the salvage amount 6 yr in the future can also be calculated using the *present worth compound amount factor* with the *uniform series capital recovery factor*

$$A_{\text{salvage}} = \left[\frac{\$4,500}{(1 + 0.07)^6} \right] \left[\frac{0.07(1 + 0.07)^6}{(1 + 0.07)^6 - 1} \right] = \$629.08$$

The annual FOG and minor maintenance cost is

$$A_{FOG} = \$12.34 \text{ per hr} \times 1{,}400 \text{ hr per yr} = \$17{,}276.00*$$

In addition to the original set of tires, two sets of replacement tires will have to be purchased, one set 2 yr into the life of the machine and a second set 4 yr into the life. To annualize the tire replacement cost, these future-point-in-time costs must be made equivalent to a present amount at time zero, and then the resulting amount is annualized across the 6-yr life of the machine. To do this, use the *present worth compound amount factor* with the *uniform series capital recovery factor*

$$A_{tires} = \left[\$3{,}200 + \frac{\$3{,}200}{(1 + 0.07)^2} + \frac{\$3{,}200}{(1 + 0.07)^4} \right] \left[\frac{0.07(1 + 0.07)^6}{(1 + 0.07)^6 - 1} \right]$$

$$= [\$3{,}200 + \$2{,}795.00 + 2{,}441.26][0.209796] = \$1{,}769.89$$

The annualized cost for major repair 3 yr into the life is

$$A_{major\ repairs} = \left[\frac{\$6{,}000}{(1 + 0.07)^3} \right] \left[\frac{0.07(1 + 0.07)^6}{(1 + 0.07)^6 - 1} \right]$$

$$= [\$4{,}897.79][0.209796] = \$1{,}027.54$$

The resulting total annual cost is

$$A_{total} = \$8{,}769.46 - \$629.08 + \$17{,}276.00 + \$1{,}769.89 + \$1{,}027.54$$

$$= \$28{,}213.81$$

The total cost per hour is

$$\text{Total cost} = \frac{\$28{,}213.81 \text{ per yr}}{1{,}400 \text{ hr per yr}} = \$20.153 \text{ per hr}$$

EVALUATING INVESTMENT ALTERNATIVES

Often in engineering economic studies, as well as in general business, an analysis is made of each financial-investment alternative under consideration to determine how best to use a company's assets. All of these analysis methods employ time value of money principles and either require an input interest rate or calculate the interest rate of the assumed cash flows.

Cost of Capital

The interest rate a company experiences is really a weighted average rate resulting from the combined cost associated with all external and internal sources of capital funds—debt (borrowing), equity (sale of stock), and internally generated (retained earnings). The cost-of-capital interest rate a company experiences is also affected by the risk associated with its business type. The market perceives

---

*This neglects interest within a given year.

the risks of the business and applies an after-tax discount rate to the future wealth it expects to derive from the firm. Therefore, the rate that banks charge for borrowed funds cannot be taken alone as the company's cost-of-capital interest rate when making discounted present worth analyses. For a complete treatment of cost of capital, see Modigliani and Miller's classic paper published in *The American Economic Review*.

Discounted Present Worth Analysis

This type of analysis involves calculating the *equivalent* present worth or present value of all the dollar amounts involved in each of the individual alternates to determine the present worth of the proposed alternates. The present worth is discounted at a predetermined rate of interest *i*, often termed the minimum attractive rate of return (MARR). The MARR is usually equal to the current *cost-of-capital* rate for the company. Example 1.4 illustrates the use of a discounted present worth analysis to evaluate three mutually exclusive investment alternatives.

EXAMPLE 1.4

Ace Builders is considering three different methods for acquiring company pickup trucks. The alternatives are

A. Purchase the trucks for $16,800 each and sell after 4 yr for an estimated $5,000 each.

B. Lease the trucks for 4 yr for $4,100 per year paid in advance at the beginning of each year. The contractor pays all operating and maintenance costs for the trucks and the leasing company retains ownership.

C. Purchase the trucks on special time payments with $4,000 down now and $4,500 per year at the end of each year for 3 yr. Assume the trucks will be sold after 4 yr for $5,000 each.

If the contractor's MARR is 8%, which alternative should be used? To solve, calculate the net present worth (NPW) of each alternative at an 8% interest rate and select the least costly alternative.

For alternative A, use the present worth compound amount factor to calculate the equivalent salvage value at time zero. Add the purchase price, which is negative because it is a cash outflow, and the equivalent salvage value, which is positive because it is a cash inflow. The result is the net present worth of alternative A.

$$NPW_A = -\$16,800 + \frac{\$5,000}{PWCAF}$$

Calculating the PWCAF with *i* equal to 8% and *n* equal to 4.

$$NPW_A = -\$16,800 + \frac{\$5,000}{1.360489} = -\$13,124.85$$

For alternative B, use the uniform series present worth factor to calculate the time zero equivalent value of the future lease payments, and add to that result the value of the initial lease payment; both of these are negative because they are cash outflows. There is no salvage in this case as the leasing company retains ownership of the trucks.

$$NPW_B = -\$4{,}100 - \$4{,}100(USPWF)$$

Calculating the USPWF with i equal to 8% and n equal to 4.

$$NPW_B = -\$4{,}100 - \$4{,}100\left[\frac{0.259712}{0.100777}\right] = -\$14{,}666.10$$

For alternative C, use the uniform series present worth factor to calculate the time zero equivalent value of the future payments and the present worth compound amount factor to calculate the equivalent salvage value at time zero. Add the three values, the initial payment and the equivalent annual payment both being negative and the salvage being positive, to arrive at the net present worth of alternate C.

$$NPW_C = -\$4{,}000 - \$4{,}500(USPWF) + \left[\frac{\$5{,}000}{PWCAF}\right]$$

$$NPW_C = -\$4{,}000 - \$4{,}500\left[\frac{0.259712}{0.100777}\right] + \left[\frac{\$5{,}000}{1.360489}\right] = -\$11{,}921.79$$

The least costly alternative is C.

Example 1.4 was simplified in two respects. First, the number of calculations required was quite small. Second, all three alternatives involved the same lives (4 yr in the example). Problems involving more data and calculations are no different in approach from Example 1.4. But when alternatives involve different lives, the analysis becomes more difficult. Obviously, if a comparison is made of one alternative having a life of 5 yr and another with a life of 10 yr, the respective discounted present worths are not directly comparable. How is such a situation handled? Two approaches are generally used.

Approach 1. Truncate (cut off) the longer-lived alternative(s) to equal the shorter-lived alternative and assume a salvage value for the unused portion of the longer-lived alternative(s). Then make the comparison on the basis of equal lives.

Approach 2. Compute the discounted present worth on the basis of the least common denominator of the different alternatives' lives.

Example 1.5 illustrates these two approaches.

EXAMPLE 1.5

A contractor is considering the purchase of a new tractor that has an expected useful life of 6 yr. This tractor costs $73,570 and should generate a net annual income of $26,000. Its likely salvage value is $8,000. Another option is to purchase a used tractor having an estimated life of 3 yr for $24,680. The used tractor would have no salvage value at the end of its useful life. The estimated net annual income for the used tractor is $12,000. If the contractor's MARR is 12%, which tractor, if either, should be chosen?

To use approach 1, a suitable salvage value must be assumed for the new tractor after 3 yr. For this example, assume a 3-yr salvage value for the new tractor of $20,000. Therefore with i equal to 12% and n equal to 3, the discounted present worth of each alternative is

$$NPW_{new} = -\$73,570 + \$26,000(USPWF) + \$20,000(PWCAF)$$

$$NPW_{new} = -\$73,570.00 + \$62,447.61 + \$14,235.60 = \$3,113.22$$

$$NPW_{used} = -\$24,680 + \$12,000(USPWF)$$

$$NPW_{used} = -\$24,680 + \$28,821.98 = \$4,141.98$$

Using approach 1, purchasing the used tractor is the better alternative. The discounted present worths of each of the alternatives using approach 2, with i equal to 12% and n now equal to 6, are

$$NPW_{new} = -\$73,570 + \$26,000(USPWF) + \$8,000(PWCAF)$$

$$NPW_{new} = -\$73,570.00 + \$106,896.59 + \$4,053.05 = \$37,379.64$$

To use a 6-yr life it is necessary to purchase a second used tractor at the end of 3 yr. The PWCAF used in this approach to account for purchasing a second tractor will be with i equal to 12% and n equal to 3. With i equal to 12% and n equal to 6, the USPWF is

$$NPW_{used} = -\$24,680 - \$24,680(PWCAF) + \$12,000(USPWF)$$

$$NPW_{used} = -\$24,680 + \$17,566.74 + \$49,336.89 = \$7,090.15$$

Using approach 2, purchasing the new tractor is the better alternative. Neither of the approaches is entirely satisfactory, and they can yield different solutions (as shown here). Which solution is the best for the company? The answer is "whichever one best fits the situation." Assuming that a high (or low) salvage value can be received by selling a piece of equipment before the end of its useful life may be very erroneous. On the other hand, assuming an equal replacement cost for a used piece of equipment 3 yr in the future may be equally erroneous. To complicate matters further, it is often appropriate to assume inflation (or deflation) on replacement of the shorter-lived alternatives and use different values for the replacement conditions. But whichever approach is taken, assumptions must be made and the analysis results will be only as good as the assumptions. Therefore, always review the analysis to ensure that the assumptions and method are *reasonable* for the specific situation. It is a good idea to try both approaches as the difference in results provides a feeling for the validity associated with the assumptions used.

Rate of Return Analysis

The *rate of return* of a proposed investment is that interest rate that makes the discounted present worth of the investment equal to zero. To calculate the rate of return, simply set up the equation and solve for i. Example 1.6 illustrates the procedure.

EXAMPLE 1.6

Here we have the same situation as in Example 1.5. A contractor is considering the purchase of a new tractor, that has an expected useful life of 6 yr. This tractor costs $73,570 and should generate a net annual income of $26,000. Its likely salvage value is $8,000. Another option is to purchase a used tractor having an estimated life of 3 yr for $24,680. The used tractor would have no salvage value at the end of its useful life. The estimated net annual income for the used tractor is $12,000. Which tractor, if either, should be chosen?

In the case of purchasing a new tractor with a 6-yr life and an $8,000 salvage value at the end of its life, the equation to solve for the rate of return is

$$NPW_{new} = -\$73,570 + \$26,000(USPWF) + \$8,000(PWCAF) = 0$$

$$-\$73,570 + \$26,000\left[\frac{(1 + i)^6 - 1}{i(1 + i)^6}\right] + \$8,000\left[\frac{1}{(1 + i)^6}\right] = 0$$

Using a spreadsheet program, a multiple trial solution can easily be found. The equation equals zero for an interest rate of 28.00%.

In the case of purchasing a used tractor with a 3-yr life and no salvage value, the equation to solve for the rate of return is

$$NPW_{used} = -\$24,680 + \$12,000(USPWF) = 0$$

$$-\$24,680 + \$12,000\left[\frac{(1 + i)^3 - 1}{i(1 + i)^3}\right] = 0$$

$$2.05667 = \frac{(1 + i)^3 - 1}{i(1 + i)^3}$$

Again using a spreadsheet program, a multiple-trial solution can be found easily. The rate-of-return interest rate for choosing the used tractor alternative is 21.54%.

The use of this approach suffers from the same problem as the discounted present worth analysis in that it ignores the difference in duration (life) of the two alternatives. In this case, while the rates of return for each alternative are correct, what happens during the *second* 3-yr period using an old tractor? If equal replacement conditions are assumed, then the new tractor yields a higher rate of return, 28.00% > 21.54%. Always remember that the predicted result is no better than the validity of the assumptions used in the analysis, and the further into the future one tries to predict, the more critical the impact of the assumption. At the same time, it is very difficult to make accurate projections of events several years in the future.

One final fact that must be considered is the magnitude of the initial investment. In this example, the initial investment required for the two alternatives is $73,570 or $24,680. While the return on the higher initial investment is greater, the company must always consider its cash flow position. Can it handle a $73,570 investment and still have the resources to pay the general operating expenses?

ESTIMATING AND PLANNING EQUIPMENT UTILIZATION

Each piece of construction equipment is specifically designed by the manufacturer to perform certain mechanical operations. The task of the project planner/estimator or the engineer on the job is to match the right machine or combination of machines to the job at hand. The proof of how well the planner understands the work and coordinates the use of the company's equipment is in the bottom line when the contract is completed—at a profit or loss!

Considering individual tasks, the quality of performance is measured by matching the equipment spread's production against its cost. Production is work done; it can be the volume or weight of material moved, the number of pieces of material cut, the distance traveled, or any similar measurement of progress. To estimate the equipment component of project cost it is necessary to first determine machine *productivity*. Productivity is governed by engineering fundamentals and management ability. Chapter 5 covers the principal engineering fundamentals that control machine productivity. Each level of productivity has a corresponding cost associated with the effort expended. The expenses that a firm experiences through machine ownership and use, and the method of analyzing such costs, are presented in Chapter 3.

Although each major type of equipment has different operational characteristics, it is not always obvious which machine is best for a particular project task. After studying the plans and specifications, visiting the project site, and performing a quantity takeoff, the planner must visualize how to best employ specific pieces of equipment to accomplish the work. Is it less expensive to make an excavation with scrapers or to top-load trucks with a dragline? Both methods will yield the required end result, but which is the most economical method of attack for the given project conditions?

To answer that question the planner develops an initial plan for employment of the scrapers and then calculates their production rate and the resulting cost. The same process is followed for the top-load operation. The type of equipment that has the lowest estimated total cost, including mobilization of the machines to the site, is selected for the job.

To perform such analyses, the planner must consider both the construction equipment and the methods of employment in relation to one another. In developing suitable equipment employment techniques, the planner must have knowledge of the material quantities involved. This book will not cover quantity takeoff per se, but that process is strongly influenced by the equipment and methods under consideration. If it is determined that different equipment and methods will be used as an excavation progresses, then it is necessary to divide the quantity take-off in a manner that is compatible with the proposed equipment utilization.

The person performing the quantity take-off must calculate the quantities so that groups of similar materials (sand, common earth, rock) are easily accessed. It is not just a question of estimating the total quantity of rock and the total quantity of soil to be excavated. All factors that affect equipment performance and choice of method, such as location of the water table, clay or sand seams, site dimensions, depth of excavations, and compaction requirements, must be considered in making the quantity take-off.

The normal operating modes of the particular machines are discussed in Chapters 4, 6 to 12, and 14 to 20 on equipment types. That presentation, though, should not blind the reader to other possible applications. The most successful construction companies are those that carefully study all possible approaches to the construction process of each individual project. These companies use project preplanning and risk identification and quantification techniques in approaching their work. No two projects are exactly alike; therefore, it is important that the planner begins each new project with a completely open mind and reviews all possible options. Additionally, machines are constantly being improved and new equipment is continually being introduced. In 1920 a four-horse team and a Fresno scraper could move an average of 4 cy (cubic yards) of earth per hour on a 400-foot (ft) haul. By 1939, crawler tractors pulling rubber-tired scrapers were moving 115 cy of earth per hour the same distance. Today bulldozers would be considered for such a short haul and, depending on their size, they can push 200 to 700 cy of earth per hour easily.

Heavy equipment is usually classified or identified by one of two methods: *functional* identification or *operational* identification. A bulldozer, used to push a stockpile of material, could be identified as a support machine for an aggregate production plant, a grouping that could also include front-end loaders. The bulldozer could, however, be *functionally* classified as an excavator. In this book, a combination of functional and operational groupings are used. The basic purpose is to explain to the reader the critical performance characteristics of a particular piece of equipment and then to describe the most common applications of that machine.

The efforts of contractors and equipment manufacturers, daring to develop new ideas, constantly pushes machine capabilities forward. As the array of useful equipment expands, the importance of careful planning and execution of construction operations also increases. New machines enable greater economies. It is the job of the estimator and the field personnel to match equipment to project situations, and that is the central focus of this book.

SUMMARY

This chapter presented a condensed overview of construction work and the risk associated with bidding work. Machine production, the amount of earth moved or concrete placed, is only one component in selecting a machine for a particular work task. It is also necessary to know the cost associated with that production. The analytical formulas for analyzing machine cost have been developed and several investment evaluation techniques presented. Critical learning objectives include:

■ An ability to appropriately apply the appropriate interest rate formula.

■ An ability to evaluate investment alternatives.

These objectives are the basis for the problems.

PROBLEMS

1.1 Solve these problems with an interest rate equal to 7% compounded annually.

    a. If $20,000 is borrowed for 5 yr, what total amount must be paid back?

    b. How much of the total amount repaid represents interest?

    (a. $28,051.03; b. $8,051.03)

1.2 A company's interest rate for acquiring outside capital is 5.5% compounded annually. If $40,000 must be borrowed for 4 yr, what is the total amount of interest that will be accrued?

1.3 A contractor is saving to purchase a $200,000 machine. How much will the company have to bank today if the interest rate is 10% and they would like to purchase the machine in 4 yr? ($136.602.69)

1.4 What amount will a company have to place in savings today to purchase a $300,000 machine 5 yr in the future? The expected interest rate is 7%.

1.5 A track dozer costs $163,000 to purchase. Fuel, oil, grease, and minor maintenance are estimated to cost $32.14 per operating hour. A major engine repair costing $12,000 will probably be required after 7,200 hr of use. The expected resale price (salvage value) is 21% of the original purchase price. The machine should last 10,800 hr. How much should the owner of the machine charge per hour of use, if it is expected that the machine will operate 1,800 hr per year? The company's cost of capital rate is 7.3%. (Total annual cost $89,533.76; hourly cost $49.741)

1.6 A machine costs $245,000 to purchase. Fuel, oil, grease, and minor maintenance are estimated to cost $47.64 per operating hour. A set of tires costs $13,700 to replace, and their estimated life is 3,100 use hours. A $15,000 major repair will probably be required after 6,200 hr of use. The machine is expected to last for 9,300 hr, after which it will be sold at a price (salvage value) equal to 17% of the original purchase price. A final set of new tires will not be purchased before the sale. How much should the owner of the machine charge per hour of use, if it is expected that the machine will operate 3,100 hr per year? The company's cost of capital rate is 8%.

1.7 A contractor is considering these three alternatives:

    a. Purchase a new microcomputer system for $15,000. The system is expected to last 6 yr with a salvage value of $1,000.

    b. Lease a new microcomputer system for $3,000 per year, payable in advance. The system should last 6 yr.

    c. Purchase a used microcomputer system for $8,200. It is expected to last 3 yr with no salvage value.

Use a common-multiple-of-lives approach. If a MARR of 8% is used, which alternative should be selected using a discounted present worth analysis? If the MARR is 12%, which alternate should be selected? (MARR = 8%: purchase $14,370; lease $14,978; used $14,709)

1.8 A contractor is considering three capital investments. Each investment is considered to have zero salvage value at the end of its life. Calculate the rate of return for each investment.

Investment	Initial capital required	Estimated annual receipts	Estimated annual disbursements	Estimated life (yr)
Crane	$300,000	$130,000	$62,000	12
Scraper	$140,000	$175,000	$126,000	6
Loader	$120,000	$56,000	$25,000	8

(Crane 21.17%; scraper 26.43%; loader 19.71%)

1.9. A contractor is considering the purchase of a fleet of trucks. Three different models have been proposed. It is assumed that each will have zero salvage value at the end of its life. Calculate the rate of return for each investment.

Investment	Initial capital required	Estimated annual receipts	Estimated annual disbursements	Estimated life (yr)
Truck A	$120,000	$97,000	$48,000	4
Truck B	$150,000	$105,000	$60,000	6
Truck C	$187,000	$115,000	$65,450	10

REFERENCES

1. *Building for Tomorrow: Global Enterprise and the U.S. Construction Industry,* National Research Council, National Academy Press, Washington, DC, 1988.

2. Collier, Courtland A., and W. B. Ledbetter, *Engineering and Economic Cost Analysis,* 2d ed., Harper & Row, New York, 1988.

3. *Davis-Bacon Operations Manual,* 2nd ed., Associated Builders and Contractors, Inc., Washington, DC, 1977.

4. Johnson, Robert W., *Capital Budgeting,* Kendall/Hunt Publishing Co., Dubuque, IA, 1977.

5. Lewellen, Wilbur G., *The Cost of Capital,* Kendall/Hunt Publishing Co., Dubuque, IA, 1976.

6. Modigliani, Franco, and Merton H. Miller, "The Cost of Capital, Corporate Finance and the Theory of Investment," *The American Economic Review,* Vol. XLVIII, No. 3, June 1958.

CONSTRUCTION-RELATED WEBSITES

Significant additional information about the construction industry can be found posted on the following websites.

Associations and Organizations

Sites about construction associations and organizations include

1. www.asce.org/, American Society of Civil Engineers
2. www.asme.org/, American Society of Mechanical Engineers
3. www.abc.org/index.html, Associated Builders and Contractors (ABC). ABC is an association devoted to protecting the merit shop construction industry. Members range from emerging companies to multi-billion-dollar international firms, and include general contractors, specialty contractors in all areas of construction, industry material and equipment suppliers, and professionals.
4. www.agc.org/, Associated General Contractors of America (AGC). The AGC is an organization of construction contractors and industry-related companies dedicated to skill, integrity, and responsibility. Operating in partnership with its chapters, the association provides a full range of services satisfying the needs and concerns of its members, thereby improving the quality of construction and protecting the public interest.
5. construction-institute.org/, Construction Industry Institute (CII). The Construction Industry Institute is a research organization with a singular mission: improving the competitiveness of the construction industry. CII is a unique consortium of leading owners and contractors who have joined together to find better ways of planning and executing capital construction programs.
6. www.nspe.org/, National Society of Professional Engineers. The National Society of Professional Engineers (NSPE) represents individual engineering professionals and licensed engineers (PEs) across all disciplines.
7. www.aiaonline.com/, American Institute of Architects
8. www.asaonline.com/, American Subcontractors Association—the construction subcontractor's online portal to advocacy, leadership, networking, and education.
9. www.nahb.com/, National Association of Home Builders. The National Association of Home Builders (NAHB) is a federation of state and local builders associations throughout the United States. The mission of this Washington, DC-based trade association is to enhance the climate for housing and the building industry, and to promote policies that will keep housing a national priority.

Codes and Regulations

Sites that provide information about codes or regulations that impact the construction industry include

1. www.osha.gov/, U.S. Department of Occupational Labor Safety & Health Administration (OSHA)
2. www.nist.gov/welcome.html, National Institute of Standards and Technology (NIST)
3. www.ansi.org/, American National Standards Institute (ANSI)
4. www.iso.ch/, International Organization for Standardization (ISO)
5. www.astm.org/, American Society for Testing & Materials (ASTM)
6. www.arnet.gov/far/, Federal Acquisition Regulations (FAR)

2

Planning for Earthwork Construction

The pace, complexity, and cost of modern construction are incompatible with trial-and-error corrections as the work proceeds. When the engineer prepares a plan and cost estimate for an earthwork project, the critical attributes that must be determined are (1) the quantities involved, basically volume or weight; (2) the haul distances; and (3) the grades for all segments of the hauls. An earthwork volume sheet allows for the systematic recording of information and making the necessary earthwork calculations. The mass diagram is an analysis tool for selecting the appropriate equipment for excavating and hauling material.

INTRODUCTION

Every construction project is a unique undertaking. Although similar work may have been performed previously, no two projects will have identical job conditions. The pace, complexity, and cost of modern construction are incompatible with trial-and-error corrections as the work proceeds. The goal of planning is to minimize the resource expenditures required to successfully complete the project, and this can only be accomplished by understanding that planning is a continuing activity that begins with preparing the bid and continues until final project acceptance by the owner. Planning involves all of these:

1. Operational plan—what is to be done and in what sequence.
2. Scheduling—when each operation is to be performed.
3. Cost estimating—what each operation is expected to cost.
4. Resource planning—what resources (manpower, equipment, materials) will be required for each operation.

   With a plan established, control is introduced to the work, and control provides feedback information (actual progress, cost, utilization of equipment) that provides the basis for updating the plan.

The planning that a project manager performs has various constraints, including

1. The requirements set forth in the contract documents must be satisfied. These requirements include the details of the physical construction, as described in the drawings and technical specifications, and the required completion dates.
2. Legal requirements (OSHA, licensing, environmental control) must be satisfied.
3. Physical and/or environmental limits of the job may necessitate off-site fabrication and material storage, or sequencing of construction operations (traffic control).
4. Climatic conditions can dictate when paving or stabilization operations can be performed, require the closure of a dam prior to flood flows, or limit earthwork operations because of moisture content and the inability to dry the material.

Preaward Planning

Preaward planning takes place up to that point in time when the contractor commits to the work via a contract. The extent of preaward planning depends on such things as

1. The time available between when plans and specifications are available and when a price must be submitted, the bidding period.
2. The amount of resources (manpower, money) the company is willing to devote to planning a job that they may never get. A company is only the successful bidder on a limited number of projects. Successful bidding is dependent on the type of project, but typically the success rate is in the range of 5 to 25%.
3. The anticipated cost (risk) of not planning. This relates to the complexity of the project, the company's "risk philosophy," and the type of contract reimbursement (cost plus, lump sum, or unit price).
4. The extent of the competition. If there are a large number of bidders, the company cannot cover unknown risks by a large contingency.

Earthwork Planning

When the engineer prepares a plan and cost estimate for an earthwork project, the critical attributes that must be determined are (1) the quantities involved, basically volume or weight; (2) the haul distances; and (3) the grades for all segments of the hauls. The relevant input for the earthwork planning will include

1. Analyzing the project requirements as set forth in the drawings and specifications. Risks are associated with subsurface conditions, specifically those inherent in the types of materials encountered and their behavior during the construction processes. "You bid it—you build it" is

not quite correct. Contractors have a right to rely on owner-provided information. Additionally, many contracts will contain a *differing site condition* clause. Material differences in conditions are applicable in either of two cases. A Type 1 differing site condition exists when actual conditions differ materially from those "indicated in the contract." A Type 2 differing site condition occurs when actual conditions differ from reasonable expectations. These clauses provide the constructor some protection from geotechnical risk but do not eliminate the responsibility of performing a thorough examination of project conditions.

2. Separating project elements into work units: stripping, soil excavation, rock excavation, embankment, waste material, etc. The take-off effort must calculate not just the total quantity of material to be handled but must segregate the total quantity of material based on factors affecting productivity. Different portions of an excavation will often have to be hauled to different embankment locations, or the project may have to be constructed in phases. The usual case with earthwork is that precision is inversely proportional to the amount of work done in making the calculations.

3. Performing detailed studies of the work to be performed. This would include production analyses based on different types of equipment spreads.

4. Supplementing the analysis of project drawings and specifications with field investigations, geologic and soil studies, and meteorological data. The contract documents will usually include geotechnical data and information that was gathered during the design phase of the project. If not included directly in the documents, this material is typically available to bidders as supplemental information. The geotechnical data are gathered to support the design effort; interpretation for design and interpretation for construction are two very different things. The designer is interested in structural capacities and the constructor is interested in how the *material will handle* during the construction processes.

5. Locating material sources or sites to use in wasting excess or unsatisfactory materials.

This chapter is devoted to the first two elements in the list; the performance of production studies for individual machine types is covered in Chapters 4, 6 to 12, and 14 to 20. Field investigations and location of material sources are project-specific field exercises. Good fieldwork requires persistence and sometimes a little detective work and creative problem solving.

GRAPHICAL PRESENTATION OF EARTHWORK

Three kinds of views are presented in the contract documents to show earthwork construction features:

1. *Plan view.* The plan view is drawn looking down on the proposed work and presents the horizontal alignment of features (see Fig. 2.1).

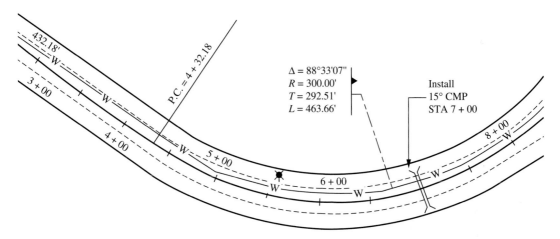

FIGURE 2.1 | Plan view of a highway project.

2. *Profile view.* The profile view is a cut view, typically along the centerline of the work. It presents the vertical alignment of features (see Fig. 2.2).

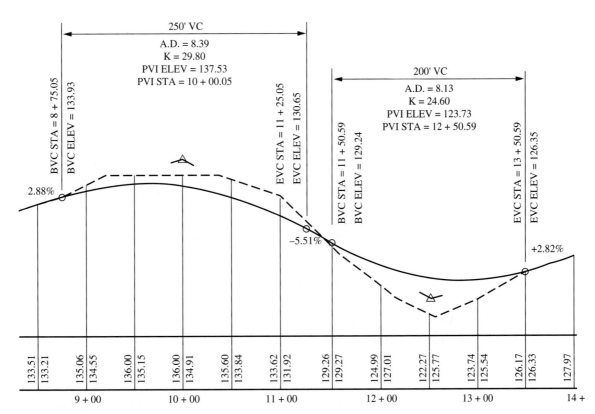

FIGURE 2.2 | Profile view of a highway project.

3. *Cross section view* (see Fig. 2.3). A view formed by a plane cutting the work at right angles to its long axis.

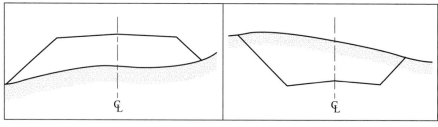

Cross section for a fill   Cross section for a cut

FIGURE 2.3 | Earthwork cross sections.

Cross Sections

In the case of a project that is linear in extent, material volumes are usually determined from cross sections. Cross sections are pictorial drawings produced from a combination of the designed project layout and measurements taken in the field at right angles to the project centerline or the centerline of a project feature, such as a drainage ditch. When the ground surface is regular, field measurements are typically taken at every full station (100 ft). When the ground is irregular, measurements must be taken at closer intervals and particularly at points of change. Typical cross sections are shown in Fig. 2.3.

The cross sections usually depict the finished subgrade elevations; this fact should always be verified. If instead the grades depicted are top of pavement, the thickness of the pavement section must be used to adjust the computations. From the cross section drawings, *end areas* can be computed using any of several methods.

EARTHWORK QUANTITIES

Earthwork computations involve calculating earthwork volumes, balancing cuts and fills, and planning the most economical material hauls. The first step in planning an earthmoving operation is estimating the quantities involved in the project. The exactness with which earthwork computations can be made is dependent on the extent and accuracy of the field measurements portrayed on the drawings.

End-Area Determination

The method chosen to compute a cross section end area will depend on the time available and the aids at hand. Most companies use commercial computer software and digitizing tablets (see the references at the end of the chapter) to arrive at cross section end areas. Other methods include the use of a planimeter, the subdivision of the area into geometric figures with definite formulas for areas (rectangles, triangles, parallelograms, and trapezoids), and the use of the trapezoidal formula.

Digitizing tablet A digitizing tablet is a board with a wire mesh grid embedded into it. When the cursor is traced over the board, a current is picked up by the board's grid, and the coordinates of the tracing device are passed on to the computer. When a plan is taped to the board and a scale is entered, the traced plan is converted to actual measurements, and then a software program computes the area, length, and volume of the traced data.

Planimeter A planimeter is a drafting instrument that is used to move a tracing point around the perimeter of the plotted area. It provides a value that is then multiplied by the square scale of the figure to calculate the figure's area. A planimeter can be used on any figure no matter how irregular the figure's shape might be.

Trapezoidal computations The mathematics of the computations is often based on breaking the drawing into small parts. The computer can very easily subdivide the drawing into a large number of strips, calculate the volume of each strip, and then sum the individual volumes to arrive at the volume of a section. If the calculations must be made by hand, the area formulas for a triangle and a trapezoid are used to compute the volume.

$$\text{Area of a triangle} = \frac{1}{2} hw \qquad \text{[2.1]}$$

where h = height of the triangle
$\quad w$ = base of the triangle

$$\text{Area of a trapezoid} = \frac{h_1 + h_2}{2} \times w \qquad \text{[2.2]}$$

where (see Fig. 2.4) $\quad w$ = distance between the two parallel sides
$\quad h_1$ and h_2 = the lengths of the two parallel sides

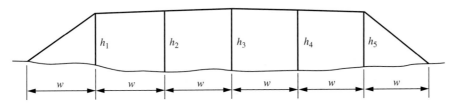

FIGURE 2.4 | Division of a cross section drawing into triangles and trapezoids.

The general trapezoidal formula for calculating area is

$$\text{Area} = \left(\frac{h_0}{2} + h_1 + h_2 + \cdots + h_{n-1} + \frac{h_n}{2} \right) \times w \qquad \text{[2.3]}$$

where (see Fig. 2.4) w = distance between the two parallel sides
$\quad h_0, \ldots, h_n$ = the lengths of the individual adjacent parallel sides

The precision achieved using this formula depends on the number of strips but is about ±0.5%.

In the case of side-hill construction, there can be both a cut area and a fill area in the same cross section. When making area computations, it is necessary to always calculate cut and fill areas separately.

Average End Area

The average-end-area method is most commonly used to determine the volume bounded by two cross sections or end areas. The principle is that the volume of the solid bounded by two parallel, or nearly parallel, cross sections is equal to the average of the two end areas times the distance between the cross sections along their centerline (see Fig. 2.5). The average-end-area formula is

$$\text{Volume [net cubic yards (cy)]} = \frac{A_1 + A_2}{2} \times \frac{L}{27} \qquad \textbf{[2.4]}$$

where (see Fig. 2.5)

   A_1 and A_2 = area in square feet (sf) of the respective end areas
   L = the length in feet between the end areas

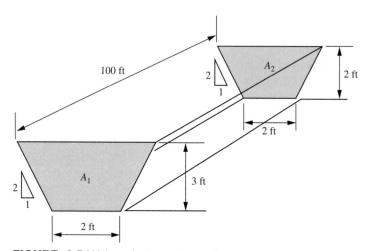

FIGURE 2.5 | Volume between two end areas.

EXAMPLE 2.1

Calculate the volume between two end areas 100 ft apart (see Fig. 2.5). End area 1 (A_1) equals 10.5 sf and end area 2 (A_2) equals 6 sf.

$$\text{Volume} = \frac{(10.5 \text{ sf} + 6 \text{ sf})}{2} \times \frac{100 \text{ ft}}{27 \text{ cf/cy}} \Rightarrow 30.6 \text{ cy}$$

where cf means cubic feet.

The principle is not altogether true because the average of the two end areas is not the arithmetic mean of many intermediate areas. The method gives volumes generally slightly in excess of the actual volumes. The precision is about $\pm 1.0\%$.

EXAMPLE 2.2

The volume between the end areas given in the table is calculated using the average-end-area method. The total volume of the section (the sum of the individual volumes) is also given.

Station	End area (sf)	Distance (ft)	Volume (cy)
150 + 00	360		
150 + 50	3,700	50	3,759
151 + 00	10,200	50	12,870
152 + 00	18,000	100	52,222
153 + 00	23,500	100	76,852
154 + 00	12,600	100	66,852
155 + 00	5,940	100	34,333
155 + 50	2,300	50	7,630
156 + 00	400	50	2,500
		Total volume	257,018

In this example, the interval between sections is 100 ft except between stations 150 + 00 and 151 + 00 and between stations 155 + 00 and 156 + 00, where the interval is 50 ft. The calculate total volume is 257,018 cy.

If the 50-ft intervals are omitted, i.e., if we assume sections were only taken on full stations, the calculations would be as shown in the table.

Station	End area (sf)	Distance (ft)	Volume (cy)
150 + 00	360		
151 + 00	10,200	100	19,556
152 + 00	18,000	100	52,222
153 + 00	23,500	100	76,852
154 + 00	12,600	100	66,852
155 + 00	5,940	100	34,333
156 + 00	400	100	741
		Total volume	250,556

The difference is 6,462 cy. This example emphasizes the importance of cross section spacing and the possible introduction of volume calculation error. The error in this case is 2.5% based on 257,018 cy of volume.

Although cross sections can be taken at any conservative interval along the centerline, judgment should be exercised, depending particularly on the irregularity of the ground and the tightness of curves. In the case of tight curves, a spacing of 25 ft is often appropriate.

Stripping

The upper layer of material encountered in an excavation is often topsoil (organic) resulting from decomposition of vegetative matter. This material is unsuitable for use in an embankment and must usually be handled in a separate excavation operation. It can be collected and wasted, or stockpiled for later use on the project to plate slopes. If the embankments are shallow, the organic material below the footprint of the fill sections must be stripped before embankment placement can commence (see Fig. 2.3). In the case of embankments over 5 ft in height, most specifications allow the organic material to remain if its thickness is only a few inches. When calculating the volume of cut sections, this stripping quantity must be subtracted from the net volume. In the case of fill sections, the quantity must be added to the calculated fill volume.

Net Volume

The computed volumes from the cross sections represent two different material states. The volumes from the fill cross sections represent compacted volume. If the volume is expressed in cubic yards, the notation is compacted cubic yards (ccy). In the case of cut sections, the volume is a natural in situ volume. The term bank volume is used to denote this in situ volume; if the volume is expressed in cubic yards the notation is bank cubic yards (bcy). If the cut and fill volumes are to be combined, they must be converted into compatible volumes. In Table 2.1 the conversion from compacted to bank cubic yards is made by dividing the compacted volume by 0.90. The theoretical basis for making soil volume conversions is explained in Chapter 4.

Earthwork Volume Sheet

An earthwork volume sheet, which can easily be constructed using a spreadsheet program, allows for the systematic recording of information and making the necessary earthwork calculations (see Table 2.1).

Stations.   Column 1 is a listing of all stations at which cross-sectional areas have been recorded. A station is a 100-ft interval. The term refers to the surveyor notation for laying out a project in the field.

Area of cut.   Column 2 is the cross-sectional area of the cut at each station. Usually this area must be computed from the project cross sections.

Area of fill.   Column 3 is the cross-sectional area of the fill at each station. Usually this area must be computed from the project cross sections. Note there can be both cut and fill at a station (see row 5, Table 2.1).

Volume of cut.   Column 4 is the volume of cut between the adjacent preceding station and the station. The average-end-area formula, Eq. [2.4], is usually used to calculate this volume. This is a *bank* volume.

Volume of fill.   Column 5 is the volume of fill between the adjacent preceding station and the station. The average-end-area formula, Eq. [2.4], is usually used to calculate this volume. This is a *compacted* volume.

TABLE 2.1 | Earthwork volume calculation sheet.

Station (1)	End-area cut (sf) (2)	End-area fill (sf) (3)	Volume of cut (bcy) (4)	Volume of fill (ccy) (5)	Stripping cut (bcy) (6)	Stripping fill (ccy) (7)	Total cut (bcy) (8)	Total fill (ccy) (9)	Adj. fill (bcy) (10)	Algebraic sum (bcy) (11)	Mass ordinate (12)
(1) 0 + 00	0	0									
(2) 0 + 50	0	115	0	106	0	18	0	124	138	−138	−138
(3) 1 + 00	0	112	0	210	0	30	0	240	267	−267	−405
(4) 2 + 00	0	54	0	307	0	44	0	351	390	−390	−796
(5) 2 + 50	64	30	59	78	26	22	59	100	111	−52	−847
(6) 3 + 00	120	0	170	28	76	0	144	28	31	114	−734
(7) 4 + 00	160	0	519	0	74	0	443	0	0	443	−291
(8) 5 + 00	317	0	883	0	60	0	809	0	0	809	518
(9) 6 + 00	51	0	681	0	21	0	621	0	0	621	1,140
(10) 6 + 50	46	6	90	6	0	0	69	6	6	63	1,202
(11) 7 + 00	0	125	43	121	0	25	43	146	163	−120	1,082
(12) 8 + 00	0	186	0	576	0	81	0	657	730	−730	352
(13) 8 + 50	0	332	0	480	0	69	0	549	610	−160	−257

Stripping volume in the cut.   Column 6 is the stripping volume of topsoil over the cut between the adjacent preceding station and the station. This volume is commonly calculated by multiplying the distance between stations or fractions of stations by the width of the cut. This provides the area of the cut footprint. The footprint area is then multiplied by an average depth of topsoil to derive the stripping volume. This represents a bank volume of cut material. Usually topsoil material is not suitable for use in the embankment. The average depth of topsoil must be determined by field investigation.

Stripping volume in the fill.   Column 7 is the stripping volume of topsoil under the fill between the adjacent preceding station and the station. This volume is commonly calculated by multiplying the distance between stations or fractions of stations by the width of the fill. This provides the area of the fill footprint. To derive the stripping volume, the area of the embankment footprint is multiplied by an average depth of topsoil. The stripping is a *bank* volume (bcy), but it also represents an additional requirement for fill material, ccy of fill.

Total volume of cut.   Column 8 is the volume of cut material available for use in embankment construction. It is derived by subtracting the cut stripping (col. 6) from the cut volume (col. 4).

Total volume of fill.   Column 9 is the total volume of fill required. It is derived by adding the fill stripping (col. 7) to the fill volume (col. 5).

Adjusted fill.   Column 10 is the total fill volume converted from compacted volume to bank volume.

Algebraic sum.   Column 11 is the difference between column 10 and column 8. This indicates the volume of material that is available (cut is positive) or required (fill is negative) within station increments after intrastation balancing.

Mass ordinate.   Column 12 is the running total of column 11 values from some point of beginning on the project profile. Then the stations being summed are excavation sections the value of this column will increase. Although summing a fill section will result in a decrease of the column 12 value, note that any material that could be used within a station length is not accounted for in the mass ordinate and therefore it is not accounted for in the mass diagram.

The mass diagram accounts only for material that must be transported beyond the limits of the two cross sections that define the volume of material. Where there is both cut and fill between a set of stations, only the excess of one over the other is used in computing the mass ordinate. Cut material between two successive stations is first used to satisfy fill requirements between those same two successive stations before there is a contribution to the mass ordinate value. Likewise if there is a greater fill requirement between two successive stations than there is cut available, the cut contribution is accounted for first. Only after all of the cut material is utilized will there be a fill contribution to the mass ordinate value. The material used between the two successive stations is considered to move at right angles to the centerline of the project and therefore is often termed *crosshaul.* The remaining material in either case represents a longitudinal haul along the length of the project.

MASS DIAGRAM

Earthmoving is basically an operation where material is removed from high spots and deposited in low spots with the "making up" of any deficit with borrow or the wasting of excess cut material. The mass diagram is an excellent method of analyzing linear earthmoving operations. It is a graphical means for measuring haul distance (stations) in terms of earthwork volume (cubic yards). A station yard (specifically a cubic yard) is a measure of work, the movement of 1 cy through a distance of one station.

On a mass diagram graph, the horizontal dimension represents the stations of a project (col. 1, Table 2.1) and the vertical dimension (col. 12, Table 2.1) represents the cumulative sum of excavation and embankment from some point of beginning on the project profile. The diagram provides information concerning

1. Quantities of materials
2. Average haul distances
3. Types of equipment that should be considered

When combined with a ground profile, the average slope of haul segments can be estimated. The mass diagram is one of the most effective tools for planning the movement of material on any project of linear extent.

Using column 1 (as the horizontal scale) and column 12 (as the vertical scale) of an earthwork volume calculation sheet, a mass diagram can be plotted. See the bottom portion of Fig. 2.6. Positive mass ordinate values are plotted

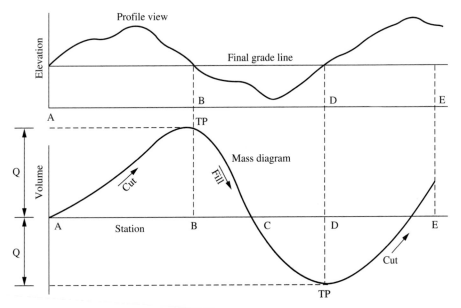

FIGURE 2.6 | Properties of a mass diagram.

above the zero datum line and negative values below. The top portion of Fig. 2.6 is the profile view of the same project.

Mass Diagram Properties

A mass diagram is a running total of the quantity of material that is surplus or deficient along the project profile. An excavation operation produces an *ascending* mass diagram curve; the excavation quantity exceeds the embankment quantity requirements. Excavation is occurring between stations A and B and between stations D and E in Fig. 2.6. The total volume of excavation between stations A and B is obtained by projecting horizontally to the vertical axis *the mass diagram line points* at stations A and B and reading the difference of the two volumes. Conversely, if the operation is a fill situation, there is a deficiency of material and a *descending* curve is produced; the embankment requirements exceed the excavation quantity being generated. Filling is occurring between stations B and D. The volume of fill can be calculated in a manner similar to the excavation calculation by a projection of the mass diagram line points to the vertical scale.

The *maximum* or *minimum points* on the mass diagram, where the curve transitions from rising to falling or from falling to rising, indicate a change from an excavation to a fill situation or vice versa. These points are referred to as *transition points.* On the ground profile, the project grade line is crossing the natural ground line (see Fig. 2.6 at stations B and D).

When the mass diagram curve crosses the datum (or zero volume) line (as at station C) exactly as much material is being excavated (between stations A and B) as is required for fill between B and C. There is no excess or deficit of material at point C in the project. The final position of the mass diagram curve above or below the datum line indicates whether the project has surplus material that must be wasted or if there is a deficiency that must be made up by borrowing material from outside the project limits. Figure 2.6 at station E indicates a waste situation, and excess material will have to be hauled off of the project.

USING THE MASS DIAGRAM

The mass diagram is an analysis tool for selecting the appropriate equipment for excavating and hauling material. The analysis is accomplished using balance lines and calculating average hauls.

Balance Lines

A balance line is a horizontal line of specific length that intersects the mass diagram in two places. The balance line can be constructed so that its length is the maximum haul distance for different types of equipment. The maximum haul distance is the limiting economical haul distance for a particular type of equipment (see Table 2.2).

TABLE 2.2 | Economical haul distances based on basic machine types.

Machine type	Economical haul distance
Large dozers, pushing material	Up to 300 ft*
Push-loaded scrapers	300 to 5,000 ft*
Trucks	Hauls greater than 5,000 ft

*The specific distance will depend on the size of the dozer or scraper.

Figure 2.7 shows a balance line drawn on a portion of a mass diagram. If this was constructed for a large push-loaded scraper, the distance between stations A and C would be 5,000 ft. Between the ends of the balance line, the cut volume generated equals the fill volume required. Between stations A and C, the amount of material the scrapers will haul is dimensioned on the vertical scale depicted by the vertical line Q. By examining either the profile view or the mass diagram, it is easy to determine the direction of haul that the cut must go to a fill location. Note the arrow on the profile view in Fig. 2.7.

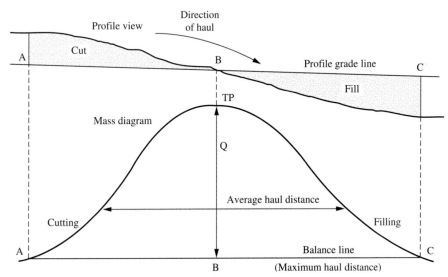

FIGURE 2.7 | Mass diagram with a balance line.

In accomplishing the balanced earthwork operation between stations A and C, some of the hauls will be short, while some will approach the maximum haul distance. The *average haul distance* is approximately the length of a horizontal line placed one-third of the distance from balance line in the direction of the high or low point of the curve. This is true in the case of a situation like that shown in Fig. 2.7 when the general shape of the mass diagram curve is a triangle. If the situation is like that shown in Fig. 2.8, where there are multiple balance lines and the area depicted is basically rectangular, the *average haul distance* is the length of a horizontal line placed midway between the balance lines. See the average haul for scrapers line shown in Fig. 2.8.

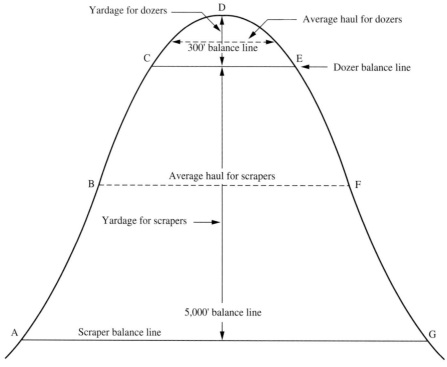

FIGURE 2.8 | Two balance lines on a mass diagram.

If the curve is above the balance line, the direction of haul is from left to right, i.e., up stationing. When the curve is below the balance line the haul is from right to left, i.e., down stationing.

Because the lengths of balance lines on a mass diagram are equal to the maximum or minimum haul distances for the balanced earthmoving operation, they should be drawn to conform to the capabilities of the particular equipment that will be utilized. The equipment will therefore operate at haul distances that are within its range of efficiency. Figure 2.8 illustrates a portion of a mass diagram on which two balance lines have been drawn. In this situation, it is planned that dozers will be used to push the short haul material. Using dozers, the excavation between stations C and D will be placed between stations D and E. Then there will be a scraper operation to excavate the material between stations A and C and haul it to fill between stations E and G.

Average Grade

When the mass diagram and the project profile are plotted one on top of the other, as shown in Figs. 2.6 and 2.9, the average haul grades of the earthmoving operations can be approximated. On the profile view, draw a horizontal line that roughly divides the cut area in half in the vertical dimension (see Fig. 2.9). Do the same for the fill area. This is a division of only that portion of the cut or fill area defined by the balance line in question. The difference in elevation between these two lines provides the vertical distance to use in calculating the average grade for the haul involving the material in the balance. The average haul distance, as determined by construction of a horizontal line on the mass diagram, is the denominator in the grade calculation.

$$\text{Average grade \%} = \frac{\text{Change in elevation}}{\text{Average haul distance}} \times 100 \qquad \textbf{[2.5]}$$

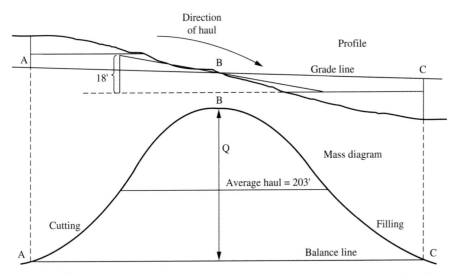

FIGURE 2.9 I Using the mass diagram and profile to determine average haul grades.

Calculate the average grade for the balance haul shown in Fig. 2.9.

$$\text{Average grade going from the cut to the fill} = \frac{-18 \text{ ft}}{203 \text{ ft}} \times 100 \Rightarrow -8.9\%$$

The return trip will be up an 8.9% grade.

Haul Distances

The mass diagram can be utilized to determine average haul distances. If the values in column 8 of Table 2.1 (volume of cut) from stations 0 + 00 to 8 + 50 are summed, the total is 2,188 bcy. This is the total volume of excavation within the project limits. If the positive values in column 11 (algebraic sum) are summed, the total is 2,049 bcy, which is the total volume of excavation that must be moved longitudinally. The difference between these two values is the crosshaul for the project, 139 bcy. Many contractors treat the crosshaul as dozer work.

Reviewing the values in column 12 of Table 2.1, we see there is a low point or valley of −847 bcy at station 2 + 50 and a high point or crest of 1,202 bcy at station 6 + 50. The sum of the absolute values of the peaks and low points equals the total excavation that must be moved longitudinally (847 + 1,202 = 2,049 bcy). The curve can have intermediate peaks and low points (see Table 2.3), and all must be accounted for when calculating the amount of material that must be moved longitudinally.

Reviewing the values in column 12 of Table 2.3, we see there is a low point or valley of −28,539 bcy at station 5 + 00, a high point or crest of −17,080 bcy at 8 + 00, and a second low point of −22,670 bcy at station 10 + 00. The sum of the absolute values of the peaks and low points equals the total excavation that must be moved longitudinally [(28,539 − 17,080) + (22,670 − 17,080) + (17,080 − 0) = 34,120 bcy]. Haul 1 involves 11,459 bcy, haul 2 is 5,590 bcy, and haul 3 is 17,080 bcy.

Graphically Determining Haul Distance   Some engineers and estimators use a graphic method to determine average haul distances. The *average haul distance line* for haul 1 is drawn below the balance line drawn through the peak at station 8 + 00. This balance line defines hauls 1, 2, and 3. It is on the vertical scale at the negative 17,080 bcy point. The low point for haul 1 is negative 28,539 bcy at station 5 + 00. Therefore, the amount of material involved is 11,459 bcy (28,539 – 17,080), as discussed previously.

An *average haul line* should be drawn through the vertical centroid of the area defined between the mass diagram curve and the balance line. In the case of haul 1, the shape of this area is roughly a triangle so the average haul line should be one-third of the vertical distance 11,459. This is based on the fact that the center of gravity of a triangle is one-third of its altitude above the base. Therefore, the average haul line for haul 1 is drawn horizontally at the −20,900 vertical scale as shown on Fig. 2.10. Scaling from the mass diagram graph (Fig. 2.10), average haul distance 1 is approximately 400 ft. The average haul line for haul 2 would be constructed in a similar manner. Haul 2 has an average haul of approximately 350 ft.

Haul 3 is slightly different because the enclosed area does not have the shape of a triangle. The centroid of this shape would be somewhere slightly closer to its base from the half-point of its height. But common practice is to use the midpoint to construct the graphical average haul line. Therefore the line is drawn at −8,540 bcy on the vertical scale and yields a distance of 12,200 ft.

TABLE 2.3 | Earthwork volume calculation sheet for the mass diagram in Fig. 2.10.

Station (1)	End-area cut (sf) (2)	End-area fill (sf) (3)	Volume of cut (bcy) (4)	Volume of fill (ccy) (5)	Stripping cut (bcy) (6)	Stripping fill (ccy) (7)	Total cut (bcy) (8)	Total fill (ccy) (9)	Adj. fill (bcy) (10)	Algebraic sum (bcy) (11) 0.90	Mass ordinate (12)
0 + 00	0	0		0							
1 + 00	0	1,700	0	3,148	0	120	0	3,268	3,631	−3,631	−3,631
2 + 00	0	3,100	0	8,889	0	120	0	9,009	10,010	−10,010	−13,641
3 + 00	0	1,500	0	8,519	0	120	0	8,639	9,598	−9,598	−23,240
4 + 00	60	600	111	3,889	60	80	51	3,969	4,410	−4,359	−27,598
5 + 00	400	200	852	1,481	80	60	772	1,541	1,713	−941	−28,539
6 + 00	1,300	30	3,148	426	110	10	3,038	436	484	2,554	−25,985
7 + 00	2,400	400	6,852	796	120	85	6,732	881	979	5,753	−20,233
8 + 00	800	850	5,926	2,315	90	100	5,836	2,415	2,683	3,153	−17,080
9 + 00	50	1,250	1,574	3,889	5	120	1,569	4,009	4,454	−2,885	−19,965
10 + 00	95	180	269	2,648	20	10	249	2,658	2,953	−2,705	−22,670
11 + 00	200	8	546	348	60	0	486	348	387	99	−22,571
12 + 00	560	0	1,407	15	65	0	1,342	15	16	1,326	−21,245
13 + 00	1,430	0	3,685	0	100	0	3,585	0	0	3,585	−17,660
14 + 00	3,580	0	9,278	0	120	0	9,158	0	0	9,158	−8,502
15 + 00	2,600	0	11,444	0	110	0	11,334	0	0	11,334	2,833

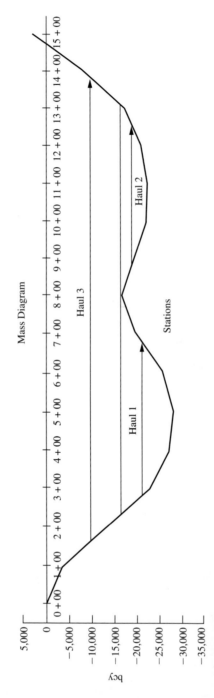

FIGURE 2.10 | Mass diagram plotted from the data in Table 2.3.

Calculating Haul Distance The average haul distance for any of the individual hauls can also be determined by calculation. Dividing the area (stations-cubic yards) that is enclosed by the balance line and the mass diagram curve by the amount of material hauled (cubic yards) will yield the average haul distance (stations).

The area (stations-cubic yards, usually referred to as station-yards) that is enclosed by the balance line, the mass diagram curve, and the zero datum line, which is haul 3, can be calculated using the trapezoidal formula, [Eq. 2.3]. The vertical at h_0 (station $0 + 00$) is 0, at h_1 is 3,631, at h_2 is 13,641, at h_3 through h_{13} is 17,080, at h_{14} is 8,502, and at h_{15} is 0. For stations $0 + 00$ and $15 + 00$, the beginning and ending stations in the equation (h_0 and h_n), the value is divided by 2. But 0 divided by 2 is still 0, so the area equals the sum of the verticals times the distance between verticals, which is 1 (one station). The sum of the verticals is 213,654, and multiplying by 1 produces an area of 213,645 station-yards (sta.-yd). There is a slight error in the computation because the mass diagram curve crosses the zero balance line somewhere between stations $14 + 00$ and $15 + 00$ and that last distance is not 100 ft. But there are no data to determine exactly where the curve and line cross, other than possibly by scaling the mass diagram.

The amount of material hauled is 17,080 bcy. Therefore, the calculated average haul for haul 3 is 12.5 stations or

$$12,500 \text{ ft} \left(\frac{213,654 \text{ station-bcy}}{17,080 \text{ bcy}} = 12.51 \text{ stations} \right).$$

Earlier the graphically determined average haul for haul 3 was 12,200 ft. This difference can be attributed to the fact that with the graphical method the *average haul line* was drawn at the vertical midpoint of the area. It is obvious from the shape of the area that the vertical centroid is slightly closer to the base (the zero datum line) of the area. Such positioning will increase the determined haul distance.

Consolidated Average Hauls

Using the individual average hauls and the quantity associated with each, a project average haul can be calculated. The calculation process is similar to that used to calculate an averaged haul for an individual haul. Consider the three hauls depicted in Fig. 2.10 and their graphically determined average haul distances. Haul 1 is 11,518 bcy with an average haul distance of 400 ft. Haul 2 is 4,590 bcy with an average haul of 350 ft, and haul 3 is 17,080 bcy with an average haul of 12,200 ft. By multiplying each haul quantity by its respective haul distance a station-yard value can be determined.

Haul 1	11,518 bcy	4.0 stations	46,072 sta.-yd
Haul 2	4,590 bcy	3.5 stations	16,065 sta.-yd
Haul 3	17,080 bcy	12.2 stations	208,376 sta.-yd
	33,188 bcy		270,513 sta.-yd

If the individual station-yard values are summed and that value divided by the total quantity moved, an average haul for the project is the result. In this case, the project average haul is 8.15 stations $\left(\dfrac{270{,}513 \text{ sta.-yd}}{33{,}188 \text{ yd}}\right)$.

If all the hauls are of about the same length, the estimator can consolidate the production calculations by using an averaging process, as just described. In the case of the data presented in Table 2.3 and Fig. 2.10, it is clear that there are two very distinct haul situations on this project. There are two short haul sections and one long haul section. These two situations will most likely each require a different number of haul units. The short haul situation will have shorter haul times and therefore will require fewer units to achieve continuous production. The long haul will require more units. Assuming the haul grades are about the same, the estimator will not calculate production for every individual part of the mass diagram. This is also driven by the practical situation of not mobilizing and demobilizing various numbers of machines for small differences in hauls.

Considering the data in Table 2.3, the estimator would most likely develop two production scenarios. The first scenario would be for the short hauls of hauls 1 and 2.

Haul 1	11,518 bcy	4.0 stations	46,072 sta.-yd
Haul 2	4,590 bcy	3.5 stations	16,065 sta.-yd
	16,108 bcy		62,137 sta.-yd
Haul 3	17,080 bcy	12.2 stations	

The average haul that results from combining these two is 3.9 stations. The second scenario would be for the long haul situation of haul 3.

There are other mass diagram applications that can be investigated. On some projects, it can prove economical to borrow material rather than undertake extremely long hauls. The same holds true for waste material. The mass diagram provides the engineer the ability to analyze such situations.

SUMMARY

The goal of planning is to minimize resource expenditures required to successfully complete the project. Three kinds of views are presented in the contract documents to show earthwork construction features: (1) plan view, (2) profile view, and (3) cross section view. Earthwork computations involve the calculation of earthwork volumes, the balancing of cuts and fills, and the planning of the most economical material hauls. Critical learning objectives that support earthwork planning include:

- Ability to calculate earthwork volume.
- Ability to adjust quantities for stripping requirements.
- Ability to construct an earthwork calculation sheet.
- Ability to construct and an understanding of the mass diagram.

These objectives are the basis for the problems that follow.

PROBLEMS

2.1   Using the average-end-area method calculate the cut and fill volumes for stations
125 + 00 through 131 + 00.

Station (1)	End-area cut (sf) (2)	End-area fill (sf) (3)	Volume of cut (bcy) (4)	Volume of fill (ccy) (5)
125 + 00	0	785	—	—
126 + 00	652	0		
127 + 00	2,150	0		
128 + 00	3,210	0		
129 + 00	1,255	147		
130 + 00	95	780		
131 + 00	0	3,666		

2.2   Using the average-end-area method calculate the cut and fill volumes for stations
19 + 00 through 24 + 00.

Station (1)	End-area cut (sf) (2)	End-area fill (sf) (3)	Volume of cut (bcy) (4)	Volume of fill (ccy) (5)
19 + 00	326	0	—	—
20 + 00	157	0		
21 + 00	44	0		
21 + 50	0	0		
22 + 00	0	147		
23 + 00	0	165		
24 + 00	0	133		

2.3   Using the average-end-area method calculate the cut and fill volumes for stations
25 + 00 through 31 + 00.

Station (1)	End-area cut (sf) (2)	End-area fill (sf) (3)	Volume of cut (bcy) (4)	Volume of fill (ccy) (5)
25 + 00	0	3,525	—	—
26 + 00	355	985		
27 + 00	786	125		
28 + 00	2,515	55		
29 + 00	1,255	23		
29 + 25	620	0		
29 + 50	25	845		
30 + 00	0	3,655		
31 + 00	0	8,560		

2.4 Complete the earthwork calculation sheet here and plot the resulting mass diagram. Divide ccy by 0.9 to convert to bcy (column 10).

Station (1)	End-area cut (sf) (2)	End-area fill (sf) (3)	Vol. of cut (bcy) (4)	Vol. of fill (ccy) (5)	Strip. cut (bcy) (6)	Strip. fill (ccy) (7)	Total cut (bcy) (8)	Total fill (ccy) (9)	Adj. fill (bcy) (10)	Algebraic sum (bcy) (11)	Mass ordinate (12)
									0.90		
0 + 00	0	0									
1 + 00	0	90			0	24					
2 + 00	0	154			0	40					
3 + 00	147	0			15	19					
4 + 00	192	0			50	0					
4 + 50	205	0			57	0					
5 + 00	179	0			21	0					
6 + 00	121	0			43	0					
6 + 50	100	0			19	0					
7 + 00	52	10			8	0					
8 + 00	0	180			10	20					
9 + 00	0	231			0	69					
10 + 00	0	285			0	18					

2.5 Complete the earthwork calculation sheet here and plot the resulting mass diagram. Divide ccy by 0.9 to convert to bcy (column 10). Calculate the average haul (trapezoidal formula) for the balances on this project. Is this a waste or borrow project?

Station (1)	End-area cut (sf) (2)	End-area fill (sf) (3)	Vol. of cut (bcy) (4)	Vol. of fill (ccy) (5)	Strip cut (bcy) (6)	Strip fill (ccy) (7)	Total cut (bcy) (8)	Total fill (ccy) (9)	Adj. fill (bcy) (10)	Algebraic sum (bcy) (11)	Mass ordinate (12)
									0.90		
10 + 00	0	0									
11 + 00	580	0			80	0					
12 + 00	2100	0			90	0					
13 + 00	4650	0			100	0					
14 + 00	6000	0			100	0					
15 + 00	3250	560			80	60					
16 + 00	1300	1620			80	80					
17 + 00	700	2450			80	85					
18 + 00	0	7800			0	100					
19 + 00	0	3620			0	90					
20 + 00	0	1980			0	80					
21 + 00	0	1310			0	80					
22 + 00	580	860			80	10					
23 + 00	1620	250			100	10					
24 + 00	3850	0			100	0					
25 + 00	2600	0			100	0					

REFERENCES

1. AGTEK, 368 Earhart Way, Livermore, CA 94550, www.agtek.com/.

2. Church, Horace K., *Excavation Handbook,* McGraw-Hill Book Co., New York, 1981.

3. InSite Software Incorporated, P.O. Box 290, 6029 E. Henrietta Rd, Rush, NY 14543, www.insitesoftware.com/.

4. Smith, Francis E., "Earthwork Volumes by Contour Method," *Journal of the Construction Division,* American Society of Civil Engineers, Vol. 102, CO1, March 1976.

5. Spectra Precision Software Inc., *Software,* 5901 Peachtree-Dunwoody Road NE, Suite 300, Atlanta, GA 30328, www.spectraprecision.com/.

6. TRAKWARE, www.trakware1.com/.

3

Equipment Cost

The only reason for purchasing equipment is to perform work that will generate a profit for the company. Data on both machine utilization and costs are the keys to making rational equipment decisions. Ownership cost is the cumulative result of those cash flows an owner experiences whether or not the machine is productively employed on a project. Operating cost is the sum of those expenses an owner experiences by working a machine on a project. The process of selecting a particular type of machine for use in constructing a project requires knowledge of the cost associated with operating the machine in the field. The economic life of equipment is of critical importance and there are three basic methods for securing a particular machine to use on a project: (1) buy, (2) rent, or (3) lease.

INTRODUCTION

Equipment cost is often one of a contractor's largest expense categories and it is a cost laced with variables and questions. To be successful, equipment owners must carefully analyze and answer two separate cost questions about their machines:

1. How much does it cost to operate the machine on a project?
2. What is the optimum economic life and the optimum manor to secure a machine?

The first question is critical to bidding and operations planning. The only reason for purchasing equipment is to perform work that will generate a profit for the company. This question seeks to identify the expense associated with productive machine work, and is commonly referred to as ownership and operating (O&O) cost. O&O cost is stated on an hourly basis (e.g., $90/hr for a dozer) because it is used in calculating the cost per unit of machine production. If a dozer can push 300 cy yd per hour and it has a $90/hr O&O cost, production cost is $0.300/cy ($90/hr ÷ 300 cy/hr). The estimator/planner can use the cost per cubic yard figure directly on unit price work. On a lump sum job, it will be necessary to multiply the cost/unit price by the estimated quantity to obtain the total amount that should be charged.

The second question seeks to identify the optimum point in time to replace a machine and the optimum way to secure a machine. This is important in that it will reduce O&O cost and thereby lower production expense, enabling a contractor to achieve a better pricing position. The process of answering this question is known as replacement analysis. A complete replacement analysis must also investigate the cost of renting or leasing a machine.

The economic analyses, which answer these two cost questions, require the input of many expense and operational factors. These input costs are discussed first and a development of the analysis procedures follows.

EQUIPMENT RECORDS

Data on both machine utilization and costs are the keys to making rational equipment decisions, but collection of individual pieces of data is only the first step. The data must be assembled and presented in usable formats. Many contractors recognize this need and strive to collect and maintain accurate equipment records for evaluating machine performance, establishing operating cost, analyzing replacement questions, and managing projects. Surveys of industry-wide practices, however, indicate that such efforts are not universal.

Realizing the advantages to be gained therefrom, owners are directing more attention to accurate record keeping. Advances in computer technology have reduced the effort required to implement record systems. Several computer companies offer record-keeping packages specifically designed for contractors. In many cases, the task is simply the retrieval of equipment cost data from existing accounting files. A warning about computer data systems is necessary. While they can turn out large stacks of printed reports, the "paper generating productivity" in itself does not guarantee that useful information is transmitted.

The structure of an equipment information system is critical, particularly if computers are used as processors. When voluminous amounts of data proliferate from a system but serve little useful purpose, the reports will only confuse management rather than provide a basis for improved practices.

Automation introduces the ability to handle more data economically and in shorter time frames, but the basic information required to make rational decisions is still the critical item. A commonly used technique in equipment costing and record keeping is the standard rate approach. Under such a system, jobs are charged a standard machine utilization rate for every hour the equipment is employed. Machine expenses are charged either directly to the piece of equipment or to separate equipment cost accounts. This method is sometimes referred to as an internal or company rental system. Such a system usually presents a fairly accurate representation of investment consumption and it properly assigns machines expenses. In the case of a company replacing machines each year and continuing in operation, this system enables a check at the end of each year on estimate rental rates as the internally generated rent should equal the expenses absorbed.

The first piece of information necessary for rational equipment analysis is not an expense but a record of the machine's use. One of the implicit assumptions of a replacement analysis is that there is a continuing need for a machine's production capability. Therefore, before beginning a replacement analysis, the disposal–replacement question must be resolved. Is this machine really necessary? A projection of the ratio between total equipment capacity and utilized capacity provides a quick guide for the dispose–replace question.

The level of detail for reporting equipment use varies. As a minimum, data should be collected on a daily basis to record whether a machine worked or was idle. A more sophisticated system will seek to identify use on an hourly basis, accounting for actual production time and categorizing idle time by classifications such as standby, down weather, and down repair. The input for either type of system is easily incorporated into regular personnel time-keeping reports, with machine time and operator time being reported together.

All owners keep records on a machine's initial purchase expense and final realized salvage value as part of the accounting data required for tax filings. Most of the information required for ownership and operating or replacement analyses is presented in the standard accounting records. Maintenance expenses can be tracked from mechanics' time sheets and parts purchase orders or from shop work orders. Service logs provide information concerning consumption of consumables. Fuel amounts can be recorded at fuel points or with automated systems. Fuel amounts should be cross-checked against the total amount purchased. When detailed and correct reporting procedures are maintained, the accuracy of equipment costs analyses is greatly enhanced.

COST OF CAPITAL

Many discussions of equipment economics include *interest* as a cost of ownership. Sometimes authors make comparisons with the interest rates that banks charge for borrowed funds or with the rate that could be earned if the funds were invested elsewhere. Such comparisons imply that these are appropriate rates to use in an equipment cost analysis. A few authors appear to have perceived the proper character of interest by realizing that a company requires capital funds for all of its operations. It is not logical to assign different interest costs to machines purchased wholly with retained earnings (cash) as opposed to those purchased with borrowed funds. A single interest rate should be determined by examination of the combined costs associated with all sources of capital funds: debt, equity, and internal.

There are two parts to the common misunderstanding concerning the proper way to account for interest. First, as just discussed, the correct interest rate should reflect the combined effect of the costs associated with all capital funds. The second error comes in trying to recoup *interest costs.* Interest is not a cost to be added together with purchase expense, taxes, insurance, etc. when calculating the total cost of a machine.

This might be easier to understand using an analogy. Consider the situation of a banker trying to decide if a loan should be granted. The question before the banker is one of *risk*: what are the chances the money will be repaid? Based on the perceived risk, a decision is made on how much return must be received to *balance* the risk. If a hundred loans are made, the banker knows some will not be repaid. The good loans have to provide the total profit margin for the bank. The good loans must carry the bad loans. A company utilizing equipment should be making a similar analysis every time a decision is made to invest in a piece of equipment.

Based on the risk of the construction business, interest is the fulcrum for determining if the value a machine will create for the company is sufficient. This is sometimes expressed as a minimum rate of return or yield. The proper interest rate will ensure that this ratio of gain in value to cost is correctly accounted for in the decision process.

The interest rate at issue is referred to in the economic literature as the *cost of capital* to the company and a market value technique has been developed for its calculation. A complete development of the market value cost-of-capital calculation is beyond the scope of this text but reference 3 at the end of this chapter is a good presentation of the subject. The resulting cost-of-capital rate is the correct interest rate to use in economic analyses of equipment decisions.

OWNERSHIP COST

GENERAL INFORMATION

Ownership cost is the cumulative result of those cash flows an owner experiences whether or not the machine is productively employed on a job. Most of these cash flows are expenses (outflows), but a few are cash inflows. The most significant cash flows affecting *ownership cost* are

1. Purchase expense.
2. Salvage value.
3. Tax saving from depreciation.
4. Major repairs and overhauls.
5. Property taxes.

6. Insurance.

7. Storage and miscellaneous.

PURCHASE EXPENSE

The cash outflow the firm experiences in acquiring ownership of a machine is the purchase expense. It is the total delivered cost (drive-away cost), including amounts for all options, shipping, and taxes, less the cost of tires if the machine has rubber tires. The machine will show as an asset on the books of the firm. The firm has exchanged money (dollars), a liquid asset, for a machine, a fixed asset with which the company hopes to generate profit. As the machine is used on projects, wear takes its toll and the machine can be thought of as being used up or consumed. This consumption reduces the machine's value because the revenue stream it can generate is likewise reduced. Normally, an owner tries to account for the decrease in value by prorating the consumption of the investment over the *service life* of the machine. This prorating is known as depreciation.

It can be argued that the amount that should be prorated is the difference between the initial acquisition expense and the expected future salvage value. Such a statement is correct to the extent of accounting for the amounts involved, but it neglects the timing of the cash flows. Therefore, it is recommended that each cash flow be treated separately to allow for a time value analysis and to allow for ease in changing assumptions during sensitivity analyses.

SALVAGE VALUE

Salvage value is the cash inflow a firm receives if a machine still has value at the time of its disposal. This revenue will occur at a future date.

Used equipment prices are difficult to predict. Machine condition, the movement of new machine prices, and the machine's possible secondary service applications affect the amount an owner can expect to receive. A machine having a diverse and layered service potential will command a higher resale value. Medium-size dozers, which often exhibit rising salvage values in later years, can have as many as seven different levels of useful life. These may range from an initial assignment as a high-production machine on a dirt spread to an infrequent land-clearing assignment by a farmer.

The effect of these relationships is illustrated by the common shapes of used equipment price curves (see Fig. 3.1). Those machines that simply wear out and have few secondary applications display a steady decline in value (see Fig. 3.1, graph a). Heavy, special-purpose machines, if maintained in good operating condition, will retain their value and produce a steady plot like Fig. 3.1b. Substantial increases in the price of new models will cause predecessor models to exhibit a convex value curve (see Fig. 3.1, graph c). Machines having multiple service life applications tend to have concave value curves (see Fig. 3.1, graph d). The tail of graph d will become very steep during periods of high inflation. Plots such as those presented in Fig. 3.1 can be helpful in making predictions about used machine value.

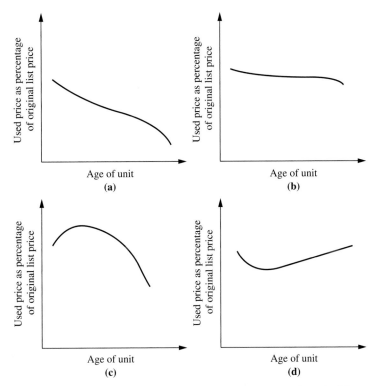

FIGURE 3.1 | Used equipment price as a percentage of original list price.

Historical resale price plots, such as those presented, provide some guidance in making salvage value predictions and can be fairly easily developed from information presented in auction price books. By studying such historical data and recognizing the effects of the economic environment, the magnitude of salvage value prediction errors can be minimized and the accuracy of an ownership cost analysis can be improved.

TAX SAVINGS FROM DEPRECIATION

The tax savings from depreciation are a phenomenon of the tax system in the United States. (This may not be an ownership cost factor under the tax laws in other countries.) Under the tax laws of the United States, depreciating a machine's loss in value with age will lessen the net cost of machine ownership. The cost saving, the prevention of a cash outflow, afforded by tax depreciation is a result of shielding the company from taxes. This is an applicable cash flow factor, however, only if a company is operating at a profit. There are carry-back features in the tax law so that the saving can be preserved even though there is a loss in any one particular year, but the long-term operating position of the company must be at a profit.

The rates at which a company can depreciate a machine are set by the revenue code. These rates usually have no relation to actual consumption of the asset (machine). Therefore, many companies keep several sets of depreciation numbers, one for depreciation tax purposes, one for corporate earnings tax accounting purposes, and one for internal and/or financial statement purposes. The first two are required by the revenue code. The last tries to accurately match the consumption of the asset based on work application and company maintenance policies.

Under the current tax laws, tax depreciation accounting no longer requires the assumption of a machine's future salvage value and useful life. The only piece of information necessary is basis. Basis refers to the cost of the machine for purposes of computing gain or loss. Basis is essential. To compute tax depreciation amounts, fixed percentages are applied to the unadjusted basis. The terminology adjusted/unadjusted refers to changing the book value of a machine by depreciation.

The tax law allows the postponement of taxation on gains derived from the exchange of like-kind depreciable property. If there is a gain realized from a like-kind exchange, the depreciation basis of the new machine is reduced by the amount of the gain. However, if the exchange involves a disposal sale to a third party and a separate acquisition of the replacement, the gain from the sale is taxed as ordinate income.

EXAMPLE 3.1

A tractor with an adjusted basis (from depreciation) of $25,000 is traded for a new tractor that has a fair market value of $400,000. A cash payment of $325,000 is made to complete the transaction. Such a transaction is a nontaxable exchange and no gain is recognized on the trade-in. The unadjusted basis of the new tractor is $350,000, even though the cash payment was $325,000 and the apparent gain in value for the traded machine was $50,000 (($400,000 − $325,000) − $25,000).

Cash payment	$325,000
Adjusted basis of the trade-in tractor	25,000
Basis of the new tractor	$350,000

If the owner had sold the old tractor to a third party for $75,000 and then purchased a new tractor for $400,000, the $50,000 profit on the third-party sale would have been taxed as ordinary income and the unadjusted basis of the new tractor would be $400,000.

The current tax depreciation law establishes depreciation percentages that can be applied each year of a machine's life. These are usually the optimum depreciation rates in terms of tax advantages. However, an owner can still utilize the straight-line method of depreciation or methods that are not expressed in terms of time duration (years). Unit-of-production would be an example of a nontime system.

Straight-Line Method of Tax Depreciation

Straight-line depreciation is easy to calculate. The annual amount of depreciation D_n, for any year n, is a constant value, and thus the book value BV_n decreases at a uniform rate over the useful life of the machine. The equations are:

$$\text{Depreciation rate, } R_n = \frac{1}{N} \qquad \text{[3.1]}$$

where N = number of years.

$$\text{Annual depreciation amount, } D_n = \text{Unadjusted basis} \times R_n$$

$$\Rightarrow \frac{\text{Unadjusted basis}}{N} \qquad \text{[3.2]}$$

$$\text{Book value year } m, BV_n = \text{Unadjusted basis} - (n \times D_n) \qquad \text{[3.3]}$$

EXAMPLE 3.2

Consider the new tractor in Example 3.1 and assume it has an estimated useful life of 5 yr. Determine the depreciation and the book value for each of the 5 yr using the straight-line method.

$$\text{Depreciation rate, } R_n = \frac{1}{5} \Rightarrow 0.2$$

$$\text{Annual depreciation amount, } D_n = \$350{,}000 \times 0.2 \Rightarrow \$70{,}000$$

The respective values are

m	BV_{n-1}	D_n	BV_n
0	$ 0	$ 0	$350,000
1	350,000	70,000	280,000
2	280,000	70,000	210,000
3	210,000	70,000	140,000
4	140,000	70,000	70,000
5	70,000	70,000	0

Tax Code Depreciation Schedules

Under the tax code, machines are classified as 3-, 5-, 10-, or 15-yr real property. Cars and light-duty trucks (under 13,000 lb unloaded) are classified as 3-yr property. Most other pieces of construction equipment are 5-yr property. The appropriate depreciation rates are given in Table 3.1.

TABLE 3.1 | Tax code specified depreciation rates.

Year of life	3-yr property	5-yr property
1	0.33	0.20
2	0.45	0.32
3	0.22	0.24
4	—	0.16
5	—	0.08

If a machine is disposed of before the depreciation process is completed no depreciation may be recovered in the year of disposal. Any gain, as measured against the depreciated value or adjusted basis, is treated as ordinary income.

EXAMPLE 3.3

A 5-yr life class machine is purchased for $125,000. It is sold in the third year after purchase for $91,000. What are the depreciation amounts and what is the book value of the machine when it is sold? Will there be income tax, if so on what amount?

$125,000 × 0.20 = $25,000   depreciation at end of the first year

$125,000 × 0.32 = $40,000   depreciation at end of second year

$65,000   total depreciation

Value when sold = $125,000 − $65,000 ⟹ $60,000

Amount of gain upon which there will be a tax = $91,000 − $60,000 ⟹ $31,000

The tax savings from depreciation are influenced by

1. Disposal method for the old machine.
2. Value received for the old machine.
3. Initial value of the replacement.
4. Class life.
5. Tax depreciation method.

Based on the relationships between these elements, three distinct situations are possible:

1. No gain on the disposal—no income tax on zero gain.
2. A gain on the disposal:
   a. Like-kind exchange—no added income tax, but basis for the new machine is adjusted.

b. Third-party sale—the gain is taxed as income, basis of new machine is fair market value paid.

3. A disposal in which a loss results—the basis of the new machine is the same as the basis of the old machine, decreased by any money received.

Assuming a corporate profit situation, the applicable tax depreciation shield formulas are

1. For a situation where there is no gain on the exchange:

$$\text{Total tax shield} = \sum_{n=1}^{N} t_c D_n \qquad \textbf{[3.4]}$$

where

n = individual yearly time periods within a life assumption of N years

t_c = corporate tax rate

D_n = annual depreciation amount in the nth time period

2. For a situation where a gain results from the exchange,
   a. Like-kind exchange—Eq. [3.4] is applicable. It must be realized that the basis of the new machine will be affected.
   b. Third-party sale.

$$\text{Total tax shield} = \left(\sum_{n=1}^{N} t_c D_n \right) - \text{gain} \times t_c \qquad \textbf{[3.5]}$$

Gain is the actual salvage amount received at the time of disposal minus the book value.

The implication of the basis is that in making analysis calculations, the actual salvage derived from the machine directly affects the depreciation saving. To perform a valid analysis, the depreciation accounting practices for tax purposes and the methods of machine disposal and acquisition the company chooses to use must be carefully examined. These dictate the appropriate calculations for the tax effects of depreciation.

MAJOR REPAIRS AND OVERHAULS

Major repairs and overhauls are included under ownership cost because they result in an extension of a machine's service life. They can be considered as an investment in a new machine. Because a machine commonly works on many different projects, considering major repairs as an ownership cost prorates these expenses to all jobs. These costs should be added to the basis of the machine and depreciated.

TAXES

In this context, taxes refer to those equipment ownership taxes that are charged by any government subdivision. They are commonly assessed at a percentage rate applied against the book value of the machine. Depending on location, property taxes can range up to about 4.5%. In many locations, there will be no property tax on equipment. Over the service life of the machine, they will decrease in magnitude as the book value decreases.

INSURANCE

Insurance, as considered here, includes the cost to cover fire, theft, and damage to the equipment. Annual rates can range from 1 to 3%. This cost can be actual premium payments to insurance companies, or it can represent allocations to a self-insurance fund maintained by the equipment owner.

STORAGE AND MISCELLANEOUS

Between jobs or during bad weather, a company will require storage facilities for its equipment. The cost of maintaining storage yards and facilities should be prorated to those machines that require such harborage. Typical expenses include space rental, utilities, and the wages for laborers or watchmen. These expenses are all combined in an overhead account and then allocated on a proportional basis to the individual machines. The rate may range from nothing to perhaps 5%.

OPERATING COST

GENERAL INFORMATION

Operating cost is the sum of those expenses an owner experiences by working a machine on a project. Typical expenses include

1. Fuel.
2. Lubricants, lube oils, filters, and grease.
3. Repairs.
4. Tires.
5. Replacement of high-wear items.

Operator wages are sometimes included under operating costs, but because of wage variance between jobs, the general practice is to keep operator wages as a separate cost category. Such a procedure aids in estimation of machine cost for bidding purposes as the differing project wage rates can readily be added to the total machine O&O cost. In applying operator cost, all benefits paid by the company must be included—direct wages, fringe benefits, insurance, etc. This is another reason wages are separated. Some benefits are on an hourly basis, some on a percentage of income, some on a percentage of income to a maximum amount, and some are paid as a fixed amount. The assumptions about project work schedule will therefore affect wage expense.

FUEL

Fuel expense is best determined by measurement on the job. Good service records tell the owner how many gallons of fuel a machine consumes over what period of time and under what job conditions. Hourly fuel consumption can then be calculated directly.

When company records are not available, manufacturer's consumption data can be used to construct fuel use estimates. The amount of fuel required to power a piece of equipment for a specific period of time depends on the brake horse-power of the machine and the work application. Therefore, most tables of hourly fuel consumption rates are divided according to the machine type and the working conditions. To calculate hourly fuel cost, a consumption rate is found in the tables (see Table 3.2) and then multiplied by the unit price of fuel.

TABLE 3.2 | Average fuel consumption—wheel loaders.

Horsepower (fwhp)	Type of utilization		
	Low (gal/hr)	Medium (gal/hr)	High (gal/hr)
90	1.5	2.4	3.3
140	2.5	4.0	5.3
220	5.0	6.8	9.4
300	6.5	8.8	11.8

Fuel consumption formulas have been published for both gasoline and diesel engines. The resulting values from such formulas must be adjusted by *time and load factors* that account for working conditions. This is because the formulas are derived assuming that the engine is operating at maximum output. Working conditions that must be considered are the percentage of an hour that the machine is actually working (*time factor*) and at what percentage of rated horsepower (*load factor*). When operating under standard conditions, a *gasoline engine* will consume approximately 0.06 gal of fuel per flywheel horsepower hour (fwhp-hr). A *diesel engine* will consume approximately 0.04 gal per fwhp-hr.

LUBRICANTS—LUBE OILS, FILTERS, AND GREASE

The cost of lubricants, filters, and grease will depend on the maintenance practices of the company and the conditions of the work location. Some companies follow the machine manufacturer's guidance concerning time periods between lubricant and filter changes. Other companies have established their own preventive maintenance change period guidelines. In either case, the hourly cost is arrived at by the operating hours between changes and a small consumption factor.

Many manufacturers provide quick cost estimating tables or rules for determining the cost of these items. Whether using manufacturer's data or past experience, notice should be taken about whether the data matches expected field conditions. If the machine is to be operated under adverse conditions, such as deep mud, water, or severe dust, the data values will have to be adjusted.

REPAIRS

Repairs, as referred to here, mean normal maintenance type repairs (see Fig. 3.2). These are the repair expenses incurred on the job site where the machine is operated and would include the costs of parts and labor. Major repairs and overhauls are accounted for as ownership cost.

Repair expenses increase with machine age. The army has found that 35% of its equipment maintenance cost is directly attributable to the oldest 10% of its equipment. Instead of applying a variable rate, an average is usually calculated by dividing the total expected repair cost, for the planned service life of the machine, by the planned operating hours. Such a policy builds up a repair reserve during a machine's early life. That reserve will then be used to cover the higher costs experienced later. As with all costs, company records are the best source of expense information. When such records are not available, manufacturers' published guidelines can be used.

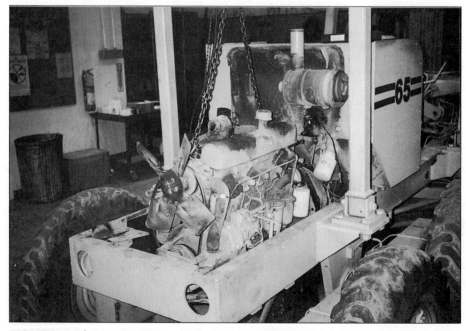

FIGURE 3.2 | Normal repairs are included in operating cost.

FIGURE 3.3 | Tires are a major operating cost.

TIRES

Tires for wheel type equipment (see Fig. 3.3) are a major operating cost because they have a short life in relation to the "iron" of a machine. Tire cost will include repair and replacement charges. These costs are very difficult to estimate because of the variability in tire wear with project site conditions and operator skill. Both tire and equipment manufacturers publish tire life guidelines based on tire type and job application. Manufacturers' suggested life periods can be used with local tire prices to obtain an hourly tire cost. It must be remembered, however, that the guidelines are based on good operating practices and do not account for abuses such as overloading haul units.

REPLACEMENT OF HIGH-WEAR ITEMS

The cost of replacing those items that have very short service lives with respect to machine service life can be a substantial operating cost. These items will differ depending on the type of machine, but typical items include cutting edges, ripper tips, bucket teeth (see Fig. 3.4), body liners, and cables. By using either

FIGURE 3.4 | Bucket teeth are a high-wear-item replacement cost.

past experience or manufacturer life estimates the cost can be calculated and converted to an hourly basis.

All machine-operating costs should be calculated per working hour so that it is easy to sum the applicable costs for a particular class of machines and obtain a total operating hour cost.

COST FOR BIDDING

GENERAL INFORMATION

The process of selecting a particular type of machine for use in constructing a project requires knowledge of the cost associated with operating the machine in the field. In selecting the proper machine, a contractor seeks to achieve unit production at a least cost. This cost for bidding is the O&O we have discussed. It is usually expressed in dollars per equipment operating hour.

OWNERSHIP EXPENSES

Depreciation is simply the prorating of a machine's consumption over its useful life. Do not confuse the depreciation discussed here with tax depreciation. Tax depreciation has nothing to do with consumption of the asset; it is simply an artificial calculation for tax code purposes.

Depreciation—Consumption of the Asset

Depreciation, as used in this section, relates to the calculations used to account for purchase expense and salvage value on an hourly basis over the service life of a machine. Because tires are a high-wear item that will be replaced many times over a machine's service life, their cost will be excluded in these calculations.

Time value method.   The depreciation portion of ownership cost can be calculated by either of two methods: time value or average annual investment. The time value method will recognize the timing of the cash flows, i.e., the purchase at time zero and the salvage at a future date. The cost of the tires is deducted from the total purchase price, which includes amounts for all options, shipping, and taxes (total cash outflow − cost of tires). A judgment about the expected service life and a corporate cost of capital rate are both necessary input parameters for the analysis. These input parameters are entered into Eq. [1.6], the uniform series capital recovery factor formula, to determine the machine's purchase price equivalent annual cost.

To account for the salvage cash inflow, Eq. [1.4], the uniform series sinking fund factor formula, is utilized. The input parameters are the estimated future salvage amount, the expected service life, and the corporate cost of capital rate.

EXAMPLE 3.4

A company having a cost of capital rate of 8% purchases a $300,000 tractor. This machine has an expected service life of 4 yr and will be utilized 2,500 hr per year. The tires on this machine cost $45,000. The estimated salvage value at the end of 4 yr is $50,000. Calculate the depreciation portion of the ownership cost for this machine using the time value method.

Initial cost	$300,000
Cost of tires	−45,000
Purchase price less tires	$255,000

Now it is necessary to calculate the uniform series required to replace a present value of $255,000. Using Eq. [1.6]:

$$A = \$255,000 \left[\frac{0.08(1 + 0.08)^4}{(1 + 0.08)^4 - 1} \right]$$

$$A = \$255,000 \times 0.3019208 \Rightarrow \$76,990 \text{ per year}$$

Next calculate the uniform series sinking fund factor to use with the salvage value. Using Eq. [1.4]:

$$A = \$50,000 \left[\frac{0.08}{(1 + 0.08)^4 - 1} \right]$$

$$A = \$50,000 \times 0.02219208 \Rightarrow \$11,096 \text{ per year}$$

Therefore using the time value method the depreciation portion of the ownership cost is

$$\frac{\$76,990/yr - \$11,096/yr}{2,500 \ hr/yr} = \$26.358/hr$$

Average annual investment method   A second approach to calculating the depreciation portion of ownership cost is the average annual investment (AAI) method.

$$AAI = \frac{P(n + 1) + S(n - 1)}{2n} \tag{3.6}$$

where

P = purchase price less the cost of the tires

S = the estimated salvage value

n = expected service life in years

The AAI is multiplied by the corporate cost of capital rate to determine the ownership cost of money portion. The straight-line depreciation of the cost of the machine less the salvage and less the cost of tires, if a wheeled machine, is then added to the cost of money part to arrive at the depreciation portion of ownership cost.

EXAMPLE 3.5

Using the same machine and company information as in Example 3.4, calculate the depreciation portion of the ownership cost using the AAI method.

$$AAI = \frac{\$255,000(4 + 1) + \$50,000(4 - 1)}{2 \times 4}$$

$$AAI = \$178,125/yr$$

$$Interest \ cost \ part = \frac{\$178,125/yr \times 8\%}{2,500 \ hr/yr} \Rightarrow \$5.700/hr$$

Straight-line depreciation part

Initial cost	$300,000
Cost of tires	−45,000
Salvage	−50,000
	$205,000

$$\frac{\$205,000}{4yr \times 2,500hr/yr} = \$20.500/hr$$

Total depreciation portion of the ownership cost using the AAI method

$$\$5.700/hr + \$20.500 = \$26.200/hr$$

For Examples 3.4 and 3.5, the difference in the calculated depreciation portion of the ownership cost is $0.158/hr ($26.358/hr − $26.200/hr). The choice of which method to use is strictly a company preference. Basically, either method is satisfactory, especially considering the impact of the unknowns concerning service life, operating hours per year, and expected future salvage. There is no single solution to calculating ownership cost. The best approach is to perform several analyses using different assumptions and to be guided by the range of solutions.

Tax Saving from Depreciation

To calculate the tax saving from depreciation, the government tax code depreciation schedules (Table 3.1) must be used. The resulting depreciation amounts are then multiplied by the company's tax rate to calculate specific savings, using Eq. [3.4] or [3.5]. The sum of the yearly saving must be divided by the total anticipated operating hours to obtain an hourly cost saving.

EXAMPLE 3.6

Using the same machine and company information as in Example 3.4, calculate the hourly tax saving resulting from depreciation. Assume that under the tax code the machine is a 5-yr type property and that there had been no gain on the exchange that procured the machine. The company's tax rate is 37%.

First, calculate the annual depreciation amounts for each of the years. In this case, the tax code depreciation rate must be used to calculate the depreciation.

Year	5-yr property rates	BV_{n-1}	D_n	BV_n
0		$ 0	$ 0	$300,000
1	0.20	300,000	60,000	240,000
2	0.32	240,000	96,000	144,000
3	0.24	144,000	72,000	72,000
4	0.16	72,000	48,000	24,000
5	0.08	24,000	24,000	0

Using Eq. [3.4], the tax shielding effect for the machine's service life would be

Year	D_n	Shielded amount*
1	$60,000	$22,200
2	96,000	35,520
3	72,000	26,640
4	48,000	17,760
	Total	$102,120

* $D_n \times 37\%$.

$$\text{Tax saving from depreciation} = \frac{\$102,120}{4 \text{ yr} \times 2,500 \text{ hr/yr}} \Rightarrow \$10.212/\text{hr}$$

Major Repairs and Overhauls

When a major repair and overhaul takes place, the machine's ownership cost will have to be recalculated. This is done by adding the cost of the overhaul to the book value at that point in time. The resulting new adjusted basis is then used in the depreciation calculated, as already described. If there are separate calculations for true depreciation and for tax depreciation, both will have to be adjusted.

Taxes, Insurance, and Storage

To calculate the taxes, insurance, and storage costs, common practice is to simply apply a percentage value to either the machine's book value or its AAI amount. The expenses incurred for these items are usually accumulated in a corporate overhead account. That value divided by the value of the equipment fleet and multiplied by 100 will provide the percentage rate to be used.

Taxes, insurance, and storage portion of ownership cost

$$\text{equals rate } (\%) \times \text{BV}_n \text{ (or AAI)} \qquad [3.7]$$

EXAMPLE 3.7

Using the same machine and company information as in Examples 3.4 and 3.5, calculate the hourly owning expense associated with taxes, insurance, and storage. Annually, the company pays as an average 1% in property taxes on equipment, 2% for insurance, and allocates 0.75% for storage expenses.

Total percentage rate for taxes, insurance, and storage = 1% + 2% + 0.75%

$$\Rightarrow 3.75\%$$

From Example 3.5 the average annual investment for the machine is $178,125/yr.

$$\text{Taxes, insurance, and storage expense} = \frac{\$178,125/\text{yr} \times 3.75\%}{2,500 \text{ hr/yr}} \Rightarrow \$2.672/\text{hr}$$

OPERATING EXPENSES

Figures based on actual company experience should be used to develop operating expenses. Many companies, however, do not keep good equipment operating and maintenance records; therefore, many operating costs are estimated as a percentage of a machine's book value. Even companies that keep records often accumulate expenses in an overhead account and then prorate the total back to individual machines using book value.

Fuel

The amount expended on fuel is a product of how a machine is used in the field and the local cost of fuel. In past years, fuel could be purchased on long-term contracts at a fixed price. Today fuel is usually offered with a *time of delivery price.* A supplier will agree to supply the fuel needs of a project, but the price will not be guaranteed for the duration of the work. Therefore, when bidding a long-duration project, the contractor must make an assessment of future fuel prices.

To calculate hourly fuel expense, a consumption rate is multiplied by the unit price of fuel. Service records are important for estimating fuel consumption.

EXAMPLE 3.8

A 220-fwhp wheel loader will be used at an asphalt plant to move aggregate from a stock pile to the cold feed hoppers. This loader is diesel powered. It is estimated that the work will be steady at an efficiency equal to a 50-min hour. The engine will work at full throttle while loading the bucket (30% of the time) and at three-quarter throttle to travel and dump. Calculate the fuel consumption using the engine consumption averages and compare the result to a medium rating in Table 3.2. If diesel costs $1.07/gal, what is the expected fuel expense?

Fuel consumption diesel engine 0.04 gal per fwhp-hr.

Load factor: Loading bucket $1.00 \times 0.30 = 0.30$

Travel and dump $\underline{0.75 \times 0.70 = 0.53}$

0.83

Time factor: 50-min hr $= 0.83$

Combined factor: $0.83 \times 0.83 = 0.69$

Fuel consumption $= 0.69 \times 0.04$ gal/fwhp-hr $\times 220$ fwhp $= 6.1$ gal/hr

Table 3.2: medium rating 200 fwhp loader $= 6.8$ gal/hr

There is considerable difference in the calculated results and those found in Table 3.2, which is why it is recommended that a company establish historical data.

Cost:

Using the formula   6.1 gal/hr $\times$ $1.07/gal $=$ $6.527/hr

Table 3.2   6.8 gal/hr $\times$ $1.07/gal $=$ $7.276/hr

Lubricants

The quantity of lubricants used by an engine will vary with the size of the engine, the capacity of the crankcase, the condition of the piston rings, and the number of hours between oil changes. For extremely dusty conditions, it may be desirable to change oil every 50 hr, but this is an unusual condition. It is common practice to change oil every 100 to 200 hr. The quantity of the oil consumed by an engine per change will include the amount added during the change plus the makeup oil between changes.

A formula that can be used to estimate the quantity of oil required is

$$\text{Quantity consumed, gph (gallons per hour)} = \frac{\text{hp} \times f \times 0.006 \text{ lb/hp-hr}}{7.4 \text{ lb/gal}} + \frac{c}{t}$$

where

    hp = rated horsepower of the engine

    c = capacity of the crankcase in gallons

    f = operating factor

    t = number of hours between oil changes

This formula contains the assumption that the quantity of oil consumed per rated horsepower hour between changes will be 0.006 lb.

EXAMPLE 3.9

Calculate the oil required, on a per hour basis, for the 220-fwhp wheel loader in Example 3.8. The operating factor will be 0.69 as calculated in that example. The crankcase capacity is 8 gal and the company has a policy to change oil every 150 hr.

$$\text{Quantity consumed (gph)} = \frac{220 \text{ fwhp} \times 0.69 \times 0.006 \text{ lb/hp} - \text{hr}}{7.4 \text{ lb/gal}} + \frac{8 \text{ gal}}{150 \text{ hr}}$$

$$= 0.18 \text{ gal/hr}$$

The cost of hydraulic oil, filters, and grease will be added to the expense of engine oil. The hourly cost of filters is simply the actual expense to purchase the filters divided by the hours between changes. If a company does not keep good machine-servicing data, it is difficult to accurately estimate the cost of hydraulic oil and grease. The usual solution is to refer to the manufacturers' published tables of average usage or expense.

Repairs

The cost of repairs is normally the largest single component of machine cost (see Table 3.3). Some general guidelines published by the Power Crane and Shovel Association (PCSA) in the past estimated repair and maintenance expenses at 80 to 95% of depreciation for crawler-mounted excavators, 80 to 85% for wheel-mounted excavators, 55% for crawler cranes, and 50% for wheel-mounted cranes. The lower figures for cranes reflect the work they perform and the intermittent nature of their use. The data assumed that half of the cost was materials and parts and half was labor, in the case of mechanical machines. For hydraulic machines, two-thirds of the cost is for materials and parts, and one-third for labor.

TABLE 3.3 | Breakdown of machine cost over its service life.

Cost category	Percentage of total cost (%)
Repair	37
Depreciation	25
Operating	23
Overhead	15

Equipment manufacturers' supply tables of average repair costs based on machine type and work application. Repair expenses will increase with machine usage (age). The repair cost to establish a machine rate for bidding should be an average rate.

Tires

Tire expenses include both tire repair and tire replacement. Tire maintenance is commonly handled as a percentage of straight-line tire depreciation. Tire hourly cost can be derived simply by dividing the cost of a set of tires by their expected life, and this is how many companies prorate this expense. A more sophisticated approach is to use a time-value calculation, recognizing that tire replacement expenses are single-point-in-time outlays that take place over the life of a wheel-type machine.

EXAMPLE 3.10

Calculate the hourly tire cost that should be part of machine operating cost if a set of tires can be expected to last 5,000 hr. Tires cost $38,580 per set of four. Tire repair cost is estimated to average 16% of the straight-line tire depreciation. The machine has a service life of 4 yr and operates 2,500 hr per yr. The company's cost of capital rate is 8%.

Not considering the time value of money:

$$\text{Tire repair cost} = \frac{\$38,580}{5,000 \text{ hr}} \times 16\% \Rightarrow \$1.235/\text{hr}$$

$$\text{Tire use cost} = \frac{\$38,580}{5,000 \text{ hr}} \Rightarrow \$7.716/\text{hr}$$

Therefore, tire operating cost is $8.951/hr ($1.235/hr + $7.716/hr).

Considering the time value of money:
Tire repair cost is the same $1.235/hr.
Will have to purchase a second set at the end of 2 yr.

$$\left(\frac{4 \text{ yr} \times 2,500 \text{ hr/yr}}{5,000 \text{ hr per set of tires}} \right) = 2 \text{ sets}$$

First set: Calculate the uniform series required to replace a present value of $38,580. Using Eq. [1.6]:

$$A = \$38,580\left[\frac{0.08(1 + 0.08)^4}{(1 + 0.08)^4 - 1}\right]$$

$$\frac{\$38,580 \times 0.301921}{2,500 \text{ hr/yr}} = \$4.659/\text{hr}$$

Second set: The second set will be purchased 2 yr in the future. Therefore what amount at time zero is equivalent to $38,580 2 yr in the future? Using the present worth compound amount factor (Eq. [1.2]), the equivalent time zero amount is calculated.

$$P = \frac{\$38,580}{(1 + 0.08)^2} \Rightarrow \$33,076$$

Calculate the uniform series required to replace a present value of $33,076.

$$A = \$33,076\left[\frac{0.08(1 + 0.08)^4}{(1 + 0.08)^4 - 1}\right]$$

$$\frac{\$33,076 \times 0.301921}{2,500 \text{ hr/yr}} = \$3.995/\text{hr}$$

Therefore, considering the time value of money, tire operating cost is $9.889/hr ($1.235/hr + $4.659/hr + $3.995/hr).

High-Wear Items

Because the cost of high-wear items is dependent on job conditions and machine application, the cost of these items is usually accounted for separate from general repairs.

EXAMPLE 3.11

A dozer equipped with a three-shank ripper will be used in a loading and ripping application. Actual ripping will take place only about 20% of total dozer operating time. A ripper shank consists of the shank itself, a ripper tip, and a shank protector. The estimated life for the ripper tip is 30 hr. The estimated life of the ripper shank protector is three times tip life. The local price for a tip is $40 and $60 for shank protectors. What hourly high-wear item charge should be added to the operating cost of a dozer in this application?

Tips: $\dfrac{30 \text{ hr}}{0.2} = 150$ hr of dozer operating time

$$\frac{3 \times \$40}{150 \text{ hr}} = \$0.800/\text{hr for tips}$$

Shank protectors: Three times tip life × 150 hr = 450 hours of dozer operating time

$$\frac{3 \times \$60}{450 \text{ hr}} = \$0.400/\text{hr for shank protectors}$$

Therefore, the cost of high-wear items is $1.200/hr ($0.800/hr tips + $0.400/hr shank protectors).

REPLACEMENT DECISIONS

GENERAL INFORMATION

A piece of equipment has two lives: (1) a physically limited working life and (2) a cost-limited economic life. Because equipment owners are in business to make money, the economic life of their equipment is of critical importance. A machine in good mechanical condition and working productively enjoys a strong bias in favor of its retention in the inventory. The equipment manager may look only at the high initial cash outflow associated with the purchase of a replacement and consequently ignore the other cost factors involved. All cost factors must be examined when considering a replacement decision. A simple example will help to illustrate the concept.

EXAMPLE 3.12

A small dozer is purchased for $106,000. A forecast of expected operating hours, salvage values, and maintenance expense is presented in the table.

Year	Operating hours	Salvage ($)	Maintenance expense ($)
1	1,850	79,500	3,340
2	1,600	63,600	3,900
3	1,400	76,320	4,460
4	1,200	74,200	5,000
5	800	63,600	6,600

A replacement analysis might look like this.

Year	1	2	3	4	5
Purchase	$106,000	$106,000	$106,000	$106,000	$106,000
Salvage	$79,500	$69,500	$76,320	$73,000	$70,000
Cost	$26,500	$36,500	$29,680	$33,800	$36,000
Cumulative operating hours	1,850	3,450	4,850	6,050	6,850
$/hr ownership	$14.32	$10.58	$6.12	$5.45	$5.26
Cumulative maintenance expense	$3,340	$7,240	$11,700	$16,700	$23,300
$/hr operating	$1.81	$2.10	$2.41	$2.76	$3.40
Total $/hr	$16.13	$12.68	$8.53	$8.21	$8.66

If an owner considers only purchase price and expected salvage, the numbers argue that the machine should not be traded (see Fig. 3.5a). However, if only operating cost is examined, the owner would want to trade the machine after the first year, as operating expenses are continually rising with usage (see Fig. 3.5b). A correct analysis of the situation requires that total machine cost be considered. So in the case of Example 3.12, the most economical service life of this machine is 4 yr, as $8.21/operating hour is the minimum total cost.

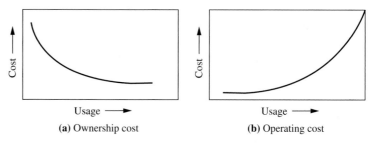

(a) Ownership cost   **(b)** Operating cost

FIGURE 3.5 | Effect of cumulative usage on cost.

The analysis is based on *cumulative* hours. This is an important point that is often missed. If the owner chooses to keep the machine 5 yr, the effective loss is $0.45 ($8.66 − $8.21) on *every operating hour,* not just the 800 hr of the last year. When the total operating hours are large, the significance of this cumulative effect can become much greater than it would appear by simply looking at the *combined cost per hour* values.

The replacement analysis should present all the cost and timing information affecting a machine or class of machines in a usable format. The format should be such that it is easy to perform sensitivity analyses to determine the correctness of the results. As described here, the model is a cost-minimization model. With such a model, the optimum economic life of a machine is that ownership time duration that results in a minimum hourly cost.

The cash flows being studied in a replacement analysis take place at different points in time, therefore the model should consider the timing effects by use of present-value techniques. The company's cost-of-capital rate is the appropriate interest rate to use in the present-value equations.

BUY, RENT, OR LEASE

GENERAL INFORMATION

There are three basic methods for securing a particular machine to use on a project: (1) *buy* (direct ownership), (2) *rent,* or (3) *lease.* Each method has inherent advantages and disadvantages. Ownership guarantees control of machine availability and mechanical condition, but it requires a continuing sequence of projects to pay for the machine. Ownership may force a company into using obsolete equipment. The calculations applicable for determining the cost of direct ownership have already been developed.

RENTAL

The rental of a machine is a short-term alternative to direct equipment ownership. With a rental, a company can pick the machine that is exactly suited for the job at hand. This is particularly advantageous if the job is of short duration or if the company does not foresee a continuing need for the particular type of machine in question. Rentals are very beneficial to a company in such situations, even though the rental charges are higher than *normal* direct ownership expense. The advantage lies in the fact that direct ownership costing assumes a continuing need for and utilization of the machine. If that assumption is not valid, a rental should be considered.

It must be remembered that rental companies only have a limited number of machines and, during peak work season, all types are not always available. Additionally, many specialized or custom machines cannot be rented.

Firms often use rentals as a way to test a machine prior to a purchase decision. A rental provides the opportunity for a company to operate a specific make or model of machine under actual project conditions. Profitability of the machine, based on the company's normal operating procedures, can then be evaluated before a major capital expenditure is approved to purchase the machine.

The general practice of the industry is to price rental rates for equipment on either a daily (8 hr), a weekly (40 hr), or a monthly (176 hr) basis. In the case of larger pieces of equipment, rentals may be available only on a monthly basis. Cost per hour usually is less for a longer-term rental (i.e., the monthly rate figured on a per hour basis would be less than the daily rate on an hourly basis).

Responsibility for repair cost is stated in the rental contract. Normally, on tractor-type equipment, the renter is responsible for all repairs. If it is a rubber-tired machine, the renting company will measure tread wear and charge the renter for tire wear. In the case of cranes and shovels, the renting company usually bears the cost of normal wear and tear. The user must provide servicing of the machine while it is being used. The renter is almost always responsible for fuel and lubrication expenses. Industry practice is that rentals are payable in advance. The renting company will require that the user furnish certificates of insurance before the machine is shipped to the job site.

Equipment cost is very sensitive to changes in use hours. Fluctuations in maintenance expenses or purchase price barely affect cost per hour. But a decrease in use hours per year can make the difference between a cost-effective machine ownership and renting. The basic cost considerations that need to be examined when considering a possible rental can be illustrated by a simple set of circumstances. Consider a small wheel loader with an ownership cost of $10.96 per hr. Assume that the cost is based on the assumption that the machine will work 2,400 hr each year of its service life. If $10.96/hr is multiplied by 2,400 hr/yr, the yearly ownership cost is found to be $26,304.

Checking with the local rental company, the construction firm receives rental quotes of $3,558 per month, $1,182 per week, and $369 per day for this size loader. By dividing with the appropriate number of hours, these rates can be expressed as hourly costs. Additionally, by dividing the calculated hourly rental rates into the construction firm's yearly ownership cost figure ($26,304) the operating hour breakeven points can be determined (see Table 3.4).

TABLE 3.4 | Rental versus ownership—operating hour breakeven points.

Rental Duration	Rate ($)	Hours	Rental rate ($/hr)	Operating hour breakeven point ($26,304/$/hr)	Operating hour breakeven point ($3,558/$/hr)
Monthly	3,558	176	20.22	1,300	—
Weekly	1,182	40	29.55	890	120
Daily	369	8	46.13	570	77

If the loader will be used for less than 1,300 hr but more than 890 hr, the construction company should consider a monthly rental instead of ownership. When the projected usage is less than 1,300 hr but greater than 120 hr, a weekly rental would be appropriate. In the case of very limited usage, that is, less than 26 hr, the daily rate is optimal.

The point is that when a company rents, it pays for the equipment only when project requirements dictate a need. The company that owns a machine must continue to make the equipment payments even when the machine lies idle. When investigating a rental the critical question is usually expected *hours of usage*.

LEASE

A lease is a long-term agreement for the use of an asset. It provides an alternative to direct ownership. During the lease term, the leasing company (lessor) always owns the equipment and the user (lessee) pays the owner to use the equipment. The lessor must retain ownership rights in order for the contract to be considered a true lease by the Internal Revenue Service. The lessor will receive lease payments in return for providing the machine. The lease payments do not have to be uniform across the lease period. The payments can be structured in the agreement to best fit the situation of the lessee or the lessor. In the lessee's case, cash flow at the beginning may be low, so the lessee wants payments that are initially low. Because of tax considerations, the lessor may agree to such a payment schedule. Lease contracts are binding legal documents, and most equipment leases are noncancelable by either party.

A lease pays for the use of a machine during the most reliable years of a machine's service life. Sometimes the advantage of a lease is that the lessor provides the management and servicing. This frees the contractor from hiring mechanics and grease men and enables the company to concentrate on the task of building.

Long-term, when used in reference to lease agreements, is a period of time that is long relative to the life of the machine in question. An agreement that is for a very short period of time, as measured against the expected machine life, is a rental. A conventional true lease may have three end-of-lease options: (1) buy the machine at fair market value, (2) renew the lease, or (3) return the equipment to the leasing company.

A lessee loses the tax depreciation shield of machine ownership but gains a tax deduction because lease payments are treated as an expense. The most important factor contributing to a decision to lease is reduced cost. Under specific conditions, the actual cost of a leased machine can be less than the ownership cost of a purchased machine. This is caused by the different tax treatments for owning and leasing an asset. An equipment user must make a careful examination of the cash flows associated with each option to determine which results in the lowest total cost.

Working capital is the cash that a firm has available to support its day-to-day operations. This *cash asset* is necessary to meet the payroll on Friday, to pay the electric bill, and to purchase fuel to keep the machines running. To be a viable business, working capital assets must be greater than the inflow of bills. A machine is an asset to the company, but it is not what the electric company will accept as payment for their bill.

A commonly cited advantage of leasing is that working capital is not tied up in equipment. This statement is only partly true. It is true that when a company borrows funds to purchase a machine, the lender normally requires that the company establish an equity position in the machine, a *down payment.* Additionally, the costs of delivery and initial servicing are not included in the loan and must be paid by the new owner. Corporate funds are therefore tied up in these up-front costs of a purchase. Leasing does not require these cash outflows and is often considered as 100% financing. However, most leases require an advance lease payment. Some even require security deposits and charge other up-front costs.

Still another argument is that because borrowed funds are not used, credit capacity is preserved. Leasing is often referred to as off-balance-sheet financing. An operating lease (used when the lessee does not ultimately want to purchase the equipment) enables leased assets to be expensed, and such assets do not appear on the balance sheet. A lease is considered an operating expense, not a liability, as is the case with a bank loan. Standards of accounting, however, require disclosure of lease obligations. It is hard to believe that *lenders* would be so naive as to not consider all of a company's fixed obligations, including both loans and leases. But the off-balance-sheet lease typically will not hurt *bonding capacity,* which is important to a company's ability to bid work.

Before entering into a contract with a construction company, most owners require that the company post a bond guaranteeing that it will complete the project. This bond is secured by a third-party surety company. The surety closely examines the construction company's financial position before issuing the bond. Based on the financial strength of the construction company, the surety typically restricts the total volume of work that the construction company may have under contract at any one time. This restriction is known as bonding capacity. It is the total dollar value of work under contract that a surety company will guarantee for a construction company.

Owners should make a careful examination of the advantages of a lease situation. The cash flows that should be considered when evaluating the cost of a lease include

1. Inflow initially of the equivalent value of the machine.
2. Outflow of the periodic lease payments.
3. Tax shielding provided by the lease payments. (This is allowed only if the agreement is a true lease. Some "lease" agreements are essentially installment sale arrangements.)
4. Loss of salvage value when the machine is returned to the lessor.

These costs all occur at different points in time so present-value computations must be made before the costs can be summed. The total present value of the lease option should be compared to the minimum ownership costs, as determined by a time-value replacement analysis. In most lease agreements, the lessee is responsible for maintenance. If, for the lease in question, maintenance expense is the same as for the case of direct ownership, then the maintenance expense factor can be dropped from the analysis. A leased machine would exhibit the same aging and the resulting reduced availability as a purchased machine.

When comparing a lease to a purchase, it is often appropriate to think of the purchase as a loan decision and the resulting consequences. The specific differences are outlined in Table 3.5.

TABLE 3.5 | Differences between a lease and a loan.

Lease	Loan
1. A lease requires no down payment and finances only the value of the equipment expected to be depleted during the lease term. The lessee usually has an option to buy the equipment for its remaining value at lease end.	1. A loan requires the end-user to invest a down payment in the equipment. The loan finances the remaining amount.
2. The leased equipment itself is usually all that is needed to secure a lease transaction.	2. A loan usually requires the borrower to pledge other assets for collateral.
3. A lease requires only a lease payment at the beginning of the first payment period, which is usually much lower than a down payment.	3. A loan usually requires two expenditures during the first payment period: a down payment at the beginning and a loan payment at the end.
4. When leases are structured as true leases, the end-user may claim the entire lease payment as a tax deduction. The equipment write-off is tied to the lease term, which can be shorter than IRS depreciation schedules, resulting in larger tax deductions each year. (Equipment financed with a conditional sale lease is treated the same as owned equipment.)	4. End-users may claim a tax deduction for a portion of the loan payment as interest and for depreciation, which is tied to IRS depreciation schedules.

SUMMARY

Equipment owners must carefully calculate machine ownership and operating cost. This cost is usually expressed in dollars per operating hour. The most significant cash flows affecting *ownership cost* are (1) purchase expense; (2) salvage value; (3) tax saving from depreciation; (4) major repairs and overhauls; and (5) property taxes, insurance, storage, and miscellaneous expenses. *Operating cost* is the sum of those expenses an owner experiences by working a machine on a project: (1) fuel; (2) lubricants, filters, and grease; (3) repairs; (4) tires; and (5) replacement of high-wear items. *Operator wages* are sometimes included under operating costs, but because of wage variances between jobs, the general practice is to keep operator wages as a separate cost category. Critical learning objectives would include:

- An ability to calculate ownership cost.
- An ability to calculate operating cost.
- An understanding of the advantages and disadvantages associated with direct ownership, renting, and leasing machines.

These objectives are the basis for the problems that follow.

PROBLEMS

3.1 What is the largest single equipment cost?

3.2 Regardless of how much a machine is used, the owner must pay owning cost. (True–False).

3.3 A machine's owning cost includes

    a. Tires

    b. Storage expenses

    c. Taxes

    d. General repairs

3.4 A tractor with an adjusted basis (from depreciation) of $55,000 is sold for $60,000, and a new tractor is purchased with a cash payment of $325,000. These are two separate transactions. What is the tax depreciation basis of the new tractor?

3.5 A tractor with an adjusted basis (from depreciation) of $65,000 is traded for a new tractor that has a fair market value of $300,000. A cash payment of $225,000 is made to complete the transaction. What is the tax depreciation basis of the new tractor?

3.6 Asphalt Pavers, Inc., purchases a loader to use at its asphalt plant. The purchase price delivered is $235,000. Tires for this machine cost $24,000. The company believes it can sell the loader after 7 yr (3,000 hr/yr) of service for $79,000. There will be no major overhauls. The company's cost of capital is 6.3%. What is the depreciation part of this machine's ownership cost? Use the time value method to calculate depreciation. ($9.623/hr)

3.7 Using the AAI method to calculate depreciation and the problem 3.6 information, what is the depreciation part of the machine's ownership cost?

3.8 Pushem Down clearing contractors purchases a dozer with a delivered price of $275,000. The company believes it can sell the used dozer after 4 yr (2,000 hr/yr) of service for $56,000. There will be no major overhauls. The company's cost of capital is 9.2%, and its tax rate is 33%. Property taxes, insurance, and storage will run 4%. What is the owning cost for the dozer? Use the time value method to calculate the depreciation portion of the ownership cost. ($29.943/hr)

3.9 Earthmovers Inc. purchases a grader to maintain haul roads. The purchase price delivered is $165,000. Tires for this machine cost $24,000. The company believes it can sell the grader after 6 yr (15,000 hr) of service for $26,000. There will be no major overhauls. The company's cost of capital is 7.3%, and its tax rate is 35%. There are no property taxes, but insurance and storage will run 3%. What is the owning cost for the grader? Use the time value method to calculate the depreciation portion of the ownership cost.

3.10 A 140-fwhp diesel-powered wheel loader will be used at an asphalt plant to move aggregate from a stock pile to the feed hoppers. The work will be steady at an efficiency equal to a 55-min hour. The engine will work at full throttle while loading the bucket (32% of the time) and at three-quarter throttle to travel and dump. Calculate the fuel consumption using the engine consumption averages and compare the result to a medium rating in Table 3.2. (4.3 gal/hr, 4.0 gal/hr)

3.11 A 60-fwhp gasoline-powered pump will be used to dewater an excavation. The work will be steady at an efficiency equal to a 60-min hour. The engine will work at half throttle. Calculate the fuel consumption using the engine consumption.

3.12 A 260-fwhp diesel-powered wheel loader will be used to load shot rock. This loader was purchased for $330,000. The estimated salvage value at the end of 4 yr is $85,000. The company's cost of capital is 8.7%. A set of tires costs $32,000. The work will be at an efficiency equal to a 45-min hour. The engine will work at full throttle while loading the bucket (33% of the time) and at three-quarter throttle to travel and dump. The crankcase capacity is 10 gal and the company has a policy to change oil every 100 hr on this job. The annual cost of repairs equals 70% of the straight-line machine depreciation. Fuel costs $1.07/gal, and oil is $2.50/gal. The cost of other lubricants and filters is $0.45/hr. Tire repair is 17% of tire depreciation. The tires should give 3,000 hr of service. The loader will work 1,500 hr/yr. In this usage, the estimated life for bucket teeth is 120 hr. The local price for a set of teeth is $640. What is the operating cost for the loader in this application? ($48.483/hr)

3.13 A 400-fwhp diesel-powered dozer will be used to support a scraper fleet. This dozer was purchased for $395,000. The estimated salvage value at the end of 4 yr is $105,000. The company's cost of capital is 7.6%. The work will be at an efficiency equal to a 50-min hour. The engine will work at full throttle while push loading the scrapers (59% of the time) and at three-quarter throttle to travel and position. The crankcase capacity is 16 gal, and the company has a policy to change oil every 150 hr. The annual cost of repairs equals 68% of the straight-line machine depreciation. Fuel cost is $1.03/gal, and oil is $2.53/gal. The cost of other lubricants and filters is $0.65/hr. The dozer will work 1,800 hr/yr. In this usage, the estimated life for cutting edges is 410 hr. The local price for a set of cutting edges is $1,300. What is the operating cost for the dozer in this application?

REFERENCES

1. *Caterpillar Performance Handbook,* Caterpillar Inc., Peoria, IL, issued annually.
2. Equipment Leasing Association of American, 4301 N. Fairfax Drive, Suite 550, Arlington, VA 22203-1627. www.elaonline.com/.

3. Lewellen, Wilbur G., *The Cost of Capital,* Kendall/Hunt Publishing Co., Dubuque, IA, 1976.

4. Modigliani, Franco, and Merton H. Miller, "The Cost of Capital, Corporate Finance and the Theory of Investment," *American Economic Review*, Vol. XLVIII, No. 3, June 1958.

5. *Rental Contract Checklist,* Associated Equipment Distributors, Inc., 615 W. 22nd Street, Oak Brook, IL 60523. www.aednet.org/.

6. Schexnayder, C. J., and Donn E. Hancher, "Inflation and Equipment Replacement Economics," *Journal of the Construction Division, Proceedings, ASCE,* Vol. 108, No. CO2, June 1982.

7. Schexnayder, C. J., and Donn E. Hancher, "Contractor Equipment Management Practices," *Journal of the Construction Division, Proceedings, ASCE,* Vol. 107, No. CO4, December 1981.

8. Schexnayder, C. J., and Donn E. Hancher, "Interest Factor in Equipment Economics," *Journal of the Construction Division, Proceedings, ASCE,* Vol. 107, No. CO4, December 1981.

9. Sunstate Equipment Co., 5425 East Washington, Phoenix, AZ 85034. www.sunstateequip.com/.

10. *The Cost of Renting Construction Equipment,* Associated Equipment Distributors, Inc., 615 W. 22nd St., Oak Brook, IL 60523. www.aednet.org/.

4

Geotechnical Materials, Compaction, and Stabilization

Knowledge of the properties, characteristics, and behavior of different soil types and aggregates is important to design and construction. The constructor is interested in how the material will handle during the construction process. Density is the most commonly used parameter for specifying construction operations because there is a direct correlation between a soil's properties and its density. The effectiveness of different compaction methods is dependent on the individual soil type being manipulated. In engineering construction, stabilization refers to when compaction is preceded by the addition and mixing of an inexpensive admixture, termed a "stabilization agent," which alters the chemical makeup of the soil, resulting in a more stable material.

GEOTECHNICAL MATERIALS

INTRODUCTION

Geotechnical materials—soils and rocks—are the principle components of many construction projects. They are used to support structures, static load; to support pavements for highways and airport runways, dynamic loads; and in dams and levees, as impoundments, to resist the passage of water. Some soils may be suitable for use in their natural state, whereas others must be excavated, processed, and compacted to meet the engineering requirements of a project. Additionally, either natural aggregate deposits or quarried rock constitute approximately 95% by weight of asphalt concrete and 75% of Portland cement concrete.

    Knowledge of the properties, characteristics, and behavior of different soil types and aggregates is important to those persons who are associated with the design or construction of projects involving these materials. Some soil and rock types are suitable for structural purposes in their natural state. In many cases, however, the locally available materials do not meet engineering requirements, and thus it is necessary to cost-effectively modify them to meet project demands. Processing can be as simple as adjusting the moisture content or mixing and blending. Because there is a direct relationship between increased density and increased strength and bearing capacity, the engineering properties of many soils can be improved simply by compaction.

GLOSSARY OF TERMS

The following glossary is used to define important terms that are used in discussing geotechnical materials, compaction, and stabilization.

Aggregate, course. Crushed rock or gravel, generally greater than 1/4 in. in size.

Aggregate, fine. The sand or fine-crushed stone used for filling voids in coarse aggregate. Generally it is less than 1/4 in. and greater than a No. 200 sieve in size.

Amplitude. The vertical distance a vibrating drum or plate is displaced from the rest position by an eccentric moment.

ASTM. American Society for Testing and Materials.

Backfill. Material used in refilling a cut or other excavation.

Bank measure. A measure of the volume of earth in its natural position before it is excavated.

Base. The layer of material, in a roadway or runway section, on which the pavement is placed. It may be of different types of materials, ranging from selected soils to crushed stone or gravel.

Binder. Fine aggregate or other materials that fill voids and hold coarse aggregate together.

Borrow pit. A pit from which fill material is mined.

Capillary. A phenomenon of a soil that enables water to be absorbed either upward or laterally.

Cohesion. The quality of some soil particles to be attracted to like particles, manifested in a tendency to stick together, as in clay.

Cohesive materials. A soil having properties of cohesion.

Compacted volume. A measurement of the volume of a soil after it has been subjected to compaction.

Grain-size curve. A graph showing the percentage by weight of soil sizes contained in a sample.

Granular material. A soil, such as sand, whose particle sizes and shapes are such that they do not stick together.

Impervious. A material that resists the flow of water through it is termed impervious.

In situ. Soil in its original or undisturbed position.

Lift. A layer of soil placed on top of previously placed embankment material. The term can be used in reference to material as spread or as compacted.

Optimum moisture content. The water content, for a given compactive effort, at which the greatest density of a soil can be obtained.

Pass. A working passage (trip) of an excavating, grading, or compaction machine.

Pavement. A layer, above the base, of rigid surfacing material that provides high bending resistance and distributes loads to the base. Pavements are usually constructed of asphalt or concrete.

Plasticity. The capability of being molded. Plastic materials do not assume their original shape after the force causing deformation is removed.

Proctor, or Proctor test. A method developed by R. R. Proctor for determining the moisture-density relationship in soils subjected to compaction.

Proctor, modified. A moisture-density test involving a higher compactive effort than the standard Proctor test.

Rock. The hard, mineral matter of the earth's crust, occurring in masses and often requiring blasting to cause breakage before excavation can be accomplished.

Shrinkage. A soil volume reduction usually occurring in fine-grained soils when they are subjected to moisture.

Soil. The loose surface material of the earth's crust, created naturally from the disintegration of rocks or decay of vegetation, that can be excavated easily using power equipment in the field.

Stabilize. To make a soil stronger, increase its strength and stiffness, and decrease its sensitivity to volume changes with changes in moisture content.

Subbase. A constructed layer of select material installed to furnish strength to the base of a road. In areas where the construction goes through marshy, swampy, unstable land it is often necessary to excavate the natural materials in the area under the roadway and replace them with more stable materials. The material used to replace the unsuitable natural soils is generally called subbase material, and when compacted it is known as the subbase.

Subgrade. The surface produced by grading native earth, or imported materials that serve as the foundation layer for a paving structural section.

Surface layer. The top layer of a road, street, parking lot, runway, etc. that covers and protects the sublayers from the action of traffic and weather. If the layer has structural strength properties, it is often referred to as a pavement layer.

PROPERTIES OF GEOTECHNICAL MATERIALS

Before discussing earth- and-rock handling techniques or analyzing problems involving these materials, it is necessary to become familiar with some of the physical properties of soils and aggregates. These properties have a direct effect on the ease or difficulty of handling the material, the selection of equipment, and equipment production rates.

Types of Geotechnical Materials

Steel and concrete are construction materials that are basically homogeneous, and uniform in composition. As such, their behavior can be predicted. Geotechnical materials are just the opposite. By nature they are heterogeneous. In their natural state, they are rarely uniform and can only be worked by comparison to a similar type material with which previous experience has been gained. To accomplish this, soil and rock types must be classified. Soils can be classified according to the sizes of the particles of which they are composed, by their physical properties, or by their behavior when their moisture content varies.

A constructor is concerned primarily with five types of soils: gravel, sand, silt, clay, and organic matter, or with a combination of these types. Different agencies and specification groups denote the sizes of these types of soil differently, causing some confusion. The following size limits represent those set forth by ASTM:

Gravel is rounded or semiround particles of rock that will pass a 3-in. and be retained on a 2.0-mm No. 10 sieve. Sizes larger than 10 in. are usually called boulders.

Sand is disintegrated rock whose particles vary in size from the lower limit of gravel 2.0 mm down to 0.074 mm (No. 200 sieve). It can be classified as coarse or fine sand, depending on the sizes of the grains. Sand is a granular noncohesive material whose particles have a bulky shape.

Silt is a material finer than sand, and thus its particles are smaller than 0.074 mm but larger than 0.005 mm. It is a noncohesive material and it has little or no strength. Silt compacts very poorly.

Clay is a cohesive material whose particles are less than 0.005 mm. The cohesion between the particles gives clays a high strength when air-dried. Clays can be subject to considerable changes in volume with variations in moisture content. They will exhibit plasticity within a range of "water contents." Clay particles are shaped like thin wafers.

Organic matter is a partly decomposed vegetable matter. It has a spongy, unstable structure that will continue to decompose and is chemically reactive. If present in soil that is to be used for construction purposes, organic matter should be removed and replaced with a more suitable soil.

Generally, the soil types are found in nature in some mixed proportions. Table 4.1 presents a classification system based on combinations of soil types.

TABLE 4.1 | Unified soil classification system

Symbol	Primary	Secondary	Supplementary
GW	Coarse-grained soils	Well-graded gravels, gravel-sand mixtures, little or no fines	Wide range of grain size
GP	Coarse-grained soils	Poorly graded gravels, gravel-sand mixtures, little or no fines	Predominantly one size or a range of intermediate sizes missing
GM	Gravel mixed with fines	Silty gravels and gravel-sand-silt mixtures— may be poorly graded	Predominantly one size or a range of intermediate sizes missing
GC	Gravel mixed with fines	Clayey gravels, gravel-sand-clay mixtures, which may be poorly graded	Plastic fines
SW	Clean sands	Well-graded sands, gravelly sands, little or no fines	Wide range in grain sizes
SP	Clean sands	Poorly graded sands, gravelly sands, little or no fines	Predominantly one size or a range of sizes with some intermediate sizes missing
SM	Sands with fines	Silty sands and sand-silt mixtures, which may be poorly graded	Nonplastic fines or fines of low plasticity
SC	Sands with fines	Clayey sands, sand-clay mixtures, which may be poorly graded	Plastic fines
ML	Fine-grained soils	Inorganic silts, clayey silts, rock flour, silty very fine sands	Plastic fines
CL	Fine-grained soils	Inorganic clays of low to medium plasticity, silty sandy or gravelly clays	Plastic fines
OL	Fine-grained soils	Organic silts and organic silt-clay of low plasticity	
MH	Fine-grained soils	Inorganic silts, clayey silts, elastic silts	
CH	Fine-grained soils	Inorganic clays of high plasticity, fat clays	
OH	Fine-grained soils	Organic clays and silty clays of medium to high plasticity	

Symbol classification

COARSE GRAINED MATERIAL
Symbol
G—Gravel grain size from 3″ to No. 4 sieve size
S—Sand grain size from No. 4 to 200 sieve size

FINE GRAINED MATERIAL
Symbol
M—Silt very fine grain size, floury appearance
C—Clay finest grain size, high dry strength— plastic
O—Organic matter partly decomposed, appears fibrous, spongy and dark in color

Subdivision
W—Well graded, little or no fines
P—Poorly graded, little or no fines
M—Concentration of silty or nonplastic fines
C—Concentration of clay or plastic fines

Subdivision
L—Low plastic material, lean soil
H—High plastic material, fat soil

Soils existing under natural conditions may not contain the relative amounts of the desired material types necessary to produce the properties required for construction purposes. For this reason, it may be necessary to obtain soils from several sources and then blend them for use in a fill.

If the material in a borrow pit consists of layers of different types of soils, the specifications for the project may require the use of equipment that will excavate vertically through the layers in order to mix the soil.

Rock was formed by one of three different means:

Igneous rocks solidified from molten masses.

Sedimentary rocks formed in layers settling out of water solutions.

Metamorphic rocks were transformed from material of the first two by heat and pressure.

Their respective formation processes will affect how rocks can be excavated and handled.

Categorization of Materials

In contract documents, excavation is typically categorized as common, rock, muck, or unclassified. *Common* refers to ordinary earth excavation, while the term *unclassified* reflects the lack of clear distinction between soil and rock. The removal of common excavation will not require the use of explosives, although tractors equipped with rippers may be used to loosen consolidated formations. The specific engineering properties of the soil—plasticity, grain-size distribution, and so on—will influence the selection of the appropriate equipment and construction methods.

In construction, *rock* is a material that cannot be removed by ordinary earth-handling equipment.* Rock must be removed by drilling and blasting or some comparable method. This normally results in considerably greater expense than earth excavation. Rock excavation involves the study of the rock type, faulting, dip and strike, and explosive characteristics as the basis for selecting material removal and aggregate production equipment.

Muck includes materials that will decay or produce subsidence in embankments. It is usually a soft organic material having a high water content. Typically, it would include such things as decaying stumps, roots, logs, and humus. These materials are hard to handle and can present special construction problems both at their point of excavation and in transportation and disposal.

You should never price an earth- or rock-handling project without first making a thorough study of the materials. In many cases, the contract documents include geotechnical information. This owner-furnished information provides a starting point for your *independent* investigation. Other good sources of preliminary information are topographic maps, agriculture maps, geologic maps, well logs, and aerial photographs. The investigation is not complete, however, until you make an on-site visit and conduct either drilling or test pit exploration.

Even though many owners provide good geotechnical data that has been put together by qualified engineers, the design engineer's primary concern has been with how well the material will perform structurally. The constructor is interested

---

*Note that this definition of rock will be affected by equipment development. Larger and heavier machines are continually changing the limits of this definition for rock.

in how the material will handle during the construction process and what volume or quantity of material is to be processed to yield the desired final structure.

Soil Weight-Volume Relationships

The primary relationships (see Fig. 4.1) are expressed as defined in Eqs. [4.1] through [4.6].

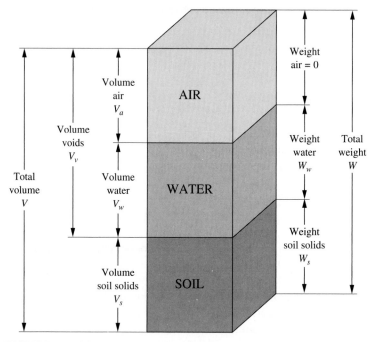

FIGURE 4.1 | Soil mass weight and volume relationships.

$$\text{Unit weight } (\gamma) = \frac{\text{total weight of soil}}{\text{total soil volume}} = \frac{W}{V} \qquad [4.1]$$

$$\text{Dry unit weight } (\gamma_d) = \frac{\text{weight of soil solids}}{\text{total soil volume}} = \frac{W_s}{V} \qquad [4.2]$$

$$\text{Water content } (\omega) = \frac{\text{weight of water in soil}}{\text{weight of soil solids}} = \frac{W_w}{W_s} \qquad [4.3]$$

$$\text{Void ratio } (e) = \frac{\text{volume of voids}}{\text{volume of soil solids}} = \frac{V_v}{V_s} \qquad [4.4]$$

$$\text{Porosity } (n) = \frac{\text{volume of voids}}{\text{total soil volume}} = \frac{V_v}{V} \qquad [4.5]$$

$$\text{Specific gravity } (G_s) = \frac{\text{weight of soil solids/volume of solids}}{\text{unit weight of water}} = \frac{W_s/V_s}{\gamma_w}$$

[4.6]

Many more formulas can be derived from these basic relationships. Two such formulas that are useful in analyzing compaction specifications are

Total soil volume (V) = volume voids (V_v) + volume solids (V_s) **[4.7]**

$$\text{Weight of solids } (W_s) = \frac{\text{weight of soil } (W)}{1 + \text{water content } (\omega)}$$

[4.8]

If unit weights are known which is the usual case, then Eq. [4.8] becomes

$$\gamma_d = \frac{\gamma}{1 + \omega}$$

[4.9]

The usefulness and application of these relationships will be shown by the problem in example 4.1.

Soil Limits

Certain limits of soil consistency—liquid limit, plastic limit—were developed to differentiate between highly plastic, slightly plastic, and nonplastic materials.

Liquid limit (LL). The water content at which a soil passes from the plastic to the liquid state is known as the liquid limit. High LL values are associated with soils of high compressibility. Typically, clays have high LL values; sandy soils have low LL values.

Plastic limit (PL). The water content at which a soil passes from the plastic to the semisolid state. The lowest water content at which a soil can be rolled into an 1/8-in. (3.2-mm)-diameter thread without crumbling.

Plasticity index (PI). The numerical difference between a soil's liquid limit and its plastic limit is its plasticity index (PI = LL − PL). Soils having high PI values are quite compressible and have high cohesion.

On many projects, the specifications will specify a certain material grada-tion, a maximum LL, and a maximum PI. The American Association of State Highway and Transportation Officials (AASHTO) system of soil classification that is the most widely used for highway construction illustrates this point (see Table 4.2).

Volumetric Measure

For bulk materials, volumetric measure varies with the material's position in the construction process (see Fig. 4.2). The same weight of a material will occupy

TABLE 4.2 | AASHTO soil classification system*

General classification	Granular materials (35% or less of total sample passing No. 200)							Silt-clay materials (more than 35% of total sample passing No. 200)			
	A-1		A-3	A-2				A-4	A-5	A-6	A-7
Group classification	A-1-a	A-1-b		A-2-4	A-2-5	A-2-6	A-2-7				A-7-5, A-7-6
Sieve analysis, percentage passing											
No. 10	50 max.										
No. 40	30 max.	50 max.	51 min.								
No. 200	15 max.	25 max.	10 max.	35 max.	35 max.	35 max.	35 max.	36 min.	36 min.	36 min.	36 min.
Characteristics of fraction passing No. 40											
Liquid limit				40 max.	41 min.	40 max.	41 min.	40 max.	41 min.	40 max.	41 min.
Plasticity index	6 max.		NP	10 max.	10 max.	11 min.	11 min.	10 max.	10 max.	11 min.	11 min.
Group index	0		0	0		4 max.		8 max.	12 max.	16 max.	20 max.

*A group index based on a formula that considers particle size, LL, and PI is given at the bottom of the table. The group index indicates the suitability of a given soil for embankment construction. A group index number of "0" indicates a good material while an index of "20" indicates a poor material.

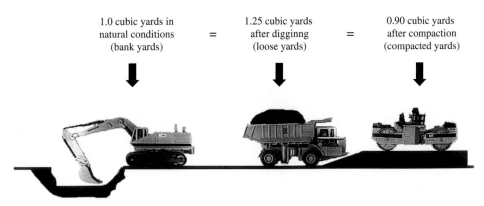

1.0 cubic yards in natural conditions (bank yards) = 1.25 cubic yards after digginng (loose yards) = 0.90 cubic yards after compaction (compacted yards)

FIGURE 4.2 | Material volume changes caused by processing.

different volumes as the material is handled on the project. Soil volume is measured in one of three states:

Bank cubic yard	1 cubic yard (cy) of material as it lies in the *natural* state, bcy
Loose cubic yard	1 cy of material after it has been disturbed by a loading process, lcy
Compacted cubic yard	1 cy of material in the compacted state, also referred to as a net in-place cubic yard, ccy

In planning or estimating a job, the engineer must use a consistent volumetric state in any set of calculations. The necessary consistency of units is achieved by use of shrinkage and swell factors. The *shrinkage factor* is the ratio of the compacted dry weight per unit volume to the bank dry weight per unit volume:

$$\text{Shrinkage factor} = \frac{\text{compacted dry unit weight}}{\text{bank dry unit weight}} \qquad \textbf{[4.10]}$$

The weight shrinkage due to compacting a fill can be expressed as a percent of the original bank measure weight:

$$\text{Shrinkage \%} = \frac{(\text{compacted dry unit weight}) - (\text{bank unit weight})}{\text{compacted unit weight}} \times 100$$

$$\textbf{[4.11]}$$

The *swell factor* is the ratio of the loose dry weight per unit volume to the bank dry weight per unit volume:

$$\text{Swell factor} = \frac{\text{loose dry unit weight}}{\text{bank dry unit weight}} \qquad \textbf{[4.12]}$$

The percent swell, expressed on a gravimetric basis, is

$$\text{Swell \%} = \left(\frac{\text{bank dry unit weight}}{\text{loose dry unit weight}} - 1 \right) \times 100 \qquad \text{[4.13]}$$

Table 4.3 gives representative swell values for different classes of earth. These values will vary with the extent of loosening and compaction. If more accurate values are desired for a specific project, tests should be made on several samples of the earth taken from different depths and different locations within the proposed cut. The test can be made by weighing a given volume of undisturbed, loose, and compacted earth.

TABLE 4.3 | Representative properties of earth and rock*

Material	Bank weight lb/cy	Bank weight kg/m³	Loose weight lb/cy	Loose weight kg/m³	Percent swell	Swell factor*
Clay, dry	2,700	1,600	2,000	1,185	35	0.74
Clay, wet	3,000	1,780	2,200	1,305	35	0.74
Earth, dry	2,800	1,660	2,240	1,325	25	0.80
Earth, wet	3,200	1,895	2,580	1,528	25	0.80
Earth and gravel	3,200	1,895	2,600	1,575	20	0.83
Gravel, dry	2,800	1,660	2,490	1,475	12	0.89
Gravel, wet	3,400	2,020	2,980	1,765	14	0.88
Limestone	4,400	2,610	2,750	1,630	60	0.63
Rock, well blasted	4,200	2,490	2,640	1,565	60	0.63
Sand, dry	2,600	1,542	2,260	1,340	15	0.87
Sand, wet	2,700	1,600	2,360	1,400	15	0.87
Shale	3,500	2,075	2,480	1,470	40	0.71

*The swell factor is equal to the loose weight divided by the bank weight per unit volume.

EXAMPLE 4.1

An earth fill, when completed, will occupy a net volume of 187,000 cy. The borrow material that will be used to construct this fill is a stiff clay. In its "bank" condition, the borrow material has a wet unit weight of 129 lb per cubic foot (cf) (γ), a water content (ω%) of 16.5%, and an in-place void ratio (e) of 0.620. The fill will be constructed in layers of 8-in. depth, loose measure, and compacted to a dry unit weight (γ_d) of 114 lb per cf at a water content of 18.3%. Compute the required volume of borrow pit excavation.

$$\text{Borrow } \gamma_d = \frac{129}{(1 + 0.165)} \Rightarrow 111 \text{ lb/cu ft}$$

Fill γ_d = 114 lb/cu ft

$$\underbrace{187,000 \text{ cy} \times \frac{27 \text{ cf}}{\text{cy}} \times \frac{114 \text{ lb}}{\text{cf}}}_{\text{Fill}} = \underbrace{x \times \frac{27 \text{ cf}}{\text{cy}} \times \frac{111 \text{ lb}}{\text{cf}}}_{\text{Borrow}}$$

$$187,000 \times \frac{114}{111} = 192,054 \ cy, \ borrow \ required$$

Note that the element 114/111 is the shrinkage factor 1.03.

The key to solving this type of problem is unit weight of the solid particles (dry weight) that make up the soil mass. In the construction process, the specifications may demand that water either be expelled or added to the soil mass. In this example, the contractor would be required to add water to the borrow to increase the moisture content from 16.5 to 18.3%. Adjusting for the extra borrow cubic yards required to make one fill cubic yard, note the water difference:

Fill		Borrow	
γ 114 × 1.183 = 135 lb/cf		129 lb/cf	
γ_d	114 lb/cf	−111 lb/cf	
		18 lb/cf	
Water	21 lb/cf	× 1.03 (shrinkage factor)	
		19 lb/cf	

To achieve the desired fill density and water content, the contractor will have to add water. This water must be hauled in by water wagon and is not part of the in-place borrow unit weight. The quantity of water that must be added is

Fill
Water content = 0.183

$$187,000 \ cy \ \times \ \frac{114 \ lb}{cf} \ \times \ 27 \ \frac{cf}{cy} \ \times \ 0.183 = 105,332,238 \ lb \ water$$

Borrow
Water content = 0.165

$$192,054 \ cy \ \times \ \frac{111 \ lb}{cf} \ \times \ 27 \ \frac{cf}{cy} \ \times \ 0.165 = \frac{94,971,663 \ lb}{10,360,575 \ lb} \ water$$

Which is 1,241,941 gal or approximately 6.5 gal per cy of borrow.

The method of soil preparation prior to compaction is an important factor whose influence on successful results is not sufficiently appreciated. This includes adding water or conversely drying the soil. The blending of the excavation material to achieve a homogeneous composition and uniform water content within a placed layer is especially important.

Constructors commonly apply what is referred to as a *swell factor* when estimating jobs. This rule-of-thumb factor should not be confused with the previously defined factors. The term swell factor is used in this case because of how the number is applied. The embankment yardage of the job is multiplied by this factor; that is, it is swelled in order to put it in the same reference units as the borrow. The job is then figured in *borrow yards*. This swell factor is strictly a guess—based on past experience with similar materials. It may, also, reflect consideration of the project design. A case in point would be when the embankment is less than 3 ft in total height, in which case more embankment material will be required to compensate for the compaction of the natural ground below the fill. The constructor would therefore apply a higher swell factor when calculating the required borrow material for fills of minimum height.

COMPACTION SPECIFICATION AND CONTROL

INTRODUCTION

Prior to preparing the specifications for a project, representative soil samples are usually collected and tested in the laboratory to determine material properties. Normal testing would include grain-size analysis, because the size of the grains and the distribution of those sizes are important properties that affect a soil's suitability.

Maximum Dry Density/Optimum Moisture

Another critical test is the laboratory-established compaction curve. From such a curve, the maximum dry unit weight (density) and the percentage of water required to achieve maximum density can be determined. This percentage of water, which corresponds to the maximum dry density (for a given compactive effort), is known as the *optimum water content*. It is the amount of water required for a given soil to reach maximum density.

Figure 4.3 shows two compaction curves based on different input energy levels. The curves are plotted in dry weight (in pounds per cubic foot) against water content (percent by dry weight). Each illustrates the effect of varying amounts of moisture on the density of a soil subjected to given compactive effort (energy input level). The two energy levels depicted are known as standard and modified Proctor tests. It will be noted that the modified Proctor (higher input energy) gives a higher density at a lower moisture content than the standard Proctor. For the material depicted by the curves in Fig. 4.3, the optimum moisture for the standard Proctor is 16% versus 12% for the modified Proctor.

The difference in optimum water content is a result of mechanical energy replacing the lubricating action of the water during the densification process. The contractor working to a modified Proctor specification (higher input energy) will have to plan to either make more passes with the compaction equipment or use heavier compaction equipment on the project. But at the same time, there will be less of a requirement to haul water and to mix water into the material.

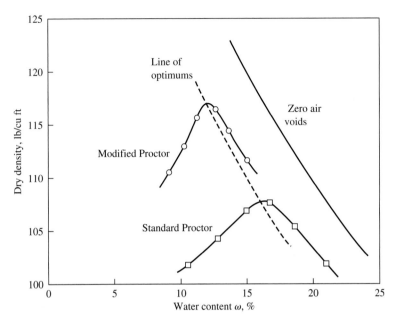

FIGURE 4.3 | Standard and modified compaction curves.

COMPACTION TESTS

The laboratory compaction test that is accepted by highway departments and other agencies is the Proctor test. For this test, a sample of soil consisting of 1/4 in. and finer material is used. The sample is placed in a steel mold in three equal layers. The cylindrical steel mold has an inside diameter of 4.0 in. and a height of 4.59 in. In the standard test, each of three equal layers are compacted by dropping a 5.5-lb rammer, with a 2-in. circular base, 25 times from a height of 12 in. above the specimen (see Fig. 4.4). The specimen is removed from the mold and the entire specimen is immediately weighed. Then a sample of the specimen is taken and weighed. That sample is dried to a constant weight to remove all moisture and weighed again so that the water content can be determined. With the water content information, the dry weight of the specimen can be determined. The test is repeated, using varying water content specimens, until the water content that produces the maximum density is determined. This test is designated as ASTM D-698 or AASHTO T99.

The modified Proctor test, designated as ASTM D-1557 or AASHTO T 180, is performed in a similar manner, except the applied energy is greater because a 10-lb rammer is dropped 18 in. on each of five equal layers (see Fig. 4.4).

Compaction Control

The specifications for a project may require a contractor to compact the soil to a 100% relative density, based on the standard Proctor test or a laboratory test at some other energy level. If the maximum laboratory dry density of the soil is

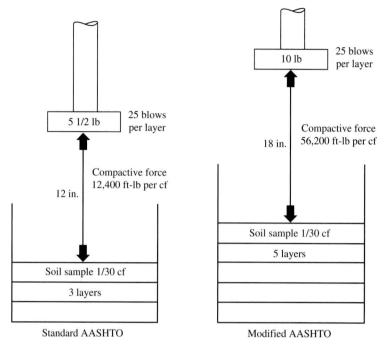

FIGURE 4.4 I Standard and modified compaction test.

determined to be 120 lb per cf, the contractor must compact the soil in the field to a density of 120 lb per cf.

Field verification tests of achieved compaction can be conducted by any of several accepted methods: sandcone, balloon, or nuclear. The first two methods are destructive tests. They involve

- Excavating a hole in the compacted fill and weighing the excavated material,
- Determining the water content of the excavated material,
- Measuring the volume of the resulting hole by use of the sandcone or a water-filled balloon, and
- Computing the density based on the obtained total weight of the excavated material and hole volume.

The dry density conversion can be made because the water content is known. Difficulties associated with such methods are that (1) it is too time consuming to conduct sufficient tests for statistical analysis, (2) there are problems with oversized particles, and (3) there is a time delay in determining the water content. Because the tests are usually conducted on each placement lift, delays in testing and acceptance can delay construction operations.

Nuclear Compaction Test

Nuclear methods are used extensively to determine the water content and density of soils. The instrument required for this test can be easily transported to the fill and placed at the desired test location, and within a few minutes, the results can be read directly from the digital display.

The device uses the Compton effect of gamma-ray scattering for density determinations and hydrogenous thermalization of fast neutrons for moisture determinations. The emitted rays enter the ground, where they are partially absorbed and partially reflected. Reflected rays pass through Geiger-Müller tubes in the surface gauge. Counts per minute are read directly on a reflected-ray counter gauge and are related to moisture and density calibration curves.

Advantages of the nuclear method when compared with other methods include the fact that it

1. Decreases the time required for a test from as much as a day to a few minutes, thereby eliminating potentially excessive construction delays, and because more samples can be taken per unit of time, the engineer is better able to characterize the achieved density.

2. Is nondestructive in that it does not require the removal of soil samples from the site of the tests.

3. Provides a means of performing density tests on soils containing large-size aggregates.

4. Reduces or eliminates, when properly calibrated and used correctly, the effect of the personal element, and possible errors. Erratic results can be easily and quickly rechecked.

Because nuclear tests are conducted with instruments that present a potential source of radiation, an operator needs to be certified and should exercise reasonable care to ensure that no harm can result from the use of the instruments. By following the instructions furnished with the instruments and by exercising proper care, exposure can be kept well below the limits set by the Nuclear Regulatory Commission (NRC). In the United States and most individual states, a license is required to own, possess, or use nuclear-type instruments.

GeoGauge

Another nondestructive device that does not require the removal of soil samples from the site of the tests is the GeoGauge. This device is very new to the field. In 1994 the Minnesota Department of Transportation tested the first prototype models in a program sponsored by the Federal Highway Administration (FHWA). Production models are currently available and each year more agencies are conducting independent field evaluations [8].

The GeoGauge is a portable instrument that provides a simple, rapid, and precise means of directly measuring lift stiffness and soil modulus. It also provides an alternative means of measuring soil density. This device imparts very small displacements to the soil ($< 1.27 \times 10^{-6}$ m or $< 0.00005''$) at 25 steady state frequencies between 100 and 196 Hz. Stiffness is determined at each frequency, and the average is displayed. The entire process takes about 1 min. If a Poisson's ratio is assumed and the gauge's physical dimensions are known, shear and Young's modulus can be derived. The gauge weighs about 10 kg ($\sim$ 22 lb), is 28 cm ($\sim$ 11'') in diameter and 25.4 cm (~ 10″) tall, and rests on the soil surface on a ring-shaped footing. Six disposable D-cell batteries power the gauge.

Laboratory versus Field

Maximum dry density is only a maximum for a specific compaction effort (input energy level) and the method by which that effort is applied. If more energy is applied in the field, a density greater than 100% of the laboratory value can be achieved. Dissimilar materials have individual curves and maximum values for the same input energy (Fig. 4.5). Well-graded sands have a higher dry density than uniform soils. As plasticity increases, the dry density of clay soils decreases.

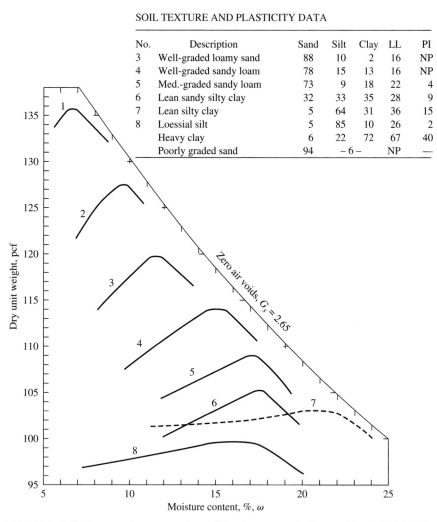

SOIL TEXTURE AND PLASTICITY DATA

No.	Description	Sand	Silt	Clay	LL	PI
3	Well-graded loamy sand	88	10	2	16	NP
4	Well-graded sandy loam	78	15	13	16	NP
5	Med.-graded sandy loam	73	9	18	22	4
6	Lean sandy silty clay	32	33	35	28	9
7	Lean silty clay	5	64	31	36	15
8	Loessial silt	5	85	10	26	2
	Heavy clay	6	22	72	67	40
	Poorly graded sand	94	– 6 –		NP	—

FIGURE 4.5 I Compaction curves for eight soils compacted according to AASHTO T99.

Source: the Highway Research Board.

It should be noted that at some point, higher water contents result in decreased density. This is because initially the water serves to "lubricate" the soil grains and helps the mechanical compaction operation move them into a compact physical arrangement. But the density of water is less than that of the soil solids, and at water contents above optimum, water is replacing soil grains in the matrix. If compaction is attempted at a water content that is much above optimum, no amount of effort will overcome these physical facts. Under such conditions, extra compactive effort will be wasted work. In fact, soils can be "over compacted." Shear planes are established and there is a large reduction in strength.

SOIL PROCESSING

The optimum water content for compaction varies from about 12 to 25% for fine-grained soils and from 7 to 12% for well-graded granular soils. Since it is difficult to attain and maintain the exact optimum water content, normal practice is to work within an acceptable moisture range. This range, which is usually ±2% of optimum, is based on attaining the maximum density with the minimum compactive effort.

Adding Water to Soil

If the water content of a soil is below the optimum moisture range, water must be added to the soil prior to compaction. When it is necessary to add water consider the

- Amount of water required.
- Rate of water application.
- Method of application.
- Effects of the climate and weather.

Water can be added to the soil at the borrow pit or in-place (at the construction site). When processing granular materials, the best results are usually obtained by adding water in-place. After water is added, it must be thoroughly and uniformly mixed with the soil.

Amount of Water Required

It is essential to determine the amount of water required to achieve a soil water content within the acceptable moisture range for compaction. The amount of water that must be added or removed is normally computed in gallons per station (100 ft of length); therefore, the volume in the following formula would normally be that for one station length. The computation is based on the dry weight of the soil and the compacted volume. This formula can be used to compute the amount of water to be added or removed from the soil:

Gallons = desired dry density pounds per cf (pcf)

$$\times \frac{(\text{desired water content }\%) - (\text{water content borrow }\%)}{100}$$

$$\times \frac{\text{compacted vol. of soil (cf)}}{8.33 \text{ lb per gal}} \qquad \text{[4.14]}$$

When having to add water, it is normally good practice to adjust the desired moisture content 2% above optimum, but this depends on the environmental conditions (temperature and wind) and the soil type. A negative answer indicates that water must be removed from the borrow material before it is compacted on the fill. The 8.33 lb per gal is the weight of a gallon of water.

EXAMPLE 4.2

Job specifications require placement of the embankment fill soil in 6-in. (compacted) lifts. The desired dry unit weight of the embankment is 120 pcf. The laboratory compaction curve indicates that the optimum water content, sometimes referred to as optimum moisture content, (OMC) of the soil is 12%. Soil tests indicated that the moisture content of the borrow material is 5%. The roadway lift to be placed is 40 ft wide. Compute the amount of water in gallons to add on a per station basis for each lift of material.

$$\text{Gallons per station} = 120 \text{ pcf} \times \frac{(12\% \text{ (OMC)} - 5\%)}{100} \times \frac{(40 \text{ ft} \times 100 \text{ ft} \times 0.5 \text{ ft})}{8.33 \text{ lb/gal}}$$

$$\text{Gallons per station} = 120 \text{ pcf} \times 0.07 \times \frac{2,000 \text{ cf}}{8.33 \text{ lb/gal}} \Rightarrow 2,017 \text{ gallons per station}$$

Application Rate

Once the total amount of water has been calculated, the application rate can be calculated. The water application rate is normally calculated in gallons per square yard, using the following formula:

Gallons per square yard =

$$\text{desired dry density of soil (pcf)} \times \frac{(\% \text{ moisture added or removed})}{100}$$

$$\times \text{ lift thickness (ft) (compacted)} \times \frac{9 \text{ sf/sy}}{8.33 \text{ lb/gal}} \quad \textbf{[4.15]}$$

EXAMPLE 4.3

Using the data from Example 4.2, determine the required application rate in gallons per square yard:

$$\text{Gallons per square yard} = 120 \text{ pcf} \times 0.07 \times 0.5 \text{ ft} \times \frac{9 \text{ sf/sy}}{8.33 \text{ lb/gal}}$$

$$\Rightarrow 4.5 \text{ gallons per sy}$$

Application Methods

Once the application rate has been calculated, the method of application must be determined. Regardless of which method of application is used, it is important to ensure that the proper application rate is achieved and that the water is uniformly distributed.

Water Distributor   On construction projects, the most common method of adding water to a soil is with a water distributor. Water distributors are designed to evenly distribute the correct amount of water over the fill. These truck-mounted or towed water distributors (see Fig. 4.6) are designed to distribute water under various pressures, or by gravity feed. Many distributors are equipped with rear-mounted spray bars. The operator can maintain the water application rate by controlling the forward speed of the vehicle.

Ponding   If time is available, water can be added to a soil by ponding or pre-wetting the area until the desired depth of penetration is achieved (see Fig. 4.7). With this method, it is difficult to control the application rate. Ponding usually requires several days to achieve a uniform moisture distribution.

Reducing the Moisture Content

As previously stated, soil that contains more water than desired (above the optimum moisture range) is correspondingly difficult to compact. Excess water makes achieving the desired density very difficult. In these cases, steps must be taken to reduce the moisture content to within the required moisture range. Drying actions may be as simple as aerating the soil. They may, however, be as

FIGURE 4.6 | Water wagon delivering water to the fill.

FIGURE 4.7 | Prewetting borrow material using a sprinkler system.

complicated as adding a soil stabilization agent that actually changes the physical properties of the soil. Lime or fly ash is the typical stabilization agent for fine-grained soils. If a high water table is causing the excess moisture, some form of subsurface drainage may be required before the soil's moisture content can be reduced.

The most common method of reducing the moisture is to scarify the soil prior to compaction. This can be accomplished with either the scarifying teeth or rippers on a motor grader (see Fig. 4.8), or by disking (see Fig. 4.9) the soil. A motor grader can also use its blade to toe the soil over into furrows to expose more material for drying.

Effects of Weather

Weather conditions substantially affect soil moisture content. Cold, rainy, cloudy, and calm weather will allow a soil to retain water. Hot, dry, sunny, and windy weather is conducive to drying the soil. In a desert climate, evaporation claims a large amount of water intended for the soil lift. Thus, for a desert project, the engineer might have to go as high as 6% above the optimum water content as a target for all water application calculations so that the actual water content will fall very near to the desired content when the material is placed and compacted.

Mixing and Blending

Whether adding water to a soil to increase the water content or adding a drying agent to reduce it, it is essential to mix the water or drying agent thoroughly and

FIGURE 4.8 I A motor grader using rear rippers to scarify material.

FIGURE 4.9 I Disk harrow used to scarify soil.

uniformly with the soil. Even if additional water is not needed, mixing may still be necessary to ensure a uniform distribution of the existing moisture. Mixing can be accomplished using motor graders, farm disks, or rotary cultivators.

Conventional motor graders can be used to mix or blend a soil additive (water or stabilizing agent) by windrowing the material from one side of the working lane to the other (see Fig. 4.10).

FIGURE 4.10 I A motor grader using its blade to mix a soil.

COMPACTION EQUIPMENT

COMPACTION OF GEOTECHNICAL MATERIALS

With time, material will settle or compact itself naturally, but the object of compaction is to achieve the required density quickly. The earliest recorded use of compaction can be found in the Roman Empire records of their road construction projects. The Romans realized that compaction would improve the engineering properties of soils; therefore, they used large cylindrical stone rollers to achieve mechanical densification of their road bases (see Fig. 4.11).

Obtaining a greater soil unit weight is not the direct objective of compaction. The reason for compaction is improve soil properties to

1. Reduce or prevent settlements.
2. Increase strength.
3. Improve bearing capacity.
4. Control volume changes.
5. Lower permeability.

Density, however, is the most commonly used parameter for specifying construction operations because there is a direct correlation between these properties and a soil's density. Construction contract documents usually call for achieving a specified density, even though one of the other soil properties is the crucial objective.

There may be other methods whereby the desired properties could be attained, but by far the most widely used method of soil strengthening is compaction of the soil at optimum moisture. The benefits of proper compaction are

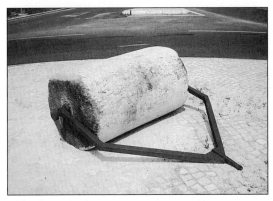

FIGURE 4.11 I A cylindrical stone roller.

enormous, far outweighing their costs. Typically, a uniform layer, or lift, of soil from 4 to 12 in. thick is compacted by means of several passes of mechanized compaction equipment.

TYPES OF COMPACTING EQUIPMENT

Applying energy to a soil by one or more of these methods will cause compaction:

1. Impact—sharp blow
2. Pressure—static weight
3. Vibration—shaking
4. Kneading—manipulation or rearranging

The effectiveness of different compaction methods is dependent on the individual soil type being manipulated. Appropriate compaction methods based on soil type are identified in Table 4.4.

TABLE 4.4 I Soil types versus the method of compaction

Material	Impact	Pressure	Vibration	Kneading
Gravel	Poor	No	Good	Very good
Sand	Poor	No	Excellent	Good
Silt	Good	Good	Poor	Excellent
Clay	Excellent with confinement	Very good	No	Good

Manufacturers have developed distinct compactors that incorporate at least one of the compaction methods and in some cases more than one into their performance capabilities. Many types of compacting equipment are available, including:

1. Sheepsfoot rollers.
2. Tamping rollers.
3. Smooth drum vibratory soil compactors.

4. Pad drum vibratory soil compactors.

5. Pneumatic-tired rollers.

Table 4.5 summarizes the principal method of compaction for the various types of compactors. There are also static, smooth steel-wheel rollers. These generally consist of two tandem drums, one in front and one behind. Steel three-wheel rollers are another version of this type of compactor, but they are not very common today. Vibratory rollers are more efficient than static steel-wheel rollers for earthwork and have largely replaced them.

TABLE 4.5 | Principal method of compaction used by various compactors

Compactor type	Impact	Pressure	Vibration	Kneading
Sheepsfoot		X		
Tamping foot	X	X		
Vibrating smooth	X		X	
Vibrating padfoot	X		X	
Pneumatic		X		X

On some projects, it may be desirable to use more than one type of equipment to attain the desired results and to achieve the greatest economy. The ultimate goal is to construct a quality embankment in the shortest time at the least cost, and that means the compaction equipment must be matched to the material. Therefore, the job should always be closely examined and samples taken of the excavation or borrow materials. The proper excavation and compaction equipment cannot be selected until the soils are identified. Table 4.6 provides guidance for selecting compaction equipment based on the type of material that must be compacted. As seen in the table, if the required density is not achieved within four to eight coverages, a different type of compactor should be considered.

TABLE 4.6 | Appropriate project compaction equipment based on material type

Material	Lift thickness (in.)	Number of passes	Compactor type	Comments
Gravel	8–12	3–5	Vib. padfoot	Foot psi 150–200
			Vib. smooth	—
			Pneumatic	Tire psi 35–130
			Sheepsfoot	Foot psi 150–200
Sand	8–10	3–5	Vib. padfoot	—
			Vib. smooth	—
			Pneumatic	Tire psi 35–65
			Smooth static	Tandem 10–15 ton
Silt	6–8	4–8	Vib. padfoot	Foot psi 200–400
			Tamping foot	—
			Pneumatic	Tire psi 35–50
			Sheepsfoot	Foot psi 200–400
Clay	4–6	4–6	Vib. padfoot	Foot psi 250–500
			Tamping foot	—
			Sheepsfoot	Foot psi 250–500

Rock fills are usually spread in 18 to 48 in. lifts. Attention to spreading the material in a uniform lift is vital to achieving density during the compaction process. Consistent spreading helps to fill voids and orients the rocks so as to provide the compaction equipment with an even surface. The largest possible smooth-drum vibratory rollers are used for deep rock lifts.

Sheepsfoot Roller

These rollers are usually found as towed drum models (see Fig. 4.12). The sheepsfoot roller is suitable for compacting all fined-grained materials, but is generally not suitable for use on cohesionless granular materials. They are steel wheels equipped with cylindrical pads (or feet) normally less than 10 in. in length. Varying the weight of the roller by the use of ballast in the drum will vary the foot-contact pressure.

The pads on a sheepsfoot drum penetrate through the top lift and actually compact the lift below. When the drum rotates the pads out of the soil, they kick up or fluff the material because of their shape. Sheepsfoot rollers can only work at speeds from 4 to 6 mph. Usually 6 to 10 passes will be needed to compact an 8-in. clay lift.

Since the sheepsfoot roller tends to aerate the soil as it compacts, it is ideally suited for working soils that have moisture contents above the acceptable moisture range. The sheepsfoot roller does not adequately compact the upper 2

FIGURE 4.12 | A two-drum towed sheepsfoot roller.

to 3 in. of a lift, and it should, therefore, be followed by a lighter pneumatic-tired or steel-wheeled roller if no succeeding lift is to be placed.

Tamping Rollers

Tamping foot compactors (Fig. 4.13) are high-speed, self-propelled, nonvibratory rollers. These rollers usually have four steel padded wheels and can be equipped with a small blade to help level the lift. The pads are tapered with an oval or rectangular face. The pad face is smaller than the base of the pad at the drum. As a tamping roller moves over the surface, the feet penetrate the soil to produce a kneading action and a pressure to mix and compact the soil from the bottom to the top of the layer. With repeated passages of the roller over the surface, the penetration of the feet decreases until the roller is said to walk out of the fill. Because the pads are tapered, a tamping foot roller can walk out of the lift without fluffing the soil. If it does not walk out, the roller is too heavy or the soil is too wet and the roller is shearing the soil.

The working speed for these rollers is in the 8- to 12-mph range. Generally two to three passes over an 8- to 12-in. lift will achieve density, but this is dependent on the size of the roller. Four passes may be necessary in poorly graded plastic silt or very fine clays. A tamping foot roller is effective on all soils except clean sand. To realize their true economical compaction potential, they need

FIGURE 4.13 I Self-propelled tamping roller with blade.

long uninterrupted passes so the roller can build up speed, which generates high production.

Like the sheepsfoot roller, tamping foot compactors do not adequately compact the upper 2 to 3 in. of a lift. Therefore, if a succeeding lift is not going to be placed, follow up with a pneumatic-tired or smooth-drum roller to complete the compaction or to seal the surface.

Vibrating Compactors

Vibration creates impact forces, and these forces result in greater compacting power than an equivalent static load. This fact is the economics behind a vibratory compactor. The impact forces are higher than the static forces because the vibrating drum converts potential energy into kinetic energy. Vibratory compactors may have one or more drums. Typically on two-drum models, one drum is powered to transmit unit propulsion. Single-drum models usually have two rubber-tire drive wheels. There are also towed vibratory compactors.

Certain types of soils such as sand, gravel, and relatively large shot rock respond quite well to compaction produced by a combination of pressure and vibration. When these materials are vibrated, the particles shift their positions and nestle more closely with adjacent particles to increase the density of the mass.

Vibrating drum rollers are actuated by an eccentric shaft that produces the vibratory action. The eccentric shaft need be only a body that rotates about an axis other than the one through the center of mass. The vibrating mass (drum) is always isolated from the main frame of the roller. Vibrations normally vary from 1,000 to 5,000 per min.

Vibration has two measurements—amplitude, which is the measurement of the movement, or throw, and frequency, which is the rate of the movement, or number of vibrations (oscillations) per second or minute (vpm). The amplitude controls the effective area, or depth to which the vibration is transmitted into the soil, while the frequency determines the number of blows or oscillations that are transmitted in a period of time.

The impacts imparted by the vibrations produce pressure waves that set the soil particles in motion, producing compaction. In compacting granular material, frequency (the number of blows in a given period) is usually the critical parameter as opposed to amplitude.

Compaction results are a function of the frequency of the blows and the force of the blows, and the time period over which the blows are applied. The frequency/time relationship accounts for the slower working speed requirement when using vibratory compactors. Working speed is important as it dictates how long a particular part of the fill is compacted. A working speed of 2 to 4 mph provides the best results when using vibratory compactors.

Smooth Drum Vibratory Soil Compactors

The smooth drum compactors, whether single- or dual-drum models, generate three compactive forces: (1) pressure, (2) impact, and (3) vibration. These rollers (see Fig. 4.14) are most effective on granular materials, with particle sizes ranging

FIGURE 4.14 I Smooth drum vibratory soil compactor.

from large rocks to fine sand. They can be used on semicohesive soils with up to about 10% of the material having a PI of 5 or greater. Large steel-drum vibratory rollers can be effective on rock lifts as thick as 3 ft.

Padded Drum Vibratory Soil Compactors

These rollers (see Fig. 4.15) are effective on soils with up to 50% of the material having a PI of 5 or greater. The pads are involuted to walk out of the lift without fluffing the soil. The typical lift thickness for padded drum units on cohesive soil is 12 to 18 in. These units are sometimes equipped with a leveling blade.

Small walk-behind vibratory rollers having widths in the range of 24 to 38 in. are available (see Fig. 4.16). These units are designed specifically for trench work or for working in confined areas. The drums of the roller extend beyond the sides of the roller body, so the compaction can be accomplished adjacent to the trench walls. Many of these small compactors can be equipped with remote control systems that permit the operator to control the roller while not having to actually enter the trench. Most remote control systems use a digitized radio frequency, this eliminates the need to have control cables dragged around the construction site.

Pneumatic-Tired Rollers

These are surface rollers that apply the principle of kneading action to effect compaction below the surface. They may be self-propelled (see Fig. 4.17) or towed.

FIGURE 4.15 | Padded drum vibratory soil compactor with a leveling blade.

FIGURE 4.16 | Padded drum walk-behind vibratory roller.

FIGURE 4.17 | Self-propelled pneumatic roller.

The small-tired units usually have two tandem axles with four to five wheels on each axle. The wheels oscillate, enabling them to follow the surface contour and reach into low areas for uniform compaction. The rear tires are spaced to travel over the surfaces between the front tires, which produces a complete coverage of the surface. The wheels may be mounted slightly out of line with the axle, giving them a weaving action (hence the name "wobbly wheel") to increase the kneading action of the soil. By adding ballast, the weight of a unit can be varied to suit the material being compacted.

Small pneumatics are not suited for high-production, thick lift embankment compaction projects. They are used on small- to medium-size soil compaction jobs, primarily on bladed granular base materials.

Large-tired rollers are available in sizes varying from 15 to 200 tons gross weight (Fig. 4.18). They utilize two or more big earth-moving tires on a single axle. The air pressure in the tires may vary from 80 to 150 psi (pounds per square inch). Because of the heavy loads and high tire pressures, they are capable of compacting all types of soils to greater depths. The expense is in propelling these large units over the lift, as they require tractors having considerable drawbar pull and traction.

These units are frequently used to proof roll roadway subgrades and airfields bases, and on earth-fill dams. On the Painted Rock Dam in Arizona, 50-ton rubber-tired rollers were used to compact sandy gravel embankment material. The pervious material was prewetted in the borrow area and then placed in the embankment in 24-in. lifts. To achieve density, four passes of the 50-ton roller were required.

FIGURE 4.18 | Fifty-ton pneumatic roller being used to proof roll a roadway subgrade.

Because the area of contact between a tire and the ground surface over which it passes varies with the air pressure in the tire, specifying the total weight or the weight per wheel is not necessarily a satisfactory method of indicating the compacting ability of a pneumatic roller. Four parameters must be known to determine the compacting ability of pneumatic rollers:

1. Wheel load.
2. Tire size.
3. Tire ply.
4. Inflation pressure.

Pneumatic-Tired Rollers with Variable Inflation Pressures

When a pneumatic-tired roller is used to compact soil through all stages of density, the first passes over a lift should be made with relatively low tire pressures to increase flotation and ground coverage. However, as the soil is compacted, the air pressure in the tires should be increased up to the maximum specified value for the final pass. Prior to the development of rollers having the capability of varying their tire pressure while in operation, it was necessary to stop the rolling and either (1) adjust the pressure in the tires, (2) vary the weight of the ballast on the roller, or (3) keep rollers of different weights and tire pressures on a project to provide units to fit the particular needs of a given compaction condition.

Several manufacturers produce rollers that are equipped to enable the operator to vary the tire pressure without stopping the machine. The first passes are made with relatively low tire pressures. As the soil is compacted, the tire pressure is increased to suit the particular conditions of the soil. The use of this type of roller usually enables adequate compaction with fewer passes than are required by constant pressure rollers.

FIGURE 4.19 | Four-sided impact compactor.

Towed Impact Compactors

Beginning in 1949, engineers in South Africa began experimenting with impact compactors or "square wheels." These compactors have used three-, four- (see Fig. 4.19), and five-sided drums. As the compactor is towed, the drum rotates, lifting itself up on edge, and then falls back to earth. The impact of the drum striking the ground provides the compactive force. Much of the impetus for this compactor design was the need to develop a high-energy compactive device that could be used for densifying materials at low moisture content in arid regions. There was also the desire to have a device that could help collapse the unstable structure of certain soils often found in arid regions.

These compactors can be used on a wide range of materials: rock, sand, gravel, silt, and clay. They will handle lifts up to 3 ft, and because they impart such high energy to the ground, density can be achieved over a wider range of moisture contents.

Compaction Wheels

To avoid the hazards of having to have men working in trenches, a compaction wheel attached to an excavator boom is often used to achieve compaction when backfilling utility trenches (see Fig. 4.20).

Manually Operated Vibratory-Plate Compactors

Figure 4.21 illustrates a self-propelled vibratory-plate compactor used for consolidating soils and asphalt concrete in locations where large units are not practical. These gasoline- or diesel-powered units are rated by centrifugal force,

FIGURE 4.20 | Compaction wheel mounted on a hydraulic excavator boom.

FIGURE 4.21 | Self-propelled vibratory-plate compactor.

FIGURE 4.22 | Manually operated rammers.

exciter revolutions per minute, depth of vibration penetration (lift), feet per minute travel, and area of coverage per hour.

Manually Operated Rammer Compactors

Gasoline-engine-driven rammers (see Fig. 4.22) are used for compacting cohesive or mixed soils in confined areas. These units range in impact from about 300 to 900 foot-pounds (ft-lb) per sec at an impact rate up to 850 per min, depending on the specific model. Performance criteria include pounds per blow, area covered per hour, and depth of compaction (lift) in inches. Rammers are self-propelled in that each blow moves them ahead slightly to contact new area.

Small compactors such as the self-propelled vibratory-plate or the rammer will provide adequate compaction if

1. Lift thickness is small (usually 3 to 4 in.),
2. Moisture content is carefully controlled, and
3. Coverages are sufficient.

The primary causes of density problems when backfilling utility trenches are (1) an inadequate number of coverages with the small equipment that must be used in the confined space, (2) lifts that are too thick, and (3) inconsistent control of moisture.

ROLLER PRODUCTION ESTIMATING

The compaction equipment used on a project must have a production capability matched to that of the excavation, hauling, and spreading equipment. Usually, excavation or hauling capability will set the expected maximum production for the job. The production formula for a compactor is

$$\text{Compacted cubic yards per hour} = \frac{16.3 \times W \times S \times L \times \text{efficiency}}{n} \qquad \textbf{[4.16]}$$

where

W = compacted width per roller pass in feet

S = average roller speed in miles per hour

L = compacted lift thickness in inches

n = number of roller passes required to achieve the required density

The computed production is in compacted cubic yards (ccy), so it will be necessary to apply a shrinkage factor to convert the production to bank cubic yards (bcy), which is how the excavation and hauling production is usually expressed.

EXAMPLE 4.4

A self-propelled tamping foot compactor will be used to compact a fill being constructed of clay material. Field tests have shown that the required density can be achieved with four passes of the roller operating at an average speed of 3 mph. The compacted lift will have a thickness of 6 in. The compacting width of this machine is 7 ft. One bank cubic yard equals 0.83 compacted cubic yards. The scraper production, estimated for the project, is 510 bcy per hour. How many rollers will be required to maintain this production? Assume a 50-min hour efficiency.

$$\text{Compacted cubic yards per hour} = \frac{16.3 \times 7 \times 3 \times 6 \times 50/60}{4} \Rightarrow 428 \text{ ccy/hr}$$

$$\frac{428 \text{ ccy per hour}}{0.83} = 516 \text{ bcy/hr}$$

$$\frac{510 \text{ bcy/hr required}}{516 \text{ bcy/hr}} = 0.99$$

Therefore only one roller will be required.

VIBROCOMPACTION METHODS

There are two distinct vibrocompaction techniques: (1) the vibrating pile method, which utilizes a top-mounted vibrator to create vibration in a vertical mode, and (2) vibroflotation equipment, which develops horizontal motion in the penetrator. Both are applicable for densification of saturated cohesionless soils having no more than 20% fines.

Vibrating Pile Method

Vibrator compaction is accomplished by the use of a vibratory pile-driving apparatus, together with an open-end tubular pile (Fig. 4.23). Using a square spacing pattern, the pile is driven into and extracted from the soil. The spacing is usually determined based on project-specific test panel construction. When the driving and extracting phases are accomplished, densification of the soil occurs both inside and outside the pile, with the concentration of vibratory energy creating extreme densification inside the pile and with densification outside the pile diminishing with distance.

FIGURE 4.23 | Vibratory pile in position to densify the soil.

Source: L. B. Foster Company.

The best pile configuration used has been an open-end 30-in. (760-mm) pipe of 3/8-in. (9.5-mm) wall thickness with 4- to 6-in. (100- to 150-mm) wide and 1/2-in. (13-mm) thick steel bands spaced 5 to 10 ft (1.5 to 3 m) apart on the outside of the pipe, together with wider driving and clamping bands at the bottom and top of the pile. Tests using other diameters have revealed that smaller diameters give less densification inside the pipe, whereas larger diameters require more vibratory energy, and a thicker and heavier pile. The pile is usually 10 to 15 ft (3 to 4.6 m) longer than the maximum penetration depth, to allow for any flexing of the pile, particularly when piles more than 50 to 55 ft (15 to 17 m) long are used. This allows for any cutoff requirements during application.

The pile is attached to the vibrator by means of a hydraulic clamp; this enables the vertical vibratory energy produced to travel to the pile material undiminished, as the pile, hydraulic clamping head, and vibrating transmission case act as a unit.

A mobile crane of sufficient size and capacity is required to handle the vibrating unit and the pile length during driving and extracting operations. An overburden of sand is required before beginning the operation to compensate for the settling that will result from the compaction. About a 12% shrinkage allowance has been satisfactory for most applications, but a hydraulic fill with a relatively low density may require a 15% allowance.

The dimensions of the spacing pattern of the piles are dependent on the required relative density of the soil. Test patterns of several different spacings should be run initially to determine the required spacing, which yields the desired density. Square patterns with spacings varying from 3 to 8 ft (0.9 to 2.4 m) have been successfully used to give the desired density. Square spacings seem to offer better results and faster operations than other patterns. Also, if additional probes are needed, they can be placed in the centers of the square patterns.

Working in saturated sands or hydraulically placed cohesionless materials, compaction by this method is very expeditious. For 50- to 55-ft piles, an average rate of about 15 pile penetrations per hour can be achieved. For projects requiring shallow piles in loose soils, the rate can be higher, whereas for projects requiring deeper piles in denser soils, the rate can be lower.

During early applications of this method, tests were conducted whereby the vibrating pile was held at full penetration depth and vibration was continued for several minutes. It was thought that such a procedure would produce additional densification. The tests revealed that little, if any, additional densification resulted from the full-depth vibration. The significant compaction is achieved during the penetration and extraction process.

On some projects in which the upper layers of soil consist of mud, muck, and silt, it will be necessary to remove all the undesirable material because it will not be densified by the vibratory method. The replacement material should be granular, cohesionless soil that will respond to vibratory densification. In some instances, it may be necessary to screen the replacement material to remove any silt or clay balls.

Vibroflotation

This method utilizes a vibrator penetrator that has water jets at both the top and bottom. As the penetrator is lowered, extension tubes are added. The vibrator penetrator requires a crane of sufficient size and capacity to handle both the penetrator and the extensions. The penetrator actually settles into the soil by its own weight as the lower water jet creates a "quick" condition. Typical penetration is 3 to 6 ft per min. To begin compaction, the lower jets are turned off, and the water is directed to the upper jets at a reduced pressure. The penetrator is then withdrawn in 1 ft per min lifts, and sand is continuously fed as backfill. The lifts method of raising the penetrator ensures compaction for the entire depth. The soil will be compacted in a radial zone extending 5 to 13 ft from the vibratory penetrator. This method is effective with greater spacing patterns than the vibrating pile method and is much more efficient in finer granular formations.

Cost Considerations

Numerous techniques and methods are available for a specific application, thus costs become an important consideration. The vibrating methods, because they are applicable to some very difficult situations, have some interesting cost considerations. An overall cost consideration should include the following items:

1. Cost of soil removal, if required.
2. Cost of replacement soil, if required.
3. Cost of necessary soil overburden, if required.

4. Cost of a testing program.
5. Cost of the vibratory compaction method.

The first four costs will be uniquely related to each project site, as each may or may not be applicable. The cost of the vibratory compaction method can vary, depending on the size of the area, the initial soil density, the required density, and the depth to be compacted. Because the mobilization and demobilization costs would be the same for a small or a large project, the unit cost for a small project would be higher. With other conditions remaining the same, the cost per unit volume should be less for projects requiring deep probes. A higher specified density, requiring closer spacing of probes, will result in a higher cost per unit of volume than the wider spacing permitted with lower density requirements.

The depth of compaction is a factor for consideration, but it is less significant than other factors because of the speed provided by vibratory device use. Overall probing time for a hole 25 ft (7.6 m) deep would not be doubled if the depth were increased to 50 ft (15 m) under the same soil conditions because the time required to move between the probes would be the same for both depths and the time required to penetrate the additional depth would be a matter of a few seconds for most projects.

The employment of this method provides these advantages:

1. An effective means for compacting a range of saturated sands.
2. An expedient method of deep compaction because of the speed of driving and extracting made available by the vibratory pile-driving device.
3. Adjustable modular spacings to adapt to final density requirements and job site conditions.
4. An effective method of compacting soil to substantial depths.
5. A compaction method where lower initial density can make the method more expedient than higher initial density.
6. A means of densification of some soils to reduce the soil liquefaction hazards of earthquakes.

DYNAMIC COMPACTION

The densification technique of repeatedly dropping a heavy weight onto the ground surface is commonly referred to as "dynamic compaction." This process has also been described as heavy tamping, impact densification, dynamic consolidation, pounding, and dynamic precompression. For either a natural soil deposit or a placed fill, the method can produce densification to depths of greater than 35 ft. Most projects have used drop weights weighing from 6 to 30 tons and typical drop heights range from 30 to 75 ft. However, on the Jackson Lake Dam project in Grand Teton National Park, a 32-ton weight from a height of 103 ft was employed.

Conventional cranes are used for drop weights of up to 20 tons and drop heights below 100 ft. The weight is attached to a single hoist line. During the drop, the hoist drum to which the line is attached is allowed to free spool, releasing the

FIGURE 4.24 | Dynamic compaction using a Lampson Thumper.

line. When heavier weights are used, specially designed dropping machines are required (see Fig. 4.24). With this densification technique, the in situ strata are compacted from the ground surface at their prevailing water contents. A possible disadvantage is that ground vibrations can be produced that travel significant distances from the impact point.

The most successful projects have been those where coarse-grained pervious soils were present. The position of the water table will have a major influence on dynamic compaction project success. It is better to be at least 6.5 ft above the water table. Operations on saturated impervious deposits have resulted in only minor improvement at high cost and should be considered ineffective.

The depth of improvement that can be achieved is a function of the weight of the tamper and the drop height:

$$D = n(W \times H)^{1/2} \qquad \textbf{[4.17]}$$

where

D = depth of improvement in meters (m)

n = an empirical coefficient, which is less than 1.0

W = weight of tamper in metric tons

H = drop height in meters

An n value of 0.5 has been suggested for many soil deposits. That value is a reasonable starting point; however, the coefficient is affected by

The type and characteristics of the material being compacted.

The applied energy.

The contact pressure of the tamper.

The influence of cable drag.

The presence of energy-absorbing layers.

SOIL STABILIZATION

GENERAL INFORMATION

Many soils are subject to differential expansion and shrinkage when they undergo changes in moisture content. Many soils also shift and rut when subjected to moving wheel loads. If pavements are to be constructed on such soils, it is usually necessary to stabilize them to reduce the volume changes and to strengthen them to the point where they can carry the imposed load, even under adverse weather and climatic conditions. In the broadest sense, *stabilization* refers to any treatment of the soil that increases its natural strength. There are two kinds of stabilization—(1) mechanical and (2) chemical. In engineering construction, however, stabilization most often refers to when compaction is preceded by the addition and mixing of an inexpensive admixture, termed a "stabilization agent," which alters the chemical makeup of the soil, resulting in a more stable material.

Stabilization can be applied in-place to a soil in its natural position, or mixing can take place on the fill. Also, stabilization can be applied in a plant, and then the blended material is transported to the job site for placement and compaction.

The two primary methods of stabilizing soils are

1. Incorporating lime or lime–fly ash into soils that have a high clay content and
2. Incorporating Portland cement (with or without fly ash) into soils that are largely granular in nature

STABILIZING SOILS WITH LIME

In general, lime reacts readily with postplastic soils containing clay, either the fine-grained clays or clay-gravel types. Such soils range in PI from 10 to 50+. Unless stabilized, these soils usually become very soft when water is introduced. The only exception would be organic soils containing more than 20% organic matter.

Soil-Lime Chemistry

In combination with compaction, soil stabilization with lime involves a chemical process whereby the soil is improved with the addition of lime. Lime, in its hydrated form $[Ca(OH)_2]$, will rapidly cause cation exchange and flocculation/agglomeration, provided it is intimately mixed with the soil. A high PI clay soil will then behave much like a material having a lower PI. This reaction begins to occur within an hour after mixing, and significant changes are realized within a very few days, depending on the PI of the soil, the temperature, and the amount of lime used. The observed effect in the field is a drying action.

Following this rapid soil improvement, a longer, slower soil improvement takes place termed "pozzolanic reaction." In this reaction, the lime chemically combines with siliceous and aluminous constituents in the soil to cement the soil together. Here some confusion exists. Some people refer to this as a "cementatious reaction," which is a term normally associated with the hydraulic action occurring between Portland cement and water, in which the two constituents chemically combine to form a hard, strong product. The confusion is increased by the fact that almost two-thirds of Portland cement is lime (CaO). But the lime in Portland cement starts out already chemically combined during manufacture with silicates and aluminates, and thus is not in an available or "free" state to combine with the clay.

The cementing reaction of the lime, as $Ca(OH)_2$, with the clay is a very slow process, quite different from the reaction of Portland cement and water, and the final form of the products is thought to be somewhat different. The slow strength with time experienced with lime stabilization of clay provides flexibility in manipulation of the soil. Lime can be added and the soil mixed and compacted, initially drying the soil and causing flocculation. Several days later, the soil can be remixed and compacted to form a dense stabilized layer that will continue to gain strength for many years. The resulting stabilized soils have been shown to be extremely durable.

Lime Stabilization Construction Procedures

Lime treatments can be characterized into three classes.

1. *Subgrade (or subbase) stabilization* includes stabilizing fine-grained soils in-place or borrow materials, which are employed as subbases.
2. *Base stabilization* includes plastic materials, such as clay-gravels which contain at least 50% coarse material retained on a No. 40 mesh screen.
3. *Lime modification* includes the upgrading of fine-grained soils with small amounts of lime, i.e., 1/2 to 3% by weight.

The distinction between modification and stabilization is that generally no credit is accorded the lime-modified layer in the structural design. It is usually used as a contractor technique to dry wet areas, to help "bridge" across underlying spongy subsoil, or to provide a working table for subsequent construction.

The basic steps in lime stabilization construction are

1. *Scarification and pulverization.* To accomplish complete stabilization, adequate pulverization of the clay fraction is essential. This is best accomplished with a rotary stabilizer (seen to the right in Fig. 4.26 later in the chapter).
2. *Lime spreading.* Dry lime should not be spread under windy conditions. Lime slurry may be prepared in a central mixing tank and spread on the grade using standard water distributor trucks.

3. *Preliminary mixing and addition of water.* During rotary mixing of the lime with the soil material, the water content should be raised to at least 5% above optimum. This may require the addition of water.

4. *Preliminary curing.* The lime-soil mixture should cure for 24 to 48 hr to permit the lime and water to break down (or mellow) the clay clods. In the case of extremely heavy clays, the curing period may extend to 7 days.

5. *Final mixing and pulverization.* During final mixing, pulverization should continue until all clods are broken down to pass a 1-in. screen and at least 60% pass a No. 4 sieve.

6. *Compaction.* The soil-lime mixture should be compacted as required by specification.

7. *Final curing.* The compacted material should be allowed to cure for 3 to 7 days prior to placing subsequent layers. Moist curing, which consists of maintaining the surface in a moist condition by light sprinkling or membrane curing, and which involves sealing the compacted layer with a bituminous material, may be used.

LIME–FLY ASH STABILIZATION

This type of stabilization, although not new, has only recently become widely used. The primary reason is that fly ash, which is the residue that would "fly" out the stack in a coal-fired power plant if it were not captured, is becoming extremely plentiful throughout the United States and, in fact, throughout many parts of the world. Fly ash is a by-product in the production of electricity from burning coal. As such, it can be a highly variable product, and its engineering usefulness can range from superior to extremely poor. Today, in the United States alone, there is in excess of 90 million tons of fly ash being collected each year [5]. The newer, more modern power plants literally pulverize the coal until almost all of it passes the No. 200 mesh sieve before it is used as fuel. The resulting fly ash is extremely fine in size (often finer than Portland cement) and contains the silicates and aluminates necessary to combine with the lime in soil stabilization. Laboratory and field results indicate that fly ash, of suitable quality, can replace a *portion* of the lime needed to stabilize a clay-type soil. Because lime is relatively expensive and fly ash is often quite inexpensive, lime–fly ash stabilization of soils is being increasingly utilized. The major drawback to the use of fly ash is that two stabilizing agents are being used instead of only one, which means more manipulation of the soil and more chance of error. There are a number of excellent references on the use of lime–fly ash as stabilization [11, 14].

CEMENT-SOIL STABILIZATION

Stabilizing soils with Portland cement is an effective method of strengthening certain soils. As long as the soils are predominately granular with only minor amounts of clay particles, the use of Portland cement has been found to be effective. A rule of thumb is that soils with PI less than about 10 are likely candidates for this type of stabilization. Soils with higher amounts of clay-sized particles are very difficult to manipulate and thoroughly mix with the cement before the cement sets. The terms "soil cement" and "cement-treated base" are often used interchangeably, and generally describe this type of stabilization. However, in some areas the term "soil cement" refers strictly to mixing and treatment of in-place soils on the grade. The term "cement-treated base" is then used to describe an aggregate/cement blend produced in a pugmill plant and hauled onto the grade. The amount of cement mixed with the soil is usually 3 to 7% by dry weight of the soil.

As discussed in connection with lime stabilization, fly ash is plentiful in many areas of the world, and it can be effectively used to replace a portion of the Portland cement in a soil-cement treatment. Replacement percentages on an equal weight basis or on a 1.25:1.0 fly ash/Portland cement replacement ratio have been used. There are a number of excellent references on this subject [10, 11, and 14].

Soil-Cement Construction Procedures

The construction methods involve scarifying the grade, spreading the Portland cement uniformly over the surface of the soil (see Fig. 4.25), then mixing it into the soil, preferably with a pulverizer-type machine, to the specified depth, followed by compaction, fine grading, and curing. If the moisture content of the soil is low, it will be necessary to add water during the mixing operation (see Fig. 4.26). The material should be compacted within 30 min after it is mixed, using either tamping foot or pneumatic-tired rollers, followed by final rolling with a smooth-wheel roller. A seal of asphalt or another acceptable material may have to be applied to the surface to retain the moisture in the mix.

FIGURE 4.25 | Flynn spreader being used to uniformly apply cement during a soil-cement stabilization project.

FIGURE 4.26 | Water truck connected to a soil stabilizer for adding water during a soil-cement mixing operation.

The nature of in-place mixed soil cement does not allow a 7 A.M. to 3 P.M. work schedule. When the cement is applied the material cannot be left overnight; you must complete the operation even if overtime work is necessary.

The Project Site   When estimating soil cement, the configuration of the areas to be treated must be considered. There are many aspects of the process that can be affected by configuration. It may be necessary to travel over treated areas to obtain water that is required to treat other areas of the project. In some instances, sites are small and it is very difficult to load and unload equipment in a safe manner. Cement dust is another issue that should be taken into consideration. Cement dust can cause damage to nearby vehicles. Always work with favorable wind conditions. Of major concern in a soil-cement operation is the amount of rock in excess of softball size. Rocks can damage the pulverizing equipment and make grading difficult. Removal of oversize rock is labor intensive, making it a critical cost consideration.

Scarifying   Scarifying, loosening the soil, can help identify problem areas. If rocks are an unforeseen problem, scarifying can bring the unacceptable rocks to the surface. A good effort at rock removal will save equipment from damage. If there are excessive amounts of rock to be removed, check the grade again. Scarifying can also reveal soft, yielding, or wet areas and hidden organic matter.

Spreading   During the spreading operations, the bulk application (spread rate) is checked and necessary adjustments made. The calibration of equipment is a crucial point in federal, state, and airport work. The spread pattern of the cement is determined by site configuration. Bulker types vary somewhat and the size of the site will determine the bulk spreader to be used. The bulker should move at a continuous rate without stopping while the product is being discharged (see Fig. 4.25).

Mixing   Mixing should be done immediately after application of the cement (see Fig. 4.26). This mix should be checked to ensure the uniformity of blend. It is important that the cement be blended thoroughly with the soil to achieve the desired finished product. The normal soil-cement mix procedure is in a down cut motion using tine-type mixer teeth (stabilization). In cohesive soils, an up cut machine with conical-type bits is better.

In the case of very dry or windy conditions, prewetting the grading can be helpful to provide an adequate material moisture content and to control dust. In extreme conditions, prewetting prior to scarifying the material may be a necessity. During the application of water, the water truck drivers should take care not to allow the discharge of water to puddle or pool. Such situations can result in soft or yielding grade conditions. These soft spots may, in turn, become areas where compaction cannot be achieved. Strict moisture control should be adhered to at all times.

Compaction   Initial compaction with a vibratory pad foot roller immediately follows the mixing operation. A roller pattern should be established at the beginning of the initial compaction effort. Normally, two to three complete roller passes will be required. The depth of mix, type of soil, and required density are factors that control the number of roller passes required. The compactor should keep pace with the mixing operation.

Following the initial compaction, the treated material should be shaped to the approximate line and grade. Special attention should always be given to grading for proper drainage. It is hard to make grade corrections to cured material. The compaction effort should continue until required density is achieved.

The fine-grading operation should follow acceptable compaction and rolling with a smooth drum vibratory or pneumatic roller. The treated material should be kept from drying at all times during rolling, shaping, and fine grading. Again, it is important that the water truck drivers exercise care not to allow puddling or pooling of water to occur.

Curing   There are several acceptable methods of curing soil cement. Some projects will specify the use of liquid asphalt curing. That method has possible environmental impacts and the material can be tracked onto adjacent roadways requiring a cleanup operation. In many areas, the effect of runoff is of great concern particularly for projects located near major waterways. The use of white curing compound has proven effective. It has to be applied at a heavy rate (0.25 to 0.30 gallons per square yard) to achieve complete coverage. This can be a time-consuming operation and requires special equipment. Therefore there is a preference for the wet cure method.

Keep the finish grade wet/damp for a period of 7 days, with as little traffic as possible being allowed to travel on the newly treated material. The water truck should be careful not to make sharp turns during the initial curing. Construction equipment, particularly track machines, should be strictly prohibited from entering the area.

During the curing operation, the treated area must be kept from freezing for 7 days. Frost should not be allowed to form on the grade during the first 48 hr of the curing process. The use of straw (4 in. minimum) and/or poly can help reduce possible freezing, depending on the projected low temperature. Temperature should always be taken into consideration at the beginning of a project. Do not begin a project without a weather projection that will allow for 7 continuous days of temperatures above freezing after completion of mixing and grading. A thin layer of straw will help reduce the possibility of frosting.

Stabilizers

Rotary cultivator stabilizers are extremely versatile pieces of equipment ideally suited for mixing, blending, and aerating soil. A stabilizer consists of a rear-mounted, removable-tine, rotating tiller blade, which is covered by a removable hood. In-place, the hood creates an enclosed mixing chamber that enhances thorough blending of the soil. The tiller blade lifts the material and throws it against the hood. The material, deflecting off of the hood, falls back onto the tiller blades for thorough blending. As the stabilizer moves forward, the material is ejected from the rear of the mixing chamber. As the material is ejected, it is struck off by the trailing edge of the hood, resulting in a fairly level working surface. With the hood removed, the blades churn the soil, exposing it to the drying action of the sun and wind. Some models are equipped with a forward-mounted spray bar that can be used to add water or stabilizing agents to the soil during the blending process. The stabilizer's use is limited to material less than 4 in. in diameter. The tines are designed to penetrate up to 10 in. below the existing surface so the unit can be used for scarifying and blending in-place (in situ) material as well as fill material.

SUMMARY

Geotechnical materials—soils and rocks—are the principle components of many construction projects. Geotechnical materials are by nature heterogeneous. In their natural state, they are rarely uniform and can only be worked by comparison to a similar type of material with which previous experience has been gained. You should never price an earth- or rock-handling project without first making a thorough study of the materials. For bulk materials, volumetric measure varies with the material's position in the construction process. Soil volume is measured in one of three states: bank cubic yard, loose cubic yard, and compacted cubic yard. The engineer must use a consistent volumetric state in any set of calculations.

A compaction curve graphically presents the maximum dry unit weight (density) and the percent water required to achieve maximum density based on a standard compactive effort (input energy). Applying energy to a soil by one or more of the following methods will cause compaction: impact, pressure, vibration, or kneading. The effectiveness of different compaction methods is dependent on the individual soil type being manipulated. The compaction equipment used on a project must have a production capability matched to that of the excavation, hauling, and spreading equipment. Usually, excavation or hauling capability will set the expected maximum production for the job.

Stabilization refers to any treatment of the soil that increases its natural strength: (1) mechanical and (2) chemical. In engineering construction, however, stabilization most often refers to when compaction is preceded by the addition and mixing of an inexpensive admixture, termed a "stabilization agent," which alters the chemical makeup of the soil, resulting in a more stable material. Stabilization agents may be applied in-place to a soil in its natural position or mixing can take place on the fill. Also, stabilization may be applied in a plant, and then the blended material is transported to the job site for placement and compaction. Critical learning objectives would include:

- An understanding of soil limits; liquid limit, plastic limit, and plasticity index.
- An ability to calculate soil volumetric changes.
- An understanding of compaction tests and specifications, and compaction curves.
- An understanding of when to use which compaction method.
- An ability to calculate estimated compaction production.
- An understanding of the construction procedures involved in either lime or cement stabilization.

These objectives are the basis for the problems that follow.

PROBLEMS

4.1 The soil borrow material, to be used to construct a highway embankment, has a mass unit weight of 96.0 pcf, a water content of 8%, and the specific gravity of the soil solids is 2.66. The specifications require that the soil be compacted to a dry unit weight of 112 lb per cf, and the water content be held to 13%. (315,000 bcy, 14.4 gal per bcy borrow, 132.3 lb/cf).

    a. How many cubic yards of borrow are required to construct an embankment having a 250,000 cy net section volume?

    b. How many gallons of water must be added per cubic yard of borrow material assuming no loss by evaporation?

    c. If the compacted fill becomes saturated at constant volume, what will be the water content and mass unit weight of the soil?

4.2 The soil borrow material to be used to construct a highway embankment has a mass unit weight of 98.0 lb per cf, a water content of 9%, and the specific gravity of the soil solids is 2.67. The specifications require that the soil be placed in the fill so the dry unit weight is 114 lb per cf and the water content be held to 12%.

    a. How many cubic yards of borrow are required to construct an embankment having an 800,000 cy net section volume?

   b. How many gallons of water must be added per cubic yard of borrow material assuming no loss evaporation?

   c. If the compacted fill becomes saturated at constant volume, what will be the water content and unit weight?

4.3 Embankment at a 12% water content is to be placed at the rate of 270 ccy per hr. The specified dry weight of the compacted fill is 2,900 lb per cy. How many gallons of water must be supplied each hour to increase the moisture content of the material from 7 to 12% by weight?

4.4 The borrow material to construct an embankment has a mass unit weight of 96.5 pcf and a water content of 8%. The specific gravity of the solids is 2.66. The contract specifications require that the soil be placed in the fill at a γ_d of 114 pcf and a water content of 10%.

   a. How many cubic yards of borrow are required to construct an embankment having a 455,000-cy net volume?

   b. How many gallons of water must be added per cubic yard of borrow material assuming no loss by evaporation?

4.5 Earth is placed in fill at the rate of 190 cy per hour, compacted measure. The placement water content is 10% and the dry weight of the compacted earth is 2,890 lb per cy. How many gallons of water must be supplied each hour to increase the moisture content of the earth from 4 to 10% by weight?

4.6 Earth—whose in situ weight is 112 lb per cf, loose weight is 95 lb per cf, and compacted weight is 120 lb per cf—is placed in a fill at the rate of 240 cy per hour, measured as compacted earth. The thickness of the compacted layers is 6 in. A tractor pulls a towed sheepsfoot roller having 5-ft-wide drums, at a speed of 2 mph. Assume a 45-min hour operating efficiency. Determine the number of drums required to provide the necessary compaction if eight drum passes are specified for each layer of earth.

REFERENCES

1. *ASTM Standards on Soil Compaction,* American Society for Testing and Materials: Philadelphia, 1992.

2. *ASTM Standards on Soil Stabilization with Admixtures,* American Society for Testing and Materials, Philadelphia, 1990.

3. Butler, R. C., "Ground Improvement Using Dynamic Compaction," *Geotechnical News,* BiTech Publishers Ltd, Vancouver, B.C., Canada, Vol. 9, No. 2, June 1991.

4. *Construction and Controlling Compaction of Earth Fills,* ASTM Special Technical Publication, 1384, D. W. Shanklin Ed., American Society for Testing and Materials, Philadelphia, April 2000.

5. Covey, James N., "An Overview of Ash Utilization in The United States," in *Proceedings of the Fly Ash Applications in 1980 Conference,* Texas A&M University, May 1980.

6. *Guide to Earthwork Construction,* State of the Art Report 8, TRB, National Research Council, Washington, D.C., 1990.

7. Holtz, R. D., *NCHRP, Synthesis of Highway Practice 147: Treatment of Problem Foundations for Highway Embankments,* TRB, National Research Council, Washington, D.C., 1989.

8. Humboldt Mfg. Co., 7300 West Agatite Ave., Norridge, Il 60656, www.hmc-hsi.com/.

9. Lukas, Robert G., *Dynamic Compaction for Highway Construction Volume I: Design and Construction Guidelines,* U.S. Department of Transportation, Federal Highway Administration, July 1986.

10. McKerall, W. C., and W. B. Ledbetter, "Variability and Control of Class C Fly Ash," *Cement, Concrete, and Aggregates,* CCAGDP, Vol. 4, No. 2, ASTM, Philadelphia, Winter 1982.

11. Meyers, J. F., R. Pichumami, and B. S. Kapples, *Fly Ash as a Highway Construction Material,* U.S. Department of Transportation, Federal Highway Administration, June 1976.

12. Monahan, E. J., *Construction of Fills,* 2nd ed., John Wiley & Sons, New York, 1993.

13. Rollings, M. P., and R. S. Rollings, *Geotechnical Materials in Construction,* McGraw-Hill, New York, 1996.

14. Terrel, R. L., *A Guide Users Manual for Soil Stabilization,* U.S. Department of Transportation, Federal Highway Administration, April 1979.

15. Wahls, H. E., *NCHRP Synthesis of Highway Practice 8: Construction of Embankments,* TRB, National Research Council, Washington, D.C., 1971.

16. Welsh, Joseph P., *Soil Improvement—A Ten Year Update, Geotechnical Special Publication No. 12,* American Society of Civil Engineers, Washington, D.C. 1987.

GEOTECHNICAL-RELATED WEBSITES

1. Federal Highway Administration, *www.fhwa.dot.gov/bridge/geo.htm.* Contains information on FHWA geotechnical program, including publications, software, and training.

2. U.S. Army Corps of Engineers, *www.erdc.usace.army.mil/.* Links are provided to the geotechnical engineering laboratory at the Waterways Experiment Station and the Cold Regions Research and Engineering Laboratory.

3. U.S. Geological Survey, *www.usgs.gov.* Provides links to geological, water, and mapping information.

4. National Resources Conservation Service, *www.nrcs.gov/TechRes.html.* Serves as the gateway to the soils database, including GIS-based polygon data.

5. United State Universities Council on Geotechnical Engineering Research (USUCGER), *www.usucger.org.* Pertinent information about geomedia research at 96 member universities in the United States is provided.

6. Transportation Research Board, *www4.national-academies.org/trb/onlinepubs.nsf.* TRB's website provides a searchable index of the board's publications and articles, which cover all modes and aspects of transportation, including geotechnical engineering.

7. Geosynthetic Institute (GSI), *www.geosynthetic-institute.org.* GSI's mission is to transfer knowledge about geosynthetics.

5

Machine Power

The constructor must select the proper equipment to relocate and/or process materials economically. The decision process for matching the best possible machine to the project task requires consideration of the mechanical capabilities of the machine. The power required is the power needed to propel the machine, and this power requirement is establish by two factors: (1) rolling resistance and (2) grade resistance. Equipment manufacturers publish performance charts for individual machine models. These charts enable the equipment planner to analyze a machine's ability to perform under a given set of job and load conditions.

GENERAL INFORMATION

On heavy construction projects, the majority of the work consists of handling and processing large quantities of bulk materials. The constructor must select the proper equipment to relocate and/or process these materials economically. The decision process for matching the best possible *machine* to the project task requires that the estimator take into account both the properties of the material to be handled and the mechanical capabilities of the machine.

When the estimator considers a construction material-handling problem, there are two primary *material* considerations: (1) total *quantity* of material and (2) *size* of the individual pieces. The quantity of material to be handled and the time constraints resulting from the project contract or weather influence the selection of equipment as to type, size, and number of machines. Larger units generally have lower unit-production cost, but there is a trade-off in higher mobilization and fixed costs. The size of the individual material pieces will affect the choice of machine size. A loader used in a quarry to move shot rock must be capable of handling the largest rock sizes produced.

Payload

The payload of hauling equipment may be expressed either *volumetrically* or *gravimetrically*. Volumetric capacity can be stated as struck or heaped, and volume can be expressed in terms of loose cubic yard (lcy), bank cubic yard (bcy), or compacted cubic yard (ccy). The payload capacity of a hauling unit is often stated by the manufacturer in terms of the volume of loose material that the unit can hold, assuming that the material is heaped in some specified angle or repose. A gravimetric capacity would represent the safe operational weight that the axles or structural frame of the machine are designed to handle.

From an economic standpoint, overloading haul trucks to improve production looks attractive and overloading by 20% might increase the haulage rate 15%, allowing for slight increases in time to load and haul. The cost per ton hauled should show a corresponding decrease, since direct labor costs will not change and fuel costs will increase only slightly. This apparently favorable situation is only temporary, however, for the advantage is being bought at the cost of premature aging of the truck and a corresponding increased replacement capital expense.

Machine Performance

"Why does the machine travel at only 12 mph when its top speed is 33 mph?" Cycle time and payload determine a machine's production rate, and machine travel speed directly affects cycle time. To answer the travel speed question, it is necessary to analyze machine power. There are three power questions that must be analyzed:

1. Required power.
2. Available power.
3. Usable power.

REQUIRED POWER

Power required is the power needed to propel the machine, and two factors establish this power requirement: (1) *rolling resistance* and (2) *grade resistance*. Therefore, power required is the power necessary to overcome the *total resistance* to machine movement, which is the sum of rolling and grade resistance.

Total resistance (TR) = Rolling resistance (RR) + Grade resistance (GR)

[5.1]

Rolling Resistance

Rolling resistance is the resistance of a level surface to constant-velocity motion across it. This is sometimes referred to as *wheel resistance* or *track resistance,* which results from friction or the flexing of the driving mechanism plus the force required to shear through or ride over the supporting surface.

This resistance varies considerably with the type and condition of the surface over which a machine moves (see Fig. 5.1). Soft earth offers a higher resistance than hard-surfaced roads such as concrete pavement. For machines that move on rubber tires, the rolling resistance varies with the size of, pressure on,

FIGURE 5.1 │ Rolling resistance varies with the condition of the surface over which a machine moves.

and the tread design of the tires. For equipment that moves on crawler tracks, such as tractors, the resistance varies primarily with the type and condition of the road surface.

A narrow-tread, high-pressure tire gives lower rolling resistance than a broad-tread, low-pressure tire on a hard-surfaced road. This is the result of the small area of contact between the tire and the road surface. If the road surface is soft and the tire tends to sink into the earth, however, a broad-tread, low-pressure tire will offer a lower rolling resistance than a narrow-tread, high-pressure tire. The reason for this condition is that the narrow tire sinks into the earth more deeply than the broad tire and thus is always having to climb out of a deeper hole that is equivalent to climbing a steeper grade.

The rolling resistance of an earth-haul road probably will not remain constant under varying climatic conditions or for varying types of soil that exist along the road. If the earth is stable, highly compacted, and well maintained by a grader, and if the moisture content is kept near optimum, it is possible to provide a surface with a rolling resistance about as low as that of concrete or asphalt. Moisture can be added, but following an extended period of rain, it may be difficult to remove the excess moisture and the haul road will become muddy, with an increase in rolling resistance. Providing good surface drainage will speed the removal of the water and should enable the road to be reconditioned quickly. For a major earthwork project, it is good economy to provide graders, water trucks, and even rollers to keep the haul road in good condition. The maintenance of low-rolling-resistance haul roads is one of the best financial investments an earthmoving contractor can make.

A tire sinks into the soil until the product of bearing area and bearing capacity is sufficient to sustain the load; then the tire is always attempting to climb out of the rut. The rolling resistance will increase about 30 lb per ton for each inch of penetration. Total rolling resistance is a function of the riding gear characteristics (independent of speed), the total weight of the vehicle, and torque. It is usually expressed as pounds of resistance per ton of vehicle weight, or as an equivalent grade resistance. For example, if a loaded truck that has a gross weight equal to 20 tons is moving over a level road whose rolling resistance is 100 lb per ton, the tractive effort required to keep the truck moving at a uniform speed will be 2,000 lb (20 tons × 100 lb/ton).

The estimation of off-road rolling resistance is based largely on empirical information, which may include experience with similar soils. Rarely are rolling resistance values based on test runs on actual terrain. Much of the actual test data available comes from performance of aircraft tire research at the U.S. Army, Waterways Experiment Station. Although it is impossible to give completely accurate values for the rolling resistances for all types of haul roads and wheels, the values given in Table 5.1 are reasonable estimates.

TABLE 5.1 | Representative rolling resistances for various types of wheels and surfaces*

| Type of surface | Steel tires, plain bearings | | Crawler type track and wheel | | Rubber tires, antifriction bearings | | | |
| | | | | | High pressure | | Low pressure | |
	lb/ton	kg/m ton	lb/ton	kg/m ton	lb/ton	kg/m ton	lb/ton	kg/m ton
Smooth concrete	40	20	55	27	35	18	45	23
Good asphalt	50–70	25–35	60–70	30–35	40–65	20–33	50–60	25–30
Earth, compacted and maintained	60–100	30–50	60–80	30–40	40–70	20–35	50–70	25–35
Earth, poorly maintained	100–150	50–75	80–110	40–55	100–140	50–70	70–100	35–50
Earth, rutted, muddy, no maintenance	200–250	100–125	140–180	70–90	180–220	90–110	150–200	75–100
Loose sand and gravel	280–320	140–160	160–200	80–100	260–290	130–145	220–260	110–130
Earth, very muddy, rutted, soft	350–400	175–200	200–240	100–120	300–400	150–200	280–340	140–170

*In pounds per ton or kilograms per metric ton of gross vehicle weight.

If desired, one can determine the rolling resistance of a haul road by towing a truck or other vehicle whose gross weight is known along a level section of the haul road at a uniform speed. The tow cable should be equipped with a dynamometer or some other device that will enable determination of the average tension in the cable. This tension is the total rolling resistance of the gross weight of the truck. The rolling resistance in pounds per gross ton will be

$$R = \frac{P}{W} \qquad [5.2]$$

where

R = rolling resistance in pounds per ton
P = total tension in tow cable in pounds
W = gross weight of truck in tons

When tire penetration is known, an approximate rolling resistance value for a wheeled vehicle can be calculated using the following formula:

$$RR = [40 + (30 \times TP)] \times GVW \qquad [5.3]$$

where

RR = rolling resistance in pounds
TP = tire penetration in inches
GVW = gross vehicle weight in tons

Grade Resistance

The force-opposing movement of a machine up a frictionless slope is known as *grade resistance*. It acts against the total weight of the machine, whether track-type or wheel-type. When a machine moves up an adverse slope (see Fig. 5.2), the power required to keep it moving increases approximately in proportion to the slope of the road. If a machine moves down a sloping road, the power required to keep it moving is reduced in proportion to the slope of the road. This is known as grade assistance.

The most common method of expressing a slope is by gradient in percent. A 1% slope is one where the surface rises or drops 1 ft vertically in a horizontal distance of 100 ft. If the slope is 5%, the surface rises or drops 5 ft per 100 ft of horizontal distance. If the surface rises, the slope is defined as plus, whereas if it drops, the slope is defined as minus. This is a physical property not affected by the type of equipment, or the condition or type of road.

FIGURE 5.2 | Off-highway truck moving up an adverse slope.

For slopes of less than 10%, the effect of grade is to increase, for a plus slope, or decrease, for a minus slope, the required tractive effort by 20 lb per gross ton of machine weight for each 1% of grade. This can be derived from elementary mechanics by calculating the required driving force.

From Fig. 5.3 the following relationships can be developed:

$$F = W \sin \alpha \qquad \textbf{[5.4]}$$

$$N = W \cos \alpha \qquad \textbf{[5.5]}$$

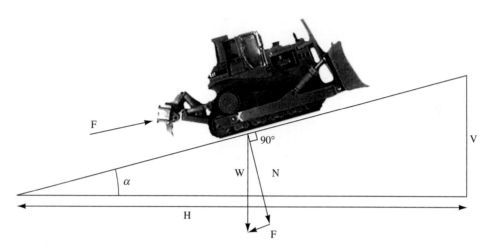

FIGURE 5.3 | Frictionless slope force relationships.

For angles less than 10°, $\sin \alpha \approx \tan \alpha$; with that substitution:

$$F = W \tan \alpha \qquad \textbf{[5.6]}$$

But

$$\tan \alpha = \frac{V}{H} = \frac{G\%}{100}$$

where $G\%$ is the gradient. Hence,

$$F = W \times \frac{G\%}{100} \qquad \textbf{[5.7]}$$

If we substitute $W = 2,000$ lb per ton, the formula reduces to

$$F = 20 \text{ lb/ton} \times G\% \qquad \textbf{[5.8]}$$

This formula is valid for a G up to about 10%, that is, the small angle assumption.

Total Resistance

Total resistance equals rolling resistance plus grade resistance or rolling resistance minus grade assistance. It can also be expressed as an *effective grade*.

Using the relationship expressed in Eq. [5.8], a rolling resistance can be equated to an equivalent gradient.

$$\frac{\text{Rolling resistance expressed in lb/ton}}{20 \text{ lb/ton}} = G\% \qquad [5.9]$$

Table 5.2 gives values for the effect of slope, expressed in pounds per gross ton or kilograms per metric ton (m ton) of weight of the vehicle.

TABLE 5.2 | The effect of grade on the tractive effort of vehicles

Slope (%)	lb/ton*	kg/m ton*	Slope (%)	lb/ton*	kg/m ton*
1	20.0	10.0	12	238.4	119.2
2	40.0	20.0	13	257.8	128.9
3	60.0	30.0	14	277.4	138.7
4	80.0	40.0	15	296.6	148.3
5	100.0	50.0	20	392.3	196.1
6	119.8	59.9	25	485.2	242.6
7	139.8	69.9	30	574.7	287.3
8	159.2	79.6	35	660.6	330.3
9	179.2	89.6	40	742.8	371.4
10	199.0	99.5	45	820.8	410.4
11	218.0	109.0	50	894.4	447.2

*Ton or metric ton of gross vehicle weight.

By combining the rolling resistance, expressed as an equivalent grade, and the grade resistance, expressed as a gradient in percent, one can express the total resistance as an effective grade. The three terms, power required, total resistance, and effective grade all mean the same thing. Power required is expressed in pounds. Total resistance is expressed in pounds or pounds per ton of machine weight, and effective grade is expressed in percent.

EXAMPLE 5.1

The haul road from the borrow pit to the fill has an adverse grade of 4%. Wheel type hauling units will be used on the job and it is expected that the haul-road rolling resistance will be 100 lb per ton. What will be the effective grade for the haul? Will the units experience the same effective grade for the return trip?

Using Eq. [5.9], we obtain

$$\text{Equivalent grade (RR)} = \frac{100 \text{ lb/ton rolling resistance}}{20 \text{ lb/ton}} = 5\%$$

$$\text{Effective grade (TR}_{\text{haul}}) = 5\% \text{ RR} + 4\% \text{ GR} = 9\%$$

$$\text{Effective grade (TR}_{\text{return}}) = 5\% \text{ RR} - 4\% \text{ GR} = 1\%$$

where

　　RR = rolling resistance
　　GR = grade resistance

Note that the effective grade is not the same for the two cases. During the haul, the unit must overcome the uphill grade; on the return the unit is aided by the downhill grade.

Haul routes   During the life of a project the haul-route grades (and, therefore, grade resistance) may remain constant. One example is trucking aggregate from a rail-yard off-load point to the concrete batch plant. In most cases, however, the haul-route grades change as the project progresses. On a linear highway project, the tops of the hills are excavated and hauled into the valleys. Early in the project, the grades are steep and reflect the existing natural ground. Over the life of the project, the grades begin to assume the final highway profile. Therefore, the estimator must first study the project's mass diagram to determine the *direction* that the material has to be moved. Then the natural ground and the final profiles depicted on the plans must be checked to determine the grades that the equipment will encounter during haul and return cycles.

Site work projects are usually not linear in extent; therefore a mass diagram is not very useful. The estimator in that case must look at the cut-and-fill areas, lay out probable haul routes, and then check the natural and finish grade contours to determine the haul-route grades.

This process of laying out haul routes is critical to machine productivity. If a route can be found that results in less grade resistance, machine travel speed can be increased, and production will likewise increase. In planning a project, a constructor should always check several haul-route options before deciding on a final construction plan.

Equipment selection is affected by travel distance because of the time factor distance introduces into the production cycle. All other factors being equal, increased travel distances will favor the use of high-speed large-capacity units. The difference between the self-loading scraper and a push-loaded scraper can be used as an illustration. The elevating scraper will load, haul, and spread without any assisting equipment, but the extra weight of the loading mechanism reduces the unit's maximum travel speed and load capacity. A scraper, which requires a push tractor to help it load, does not have to expend power to haul a loading mechanism with it on every cycle. It will be more efficient in long-haul situations as it does not have to expend fuel transporting extra machine weight.

AVAILABLE POWER

There are two factors that determine available power: (1) horsepower and (2) speed. Horsepower is the time rate of doing work and is a constant value for any given machine. Since horsepower is a machine specific constant, available pounds pull or push will change as machine speed is varied. Travel fast and pulling ability will be low; travel slow and the machine has the ability to exert a high pulling force.

Horsepower

Internal combustion engines power most construction equipment. Because diesel engines perform better under heavy-duty applications than gasoline engines, diesel-powered machines are the workhorses of the construction industry. The characteristics that control the performance differences of these two engines are

> *Carburetor.* Used on gasoline engines, is an efficient method of regulating fuel.
>
> *Injector.* Used on diesel engines, is a better method of regulating fuel.
>
> *Ignition System.* Gasoline engines use spark-ignition; diesel engines meter fuel and air for compression-ignition.

Additionally, diesel engines have longer service lives and lower fuel consumption, and diesel fuel presents less of a fire hazard.

No matter which type of engine serves as the power source, the mechanics of energy transmission are the same. The engine develops a piston force F_p, which acts on a crankshaft having a radius r, producing a crankshaft torque T_g at a governed speed N_g.

$$T_g = F_p \times r \qquad \qquad [5.10]$$

The output of the engine at the flywheel at rated revolutions per minute (rpm) and under environmental conditions of testing (temperature and altitude) is known as flywheel horsepower (fwhp). This output can be measured either by friction belt or by brake, hence the names *belt horsepower* or *brake horsepower* (bhp).

$$\text{fwhp} = \frac{2\pi N_g F_p r}{33,000} = \frac{2\pi N_g T_g}{33,000} \qquad \qquad [5.11]$$

where N_g is in rpm, F_p is in pounds, r is in feet, and T_g is in lb-ft.

Typically, manufacturers rate machine horsepower at a specified rpm based on the Society of Automotive Engineers (SAE) standardized rating procedure. Under the SAE standard, horsepower will be listed as either gross or flywheel (also often listed as net horsepower). Flywheel horsepower can be considered as *usable* horsepower. It is the power available to operate a machine after deducting for power losses in the engine. If machine horsepower is listed without reference to the SAE standard it could be a *maximum,* which would be substantially higher than flywheel. When the SAE standard is not cited, always check with the manufacturer to determine the basis of the rating. For applications requiring continuous high power output, an engine that develops the required horsepower at a lower rpm will be under less strain. Such an engine may cost more to purchase, but it should last longer and require fewer repairs.

The power output from the engine, fwhp, becomes the power input to the transmission system. This system consists of the drive shaft, a transmission, planetary gears, drive axles, and drive wheels (see Fig. 5.4).

When analyzing a piece of equipment, we are interested in the usable force developed at the point of contact between the tire and the ground (*rimpull*) for a wheel machine. In the case of a track machine, the force in question is that available at the drawbar (*drawbar pull*). The difference in the name is a matter of

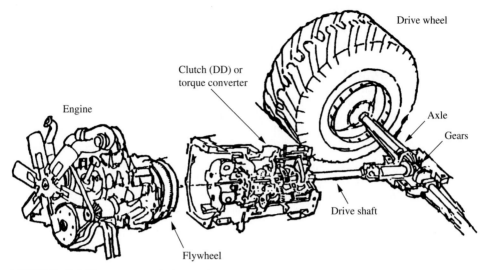

FIGURE 5.4 | Power transmission system.

convention; both rimpull and drawbar pull are measured in the same units, pounds pull.

In the mechanical process of developing rimpull or drawbar pull, there are power losses. For any specified gear or speed torque position on a torque converter

$$\text{Usable horsepower} = \text{fwhp} \times \frac{E\%}{100} \qquad \textbf{[5.12]}$$

where $E\%$ is the efficiency of the power transmission.

There are two methods for arriving at a machine's developed output force, F_w (force at the wheel):

1. If the whole-body velocity of the machine when operating at governed engine speed N_g is known for a specific gear, the relationship is

$$F_w = \frac{33{,}000 \times \text{fwhp} \times (E\%/100)}{v} \qquad \textbf{[5.13]}$$

where v is the velocity in feet per minute (fpm).

2. If the transmission gear ratio and the rolling radius of the wheel are known, v can be computed and then F_w by Eq. [5.13]. This assumes that there is no slippage in the gear train.

$$N \text{ drive axle} = N_g \times \text{gear ratio} \qquad \textbf{[5.14]}$$

where N drive axle is in rpm:

$$v = 2\pi \times R \text{ drive wheel} \times N \text{ drive axle} \qquad \textbf{[5.15]}$$

where R drive wheel is the radius of the drive wheel.

Normally, the F_w and v are measured and then usable horsepower and, ultimately, $E\%$ are backfigured. This mechanical efficiency, $E\%$, is approximately 90 for direct-drive machines and approximately 80 for torque-converter drives.

Rimpull

Rimpull is a term that is used to designate the tractive force between the rubber tires of driving wheels and the surface on which they travel. If the coefficient of traction is high enough to eliminate tire slippage, the maximum rimpull is a function of the power of the engine and the gear ratios between the engine and the driving wheels. If the driving wheels slip on the haul surface, the maximum effective rimpull will be equal to the total pressure between the tires and the surface multiplied by the coefficient of traction. Rimpull is expressed in pounds.

If the rimpull of a vehicle is not known, it can be determined from the equation

$$\text{Rimpull} = \frac{375 \times \text{hp} \times \text{efficiency}}{\text{speed (mph)}} \text{ lb} \qquad \textbf{[5.16]}$$

This is the formulation of the available power factors horsepower and speed.

The efficiency of most tractors and trucks will range from 0.80 to 0.85. For a rubber-tired tractor with a 140-hp engine and a maximum speed of 3.3 mph in first gear, the rimpull will be

$$\text{Rimpull} = \frac{375 \times 140 \times 0.85}{3.3} = 13,523 \text{ lb}$$

The maximum rimpull in all gear ranges for this tractor will be as shown in the table.

Gear	Speed (mph)	Rimpull (lb)
First	3.3	13,523
Second	7.1	6,285
Third	12.9	3,542
Fourth	21.5	2,076
Fifth	33.9	1,316

In computing the pull that a tractor can exert on a towed load, it is necessary to deduct from the rimpull of the tractor the force required to overcome the rolling and grade resistance. For example, consider a tractor whose maximum rimpull in the first gear is 13,730 lb, weighs 12.4 tons, and is operated up a haul road with a slope of 2% and a rolling resistance of 100 lb per ton. Given these conditions, the pull available for towing a load will be determined as follows:

Max rimpull = 13,730 lb

Pull required to overcome grade, 12.4 tn × (20 lb/tn × 2%) = 496 lb

Pull required to overcome rolling resistance, 12.4 tn × 100 lb/tn = 1,240 lb

Total pull to be deducted, 496 lb + 1,240 lb = 1,736 lb

Pull available for towing a load (13,730 lb − 1,736 lb) = 11,995 lb

Drawbar Pull

The available pull that a crawler tractor can exert on a load that is being towed is referred to as the *drawbar pull* of the tractor. The pull is expressed in pounds. From the total pulling effort of an engine, the pull required to move the tractor over a level haul road must be deducted before the drawbar pull can be determined. If a crawler tractor tows a load up a slope, its drawbar pull will be reduced by 20 lb for each ton of weight of the tractor for each 1% slope.

The performance of crawler tractors, as reported in the specifications supplied by the manufacturer, is usually based on the Nebraska tests. In testing a tractor to determine its maximum drawbar pull at each of the available speeds, the haul road is calculated to have a rolling resistance of 110 lb per ton. If a tractor is used on a haul road whose rolling resistance is higher or lower than 110 lb per ton, the drawbar pull will be reduced or increased, respectively, by an amount equal to the weight of the tractor in tons multiplied by the variation of the haul road from 110 lb per ton.

EXAMPLE 5.2

A tractor whose weight is 15 tons has a drawbar pull of 5,685 lb in the sixth gear when operated on a level road having a rolling resistance of 110 lb per ton. If the tractor is operated on a level road having a rolling resistance of 180 lb per ton, the drawbar pull will be reduced by 15 tons × (180 lb/tn − 110 lb/tn) = 1,050 lb. Thus, the effective drawbar pull will be 5,685 − 1,050 = 4,635 lb.

The drawbar pull of a crawler tractor will vary indirectly with the speed of each gear. It is highest in the first gear and lowest in the top gear. The specifications supplied by the manufacturer should give the maximum speed and drawbar pull for each of the gears.

USABLE POWER

Usable power depends on project conditions, primarily haul-road surface condition and type, and altitude and temperature. Underfoot conditions determine how much of the available power can be transferred to the surface to propel the machine. As altitude increases, the air becomes less dense. Above 3,000 ft, the decrease in air density may cause a reduction in horsepower output of some engines. Manufacturers provide charts detailing appropriate altitude power reductions (see Table 5.4 later in the chapter). Temperature will also affect engine output.

Coefficient of Traction

The total energy of an engine in any unit of equipment designed primarily for pulling a load can be converted into tractive effort only if sufficient traction can be developed between the driving wheels or tracks and the haul surface. If there is insufficient traction, the full available power of the engine cannot be used, as the wheels of tracks will slip on the surface.

The *coefficient of traction* can be defined as the factor by which the total weight on the drive wheels or tracks should be multiplied to determine the maximum possible tractive force between the wheels or tracks and the surface just before slipping will occur.

Usable force = coefficient of traction × wt. on powered running gear   **[5.17]**

The power that can be developed at the interface between running gear and the haul-road surface is often limited by *traction.* The factors controlling usable horsepower are the weight on the powered running gear (drive wheels for wheel type, total weight for track type—see Fig. 5.5), the characteristics of the running gear, and the characteristics of the travel surface.

FOR TRACK-TYPE TRACTOR

Use total tractor weight.

FOR 4-WHEEL TRACTOR

Use weight on drivers shown on spec sheet or approximately 40% of vehicle gross weight.

FOR 2-WHEEL TRACTOR

Use weight on drivers shown on spec sheet or approximately 50% of vehicle gross weight.

FIGURE 5.5 | Weight distribution on powered running gear.

The coefficient of traction between rubber tires and road surfaces will vary with the type of tread on the tires and with the road surface. For crawler tracts, it will vary with the design of the grosser and the road surface. These variations are such that exact values cannot be given. Table 5.3 gives approximate values, for the coefficient of traction between rubber tires or crawler tracks and road surfaces that are sufficiently accurate for most estimating purposes.

TABLE 5.3 | Coefficients of traction for various road surfaces

Surface	Rubber tires	Crawler tracks
Dry, rough concrete	0.80–1.00	0.45
Dry, clay loam	0.50–0.70	0.90
Wet, clay loam	0.40–0.50	0.70
Wet sand and gravel	0.30–0.40	0.35
Loose, dry sand	0.20–0.30	0.30
Dry snow	0.20	0.15–0.35
Ice	0.10	0.10–0.25

EXAMPLE 5.3

Assume that the rubber-tired tractor has a total weight of 18,000 lb on the two driving wheels. The maximum rimpull in low gear is 9,000 lb. If the tractor is operating in wet sand, with a coefficient of traction of 0.30, the maximum possible rimpull prior to slippage of the tires will be

$$0.30 \times 18{,}000 \text{ lb} = 5{,}400 \text{ lb}$$

Regardless of the horsepower of the engine, not more than 5,400 lb of tractive effort can be used because of the slippage of the wheels. If the same tractor is operating on dry clay, with a coefficient of traction of 0.60, the maximum possible rimpull prior to slippage of the wheels will be

$$0.60 \times 18{,}000 \text{ lb} = 10{,}800 \text{ lb}$$

For this surface, the engine will not be able to cause the tires to slip. Thus, the full power of the engine can be used.

EXAMPLE 5.4

A wheel tractor scraper is used on a road project. When the project initially begins, the scraper will experience high rolling and grade resistance at one work area. The rimpull required to maneuver in this work area is 42,000 lb. In the fully loaded condition, 52% of the total vehicle weight is on the drive wheels. The fully loaded vehicle weight is 230,880 lb. What minimum value of coefficient of traction between the scraper wheels and the traveling surface is needed to maintain maximum possible travel speed?

$$\text{Weight on the drive wheels} = 0.52 \times 230{,}880 \text{ lb} = 120{,}058 \text{ lb}$$

$$\text{Minimum required coefficient of traction} = \frac{42{,}000 \text{ lb}}{120{,}058 \text{ lb}} = 0.35$$

Altitude Effect on Usable Power

When a manufacturer provides a flywheel horsepower rating it is based on tests conducted at standard conditions, a temperature of 60° and sea-level barometric pressure, 29.92 in. of mercury (in. Hg). For naturally aspirated engines, operation at altitudes above sea level will cause a decrease in available engine power. This power decrease is caused by the decrease in air density associated with increased altitude. Air density, in turn, affects the fuel-to-air ratio during combustion in the engine's pistons.

The effect of the loss in power due to altitude can be eliminated by the installation of a supercharger. This is a mechanical unit that will increase the pressure of the air supplied to the engine, thus enabling sea-level performance at any altitude.

If equipment is to be used at high altitudes for long periods of time, the increased performance probably will more than pay for the installed cost of a supercharger.

For specific machine applications, the manufacturer's performance data should be consulted. Table 5.4 presents data for selected Caterpillar Inc. machines.

TABLE 5.4 | Percent flywheel horsepower available for select Caterpillar machines at specified altitudes

Model	0–2,500 ft (0–760 m)	2,500–5,000 ft (760–1,500 m)	5,000–7,500 ft (1,500–2,300 m)	7,500–10,000 ft (2,300–3,000 m)	10,000–12,500 ft (3,000–3,800 m)
Tractors					
D6D, D6E	100	100	100	100	94
D7G	100	100	100	94	86
D8L	100	100	100	100	93
D8N	100	100	100	100	98
D9N	100	100	100	96	89
D10N	100	100	100	94	87
Graders					
120G	100	100	100	100	96
12G	100	100	96	90	84
140G	100	100	100	100	94
14G	100	100	100	94	87
16G	100	100	100	100	100
Excavators					
214B	100	100	100	100	92
235D	100	100	100	98	91
245D	100	100	100	94	87
Scrapers					
615C	100	100	95	88	81
621E	100	100	94	87	80
623E	100	100	94	87	80
631E	100	100	96	88	82
Trucks					
769C	100	100	100	97	89
773B	100	100	100	100	96
Loaders					
966E	100	100	100	93	86
988B	100	100	100	100	93

Source: Caterpillar Inc.

POWER OUTPUT AND TORQUE

Figure 5.6 shows the typical curves for brake horsepower and torque as an engine increases its crankshaft speed to the governed rpm value. The important feature of this plot is the shape of the torque curve. Maximum torque is not obtained at maximum rpm. This provides the engine with a power reserve. When a machine is subjected to a momentary overload, the rpm drops but the torque goes up, keeping the engine from stalling. This is commonly referred to as "lugging" the engine.

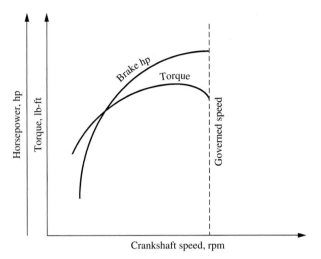

FIGURE 5.6 | Engine speed–power relationships.

Machines can be purchased with either a direct drive (standard) or torque-converter drive. With a direct-drive machine, the operator must manually shift gears to match engine output to the resisting load. The difference in power available when considering maximum torque and torque at governed speed is the machine's operating range for a given gear. In those applications where load is constantly changing, the operation of a direct-drive machine requires a skilled operator. Operator skill is a significant factor that controls the amount of wear and tear a direct-drive machine will experience. Operators of direct-drive machines will be subjected to more operator fatigue than those on power-shift models. This fatigue factor will, in turn, affect machine productivity.

A *torque converter* is a device that adjusts power output to match the load. This adjustment is accomplished hydraulically by a fluid coupling. As a machine begins to accelerate, the engine rpm will quickly reach the governed crankshaft speed and the torque converter will automatically multiply the engine torque to provide the required acceleration force. In this process, there are losses due to hydraulic inefficiencies. If the machine is operating under constant load and at a steady whole-body speed, no torque multiplying is necessary. At that point, the transmission of engine torque can be made nearly as efficient as a direct drive transmission by locking ("lock-up") the torque converter pump and transmission together.

PERFORMANCE CHARTS

Equipment manufacturers publish performance charts for individual machine models. These charts enable the equipment estimator/planner to analyze a machine's ability to perform under a given set of job and load conditions. The performance chart is a graphical representation of the power and corresponding speed the engine and transmission can deliver. The load condition is stated as either rimpull or drawbar pull. It should be noted that the drawbar pull/rimpull–speed relationship is inverse since vehicle speed increases as pull decreases.

Drawbar Pull Performance Chart

In the case of the track machine whose drawbar pull performance chart is shown in Fig. 5.7, the available power ranges from 0 to 56,000 lb (the vertical scale) and the speed ranges from 0 to 6.5 mph (the horizontal scale).

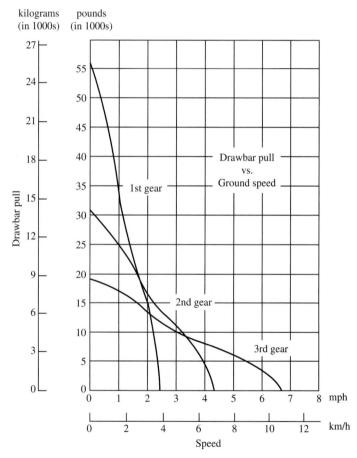

FIGURE 5.7 | Drawbar pull performance chart.
Reprinted courtesy of Caterpillar Inc.

Assuming the power required for a certain application is 25,000 lb, this machine would travel efficiently at a speed of approximately 1.5 mph in first gear. This is found by first moving horizontally across the chart at the 25,000-lb mark on the vertical scale and intersecting the first-gear curve. At the intersection point, move vertically downward to find the speed in mph on the bottom horizontal scale.

Rimpull Performance Charts

Each manufacturer has a slightly different graphical layout for presenting performance chart information. However, the procedures for reading a performance chart are basically the same. The steps described here are based on the chart shown in Fig. 5.8.

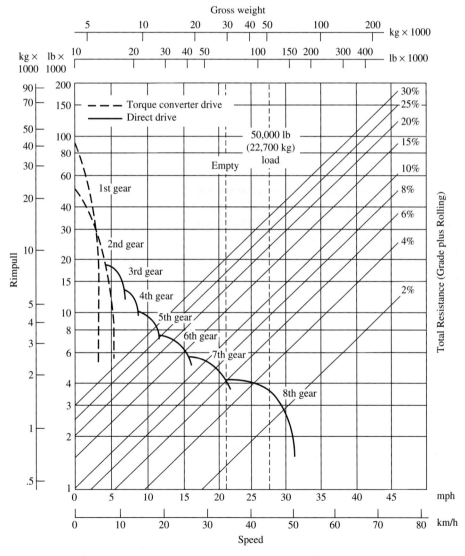

FIGURE 5.8 | Rimpull performance chart.

Reprinted courtesy of Caterpillar Inc.

Required power/total resistance   The charts are presented so that speed can be determined in two different ways, based on whether the required power was calculated in terms of force or effective grade. Assuming that the required power has been calculated, the procedures to determine speed are

1. Ensure that the proposed machine has the same engine, gear ratios, and tire size as those identified for the machine on the chart. If the gear ratios or rolling radius of a machine is changed, the performance curve will shift along both the rimpull and speed axes.
2. Estimate a power required or total resistance (rolling resistance plus grade resistance) based on the probable job conditions.
3. Extend a line horizontally to the right from the power requirement found on the left vertical scale. The point of intersection of this horizontal line with a gear curve defines the operating relationship between horsepower and speed.
4. From the point at which this horizontal line intersects the gear curve, draw a line vertically to the bottom x axis, which indicates the speed in mph. This yields the vehicle speed for the assumed job conditions. Sometimes the horizontal line from the power requirement will intersect the gear range curve at two points (see Fig. 5.9). In such a case, the speed can be interpreted in two ways.

A guide in determining the appropriate speed is

■ If the required rimpull is *less* than that required on the previous stretch of haul, use the higher gear and speed.
■ If the required rimpull is *greater* than that required on the previous stretch of haul, use the lower gear and speed.

Effective grade resistance   Assuming that an effective grade has been calculated, the procedures (many of these steps are the same as in the required power/total resistance case) to determine speed are

1. Ensure that the proposed machine has the same engine, gear ratios, and tire size as those identified for the machine on the chart.
2. Determine the machine weight both when the machine is empty and loaded. The empty weight is the operating weight and should include coolants, lubricants, full fuel tanks, and operator. Loaded weight depends on the density of the loaded material and the proposed load size. These two weights, empty and loaded, are often referred to as the *net vehicle weight* (NVW) and *the gross vehicle weight* (GVW), respectively. The net or empty weight is usually marked on the chart. Likewise the gross weight, based on the gravimetric capacity of the machine, will usually be indicated.
3. Based on the probable job conditions, estimate a total resistance (the sum of rolling plus grade resistance both expressed as percent grade). The

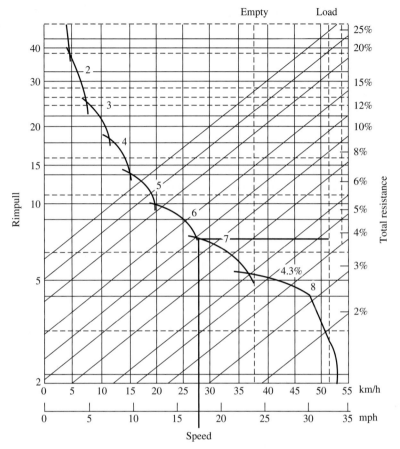

FIGURE 5.9 | Rimpull performance chart—gear effect closeup.

Reprinted courtesy of Caterpillar Inc.

intersection of the vertical vehicle weight line and the diagonal total resistance line establishes the conditions under which the machine will be operated and correspondingly the power requirement.

4. Extend a line horizontally from the vehicle-weight total-resistance intersection point. The point of intersection of this horizontal line with a gear curve defines the operating relationship between horsepower and speed.

5. From the point at which the horizontal line intersects the gear range curve, draw a line vertically to the bottom x axis, which indicates the speed in mph. This yields the vehicle speed for the assumed job conditions.

Performance charts are established assuming machine operation under standard conditions. When the machine is utilized under a differing set of conditions, the rimpull force and speed must be appropriately adjusted. Operation at higher altitudes will require a percentage derating in rimpull that is approximately equal to the percentage loss in flywheel horsepower.

Retarder Performance Chart

When operating on steep downgrades, a machine's speed may have to be limited for safety reasons. A retarder is a dynamic speed control device. By the use of an oil-filled chamber between the torque converter and the transmission, machine speed is retarded. The retarder will not stop the machine; rather it provides speed control for long downhill hauls, reducing wear on the service brake. Figure 5.10 presents a retarder chart for a wheel-tractor scraper.

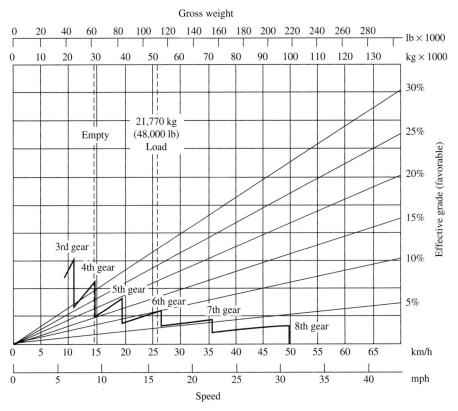

FIGURE 5.10 | Retarder performance chart.

Reprinted courtesy of Caterpillar Inc.

The retarder performance chart (see Fig. 5.10) identifies the speed that can be maintained when a vehicle is descending a grade having a slope such that the magnitude of the grade resistance is greater than the resisting rolling resistance.

This retarder-controlled speed is steady state, and use of the service brake will not be necessary to prevent acceleration.

A retarder performance chart is read in a manner similar to that already described, remembering that the total resistance (*effective grade*) values are actually negative numbers. As with the rimpull chart, the horizontal line can intersect more than one gear. In a particular gear, the vertical portion of the retarder curve indicates maximum retarder effort and resulting speed. If haul conditions dictate, the operator will shift into a lower gear and a lower speed would be applicable. Many times the decision as to which speed to select is answered by the question: "How much effort will be expended in haul route maintenance?" Route smoothness is often the controlling factor affecting higher operating speeds.

EXAMPLE 5.5

A contractor proposes to use scrapers on an embankment job. The performance characteristics of the machines are shown in Figs. 5.8 and 5.10. The scrapers have a rated capacity of 14-cy struck. Operating weight empty is 69,000 lb. Loaded weight distribution is 53% on the drive wheels.

The contractor believes that the average load for the material that must be hauled will be 15.2 bcy. The haul from an excavation area is a uniform adverse gradient of 5.0% with a rolling resistance of 60 lb per ton. The material to be excavated and transported is a common earth with a bank unit weight of 3,200 lb per bcy.

a.   Calculate the maximum travel speeds that can be expected.

Machine weight:

Empty operating weight	69,000 lb
Payload weight, 15.2 bcy × 3,200 lb per bcy	48,640 lb
Total loaded weight =	117,640 lb

Haul Conditions:

	Loaded (haul) (%)	**Empty (return) (%)**
Grade resistance	5.0	−5.0
Rolling resistance 60/20	3.0	3.0
Total resistance	8.0	−2.0

Loaded speed (haul): Using Fig. 5.8, enter the upper horizontal scale at 117,640 lb, the total loaded weight, and move down to the intersection with the 8.0% total resistance line. The line intersects fifth gear. By drawing a vertical line from the fifth-gear intersect point to the lower horizontal scale, the determined scraper speed is approximately 11 mph.

Empty Speed (return): Using Fig. 5.10, enter the upper horizontal scale at 69,000 lb, the empty operating weight. Because this is a commonly used weight, it is marked on the chart as a dashed line. The intersection with the 2.0% effective resistance line defines the point from which to construct the horizontal line. The horizontal line intersects eighth gear, and the corresponding speed is 31 mph.

b.   If the job is at elevation 12,500 ft, what will be the operating speeds when consideration of the nonstandard pressure is included in the analysis? Operating at an altitude of 12,500 ft the manufacturer has reported that these scrapers, which have turbocharger engines, can deliver 82% of rated flywheel horsepower.

At an altitude of 12,500 ft, the rimpull that is necessary to overcome a total resistance of 8% must be adjusted for the altitude derating.

$$\text{Altitude adjusted effective resistance: } \frac{8.0}{0.82} = 9.8\%$$

Now proceed as in part (a) by locating the intersection of the total loaded weight line and the altitude adjusted total resistance line. Constructing a horizontal line from the 9.8% altitude adjusted effective resistance diagonal yields an intersection with the fourth-gear curve and, by drawing a vertical line at that point, a speed of approximately 8 mph.

In this problem, the nonstandard altitude would not affect the empty (return) speed because the machine does not require rimpull force for motion. The retarder controls the machine's downhill momentum. Therefore, the empty (return) speed is still 31 mph at 12,500-ft altitude.

SUMMARY

The payload of hauling equipment can be expressed either *volumetrically* or *gravimetrically*. The power required is the power necessary to overcome the *total resistance* to machine movement, which is the sum of rolling and grade resistance. Rolling resistance is the resistance of a level surface to constant-velocity motion across it. The force-opposing movement of a machine up a frictionless slope is known as *grade resistance*.

The *coefficient of traction* is the factor by which the total weight on the drive wheels or tracks should be multiplied to determine the maximum possible tractive force between the wheels or tracks and the surface just before slipping will occur. A performance chart is a graphical representation of power and corresponding speed the engine and transmission can deliver. Critical learning objectives would include:

■   An ability to calculate vehicle weight.
■   An ability to determine rolling resistance based on anticipated haul-road conditions.
■   An ability to calculate grade resistance.
■   An ability to use performance charts to determine machine speed.

These objectives are the basis for the problems that follow.

PROBLEMS

5.1 A four-wheel tractor whose operating weight is 48,000 lb is pulled up a road whose slope is +4% at a uniform speed. If the average tension in the towing cable is 4,680 lb, what is the rolling resistance of the road? (115 lb/ton)

5.2 A wheel-type tractor-pulled scraper having a combined weight of 172,000 is push-loaded down a 6% slope by a crawler tractor whose weight is 70,000 lb. What is the equivalent gain in loading force for the tractor and scraper resulting from loading the scraper down slope instead of up slope?

5.3 Consider a wheel-type tractor-pulled scraper whose gross weight is 94,000 lb, including tractor, scraper, and its load. What is the equivalent gain, in horse-power, resulting from operating this vehicle down a 4% slope instead of up the same slope at a speed of 12 mph? One horsepower equals 33,000 ft-lb of work per minute. (241 hp)

5.4 A wheel-type tractor with a 210-hp engine has a maximum speed of 4.65 mph in first gear. Determine the maximum rimpull of the tractor in each of the indicated gears if the efficiency is 90%.

Gear	Speed (mph)
First	4.65
Second	7.60
Third	11.50
Fourth	17.50
Fifth	26.80

5.5 If the tractor of Problem 5.4 weighs 21.4 tons and is operated over a haul road whose slope is +3% with rolling resistance of 80 lb per ton, determine the maximum external pull by the tractor in each of five gears. (12,246 lb; 6,330 lb; 3,167 lb; 1,077 lb; none)

5.6 If the tractor of Problems 5.4 and 5.5 is operated down a 4% slope whose rolling resistance is 80 lb per ton, determine the maximum external pull by the tractor in each of the five gears.

5.7 A wheel-type tractor, operating in its second-gear range and at its full rated rpm, is observed to maintain a steady speed of 1.50 mph when operating under the conditions described herein. Ambient air temperature is 60°F, altitude is sea level. The tractor is climbing a uniform 6.5% slope with a rolling resistance of 65 lb per ton, and it is towing a pneumatic-tired trailer loaded with fill material. The two-axle tractor has a total weight of 60,000 lb, 55% of which is distributed to the power axle. The loaded trailer has a weight of 75,000 lb.

   a.   For the environmental conditions just described, the manufacturer rates the tractive effort of the new tractor at 60 rimpull horsepower. What percentage of this rated rimpull does the tractor actually develop? In performing your calculations, it will be acceptable to assume that the component of weight normal to the traveling surface is equal to the weight itself (i.e., cos 0 taken as 1.00, where 0° is the angle of total resistance). The "20 lb of rimpull required ton of weight per % of slope" approximation will be acceptable. You may also disregard the power required to overcome wind resistance and to provide acceleration. Assume traction is not a limiting factor. (87.7%; 0.4)

b.   What is the value of the coefficient of traction if the drive wheels of the tractor are at the point of incipient slippage for the conditions just described?

5.8   A wheel-type tractor unit, operating in its fourth-gear range and at its full rated rpm, is observed to maintain a steady speed of 7.50 mph when operating under the conditions described herein. Ambient air temperature is 60°F. Altitude is sea level. The tractor is climbing a uniform 5.5% slope with a rolling resistance of 55 lb per ton, and it is towing a pneumatic-tired trailer loaded with fill material. The single-axle tractor has an operating weight of 66,000 lb. The loaded trailer has a weight of 48,000 lb. The weight distribution for the combined tractor trailer unit is 55% to the drive axle and 45% to the rear axle.

a.   For the environmental conditions just described, the manufacturer rates the tractive effort of the new tractor at 330 rimpull horsepower. What percentage of this rated rimpull hp does the tractor actually develop? In performing your calculations, it will be acceptable to assume that the component of weight normal to the traveling surface is equal to the weight itself (i.e., cos 0 taken as 1.00, where 0° is the angle of total resistance). The "20 lb of rimpull required per ton of weight per % of slope" approximation will be acceptable. You may also disregard the power required to overcome wind resistance and to provide acceleration. Assume traction is not a limiting factor.

b.   What is the value of the coefficient of traction if the drive wheels of the tractor are at the point of incipient slippage for the conditions just described?

5.9   A wheel tractor-scraper is operating on a level grade. Assume no power derating is required for equipment condition, altitude, temperature, etc. Use equipment data from Fig. 5.8.

a.   Disregarding traction limitations, what is the maximum value of rolling resistance (in pounds per ton) over which the fully loaded unit can maintain a speed of 20 mph?

b.   What minimum value of coefficient of traction between the tractor wheels and the traveling surface is needed to satisfy the requirements of part (a)? For the fully loaded condition, 67% of the weight is distributed to the drive axle. Operating weight of the empty scraper is 70,000 lb.

5.10   A wheeled tractor with high-pressure tires and weighing 81,000 lb is pulled up a 5% slope at a uniform speed. If the tension in the tow cable is 15,200 lb, what is the rolling resistance of the road? What type of surface would this be?

5.11   A wheel tractor-scraper is operating on a 4% adverse grade. Assume that no power derating is required for equipment condition, altitude, and temperature. Use equipment data from Fig. 5.8. Disregarding traction limitations, what is the maximum value of rolling resistance (in pounds per ton) over which the empty unit can maintain a speed of 14 mph?

REFERENCES

1. *Caterpillar Performance Handbook,* Caterpillar Inc., Peoria, Ill. Published annually. (www.cat.com.)

2. *Handbook of Earthmoving,* Caterpillar Tractor Inc., Peoria, Ill., 1981.

3. Nowatzki, E. A., L. L. Karafiath, and R. L. Wade, "The Use of Mobility Analyses for Selection of Heavy Mining Equipment," *Proceedings of the Specialty Conference on Construction Equipment & Techniques for the Eighties,* American Society of Engineers, New York, 1982, pp. 130–155.

4. Schexnayder, Cliff, Sandra L. Weber, and Brentwood T. Brooks, "Effect of Truck Payload Weight on Production," *Journal of Construction Engineering and Management, ASCE,* Vol. 125, No. 1, January–February 1999, pp. 1–7.

5. Sullivan, R. J., "Mobility Prediction in Earthmoving," *Proceedings 7th International Conference for Terrain Vehicle Systems,* Vol. 1, August 1981.

6

Dozers

A dozer is a tractor-power unit that has a blade attached to the machine's front. It is designed to provide tractive power for drawbar work. A dozer has no set volumetric capacity. The amount of material the dozer moves is dependent on the quantity that will remain in front of the blade during the push. Crawler dozers equipped with special clearing blades are excellent machines for land clearing. Heavy ripping of rock is accomplished by crawler dozers equipped with rear-mounted rippers because of the power and tractive force that they can develop.

DESCRIPTION

Dozers are self-contained units equipped with a blade. They are designed to provide tractive power for drawbar work. Dozers may be either tracklaying crawler or wheel-type machines. Consistent with their purpose, as a unit for drawbar work, they are low-center-of-gravity machines. This is a prerequisite of an effective machine. The larger the difference between the line-of-force transmission from the machine and the line-of-resisting force, the less effective the utilization of developed power. Dozers are used for dozing (pushing materials), land clearing, ripping, assisting scrapers in loading, and towing other pieces of construction equipment. They can be equipped with either a rear-mounted winch or a ripper. For long moves between projects or within a project, the track dozer should be transported. Moving them under their own power, even at slow speeds, increases track wear and shortens the machine's operational life.

PERFORMANCE CHARACTERISTICS OF DOZERS

Dozers are classified on the basis of running gear:

1. Crawler (tracklaying) type (see Fig. 6.1).
2. Wheel type (see Fig. 6.2).

FIGURE 6.1 | Crawler dozer.

FIGURE 6.2 | Wheel dozer.

Crawler dozers are actually tracklaying machines. They have a continuous track of linked shoes (see Fig. 6.3) that moves in the horizontal plane across fixed rollers. At the rear of the machine, the track passes over a vertically mounted sprocket drive wheel. As the sprocket turns, it forces the track forward or back, imparting motion to the dozer. In the front of the machine, the track passes over a vertically mounted idler wheel that is connected to a recoil device

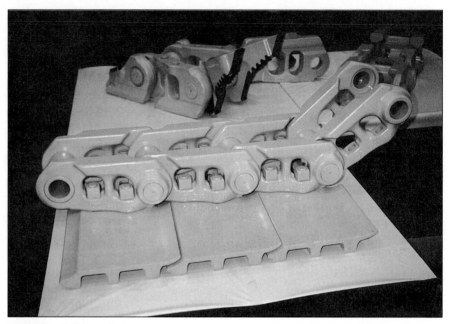

FIGURE 6.3 | Track shoes.

having adjustable tension. The idler wheel maintains the proper tension in the track and enables it to absorb heavy shocks. The linked shoes are made of heat-treated steel designed to resist wear and abrasion. There are several companies that now offer tracks having rubber-covered steel shoes.

As discussed in Chapter 4, the usable force available to perform work is often limited by traction. This limitation is dependent on two factors:

1. Coefficient of traction of the surface being traversed.
2. Weight carried by the drive wheels.

Sometimes users weight the tires of wheel-type dozers to overcome tractive-power limitations. A mixture of calcium chloride and water is recommended as tire ballast. Care must be taken to ensure that the new weight distribution is equal between all drive wheels.

Traction or floatation requirements can be met by proper undercarriage or tire selection. A standard crawler dozer undercarriage is appropriate for general work in rock to moderately soft ground. Typical ground pressure for a crawler dozer with a standard undercarriage is about 6 to 9 psi [41–62 kilopascals (kPa)]. The low-ground-pressure (LGP) undercarriage configuration is for soft ground conditions. The ground pressure exerted by a crawler dozer with an LGP under-carriage is about 3 to 4 psi (21–28 kPa). LGP machines should not be used in hard or rocky conditions, as such a practice will reduce undercarriage life. There are extralong (XL) undercarriages available for machines dedicated to finish work.

In the case of wheel machines, wider tires provide greater contact area and increase floatation. It must be remembered, however, that rimpull charts are based on standard equipment, including tires. Larger tires will reduce developed rimpull.

The crawler- (tracklaying-)type unit is designed for those jobs requiring high tractive effort. No other piece of equipment can provide the power, traction, and floatation needed in such a variety of working conditions. A crawler dozer can operate on slopes as steep as 45°.

Both track- and wheel-type dozers are rated by flywheel power (fwhp) and weight. Normally the weight is an operating weight and includes lubricants, coolants, a full fuel tank, a blade, hydraulic fluid, the OSHA rollover protective standards canopy (ROPS), and an operator. Dozer weight is important on many projects because the maximum tractive effort that a unit can provide is limited to the product of the weight times the coefficient of traction for the unit and the particular ground surface, regardless of the power supplied by the engine. Table 5.3 gives the coefficients of traction for various surfaces.

An advantage of a wheel-type dozer as compared with a crawler dozer is the higher speed possible with the former machine—in excess of 30 mph for some models. To attain a higher speed, however, a wheel dozer must sacrifice pulling effort. Also, because of the lower coefficient of traction between rubber tires and some ground surfaces, the wheel dozer may slip its wheels before developing its rated pulling effort. Table 6.1 provides a comparison of crawler dozer and wheel dozer utilization.

TABLE 6.1 | Dozer-type utilization comparison

Wheel dozer	Crawler dozer
Good on firm soils and concrete and abrasive soils that have no sharp-edged pieces	Can work on a variety of soils; sharp-edged pieces not as destructive to dozer though fine sand will increase running gear wear
Best for level and downhill work	Can work over almost any terrain
Wet weather causing soft and slick surface conditions will slow or stop operation	Can work on soft ground and over mud-slick surfaces; will exert very low ground pressures with special low-ground-pressure undercarriage and track configuration
The concentrated wheel load will provide compaction and kneading action to ground surface	
Good for long travel distances	Good for short work distances
Best in handling loose soils	Can handle tight soils
Fast return speeds, 8–26 mph	Slow return speeds, 5–10 mph
Can only handle moderate blade loads	Can push large blade loads

Internal combustion engines are used to power most dozers, with diesel engines being the most common primary power units. Gasoline engines are used in some smaller machines. There are electric- and air-powered dozers available for tunnel work.

Because the crankshaft rotation derived from the engine is usually too fast and does not have sufficient force (torque), machines have transmissions that reduce the rotational speed of the crankshaft and increase the force available to do work. Transmissions provide the operator with the ability to change the machine's speed-power ratio so that it matches the work requirements. Manufacturers provide dozers with a variety of transmissions, but primarily the options are

■ Direct drive.
■ Torque converter and power-shift transmission.

Some less-than-100-hp dozers are available with hydrostatic powertrains. The smaller, less-than-300-hp, diesel-powered machines are commonly available with either direct- or power-shift-type transmissions. Larger dozers are always equipped with power-shift transmissions.

Crawler Dozers with Direct Drive

The term direct drive means the power is transmitted straight through the transmission as if there was a single shaft. This is usually what happens when the transmission is in its highest gear. In all other gears, mechanical elements match speed and torque. Direct drive dozers are superior when the work involves constant loading conditions. A job where full blade loads must be pushed long distances would be an appropriate application of a direct drive machine.

Some manufacturers' specifications list two sets of drawbar pulls—rated and maximum for direct drive dozers. The rated value is the drawbar pull that can be sustained for continuous operation. The maximum drawbar pull is the pull that the dozer can exert for a short period while lugging the engine, such as when passing over a soft spot in the ground that requires a temporary higher tractive effort. Thus, the rated pull should be used for continuous operation. Available drawbar pull is subject to the limitation imposed by the traction that can be developed between the tracts and the ground.

Crawler Dozers with Torque Converter and Power-Shift Transmissions

Transmissions that can be shifted while transmitting full engine power are known as power shift. These transmissions are teamed with torque converters to absorb drive train shock loads caused by changes in gear ratios. A power-shift transmission provides an efficient flow of power from the engine to the tracks and gives superior performance in applications involving variable load conditions. Figure 6.4 illustrates the performance curves for a track-type dozer equipped with a power-shift transmission.

Crawler Dozers with Hydrostatic Powertrains

Confined oil under pressure is an effective means of power transmission. A hydrostatic powertrain offers an infinitely variable speed range with constant power to both tracks. This type of powertrain improves machine controllability and increases operational efficiency. Hydrostatic powertrain transmissions are available on some small-size dozers.

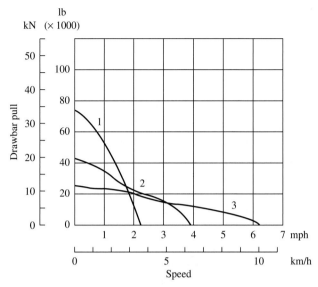

FIGURE 6.4 | Performance chart for a 200-hp 45,560-lb track-type dozer with a power shift.

Reprinted courtesy of Caterpillar Inc.

Wheel Dozers

Most wheel dozers are equipped with torque converters and power-shift transmissions. Figure 6.5 illustrates the performance curves for a wheel dozer equipped with a power-shift transmission. Wheel dozers exert comparatively high ground pressures, 25 to 35 psi (172–241 kPa).

Comparison of Performance

Heed the cautionary note on Fig. 6.5 that usable pull/rimpull will depend on the weight and traction of the fully equipped dozer. This is a warning that, even though the engine can develop a certain drawbar pull or rimpull force, all of that pull may not be available to do work. The caution is a restatement of Eq. [5.17]. If the project working surface is dry clay loam, Table 5.3 provides the following coefficient of traction factors:

Rubber tires   0.50–0.70
Track   0.90

Using the factor for tracks, 0.90, and considering a track-type dozer with a power shift (see Fig. 6.4), the usable drawbar pull is found to be

$$45,560 \text{ lb} \times 0.90 = 41,004 \text{ lb}$$

Now consider a wheel-type dozer (Fig. 6.5):

$$45,370 \text{ lb} \times 0.60 = 27,222 \text{ lb}$$

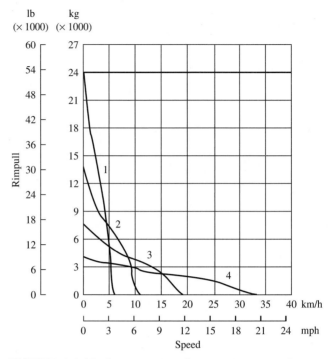

FIGURE 6.5 | Performance chart for a 216-hp 45,370-lb wheel dozer with power-shift transmission. Usable rimpull depends on traction and the weight of dozer.

Reprinted courtesy of Caterpillar Inc.

The two machines have approximately the same operating weight and flywheel power; yet, because of the effect of traction, the track machine can supply one and a half times the *usable* power.

In the case of most soil conditions, the coefficient of traction for wheels is less than that of tracks. Therefore, a wheel-type dozer must be considerably heavier (approximately 50%) than a crawler dozer to develop the same amount of usable force.

As the weight of a wheel-type dozer is increased, a larger engine will be required to maintain the weight-to-horsepower ratio. There is a limit to the weight that can be added to the wheel-type dozer and still have a machine with a speed and mobility advantage over track dozers.

PUSHING MATERIAL

GENERAL INFORMATION

A *dozer* is a tractor power unit that has a blade attached to the machine's front. The blade is used to push, shear, cut, and roll material ahead of the dozer. Dozers are effective and versatile earthmoving machines.

Dozers are used as both support and as production machines on many construction projects. They may be used for operations such as

1. Moving earth or rock for short haul (push) distances, up to 300 ft (91 m) in the case of large dozers.
2. Spreading earth or rock fills.
3. Back-filling trenches.
4. Opening up pilot roads through mountains or rocky terrain.
5. Clearing the floors of borrow and quarry pits.
6. Helping load tractor-pulled scrapers.
7. Clearing land of timber, stumps, and root mat.

BLADES

A dozer blade consists of a moldboard with replaceable cutting edges and side bits. Push arms and tilt cylinders or a C-frame connect the blade to the dozer. Blades vary in size and design based on specific work applications. The hardened-steel cutting edges and side bits are bolted on because they receive most of the abrasion and wear out rapidly. The bolted connection enables easy replacement. The design of some machines enables either end of the blade to be raised or lowered in the vertical plane of the blade, *tilt.* The top of the blade can be pitched forward or backward varying the angle of attack of the cutting edge, *pitch.* Blades mounted on a C-frame can be turned from the direction of travel, *angling.* These features are not applicable to all blades, but any two of these may be incorporated in a single mount. Figure 6.6 illustrates tilt, pitch, and angling.

> *Tilt.* This movement is within the vertical plane of the blade. Tilting enables concentration of dozer driving power on a limited portion of the blade's length.
>
> *Pitch.* This is a pivotal movement about the point of connection between the dozer and blade. When the top of the blade is pitched forward, the bottom edge moves back, and this increases the angle of cutting edge attack.
>
> *Angling.* Turning the blade so that it is not perpendicular to the direction of the dozer's travel is known as angling. Angling causes the pushed material to roll off the trailing end of the blade. This procedure of rolling material off one end of the blade is called *side casting.*

Blade Performance

A dozer's pushing potential is measured by two standard ratios:

- Horsepower per foot of cutting edge.
- Horsepower per loose cubic yard of material retained in front of the blade.

Horsepower per foot (hp/ft) provides an indication of the ability of the blade to penetrate and obtain a load. The higher this ratio the more aggressive the blade.

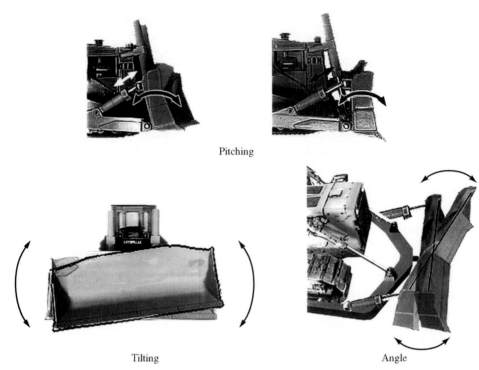

Pitching

Tilting

Angle

FIGURE 6.6 I Bulldozer blade adjustments—tilt, pitch, and angle.

Horsepower per loose cubic yard (hp/lcy) measures the blade's ability to push a load. A higher ratio means that the dozer can push a load at a greater speed.

The blade is raised or lowered by hydraulic rams, therefore a positive downward force can be exerted. Additionally, basic earthmoving blades are curved in the vertical plane in the shape of a flattened C. When the blade is pushed down, the edge cuts into the earth. As the dozer moves forward, the cut material is pushed up the face of the blade. The upper part of the flattened C rolls this material forward. The total effect is to "boil" the pushed material over and over in front of the blade. The flattened C shape provides the necessary cutting angle for the edge and at the beginning of the pass the weight of the cut material on the lower half of the C helps achieve edge penetration. As the push progresses, the load in front of the blade passes the midpoint of the C and begins to exert an upward force on the blade. This "floats" the blade reducing the penetration of the cutting edge.

Many different special application blades may be attached to a dozer (see Fig. 6.7), but basically only five blades are common to earthwork: (1) the *straight* "S" blade, (2) the *angle* "A" blade, (3) the *universal* "U" blade, (4) the semi-U "SU" blade, and (5) the *cushion* "C" blade.

Straight blades "S." The straight blade is designed for short- and medium-distance passes, such as backfilling, grading, and spreading fill material. These blades have no curvature in their length and are mounted in a fixed

Straight blade

Angle blade

Universal blade

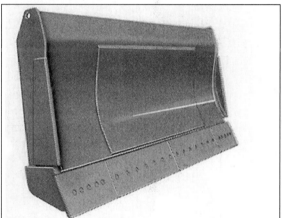

Cushion blade

FIGURE 6.7 | Common earthmoving dozer blades.

position, perpendicular to the dozer's line of travel. Generally, a straight blade is heavy duty and normally it can be tilted, within a 10° arc, increasing penetration for cutting or decreasing penetration for backdragging. It may be equipped to pitch. The ability to pitch means that the operator can set the cutting edge to dig hard materials or move the edge's plane of attack to ease the drifting of light materials.

Angle blades "A." An angle blade is wider by 1–2 ft than an S blade. It can be angled up to a maximum of 25° left or right of perpendicular to the dozer, or used as a straight blade. An angle blade can be tilted, but because it is attached to the dozer by a C-frame mount, it cannot be pitched. The angle blade is very effective for side casting material particularly for backfilling or making sidehill cuts.

Universal blades "U." This blade is wider than a straight blade and the outside edges are canted forward about 25°. The canting of the edges reduces the spillage of loose material making the U blade efficient for moving large loads over long distances. The hp/ft ratio is lower for the U than the S blade mounted on a similar dozer. Penetration is not a prime objective of the U blade, as this ratio relationship indicates. The U blade's hp/lcy ratio is lower than that of an S blade. This denotes that the blade is best suited for lighter materials. Typical usages are working stockpiles and drifting loose or noncohesive materials.

Semi-U blades "SU." This blade combines the characteristics of the S- and U-blade designs. It has increased capacity by the addition of short wings.

Cushion blades "C." Cushion blades are mounted on large dozers that are used primarily for push-loading scrapers. The C blade is shorter than the S blade so as to avoid pushing the blade into and cutting the rear tires of the scraper while push-loading. The shorter length also facilitates maneuvering into position behind the scrapers. Rubber cushions and springs in the mounting enable the dozer to absorb the impact of contacting the scraper push block. By using a cushion blade instead of a "pusher block" to push scrapers, the dozer can clean up the cut area and increase the total fleet production. It is a blade of limited utility in pushing material and should not be used for production dozing. It cannot be tilted, pitched, or angled.

PROJECT EMPLOYMENT

Stripping

Dozers are excellent machines for *stripping,* which is the removal of a thin layer of material. On most projects, this is a term used to describe the removal of top-soil. As with all dozer earthmoving operations, stripping should be conducted in such a manner that push distances are minimized. Dozers are economical machines for moving material only about 300 ft in the case of large machines. The economical push distance decreases as dozer size decreases, but economical push distance also depends on the material being handled. A material exhibiting cohesion (clay content) is easier to push than a granular material (sand), which tends to run in front of the blade. In situations where material must be moved a distance greater than 300 ft, scrapers should be considered. Dozers can be very effective support machines for long-haul stripping situations when used to create windrows of stripped material that can be easily picked up by the scrapers.

Sidehill Cuts

It is very difficult to develop the initial working table for excavations made on steep ground. Usually, the excavated material from such a cut is pushed over the side of the hill. The first passes are made perpendicular to the long direction of the project. Starting on the uphill side, short passes are made to push the material across the centerline and over the side. Pushing downhill takes advantage of gravity. Because these perpendicular passes are short, the dozer usually is not able to develop a full-blade load. Therefore, once a bench is established, the dozer should push in the long direction of the project, develop a full-blade load, and then use turns to push the material over the side.

Ditching

A dozer can be used to accomplish ditching, but this is practical only for very rough ditch sections. Small shallow ditches are usually cut with a motor grader. Large deep ditches are either cut with excavators or, if the cut is made before water enters the ditch, scrapers can be used. A dozer will follow the scrapers and perform the final dressing of the slopes.

If a dozer is used to cut rough ditches, the machine pushes the material out of the cut by working perpendicular to the line of the ditch.

Backfilling

A dozer can efficiently accomplish backfilling by drifting material sideways with an angle blade. This enables forward motion parallel to the excavation. If a straight blade is used, the dozer will approach the excavation at a slight angle and then, at the end of the pass, turn in toward the excavation. No part of the tracks should hang over the edge of the excavation.

Caution must be exercised in making the initial pass completely across pipes and culverts. As a minimum, 12 in. of material should cover the pipe or structure before accomplishing a crossing. The diameter of the pipe, the pipe type, the distance between the sidewalls of the excavation, and the number of lines of pipe in the excavation dictate the minimum required cover. Larger diameter pipe, larger excavation widths, and multiple lines of pipe are factors that all dictate more cover before crossing the structure.

Rocks or Frozen Ground

With proper attack techniques, a dozer can move rocks or frozen ground. In both cases, the blade must be worked under the material to be moved. This can be accomplished by tilting the corner of the blade. To maximize the driving force of the blade, hook only the tilted end under the rock or ground. It may be necessary to use the blade as a pry bar to lift the rock. Once the blade is in contact beneath the rock and the dozer is driving forward, the operator lifts the blade to pry up the rock.

Weak formations of soft rocks, such as shale and sandstone, can be attacked in a similar manner. Work under the outcrop and lift. Once a plane of weakness slides, a track machine can often crush the material by running over it. It should be remembered that dozer work in rocky areas increases track wear.

Spreading

The spreading of material dumped by trucks or scrapers is a common dozer task. Ordinarily, project specifications state a maximum loose lift thickness. Even when lift thickness limits are not stated in the contract specifications, density requirements and proposed compaction equipment will force the contractor to control the height of each lift. Uniform spreading is accomplished with a dozer by keeping the blade straight and at the desired height above the previously placed fill surface. The dumped material is forced directly under the blade's cutting edge. Fairly uniform spreading can be achieved, even by semiskilled operators, if two complete passes are made across the dump area, with the second pass made perpendicular to the first. Today, laser blade controls are available for this type of work (see Fig. 6.8).

Slot Dozing

Slot dozing is the technique whereby the blade end spillage from the first pass or the sidewalls from previous cuts are used to hold material in front of the dozer blade on subsequent passes. When employing this method to increase production, align cuts parallel, leaving a narrow uncut section between slots. Then, remove the uncut sections by normal dozing. The technique prevents spillage at each end of the blade and usually increases production by about 20%. The production increase is highly dependent on the slope of the push and the type of material being pushed.

Blade-to-Blade Dozing

Another technique used to increase bulldozer production is blade-to-blade dozing (see Fig. 6.9). The technique is sometimes referred to as *side-by-side dozing*.

FIGURE 6.8 I Spreading using a laser blade control.

FIGURE 6.9 | Blade-to-blade dozing used to increase production by minimizing spillage.

As the names imply, two machines maneuver so that their blades are right next to each other during the pushing phase of the production cycle. This reduces the side spillage of each machine by 50%. The extra time necessary to position the machines together increases that phase of the cycle. Therefore, the technique is not effective on pushes of less than 50 ft because of the excess maneuver time required. When machines operate simultaneously, delay to one machine is in effect a double delay. The combination of less spillage but increased maneuver time tends to make the total increase in production for this technique somewhere between 15 and 25%.

DOZER PRODUCTION ESTIMATING

A dozer has no set volumetric capacity. There is no hopper or bowl to load; instead the amount of material the dozer moves is dependent on the quantity that will remain in front of the blade during the push. The factors that control dozer production rates are

1. Blade type.
2. Type and condition of material.
3. Cycle time.

Blade Type

The description of blade types was already presented. An important production estimation characteristic of blades stated there was that straight blades roll material in front of the blade, whereas universal and semi-U blades control side spillage holding the material within the blade. Because the U and SU blades force the material to move to the center, there is a greater degree of swelling. The U or SU blade's quantity of loose material will be greater than that of the S blade. But the ratio of this difference is not the same when considering bank yards. This is because the factor to convert loose cubic yards to bank cubic yards for the universal type blades is not the same as that for a straight blade. The U or SU blade's boiling effect causes the difference.

The same type of blade comes in different sizes to fit different size dozers. Blade capacity then is a function of blade type and physical size. Manufacturers' specification sheets will provide the necessary information concerning blade dimensions.

Type and Condition of Material

The type and condition of the material being handled affects the shape of the pushed mass in front of the blade. Cohesive materials (clays) will "boil" and heap. Materials that exhibit a slippery quality or those that have a high mica content will ride over the ground and swell out. Cohesionless materials (sands) are known as "dead" materials because they do not exhibit heap or swell properties. Figure 6.10 illustrates these material attributes.

Clay material boiling in front of the blade.

Cohesionless, sandy loam in front of the blade.
FIGURE 6.10 I Bulking attributes of materials when being pushed.

Blade Load

The load a blade will carry can be estimated by several methods:

1. Manufacturer's blade rating.
2. Previous experience (similar material, equipment, and work conditions).
3. Field measurements.

Manufacturer's blade ratings   Manufacturers may provide a blade rating based on SAE practice J1265.

$$V_s = 0.8WH^2 \qquad\qquad \textbf{[6.1]}$$

$$V_u = V_s + ZH(W - Z)\tan x° \qquad\qquad \textbf{[6.2]}$$

where

V_s = capacity of straight or angle blade, in lcy
V_u = capacity of universal blade, in lcy
W = blade width, in yards, exclusive of end bits
H = effective blade height, in yards
Z = wing length measured parallel to the blade width, in yards
x = wing angle

Previous experience   Properly documented past experience is an excellent estimating method. Documentation requires that the excavated area be cross-sectioned to determine the total volume of material moved and that the number of dozer cycles be recorded. Production studies can also be made based on the weight of the material moved. In the case of dozers, the mechanics of weighing the material are normally harder to accomplish than surveying the volume.

Field measurement   A procedure for measuring blade loads follows:

1. Obtain a normal blade load:
   a. The dozer pushes a normal blade load onto a level area.
   b. Stop the dozer's forward motion. While raising the blade move forward slightly to create a symmetrical pile.
   c. Reverse and move away from the pile.
2. Measurement (see Fig. 6.11):
   d. Measure the height (H) of the pile at the inside edge of each track.
   e. Measure the width (W) of the pile at the inside edge of each track.
   f. Measure the greatest length (L) of the pile. This will not necessarily be at the middle.
3. Computation: Average both the two-height and the two-width measurements. If the measurements are in feet, the blade load in lcy is calculated by the formula

$$\text{Blade load (lcy)} = 0.0139HWL \qquad\qquad \textbf{[6.3]}$$

Front view

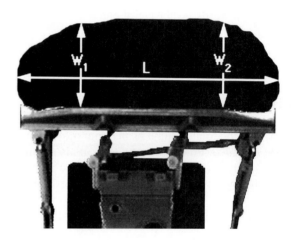

Top view

FIGURE 6.11 I Measurement of blade loads.

Cycle Time

The sum of the time required to push, backtrack, and maneuver into position to push represents the complete dozer cycle. The time required to push and backtrack can be calculated for each dozing situation considering the travel distance and obtaining a speed from the machine's performance chart.

Dozing, however, is generally performed at slow speed, 1.5 to 2 mph. The lower figure is appropriate for very heavy cohesive materials.

Return speed is usually the maximum that can be attained in the distance available. When using performance charts to determine possible speeds, remember the chart identifies instantaneous speeds. In calculating cycle duration, the estimator must use an average speed that accounts for the time required to accelerate to the attainable speed as indicated by the chart. Usually the operator cannot shift the machine past second gear in the case of distances that are less than 100 ft. If the distance is greater than 100 ft and the ground conditions are relatively smooth and level, maximum machine speed may be obtained. Maneuver time for power-shift dozers is about 0.05 min.

Production

The formula to calculate dozer production in loose cubic yards per a 60-min hour is presented below:

Production (lcy per hour)

$$= \frac{60 \text{ min} \times \text{blade load}}{\text{push time (min)} + \text{return time (min)} + \text{maneuver time (min)}} \quad \textbf{[6.4]}$$

EXAMPLE 6.1

A track-type dozer equipped with a power shift (see Fig. 6.4) can push an average blade load of 6.15 lcy. The material being pushed is silty sand. The average push distance is 90 ft. What production can be expected in loose cubic yards?

Push time: 2 mph average speed (sandy material):

$$\text{Push time} = \frac{90 \text{ ft}}{5,290 \text{ ft/mi}} \times \frac{1}{2 \text{ mph}} \times 60 \text{ min/hr} = 0.51 \text{ min}$$

Return time: Figure 6.4, second gear because less than 100 ft
Maximum speed 4 mph:

$$\text{Return time} = \frac{90 \text{ ft}}{5,290 \text{ ft/mi}} \times \frac{1}{4 \text{ mph}} \times 60 \text{ min/hr} = 0.26 \text{ min}$$

The chart provides information based on a *steady state velocity*. The dozer must accelerate to attain that velocity. Therefore, when using such speed data, it is always necessary to make an allowance for acceleration time. Because the change in speed is very small, in this example an allowance of 0.05 min is made for acceleration time.

Return time (0.26 + 0.05) = 0.31 min

Maneuver time = 0.05 min

$$\text{Production} = \frac{60 \text{ min} \times 6.15 \text{ lcy}}{0.51 \text{ min} + 0.31 \text{ min} + 0.05 \text{ min}} = 424 \text{ lcy/hr}$$

This production is based on working 60 min per hour, or an ideal condition. How well the work is managed in the field, the condition of the equipment, and the difficulty of the work are factors that will affect job efficiency. The efficiency of an operation is usually accounted for by reducing the number of minutes worked per hour. The efficiency factor is then expressed as working minutes per hour, for example, a 50-min hour or a 0.83 efficiency factor.

If it is necessary to calculate the production in terms of bank cubic yards (bcy), the swell factor Eq. [4.12], or the percent swell Eq. [4.13], can be used to make the conversion.

EXAMPLE 6.2

Assume a percent swell of 0.25 for the silty sand of the previous example and a job efficiency equal to a 50-min hour. What is the actual production that can be expected in bank cubic yards?

$$\text{Production} = \frac{424 \text{ lcy}}{1.25} \times \frac{50 \text{ min}}{60 \text{ min}} = 283 \text{ bcy/hr} \Rightarrow 280 \text{ bcy/hr}$$

The final step is to compute the unit cost for pushing the material. A large machine should be able to push more material per hour than a small one. However, the cost to operate a large machine will be greater than the cost to operate a small one. The ratio of *cost to operate to the amount of material moved* determines the most economical machine for the job. That ratio is the cost figure that is used in bidding unit price work.

EXAMPLE 6.3

The machine in Example 6.2 has an owning and operating cost of $40.50 per hour. Operators in the area where the proposed work will be performed are paid a wage of $15.50 per hour. What is the unit cost for pushing the silty sand?

$$\text{Unit cost} = \frac{\$40.50 \text{ per hour} + \$15.50 \text{ per hour}}{280 \text{ bcy/hr}} = \$0.200 \text{ per bcy}$$

Production Formulas

Equipment manufacturers have developed production formulas for use in estimating the amount of material bulldozers can push. Equation [6.5] is a *rule-of-thumb* formula proposed by International Harvester (IH). This formula equates the horsepower for a power-shift crawler dozer to lcy production.

$$\text{Production (lcy per 60-min hr)} = \frac{\text{net hp} \times 330}{(D + 50)} \qquad \textbf{[6.5]}$$

where

net hp = net horsepower at the flywheel for a power-shift crawler dozer

D = one-way push distance, in feet

EXAMPLE 6.4

The power-shift dozer whose characteristics are shown in Fig. 6.4 will be used to push material 90 ft. Use the IH formula to calculate the lcy production that can be expected for this operation.

From Fig. 6.4, net hp = 200.

$$\frac{200 \times 330}{(90 + 50)} = 471 \text{ lcy per 60-min hr}$$

Again, the actual production will be less, as a 60-min hour represents an ideal situation.

Production Curves

Figures 6.12 and 6.13 present production curves for estimating the amount of material Caterpillar dozers can push. These are published in the *Caterpillar Performance Handbook*. A production number taken from the Caterpillar curves is a maximum value in lcy per hour based on a set of "ideal conditions":

1. A 60-min hour (100% efficiency).
2. Power-shift machines with 0.05-min fixed time.
3. The machine cuts for 50 ft, then drifts the blade load to dump over a high wall.
4. A soil density of 2,300 lb per lcy.

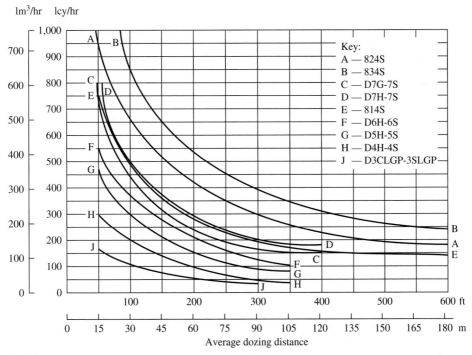

FIGURE 6.12 | Dozing production estimating curves for Caterpillar D3, D4, D5, D6, D7, 814, 824, and 834 dozers equipped with straight blades.

Reprinted courtesy of Caterpillar Inc.

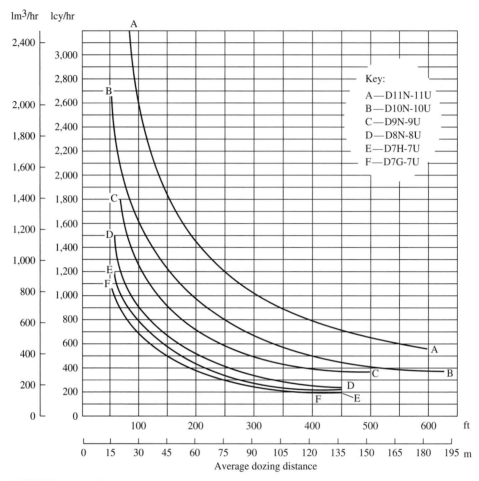

FIGURE 6.13 | Dozing production estimating curves for Caterpillar D7 through D11 dozers equipped with universal blades.

Reprinted courtesy of Caterpillar Inc.

5. Coefficient of traction:
   a. Track machines—0.5 or better
   b. Wheel machines—0.4 or better*
6. The use of hydraulic-controlled blades.

To calculate field production rates, the curve values must be adjusted by the expected job conditions. Table 6.2 lists the correction factors to use with the Caterpillar dozer production curves. The formula for production calculation is:

---

* Poor traction affects both track and wheel machines, causing smaller blade loads. Wheel dozers, however, are affected more severely. There are no fixed rules to predict the resulting production loss caused by poor traction. A rough rule of thumb for wheel-dozer production loss is a 4% decrease for each 0.01 decrease in the coefficient of traction below 0.40.

$$\text{Production (lcy/hour)} = \text{maximum production from curve}$$
$$\times \text{ product of the correction factors} \qquad \textbf{[6.6]}$$

TABLE 6.2 | Caterpillar job condition correction factors for estimating dozer production

	Track-type tractor	Wheel-type tractor
Operator		
Excellent	1.00	1.00
Average	0.75	0.60
Poor	0.60	0.50
Material		
Loose stockpile	1.20	1.20
Hard to cut; frozen		
with tilt cylinder	0.80	0.75
without tilt cylinder	0.70	—
cable controlled blade	0.60	—
Hard to drift; "dead" (dry, noncohesive	0.80	0.80
material) or very sticky material		
Rock, ripped or blasted	0.60–0.80	—
Slot dozing	1.20	1.20
Side-by-side dozing	1.15–1.25	1.15–1.25
Visibility		
Dust, rain, snow, fog, or darkness	0.80	0.70
Job efficiency		
50 min/hr	0.83	0.83
40 min/hr	0.67	0.67
Direct drive transmission		
(0.1-min fixed time)	0.80	—
Bulldozer*		
Adjust based on SAE capacity relative		
to the base blade used in the estimated		
dozing production graphs		
Grades—see the graph		

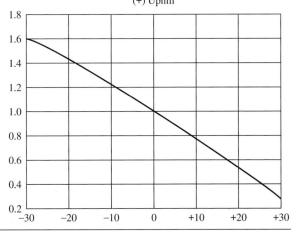

% Grade vs. Dozing Factor
(−) Downhill
(+) Uphill

*Note: Angling blades and cushion blades are not considered production-dozing tools. Depending on job conditions, the A blade and C blade will average 50–75% of straight-blade production.

Reprinted courtesy of Caterpillar Inc.

EXAMPLE 6.5

A D7G crawler dozer with a straight blade is to be used in a slot-dozing operation. The material is a dry, noncohesive silty sand and it is to be moved a distance of 300 ft from the beginning of the cut. Dozing is downhill on a 10% grade. The operator will have average skill, the dozer will have a power-shift transmission, and both visibility and traction should be satisfactory. The material weighs 108 pcf (lb per cu ft) in the bank state and is estimated to swell 12% in the loose state. Job efficiency is assumed to be equivalent to a 50-min hour. Calculate the direct cost of the proposed earthmoving operation in dollars per bcy. Assume that the owning and operating (O&O) cost for the dozer is $32.50 per hour and the operator's wage is $10.85 per hour.

Step 1. *Ideal maximum production.* Determine the ideal maximum production from the appropriate curve based on the model dozer and blade being used. Find the dozing distance on the bottom horizontal scale of the proper figure. Draw a vertical line upward until it intersects the production curve for the dozer under consideration, then construct a line horizontally to the vertical scale on the left side of the figure. Read from the intersect point on vertical scale the maximum production in lcy per hr.

D7 with straight blade: From Fig. 6.12, ideal production for 300-ft push is 170 lcy per hour.

Step 2. *Material-weight correction factor.* If the actual unit weight of the material to be pushed is not available from soil investigations, the average values found in Table 4.3 can be used. Divide 2,300 lb/lcy by the lcy weight of the material being pushed in order to determine the correction factor.

Bank weight for this project is given as 108 pcf; therefore,

$$108 \text{ lb/cu ft} \times 27 \text{ cu ft/cu yd} = 2,916 \text{ lb/bcy}$$

Swell is 12%, therefore the loose weight of the material being pushed is

$$\frac{2,916}{1.12} = 2,604 \text{ lb/lcy}$$

Standard condition is 2,300 lb/lcy

$$\text{Material-weight correction} = \frac{2,300 \text{ lb/lcy}}{2,604 \text{ lb/lcy}} = 0.88$$

Step 3. *Determine the operator correction factor* (see Table 6.2) based on the skill of the dozer operator.

Operator   0.75 (average skill, track-type tractor)

Step 4. *Material-type correction factor.* Dozer blades are designed to cut material and give it a rolling effect in front of the blade. The normal condition is a production factor of 1, however, some materials do not behave in the ideal manner and a correction factor must be applied (see Table 6.2).

Material (type)   0.80 (dry, noncohesive)

Step 5. *Operating-technique correction factor.* In the case of a dozer operating alone the factor is 1, see Table 6.2. In the case of slot or side-by-side dozing:

Operating techniques   1.20 (Slot dozing)

Step 6. *Visibility correction factor.* In the case of good visibility use 1, see Table 6.2 for factors based on other conditions.

Visibility   1.00

Step 7. *Efficiency factor.* See Table 6.2 or use the assumed number of operating minutes per hour divided by 60 min.

Job efficiency   0.83 (50-min hour)

Step 8. *Machine transmission factor.* See Table 6.2.

Transmission   1.00

Step 9. *Blade adjustment factor.* See the note at the bottom of Table 6.2.

Blade   1.00

Step 10. *Grade correction factor:* ($-$) favorable or ($+$) unfavorable. Find the percent grade on the bottom of the horizontal scale (Table 6.2 graph). Move up vertically and intersect the grade correction curve, then move left horizontally and locate the grade correction factor on the vertical scale.

Grade   1.24 ($-$10% grade)

Step 11. Determine the product of the correction factors.

Product, correction factors

$$= 0.88 \times 0.75 \times 0.80 \times 1.20 \times 1.0 \times 0.83 \times 1.0 \times 1.0 \times 1.24 = 0.652$$

Step 12. Determine the dozer production.

$$\text{Production} = 170 \text{ lcy/hr} \times 0.652 = 111 \text{ lcy/hr}$$

Step 13. Determine the material conversion, if required. The average values that are found in Table 4.3 can be used if no project specific data is available. Note that this conversion does not change the dozer production only how production is started.

$$\frac{111 \text{ lcy/hr}}{1.12} = 99 \text{ bcy/hr}$$

Step 14. Determine the total cost to operate the dozer.

Cost:

O&O	$32.50 per hour
Operator	$10.85
Total	$43.35 per hour

Step 15. Determine the direct unit production cost.

$$\text{Direct production cost} = \frac{\$43.35 \text{ per hour}}{99 \text{ bcy/hr}} = \$0.438 \text{ per bcy}$$

LAND CLEARING

LAND-CLEARING OPERATIONS

Crawler dozers equipped with special clearing blades are excellent machines for land clearing. Clearing of vegetation and trees is usually necessary before undertaking earthmoving operations. Trees, brush, and even grass and weeds make material handling very difficult. If these organic materials are allowed to become mixed into an embankment fill, their decay over time will cause settlement of the fill.

Clearing land can be divided into several operations, depending on the type of vegetation, the condition of the soil and topography, the amount of clearing required, and the purpose for which the clearing is done:

1. Removing all trees and stumps, including roots.
2. Removing all vegetation above the surface of the ground only, leaving stumps and roots in the ground.
3. Disposing of vegetation by stacking and burning.
4. Knocking all vegetation down, then chopping or crushing it to or into the surface of the ground, or burning it later.
5. Killing or retarding the growth of brush by cutting the roots below the surface of the ground.

The project specifications will dictate the proper clearing techniques.

TYPES OF EQUIPMENT USED

Several types of equipment are used for clearing land, with varying degrees of success:

1. Crawler dozers with earthmoving blades.
2. Crawler dozers with special clearing blades.
3. Crawler dozers with clearing rakes.

Crawler Dozers with Earthmoving Blades

Once dozers with earthmoving blades were used extensively to clear land. There are at least two valid objections to the use of dozers with earthmoving blades. Prior to felling large trees, they must excavate earth from around the tree and cut the main roots, which leaves objectionable holes in the ground and requires considerable time. Also, when stacking the felled trees and other vegetation, a considerable amount of earth is transported to the piles, which makes burning difficult.

Crawler Dozers with Special Clearing Blades

There are blades specially designed for use in felling trees. There are two basic types of clearing blades: (1) the single-angle with projecting stinger and (2) the V blade. The single-angle blade with stinger is often referred to as a *"K/G"*

FIGURE 6.14 | The Rome K/G clearing blade.

blade (Fig. 6.14). This name comes from the highly successful *Rome K/G clearing blade,* manufactured by the Rome Plow Company.

The major components of a single-angle clearing blade are the stinger, web, cutting edge, and guide bar. The stinger is a protruding vertical knife. It is designed to be used as a knife to cut and split trees, stumps, and roots. The stinger is sharp vertically, while the web is sharp horizontally; together they cut the tree in both planes simultaneously (see Fig. 6.14).

The blade can be tilted and is mounted at a 30° angle with the stinger forward. The guide bar serves to guide the cut material forward and to the side of the tractor.

The V clearing blade is shaped as the name implies. The point of the V is to the front, enabling debris to be cast off to each side. There is a stinger that protrudes from the lead point. The cutting edges down each side of the V are serrated. As with the single-angle blade, the design is such that vegetation is sheared off at ground level. Large hardwood trees are rammed in their middle and split by the stinger. Simultaneously, the two halves are then sheared off at ground level by the cutting edges on each side of the stinger.

Both types of clearing blades are most efficient when the dozer is operating on level ground and the cutting edges can maintain good contact with the ground surface. It is easier to work with soil types that hold the vegetation's root structure when the trunks are sheared. Large rocks will slow production by damaging the cutting edges.

Crawler Dozers with Clearing Rakes

A *rake* is a frame with multiple vertical teeth or tines mounted in the place of a solid-faced blade. Clearing rakes are used to grub and to pile trees after the

FIGURE 6.15 | Crawler-dozer-mounted land-clearing rakes.

clearing blades have worked an area (see Fig. 6.15). Like earthmoving blades, the teeth of a rake are curved in the vertical plane in the shape of a "C" to get under roots, rocks, and boulders. As the dozer moves forward, it forces the teeth of the rake below the ground surface. The teeth will catch the belowground roots and surface brush left from the felling operation, while allowing the soil to pass through. Rakes have an upward extension brush guard that can be either solid plate or opened spaced ribs. Some rakes have a steel center plate to protect the dozer's radiator.

The size, weight, and spacing of the rake's teeth depend on the intended application. Rakes used to grub-out stumps and heavy roots must have teeth of sufficient strength so that a single tooth can take the push of the dozer at full power. Lighter rakes having smaller and closer-spaced teeth are used for finish raking and to clear light root systems and small branches on the ground.

Rakes are also used to push, shake, and turn piles of trees and vegetation before and during burning operations. These tasks shake the dirt out of the piles and improve burning.

Disposal of Trees, Brush, Stumps, and Root Debris

When debris is to be disposed of by burning, it should be piled into stacks or rows (these piled-up or created rows are sometimes referred to as windrows), with a minimum amount of soil. Shaking a rake while it is moving the debris will reduce the amount of soil in the pile.

If the debris is burned while the moisture content is high, it may be necessary to provide an external source of fuel, such as diesel, to start combustion. A burner consists of a gasoline-engine-driven pump and a propeller, and is capable of maintaining a fire even under adverse conditions. The liquid fuel is blown as a stream into the pile of material while the propeller furnishes a supply of air to assure vigorous burning. Once combustion is started the fuel is turned off.

In many urban areas today, the burning of debris is restricted. Therefore, it is becoming common to chip brush and trees. Mixed chips, bark, and clean wood can be sold as mulch or used as boiler fuel. Clean wood chips without bark are used to manufacture pressed boards.

CLEARING TECHNIQUES

Applicable clearing techniques vary with the type of vegetation being cleared, the ground's soil type, and the soil's moisture condition. The job specifications and the nature of the field conditions will dictate the proper techniques to be used on a project. Job specifications may allow shearing of the vegetation and trees at ground level or the contract may require grubbing, which is the removal of stumps and roots from below the ground.

Shearing at Ground Level and Piling Operations

Techniques for conducting shearing and piling operations are

1. Brush and small trees can be removed by advancing the dozer with the blade slightly below ground level. The blade cuts, breaks off, or uproots most of the trees and bends the rest for removal on the return pass. A medium-size dozer can clear and pile 0.25 acres of brush or small trees per hour.

2. Small to medium-diameter (7–12 in. in diameter) vegetation through which the dozer can move almost continuously may be attacked either from the outside or from the center of the area. In both types of operations, production is best if long areas having widths of between 200 and 400 ft can be worked. The angle-type clearing blade is very efficient for these conditions.

When working in toward the center of an area, the operator rolls the cut material to the outside and away from the uncut area. After cutting, windrowing is accomplished by working from midwidth and pushing the downed material to the borders of the long dimension (see Fig. 6.16).

If the shearing is accomplished from the inside toward the outside, the cut material is rolled to the center, which becomes the site for the windrow (see Fig. 6.17).

3. Large trees (12–30 in. in diameter) are first cut 2 to 4 ft above ground level by holding the clearing blade in a raised position and driving the stinger into the tree. The vertical edge of the stinger splits the tree vertically. After the tree is felled, the horizontal edge of the stinger is used to cut the exposed stump at ground level. The time duration for a medium dozer to clear and pile large trees can be from 5 to 20 min per tree.

4. In the case of steep slopes or a significant number of large trees, cutting and windrowing is normally accomplished in one operation. Once a blade load

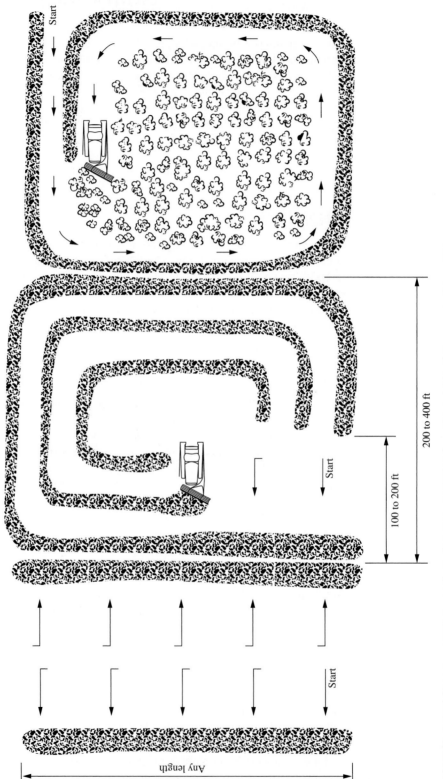

FIGURE 6.16 | Shearing small- to medium-diameter vegetation from the outside of the area to the center.

Reprinted courtesy of Caterpillar Inc.

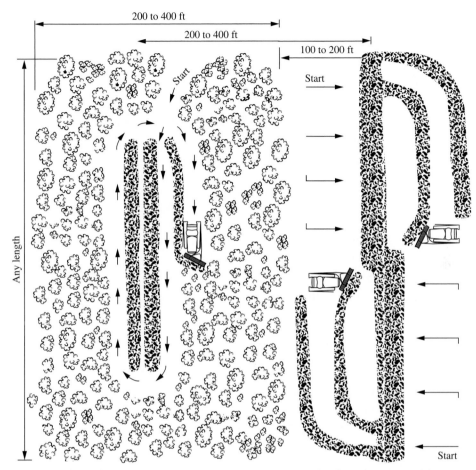

FIGURE 6.17 | Shearing small- to medium-diameter vegetation from the center of the area to the outside.

Reprinted courtesy of Caterpillar Inc.

of cut material is obtained, the dozer is turned 90° and the material deposited in the windrow. All cutting is performed in the direction of the ground's contour, and the windrow is established on the downhill side so that the dozer does not have to work against the natural grade.

Removal of Trees and Stumps to Below Ground Level

Stumps can be removed during the clearing operation by the following techniques:

1. To remove stumps, the stinger of the clearing blade is used to cut the roots below ground level, then the stump is pushed out of the ground with the edge of the blade. The stinger is designed to be used as a knife or spear—not as a lever. The edge of the blade should be used for pushing.

2. Extremely large or hardwood trees do not split readily. These are removed by first cutting the lateral roots around the tree and then pushing the tree with the blade raised high to provide leverage. In the case of very large trees, it may be necessary to build an earth ramp on one side of the tree to gain greater pushing leverage.

LAND-CLEARING PRODUCTION ESTIMATING

Typically, land clearing of timber is performed with crawler dozers having between 160 and 460 hp. The speed at which the dozer can move through the vegetation will depend on the nature of the growth and the size of the machine. It is best to estimate land clearing using historical data from similar projects. When data from past projects are not available, the estimator can utilize the formula presented in this section as a rough guide for probable production rates. However, production rates calculated strictly from the formula should be used with caution.

Critical factors to consider when estimating land clearing include

1. Nature of the vegetation:
   a. Number of trees.
   b. Size of the trees.
   c. Density of the trees.
   d. Type of trees.
   e. Root systems.
   f. Undergrowth and vines.
2. Soil condition and bearing capacity:
   a. Type of soil (cohesive, noncohesive).
   b. Moisture content of the soil.
   c. Depth to the water table.
   d. Presence of rock.
3. Topography—grade and terrain, hills, swamps, etc.
4. Climate and rainfall:
   a. Temperature, humidity, wind.
   b. Amount of rainfall.
   c. Amount of rainfall per rain.
   d. Days between rainfalls.
   e. Days of sunshine.
5. Job specifications:
   a. Clear (shear only).
   b. Clear and grub.
   c. Size of project.
   d. Time available to perform the clearing.

The estimator should always walk the project site prior to preparing an estimate. This is necessary to obtain information needed to properly evaluate the appropriate factors and to develop a complete understanding of the project requirements and possible variations from the clearing formula assumptions.

Constant Speed Clearing

When there is only light vegetation and it is possible to clear at a constant speed, production can be estimated by the dozer speed and the width of the pass:

Production (acre/hr)

$$= \frac{\text{width of cut (ft)} \times \text{speed (mph)} \times 5{,}280 \text{ ft/mi} \times \text{efficiency}}{43{,}560 \text{ sf/acre}} \quad \textbf{[6.7]}$$

The American Society of Agricultural Engineers' formula for estimating land-clearing production at constant speed is based on a 49.5-min hour, which is a 0.825 efficiency. Formula [6.7] then reduces to

$$\text{Production (acre/hr)} = \frac{\text{width of cut (ft)} \times \text{speed (mph)}}{10} \quad \textbf{[6.8]}$$

The width of cut is the resulting cleared width, measured perpendicular to the direction of dozer travel. With an angled blade, it is obvious that this is not equal to the width of the blade. Even when working with a straight blade it may not be the same as the blade width. The width of cut should be determined by field measurement, but it may have to be estimated.

EXAMPLE 6.6

A 200-hp crawler dozer will be used to clear small trees and brush from a 10-acre site. By operating in first gear, the dozer should be able to maintain a continuous forward speed of 0.9 mph. An angled clearing blade will be used, and from past experience the average resulting clear width will be 8 ft. Assuming normal efficiency, how long will it take to knock down the vegetation?

Using Eq. [6.8], we have

$$\frac{8 \text{ ft} \times 0.9 \text{ mph}}{10} = 0.72 \text{ acre/hr}$$

$$\frac{10 \text{ acres}}{0.72 \text{ acre/hr}} = 13.9 \text{ hr} \Rightarrow 14 \text{ hr}$$

Quick Estimating Method

Table 6.3 provides quick estimate production factors for typical area clearing conditions. This quick method should be used only when a detailed reconnaissance and tree count are not possible. The time required to clear a specific acreage, expressed in acres, can be calculated using the following formula:

$$\text{Total time (hr)} = \frac{\text{quick estimating factor} \times \text{area (acres)}}{\text{efficiency factor}} \quad \textbf{[6.9]}$$

TABLE 6.3 | Quick estimating factors for area clearing

	Equipment hours per acre		
Equipment type	Light (12 in. or less)	Medium (12 to 18 in.*)	Heavy (18 in.*)
200-hp dozer w/ clearing blade	0.53	1.07	1.73
300-hp dozer w/ clearing blade	0.40	0.67	1.07

*Maximum tree size.

Note: These clearing rates are averages for tree counts of 50 trees per acre. Adverse conditions (slopes, rocks, soft ground) can reduce these rates significantly.

EXAMPLE 6.7

Determine the time required to clear an area that is 500 ft wide by one-half mile long. Two 300-hp crawler dozers with clearing blades are available for the task. The largest trees in the area are 14 in. in diameter. The ground is fairly level and the expected efficiency is 50 min per hr.

$$\text{Total area in acres} = \frac{500 \text{ ft} \times (0.5 \times 5{,}280 \text{ ft/mile})}{43{,}560 \text{ sf/acre}} \Rightarrow 30.3 \text{ acres}$$

0.67 acres per hour, from Table 6.3, for 14 in.-diameter trees.
    Using Eq. (6.9),

$$\text{Total time} = \frac{0.67 \text{ hr/acre} \times 30.3 \text{ acres}}{(50/60) \times 2 \text{ tractors}} \Rightarrow 12.2 \text{ hours}$$

Tree Count Method—Cutting and Piling-Up Production

Rome Industries has developed a formula for estimating cutting and piling-up production. (Refer to the *Caterpillar Performance Handbook*.) The Rome formula and tables of constants provide guidance for variable speed operations, but use of the results should be tempered with field experience.

To develop the necessary input data for the Rome formula, the estimator must make a field survey of the area to be cleared and collect information on the following items:

1. Density of vegetation *less than 12 in.* in diameter:
   Dense—600 trees per acre.
   Medium—400 to 600 trees per acre.
   Light—fewer than 400 trees per acre.
2. Presence of hardwoods expressed in percent.
3. Presence of heavy vines.
4. Average number of trees per acre in each of the following size ranges:
   Less than 1 ft in diameter.
   1 to 2 ft in diameter.
   2 to 3 ft in diameter.

3 to 4 ft in diameter.

4 to 6 ft in diameter.

The diameter of the tree is taken at breast-height or 4.5 ft above the ground. In the case of a tree having a large buttress, the measurement should be taken where the trunk begins to run straight and true.

5. Sum of diameter of all trees per acre above 6 ft in diameter at ground level.

Once the field information is collected, the estimator can enter the table of production factors for cutting (see Table 6.4) to determine the time factors that should be used in the Rome cutting formula, Eq. [6.10]. The formula is based on the assumptions that a power shift dozer is being used, the ground is reasonably level (less than 10% grades), and the machine has good footing.

TABLE 6.4 I Production factors for felling with Rome K/G blades*

Tractor (hp)	Base minutes per acre* (B)	Diameter range				
		1–2 ft (M_1)	2–3 ft (M_2)	3–4 ft (M_3)	4–6 ft (M_4)	Above 6 ft (F)
165	34.41	0.7	3.4	6.8	—	—
215	23.48	0.5	1.7	3.6	10.2	3.3
335	18.22	0.2	1.3	2.2	6.0	1.8
460	15.79	0.1	0.4	1.3	3.0	1.0

*Based on power-shift tractors working on reasonably level terrain (10% maximum grade) with good footing and no stones and an average mix of soft- and hardwoods.

Reprinted courtesy of Caterpillar Inc.

Time (min) per acre for cutting

$$= H[A(B) + M_1 N_1 + M_2 N_2 + M_3 N_3 + M_4 N_4 + DF] \qquad \textbf{[6.10]}$$

where H = hardwood factor affecting total time

Hardwoods affect overall time as follows:

75 to 100% hardwoods; add 30% to total time ($H = 1.3$)

25 to 75% hardwoods; no change ($H = 1.0$)

0 to 25% hardwoods; reduce total time 30% ($H = 0.7$)

A = tree density and presence of vines effect on base time

The density of undergrowth material less than 1 ft in diameter and the presence of vines affect base time.

Dense: greater than 600 trees per acre; add 100% to base time ($A = 2.0$)

Medium: 400 to 600 trees per acre; no change ($A = 1.0$)

Light: less than 400 trees per acre; reduce base time 30% ($A = 0.7$)

Presence of heavy vines; add 100% to base time ($A = 2.0$)

B = base time for each dozer size per acre

M = minutes per tree in each diameter range

N = number of trees per acre in each diameter range, from field survey

D = sum of diameter in foot increments of all trees per acre above 6 ft in diameter at ground level, from field survey

F = minutes per foot of diameter for trees above 6 ft in diameter

Where the job specifications require removal of trees and grubbing of the roots and stumps greater than 1 ft in diameter in one operation, increase total time per acre by 25%. When the specifications require removal of stumps in a separate operation, increase the time per acre by 50%. Note that this tree count method has no correction for efficiency. The time values in Tables 6.4 and 6.5 are based on normal efficiency.

<hr>

EXAMPLE 6.8

Estimate the rate at which a 215-hp dozer equipped with a K/G blade can fell the vegetation on a highway project. Highway specifications require grubbing of stumps resulting from trees greater than 12 in. in diameter. Felling and grubbing will be performed in one operation.

The site is reasonably level terrain with firm ground and less than 25% hardwood. The field survey gathered the following tree counts:

Average number of trees per acre, 700

1 to 2 ft in diameter,   100 trees

2 to 3 ft in diameter,   10 trees

3 to 4 ft in diameter,   2 trees

4 to 6 ft in diameter,   0 trees

Sum of diameter increments above 6 ft, none.
The necessary input values for formula [6.10] are

$H = 0.7$ less than 25% hardwoods

$A = 2.0$ dense, > 600 trees per acre

From Table 6.5 for a 215-hp dozer:

$$B = 23.48,$$

$$M_1 = 0.5, M_2 = 1.7, M_3 = 3.6, M_4 = 10.2, \text{ and } F = 3.3$$

Time per acre = 0.7[2.0(23.48) + 0.5(100) + 1.7(10) + 3.6(2) + 10.2(0) + 0(3.3)]
Time per acre = 84.8 min per acre

Because the operation will include grubbing, the time must be increased by 25%.

Time per acre = 84.8 min per acre × 1.25 = 106.0 min per acre

Clearing rates are often expressed in acres per hour, so for this example the rate would be 60 min per hr/106 min per acre = 0.57 acres per hour.

Table 6.4 and Eq. [6.10] are used to calculate the time for a cutting operation. Usually the cut material must be piled for burning or so that it can be easily picked up and hauled away. Rome Industries has developed a separate formula and set of constants for estimating piling-up production rates.

Time (min) per acre for piling-up

$$= B + M_1N_1 + M_2N_2 + M_3N_3 + M_4N_4 + DF \qquad \textbf{[6.11]}$$

The factors have the same definitions as when used previously in Eq. [6.10], but their values must be determined from Table 6.5 for input into the piling-up Eq. [6.11]. Piling-up grubbed vegetation increases the total piling-up time by 25%.

TABLE 6.5 | Production factors for piling-up in windrows*

Tractor (hp)	Base minutes per acre* (B)	Diameter range				
		1–2 ft (M₁)	2–3 ft (M₂)	3–4 ft (M₃)	4–6 ft (M₄)	Above 6 ft (F)
165	63.56	0.5	1.0	4.2	—	—
215	50.61	0.4	0.7	2.5	5.0	—
335	44.94	0.1	0.5	1.8	3.6	0.9
460	39.27	0.08	0.1	1.2	2.1	0.3

*May be used with most types of raking tools and angled shearing blades. Windrows to be spaced approximately 200 ft apart.

Reprinted courtesy of Caterpillar Inc.

EXAMPLE 6.9

Consider that the vegetation cut in Example 6.8 must be piled. What is the estimated rate at which piling-up can be accomplished?

Time per acre = 50.61 + 0.4(100) + 0.7(10) + 2.5(2) + 5.0(0) + 0(0)

= 103 min per acre

Because the operation will include grubbing, the time must be increased by 25%.

103 min per acre × 1.25 = 129 min per acre or 0.47 acre piled-up per hr

Safety

Never operate clearing dozers close together, as falling trees may strike a neighboring machine. When pushing a tree over, do not follow too closely, because as the tree falls, its stump and root mass may catch under the front of the dozer. The dozer's belly pan should be cleaned often of accumulated debris to prevent fires in the engine compartment.

FIGURE 6.18 | Dozer-mounted hydraulically operated triple-shank ripper.

RIPPING ROCK

GENERAL INFORMATION

The ripper is a relatively narrow profile implement. It penetrates the earth and is pulled to loosen and split hard ground, weak rock, or old pavements and bases. Heavy ripping is accomplished by crawler dozers equipped with rear mounted rippers because of the power and tractive force available from such machines (see Fig. 6.18). Motor graders can, also, be equipped with rippers for light duty applications.

Although rock has been ripped with varying degrees of success for many years, recent developments in methods, equipment, and knowledge have greatly increased the extent of ripping. Rock that was once considered to be unrippable is now ripped with relative ease, and at cost reductions—including ripping and hauling with scrappers—amounting to as much as 50% when compared with the cost of drilling, blasting, loading with loaders, and hauling with trucks.

The major developments responsible for the increase in ripping rock include

1. Heavier and more powerful dozers.
2. Improvements in the sizes and performance of rippers, to include development of impact rippers.
3. Better instruments for determining the rippability of rocks.
4. Improved techniques in using instruments and equipment.

DETERMINING THE RIPPABILITY OF ROCK

Prior to selecting the method of excavating and hauling rock, it is desirable to determine if the rock can be ripped or if it will be necessary to drill and blast.

When investigating a project, many contractors ask the question "Do I have to blast?" The question that should be investigated is "Can I rip?" Evaluating rippability of rock involves study of the rock type and a determination of the rock's density. Igneous rocks, such as granites and basaltic types, are normally impossible to rip, because they lack stratification and cleavage planes and are very hard. Sedimentary rocks have a layered structure caused by the manner in which they were formed. This characteristic makes them easier to rip. The metamorphic rocks, such as gneiss, quartzite, schist, and slate, being a changed form from either igneous or sedimentary, vary in rippability with their degree of lamination or cleavage.

Physical characteristics that favor ripping are

1. Fractures, faults, and joints all act as planes of weakness facilitating ripping.
2. Weathering, the greater the degree of weathering the more easily the rock is ripped.
3. Brittleness and crystalline structure.
4. High degree of stratification or lamination offers good possibilities for ripping.
5. Large grain size, coarse-grained rocks rip more easily than fine-grained rocks.

Because the rippability of most types of rocks is related to the speed at which sound waves travel through rock, it is possible to use refraction seismographic methods to determine with reasonable accuracy if a rock can be ripped. Rocks that propagate sound waves at low velocities, less than 7,000 ft/sec, are rippable, whereas rocks that propagate waves at high velocities, 10,000 ft/sec or greater, are not rippable. Rocks having intermediate velocities are classified as marginal.

Fig. 6.19 indicates rippability for a specific size dozer based on velocity ranges for various types of soil and rocks encountered on construction projects. The indication that a rock may be rippable, marginal, or nonrippable is based on using multi- or single-shank rippers mounted on a crawler dozer. The information appearing in the figure should be used only as a guide and should be supplemented with other data such as boring logs or core samples. The decision to rip or not to rip rock should be based on the relative costs of excavating, using the methods under consideration, and equipment available. Field tests may be necessary to determine if a given rock can be ripped economically.

DETERMINING THE SPEED OF SOUND WAVES IN ROCK

A refraction seismograph can be used to determine the thickness and degree of consolidation of rock layers at or near the ground surface. The paths followed by sound waves from a wave-generating source through a formation to detecting instruments are illustrated in Fig. 6.20.

A geophone, that is, a sound sensor, is driven into the ground at station 0. Equally spaced points 1, 2, 3, and so on are located along a line, as indicated in

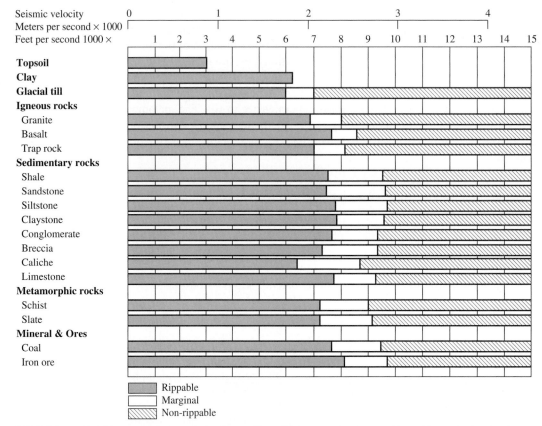

FIGURE 6.19 | Ripper performance for Caterpillar 370-hp crawler dozer with multi- or single-shank rippers. Estimated by seismic wave velocities.

Reprinted courtesy of Caterpillar Inc.

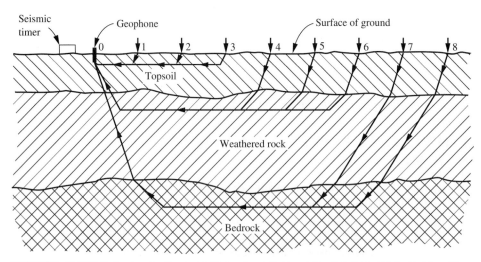

FIGURE 6.20 | Path of seismic waves through earth layers having increased density with depth.

Fig. 6.20. A wire is connected from the geophone to the seismic timer (see Fig. 6.21), and another wire is connected from the timer to a sledgehammer, or other impact-producing tool. A steel plate is placed on the ground at the sensing stations, in successive order. When the hammer strikes the steel plate, a switch closes instantly to send an electric signal, which starts the timer. At the same instant, the blow from the hammer sends sound waves into the formation, which travel to the geophone. On the receipt of the first wave, the geophone signals the timer to stop recording elapsed time. With the distance and time known, the velocity of a wave can be determined.

FIGURE 6.21 | Refractive seismograph recording instrument and geophones.

As the distance from the geophone to the wave source, namely, the striking hammer, is increased, waves will enter the lower and denser formation, through which they will travel at a higher speed than through the topsoil. The waves that travel through the denser formation will reach the geophone before the waves traveling through the topsoil arrive. The velocity through the denser formation is a higher value, and it will remain approximately constant as long as the waves travel through a formation of uniform density.

A plot such as that shown in Fig. 6.22 yields wave velocity simply by measuring the distance and time associated with each layer and dividing the distance by the corresponding time:

$$\text{Velocity, } V_i = \frac{\text{horizontal distance, } L_i}{\text{time to travel distance } L_i} \qquad [6.12]$$

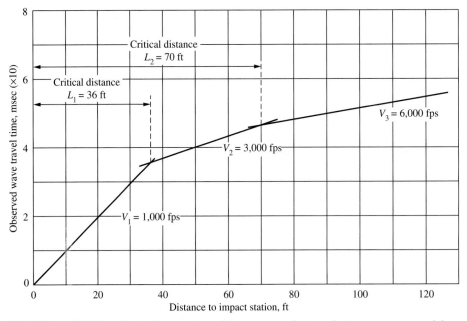

FIGURE 6.22 | Plot of seismic wave travel time versus distance between source and the geophone.

The velocity remains constant as the wave passes through material of uniform density. Breakpoints in the plot indicate changes in material density, and therefore velocity.

The depth to the surface separating the two strata depends on the critical distance and the velocities in the two materials. It can be computed from the equation:

$$D_1 = \frac{L_1}{2}\sqrt{\frac{V_2 - V_1}{V_2 + V_1}} \qquad [6.13]$$

where

D = depth in feet

L_1 = critical distance in feet

V_1 = velocity of wave in top stratum in fps

V_2 = velocity of wave in lower stratum in fps

Solving for D_1 from Fig. 6.22 gives

$$D_1 = \frac{36}{2}\sqrt{\frac{3,000 - 1,000}{3,000 + 1,000}} = 13 \text{ ft}$$

Thus, the topsoil has an apparent depth of 13 ft.

Equation [6.14] can be used to determine the apparent thickness of the second stratum.

$$D_2 = \frac{L_2}{2}\sqrt{\frac{V_3 - V_2}{V_3 + V_2}} + D_1\left[1 - \frac{V_2\sqrt{V_3^2 - V_1^2} - V_3\sqrt{V_2^2 - V_1^2}}{V_1\sqrt{V_3^2 - V_2^2}}\right] \qquad \textbf{[6.14]}$$

where L_2 is 70 ft, D_1 is 13 ft, V_1 is 1,000 fps, V_2 is 3,000 fps, and V_3 is 6,000 fps. Solving, we obtain

$$D_2 = 31 \text{ ft}$$

Ripping is normally an operation involving the top strata of a formation. Determination of the depth of the upper two layers and the velocity of the upper three layers will provide the necessary data to estimate most construction jobs. The procedure can become very complicated, however, when dipping strata are encountered. The investigation of complicated formations may require, in addition to seismic studies, the core drilling of samples and the excavation of test pits.

RIPPER ATTACHMENTS

Ripper attachments for dozers are generally rear mounted. The mounting may be fixed radial, fixed parallelogram, or parallelogram linkage with hydraulically variable pitch (see Fig. 6.23). The vertical piece that is forced down into the material to be ripped is known as a *shank*. A ripper tip (a tooth, point, or tap) is fixed to the lower, cutting end of the shank. The tip is detachable for easy replacement, as it is the high-wear surface of the ripper. A tip can have a service life from only 30 min. to 1000 hr, depending on the abrasive characteristics of the material being ripped.

Both straight and curved shanks are available. Straight shanks are used for massive or blocky formations. Curved shanks are used for bedded or laminated rocks or for pavements, where a lifting action will help shatter the material.

With the radial-type ripper linkage, the beam of the ripper pivots on link arms about its point of attachment to the dozer, therefore, the angle of tip attack varies with the depth the shank is depressed. This may make it difficult to achieve penetration in tough materials. The shank may, also, tend to "dig itself in."

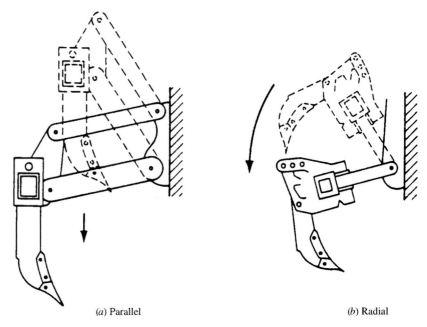

(*a*) Parallel   (*b*) Radial

FIGURE 6.23 | Types of linkages used to mount rippers on dozers.

The parallelogram-type ripper maintains the shank in a vertical position and keeps the tip at a constant angle. The adjustable parallelogram-type rippers enable the tip angle to be controlled by the operator.

Heavy-duty rippers are available in either single-shank or multishank models. The multishank models are available with up to five shanks in the case of smaller dozers and up to three shanks for larger machines. The shanks are pinned into position on the ripper frame. This enables ease of removal so that shank arrangement can be matched to material and project requirements. In the case of very heavy work (i.e., high seismic velocity material) a single center-mounted shank will maximize production. The use of multiple shanks will produce more uniform breakage if full penetration can be obtained. A single shank often rolls individual oversized pieces to the side.

The effectiveness of a ripper depends on

1. Down pressure at the ripper tip.
2. The dozer's usable power to advance the tip; a function of power available, dozer weight, and coefficient of traction.
3. Properties of the material being ripped; laminated, faulted, weathered, and so on.

A rule of thumb in sizing a dozer for a ripping operation is that it must have one flywheel horsepower per 100 lb of down pressure on the ripper and it must have 3 lb of machine weight per lb of down pressure to ensure adequate traction.

The number of shanks used depends on the size of the dozer, the depth of penetration desired, the resistance of the material being ripped, and the degree of breakage of the material desired. If the material is to be excavated by scrapers, it should be broken into particles that can be loaded into scrapers, usually of not more than 24 to 30 in. maximum sizes. Two shanks can be effective in softer, easily fractured materials that are to be scraper-loaded. Three shanks should be used only in very easy to rip material such as hardpan or some shales. Only a field test conducted at the project site will demonstrate which method, depth, and degree of breakage are appropriate and economical.

RIPPING PRODUCTION ESTIMATES

Although the cost of excavating rock by ripping and scraper loading is considerably higher than for earth that requires no ripping, it may be much less expensive than using an alternate method, such as drilling, blasting, excavator loading, and truck hauling. Estimating ripping production is best accomplished by working a test section and conducting a study of the operational methods, and determining production by weight of ripper material. However, the opportunity to conduct such field tests is often nonexistent; therefore, most initial estimates are based on equipment manufacturers' production charts.

Quick Method

By field timing several passes of a ripper over a measured distance, an approximate production rate can be determined. The timed duration should include the turnaround time at the end of each pass. An average cycle time can be determined from the timed cycles. The quantity (volume) is determined by measuring the length, width, and depth of the ripped area. Experience has shown that the production rate calculated by this method is about 20% higher than an accurately cross-sectioned study. Therefore, the formula for estimated ripping production is

$$\text{Ripping production (bcy/hr)} = \frac{\text{measured volume (bcy)}}{1.2 \times \text{average time (hr)}} \qquad \textbf{[6.15]}$$

Seismic-Velocity Method

Manufacturers have developed relationships between seismic wave velocities and rippability by conducting field tests on a variety of materials (see Fig. 6.19). Ripping performance charts such as those in Fig. 6.19 enable the estimator to make an initial determination of equipment that may be suitable to rip a material based on general rock-type classifications. After the initial determination of applicable machines is made, production rates for the particular machines are calculated from production charts.

Ripping production charts (see Fig. 6.24) developed from field tests conducted in a variety of materials have been published. However, because of the extreme variations possible among materials of a specific classification, a great deal of judgment is necessary when using such charts. The production rates obtained from the charts must be adjusted to reflect the actual field conditions of the project.

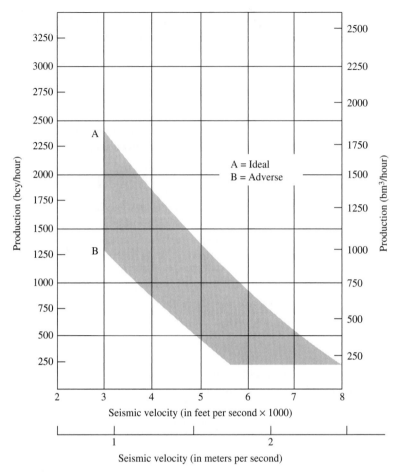

FIGURE 6.24 I Ripper production Caterpillar 370-hp crawler dozer with a single shank.

Reprinted courtesy of Caterpillar Inc.

Owning and operating (O&O) costs of a dozer will increase for machines consistently used in ripping operations. Caterpillar warns that normal O&O costs must be increased by 30 to 40% if the machine is used in heavy ripping applications.

EXAMPLE 6.10

A contractor encounters a shale formation at shallow depth in a cut section of a project. Seismographic tests indicate a seismic velocity of 7,000 fps (feet per second) for the shale. On this basis, it is proposed to rip the material with a 370-hp dozer. Estimate the production in bank cubic yards (bcy) for full-time ripping, with efficiency based on a 45-min hour (a typical factor for ripping operations). Assume that the ripper is equipped with a single shank and that ripping conditions are ideal. The "normal" O&O cost excluding operator for the dozer-ripper combination is $86 per hour. Operator wages are $9.50 per hour. What is the estimated ripping production cost in dollars per bcy?

Step 1. From Fig. 6.19, ripper performance is

370 hp—Rippable in shale having a seismic velocity up to 7,500 fps

The dozer is applicable for this situation according to the chart, but it is at the limit of its capability. Therefore, the contractor should consider a larger machine.

Step 2. Using the 370-hp-dozer ripper production chart (see Fig. 6.24) for a seismic velocity of 7,000 fps and ideal conditions, results in

$$\text{Ideal production 370-hp dozer} = 560 \text{ bcy/hr}$$

$$\text{Adjusted production} = 560 \times \frac{45}{60} = 420 \text{ bcy/hr}$$

Step 3. Increase normal O&O cost because of ripping application:

$$\$86.50 \text{ per hour} \times 1.35 = \$116.10 \text{ per hour}$$

$$\text{Total cost with operator: } \$116.10 + \$9.50 = \$125.60 \text{ per hour}$$

$$\text{Production cost} = \frac{\$125.60 \text{ per hour}}{420 \text{ bcy/hr}} = \$0.299 \text{ per bcy}$$

Operating Techniques

Ripping should be done at the maximum penetration depth that traction will allow, but it should also be accomplished at a uniform depth. For the most economical production, ripping is performed in low gear at low speed, typically 1 to $1\frac{1}{2}$ mph. At speeds only slightly higher, there can be a dramatic increase in operating cost from undercarriage ripper tip wear. When ripping to load scrapers it is best to rip in the same direction that the scrapers load. When removing the ripped material, always leave a "cushion" 4 to 6 in. in depth of loose material. The "cushion" will create better underfoot conditions for the dozer doing the ripping and reduce track wear.

Take advantage of gravity and rip downhill when possible. Cross ripping will make the pit rougher, increase scraper tire wear, and require twice as many passes. Cross ripping does, however, help to break up "hard spots" or material that comes loose in large slabs.

SUMMARY

Dozers are used for dozing (pushing materials), land clearing, ripping, assisting scrapers in loading, and towing other pieces of construction equipment. The factors that control dozer production rates are (1) blade type, (2) type and condition of material, and (3) cycle time. Manufacturers provide production curves for estimating the amount of material their dozers can push. The curves provide a maximum value in lcy per hour based on a set of "ideal conditions."

Crawler dozers equipped with special clearing blades are excellent machines for land clearing. Typically, land clearing of timber is performed with crawler dozers having between 160 and 460 hp. It is best to estimate land clearing by using historical data from similar projects. If historical data is not available, Rome Industries has developed a formula for estimating cutting and piling-up production.

Heavy ripping of rock is accomplished by crawler dozers equipped with rear-mounted rippers because of the power and tractive force available from such machines. A refraction seismograph can be used to determine the thickness and degree of consolidation of rock layers at or near the ground surface. Manufacturers have developed relationships between seismic wave velocities and rippability. Additionally, there are manufacturer ripping production charts that have been developed from field tests. Critical learning objectives would include:

- An ability to calculate dozer pushing production.
- An ability to calculate dozer production for clearing operations.
- An ability to calculate dozer production for ripping operations.

These objectives are the basis for the problems that follow.

PROBLEMS

6.1   A contractor wants to investigate using a 6S blade on the D6H dozer. The material is a dry noncohesive silty sand, and it is to be moved a distance of 200 ft from the beginning of the cut. The dozing is uphill on a 2% grade. The operator will have average skill, the dozer has a power-shift transmission, and both visibility and traction are assumed to be satisfactory. The material weighs 110 lb per cf (pcf) in the bank state. Job efficiency is assumed to be equivalent to a 50-min hour. Calculate the direct cost of the proposed earthmoving operation in dollars per bcy. Assume that the O&O cost for the dozer is $54.00 per hour and the operator's wage is $12.00 per hour plus 35% for fringes, workman's compensation, and so on. ($0.900 per bcy)

6.2   A contractor wants to investigate the differences in using a 7S blade on a D7H dozer or an S blade on an 824 dozer. The material is a dry clay, and it is to be moved a distance of 200 ft from the beginning of the cut. The dozing is uphill on a 5% grade. The operator will have average skill, the dozer has a power-shift transmission, and traction is assumed to be satisfactory. The site will be dusty. The material weighs 98 pcf in the bank state. Job efficiency is assumed to be equivalent to a 50-min hour. Calculate the direct cost of the proposed earthmoving operation in dollars per bcy. Assume that the O&O cost for the D7 dozer is $76.00 per hour and $73.00 per hour for the 824. The operator's wage is $16.00 per hour plus 35% for fringes, workman's compensation, and other benefits.

a.   Which machine would you use on this job?

b.   It rains and the coefficient of traction for the 824 is now 0.37. The coefficient of traction for the D7 is 0.55. Which machine would you use on this job?

6.3   A crawler dozer with a universal blade is to be used in a slot-dozing operation. The material is a dry, noncohesive silty sand, and it is to be moved a distance of 350 ft from the beginning of the cut. The dozing is downhill on a 10% grade. The operator will have average skill, the dozer has a power-shift transmission, and both visibility and traction are assumed to be satisfactory. The material weighs 108 pcf in the bank state, and will swell 10% when excavated by dozing. Job efficiency is assumed to be equivalent to a 50-min hour.

a. Assuming a Caterpillar D8 is used, calculate the direct cost of the proposed earthmoving operation in dollars per bcy. Assume that the O&O cost for the dozer is $64.80 per hour and the operator's wage is $10.85 per hour plus 36% for fringes, workman's compensation, and so on.

b. Assume that a CAT 834 wheel-type dozer with a straight blade is substituted for the D8. (The 834 has 58% more flywheel horsepower than the D8.) All other conditions of the job are unchanged, except that the O&O cost for the 834 is $69.95 per hour.

c. Assume that the 834 dozer, when operating on this particular soil, will have a coefficient of traction of 0.34. Estimate the unit production cost in dollars per bcy, with all other job conditions remaining unchanged.

(a. unit production cost = $0.479/bcy; b. unit production cost = $0.522/bcy; c. 834 unit production cost w/ traction of 0.34% = $0.687 per bcy)

6.4 Two CAT D-10 (power-shift) dozers are to be used in a side-by-side pushing operation. The dozers are equipped with universal blades. The material, dry and noncohesive, weighs 102 pcf in the bank state. It is estimated that the material will swell 8%, from bank to loose state. The center of mass to center of mass-pushing distance is 380 ft downhill on a 5% grade. The operators have average skill and the job will be performed in dusty conditions. Job efficiency can be assumed to be equivalent to a 50-min hour.

Calculate the direct cost of the proposed earthmoving operation in dollars per bcy. The company's normal O&O cost for these machines is $93.82 per hour and the operator's wage is $12.18 per hour plus 42% for fringes, insurance, workman's compensation, and so on.

6.5 A CAT 824C with a straight blade is to be used to push a material that weighs 125 pcf in the bank state. It is estimated that the material will swell 20%, from the bank to loose state. This cohesive soil is to be moved an average distance of 100 ft up a 12% grade. The operator is a new hire who has poor skills. Job efficiency is estimated to be equivalent to a 45-min hour. Assume that the dozer will have a 0.36 coefficient of traction. Calculate the direct cost of the proposed earthmoving operation in dollars per bcy. The O&O cost for the dozer is $47.30 per hour and the operator's wage is $8.76 per hour plus 34% for fringes, insurance, workman's compensation, and so on.

6.6 Given the geophone data shown here, calculate the seismic wave velocity in each layer. Find the bottom depth of each layer except the last.

Distance (ft)	15	30	45	60	75	90	105	125	145
Time (msec)	7.5	15.0	22.5	26.3	30.0	33.8	37.5	40.4	43.2

6.7 An earthmoving contractor encounters a trap-rock formation at a shallow depth in a rock cut. He performs seismographic tests, which indicate a seismic velocity of 6,500 fps for the material. On this basis, he proposes to rip the material with a 370-hp dozer. Estimate the production in bcy, for full-time ripping, with efficiency based on a 50-min hour. Assume that the dozer is equipped with a single shank and that ripping conditions are *average* (i.e., intermediate between extreme conditions). The *normal* O&O cost per hour, including operator for a 370-hp dozer with ripper, is $95.82. This work would be classified as a *heavy ripping* application. Using your estimate of hourly production, estimate the ripping unit production cost in dollars per bcy.

6.8   A limestone formation with an average depth of 5 ft is exposed over the full 80-ft width and 23,700-ft length of a highway cut. Preliminary seismic investigations indicate that the rock layer has a seismic velocity of 5,500 fps. The contractor proposes to loosen this rock using a single-shank ripper on a power-shift, track-type dozer. The job conditions relative to lamination, and other rock characteristics are *average*, or midway between *adverse* and *ideal*. A 40-min hour working efficiency is possible. The contractor wishes to explore the use of two 370-hp dozers. O&O costs are $126.00 per hour, including the rippers, for the ripping operations. The operator's wage is $14.30 per hour. What is the unit cost to rip this rock? What will be the total cost of the work and how many hours will the project require? (unit cost $0.439 per bcy, total cost $153,997, time with two dozers 405 hours)

6.9   A limestone formation with an average depth of 2.4 ft is exposed over the full 80-ft width and 23,700-ft length of a highway cut. Preliminary seismic investigations indicate that the rock layer has a seismic velocity of 5,500 fps.

The contractor proposes to loosen this rock using a single-shank ripper on a power-shift, track-type dozer. It is estimated that job conditions relative to lamination, and other rock characteristics are *average*, or midway between *adverse* and *ideal*. A 40-min hour working efficiency should be obtainable while the ripper is actually in production, but it is estimated that the ripper will be out of service for repairs for 5% of the total hours that the ripper could be used on the job (i.e., availability of 0.95). It is further estimated that the ripper will be out of service due to adverse weather for 20% of the same total hours.

The contractor wishes to explore the use of two 370-hp dozers for the ripping operations. O&O costs are estimated at $79.82 per working hour for each 370-hp dozer, including rippers. Each machine is to be charged against the job at 60% of its normal O&O rate during working hours, when the equipment is not used because of adverse weather or scheduling problems, but it will not be charged against the job when the equipment is down for repair.

The operator's wage, paid during hours worked and hours when the equipment is being repaired, is $14.30 per hour. Associated charges against the ripping operation for grade foreman, mechanic, and so on are estimated to be $25 per nominal (i.e., regularly scheduled, including downtime for repairs or adverse weather) working hours. Move-in and move-out costs combined are estimated to be $200 per unit. Calculate the total time in weeks, based on a normal 40-hour workweek, to loosen the rock cut, and calculate the total loosening cost, in dollars per bcy.

6.10   A contract is being bid to clear and grub 154 acres. You are considering the cost associated with using a 335-hp dozer. The dozer's O&O cost is $63.96 per hour. The wage determination for operators is $10.85 per hour with fringes. Project overhead will run $357 per workday. Assume a 10-hour day.

A field engineer has made a site visit and provided you with the following information. The site is reasonably level and the machines will have no problem with bearing or traction. About 10% of the trees are pines and the rest are oak. The total number of trees per acre is about 300; of that number

210 trees per acre are 1 to 2 ft in diameter.

9 trees per acre are 2 to 3 ft in diameter.

1 tree per acre is 3 to 4 ft in diameter.

You plan to clear and grub in one operation, and then to follow with piling-up in windrows and burning. Burning will require that the dozer be employed twice the time expected for piling-up alone.

a. What is the estimated clearing and grubbing production rate?

b. What is the cost associated with the total operation, including overhead, when only one 335-hp dozer is used?

c. Is it cheaper to employ two 335-hp dozers?

(production 0.205 acre/hr, total cost $109,870; cheaper to use two dozers, $96,661)

6.11 A contract is being bid to clear and grub 193 acres. You are considering the cost associated with using a 460-hp dozer. The dozer's O&O cost is $96.96 per hour. The wage determination for heavy-duty operators is $13.85 per hour, plus 30% for fringes. Project overhead will run $957 per workday. Assume a 10-hour workday.

A field engineer has made a site visit and provided you with the following information. The site is reasonably level and the machines will have no problem with bearing or traction. About 80% of the trees are pines and the rest are oak. The total number of trees per acre is about 500; of these

200 trees per acre are 1 to 2 ft in diameter.

50 trees per acre are 2 to 3 ft in diameter.

20 trees per acre are 3 to 4 ft in diameter.

You plan to clear and grub in one operation, and then to follow with piling-up in windrows and burning. Burning will require that the dozer be employed twice the time expected for piling-up alone.

a. What is the estimated production rate for clearing, grubbing, piling, and burning?

b. What is the cost associated with the total operation, including overhead, when only one 460-hp dozer is used?

c. Is it cheaper to employ two, three, or four of the 460-hp dozers?

6.12 A contract is being bid to clear and grub 147 acres. You are considering the cost associated with using either of two different power-shift dozers. One dozer is a 215-hp model and the other is rated at 335 hp. The 215-hp dozer has an O&O cost, without operator, of $43 per hour and the 335-hp dozer's O&O is $63.96 per hour. The wage determination for heavy-duty operators is $10.85 per hour. Project overhead will run $357 per workday. Assume a 10-hour day.

A field engineer has made a site visit and provided you with the following information. The site is reasonably level and the machines will have no problem with bearing or traction. About 10% of the trees are pines and the rest are oak. There are about

210 trees per acre 1 to 2 ft in diameter.

9 trees per acre 2 to 3 ft in diameter.

1 tree per acre 3 to 4 ft in diameter.

You plan to clear and grub in one operation, and then to follow with piling-up in windrows and burning. Burning will require that the dozer be employed twice the time expected for piling-up alone.

a. What is the estimated clearing and grubbing production rate for each dozer?

b. What is the cost associated with the total operation, including overhead, when only one 215-hp dozer is used?

c. What is the cost associated with the total operation, including overhead, when only one 335-hp dozer is used?

d. Is it cheaper to employ three 215-hp dozers or two 335-hp dozers?

REFERENCES

1. *Caterpillar Performance Handbook,* Caterpillar Inc., Peoria, Ill.

2. Church, Horace K., "433 Seismic Excavation Studies: What They Tell About Rippability," *Roads and Streets,* Vol. 115, pp. 86–92, January 1972.

3. *Handbook of Ripping,* 7th ed., Caterpillar Tractor Co., Peoria, Ill., January 1983.

4. *Land Clearing,* Caterpillar Tractor Co., Peoria, Ill.

5. Schexnayder, Cliff, discussion of "Rational Equipment Selection Method Based on Soil Conditions," by Ali Touran, Thomas C. Sheahan, and Emre Ozcan, *Journal of Construction Engineering and Management, ASCE,* Vol. 124, No. 6, November–December 1998, p. 506.

6. *The Rome K/G Clearing Blade, Operator's Handbook,* Rome Plow Company, Cedartown, Ga.

7. Touran, Ali, Thomas C. Sheahan, and Emre Ozcan, "Rational Equipment Selection Method Based on Soil Conditions," *Journal of Construction Engineering and Management, ASCE*, Vol. 123, No. 1, pp. 85–88. 1997.

7

Scrapers

The key to a pusher-scraper spread's economy is that both the pusher and the scraper share in the work of obtaining the load. The production cycle for a scraper consists of six operations: (1) loading, (2) haul travel, (3) dumping and spreading, (4) turning, (5) return travel, and (6) turning and positioning to pick up another load. A systematic analysis of the scraper production cycle is the basic approach to identifying the most economical equipment employment. Scrapers are best suited for haul distances greater than 500 ft but less than 3,000, although with larger units the maximum distance can approach a mile.

GENERAL INFORMATION

Tractor-pulled scrapers are designed to load, haul, and dump loose material. The greatest advantage of tractor-scraper combinations is their versatility. The key to a pusher-scraper spread's economy is that both the pusher and the scraper share in the work of obtaining the load. Scrapers can be used in a wide range of material types (including shot rock) and are economical over a wide range of haul lengths and haul conditions. To the extent that they can self-load, they are not dependent on other equipment. If one machine in the spread experiences a temporary breakdown, it will not shut down the job, as would be the case for a machine that is used exclusively for loading. If the loader breaks down, the entire job must stop until repairs can be made. Scrapers are available with loose-heaped capacities up to about 44 cy, although in the past a few machines as large as 100 cy were offered.

Since scrapers are a compromise between machines designed exclusively for either loading or hauling, they are not superior to function-specific equipment in either hauling or loading. Excavators, such as hydraulic hoes and front shovels, or loaders will usually surpass scrapers in loading. Trucks may surpass them in hauling, especially over long distances. However, for off-highway situations having hauls of less than a mile, a scraper's ability to both load and haul gives it an advantage. Additionally, the ability of these machines to deposit their loads in layers of uniform thickness facilitates compaction operations.

SCRAPER TYPES

There are several types of scrapers, primarily classified according to the number of powered axles or by the method of loading. These are all wheel-tractor pulled machines. In the past, crawler-tractor "towed" two-axle scraper bowls were manufactured. They proved effective in short-haul situations, less than 600 ft one way. Another machine of the past is the two-axle pulling tractor that can now be found only in much older spreads. Machines currently available include:

1. Pusher-loaded (conventional)
    a. Single-powered axle (see Fig. 7.1)
    b. Tandem-powered axles (see Fig. 7.2)
2. Self-loading
    a. Push-pull, tandem-powered axles (see Fig. 7.2)
    b. Elevating (see Fig. 7.3)
    c. Auger (see Fig. 7.4)

Pusher-Loaded Scrapers

The wheel-type tractor scraper has the potential for high travel speeds on favorable haul roads. Many models can achieve speeds up to about 33 mph when fully loaded. This extends the economic haul distance of the units. However, these units are at a disadvantage when it comes to individually providing the high tractive effort required for economical loading. For the single-powered axle scraper

FIGURE 7.1 I Single-powered axle wheel-tractor scraper.

FIGURE 7.2 | Push-pull, tandem-powered axles, wheel-tractor scraper.

FIGURE 7.3 | Elevating wheel-tractor scraper.

(see Fig. 7.1), only a portion, on the order of 50 to 55% of the total loaded weight, bears on the drive wheels.

Additionally, in most materials, the coefficient of traction for rubber tires is less than for tracks. Therefore, it is necessary to supplement the loading power of these scrapers. The external source of loading power is usually a crawler-tractor pusher. Loading costs are still relatively low because both the scraper and the pusher share in providing the total power required to obtain a full load. Even tandem-powered units normally require help loading (see Fig. 7.2). The tandem-powered units have an initial cost that is about 25% more than that for a single-powered axle scraper. For that reason, they are commonly considered a specialized

FIGURE 7.4 | Auger wheel-tractor scraper.

unit; good for opening up a job, working extremely adverse grades, or working in soft ground conditions.

Single-powered axle pusher-loaded wheel-tractor scrapers become uneconomical when

■ Haul grades are greater than 5% and
■ Return grades are greater than 12%.

Push-Pull Scrapers

These are basically tandem-powered-axle units having a cushioned push block and bail mounted on the front (see Figs. 7.2 and 7.5), and a hook on the rear above the usual back push block. These features enable two scrapers to assist one another during loading by hooking together. The trailing scraper pushes the lead scraper as it loads. Then the lead scraper pulls the trailing scraper to assist it in loading. This feature allows two scrapers to work without assistance from a push tractor. They can also function individually with a pusher.

When used in rock or abrasive materials, tire wear will increase because of more slippage from the four-wheel drive action.

Elevating Scrapers

This is a completely self-contained loading and hauling scraper (see Fig. 7.3). A chain elevator serves as the loading mechanism. The disadvantage of this machine is that the weight of the elevator loading assembly is deadweight during the haul cycle. Such scrapers are economical in short-haul situations where the ratio of haul time to load time remains low. Elevating scrapers are also used for utility work dressing-up behind high-production spreads or shifting material

FIGURE 7.5 | Hook on the rear and cushioned-push block and bail on the front of two push-pull wheel-tractor scrapers.

during fine-grading operations. They are very good in small quantity situations. No pusher is required, so there is never a mismatch between pusher and the number of scrapers. Because of the elevator mechanism, they cannot handle rock or material containing rocks.

Auger Scrapers

This is another completely self-contained loading and hauling scraper. Auger scrapers can self-load in difficult conditions, such as laminated rock or granular materials.

An independent hydrostatic system powers the auger that is located in the center of the bowl. The rotating auger lifts the material off the scraper cutting edge and carries it to the top of the load, creating a void that allows new material to easily enter the bowl. This action reduces the cutting edge resistance, enabling the wheel-tractor scraper to continue moving through the cut. Auger scrapers are available in both single- and tandem-powered configurations. The auger adds nonload weight to the scraper during the travel cycle. These scrapers are more costly to own and operate than a conventional single- or tandem-powered machine.

In a soft limestone mining operation, 44-cy capacity tandem-powered auger scrapers making 3- to 5-in.-deep cuts achieved an average load time of about $1\frac{1}{2}$ min. Another advantage reported was the elimination of dust clouds that were created by other type scrapers when used in this application [4].

Volume of a Scraper

The volumetric load of a scraper may be specified as either the struck or heaped capacity of the bowl expressed in cubic yards. The struck capacity is the volume that a scraper would hold if the top of the material was struck off even at the top of the bowl. In specifying the heaped capacity of a scraper, manufacturers usually specify the slope of the material above the sides of the bowl with the designation SAE. The Society of Automotive Engineers (SAE) specifies a repose slope of 1:1 for scrapers. The SAE standard for other haul units and loader buckets is 2:1, as will be discussed in Chapters 8 and 10. Remember that the actual slope will vary with the type of material handled. In practice, both of these volumes represent loose cubic yards (lcy) of material because of how a scraper loads.

The capacity of a scraper, expressed in cubic yards bank measure (bcy), can be approximated by multiplying the loose volume in the scraper by an appropriate swell factor (see Table 4.3). Because of the compacting effect on the material in a push-loaded scraper, resulting from the pressure required to force additional material into the bowl, the swell is usually less than for material dropped into a truck by a hydraulic excavator. Tests indicate that the swell factors in Table 4.3 should be increased by approximately 10% for material push-loaded into a scraper. When computing the bank measure volume for an elevating scraper, no correction is required for the factors in Table 4.3.

EXAMPLE 7.1

If a push-loaded scraper hauls a heaped load measuring 22.5 cy and the appropriate swell factor from Table 4.3 is 0.8, the calculated bank measure volume will be

$$22.5 \text{ cy} \times 0.8 \times 1.1 = 19.8 \text{ bcy}$$

The use of rated volumetric capacity and swell values from tables provides an estimate of the scraper load. This is satisfactory for small jobs or if the estimator has developed a set of swell values for the materials commonly encountered in the work area. However, over time, actual field weights of loaded scrapers should be obtained for a variety of materials (see Fig. 7.6). From such weights, the average carrying capacity of those scrapers actually in the fleet can be calculated directly by the methods outlined in Chapter 4. This can be important on large jobs where the significance of small differences between table values and actual measured values can have large cost effects.

SCRAPER OPERATION

The basic operating parts of a scraper are (see Fig. 7.7):

- *Bowl.* The bowl is the loading and carrying component of a scraper. It has a cutting edge that extends across the front bottom edge. The bowl is lowered for loading and raised during travel.

- *Apron.* The apron is the front wall of the bowl. It is independent of the bowl. It is raised during the loading and dumping operations to allow the

FIGURE 7.6 | Weighing a loaded scraper in the field.

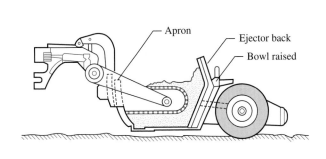

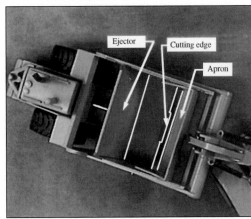

FIGURE 7.7 | Scraper bowl, apron, and ejector: cutaway view and overhead view.

material to flow into or out of the bowl. The apron is lowered during hauling to prevent material spillage.

■ *Ejector.* The ejector is the rear wall of the bowl. The ejector is in the rear position during loading and hauling. During spreading, the ejector is activated to move forward, providing positive discharge of the material in the bowl.

A scraper is loaded by lowering the front end of the bowl until the cutting edge that is attached to and extends across the width of the bowl enters the ground. At the same time, the front apron is raised to provide an open slot through which the earth can flow into the bowl. As the scraper moves forward, a horizontal strip of material is forced into the bowl. This is continued until the bowl is filled, at which point the bowl (cutting edge) is raised and the apron lowered to prevent spillage during the haul. During the haul, the bowl should be kept just high enough to clear the ground. Keeping the bowl low during travel enhances stability at high speeds and on uneven haul roads.

The dumping operation consists of lowering the cutting edge to the desired height above the fill, raising the apron, and then forcing the material out by means of a movable ejector mounted at the rear of the bowl.

The elevating scraper is equipped with horizontal flights operated by two endless elevator chains, to which the ends of the flights are connected. As the scraper moves forward with its cutting edge digging into the earth, the flights rake the material upward into the bowl. The pulverizing action of the flights permits a complete filling of the bowl, and enhances uniform spreading of the fill. An elevating scraper also has an ejector for discharging the material. Additionally, the elevator reverses to aid ejection of material.

SCRAPER PERFORMANCE CHARTS

Manufacturers of scrapers provide specifications and performance charts for each of their units. These charts contain information that may be used to analyze the performance of a scraper under various operating conditions. Table 7.1 gives the specifications for a Caterpillar 631E single-powered axle scraper. Figures 7.8 and 7.9 present the performance charts for this particular scraper.

It should be noted that owners sometimes add sideboards to the bowls of their scrapers. This practice will increase the volumetric load the machine can

TABLE 7.1 I Specifications for a Caterpillar 631E scraper

Engine: flywheel power 450			
Transmission: semiautomatic power shift, eight speeds			
Capacity of scraper:	Struck		21 cy
	Heaped		31 cy
Weight distribution:	Empty	Drive axle	67%
		Rear axle	33%
	Loaded	Drive axle	53%
		Rear axle	47%
Operating weight:	Empty		96,880 lb*
Rated load:			75,000 lb
Top speed:	Loaded		33 mph

*Includes coolant, lubricants, full fuel tank, ROPS canopy, and operator.

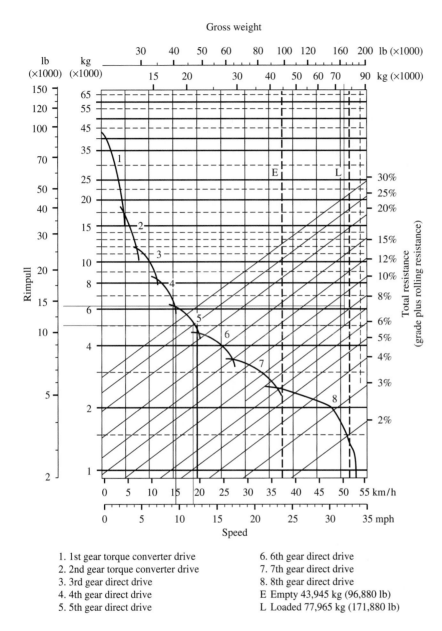

Gross weight

FIGURE **7.8** | Rimpull performance chart Caterpillar 631E scraper.

Reprinted courtesy of Caterpillar Inc.

1. 1st gear torque converter drive
2. 2nd gear torque converter drive
3. 3rd gear direct drive
4. 4th gear direct drive
5. 5th gear direct drive

6. 6th gear direct drive
7. 7th gear direct drive
8. 8th gear direct drive
E Empty 43,945 kg (96,880 lb)
L Loaded 77,965 kg (171,880 lb)

carry. If the boards cause the volumetric load of the scraper, whose specifications are given in Table 7.1, to be increased to 25.8 bcy, the weight the scraper
is carrying will also increase. This scraper has a rated payload of 75,000 lb. In
this sideboarding case, if the unit weight of the material being hauled is 3,000 lb

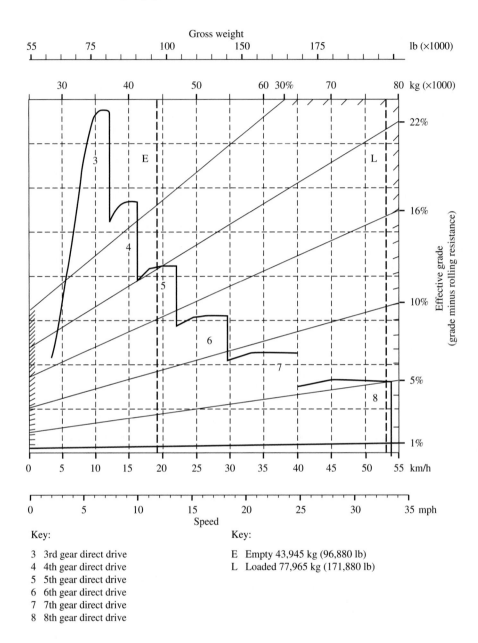

FIGURE 7.9 I Retarding performance chart Caterpillar 631E scraper.

Reprinted courtesy of Caterpillar Inc.

per bcy, the weight of the load would be 77,400 lb (25.8 bcy × 3,000 lb per bcy). This is greater than the rated load. The effect will be a few more yards per load, but in time, maintenance costs will increase because of overloading the machine. A material having a higher unit weight would aggravate this effect. However, if the machine was to be operated in lightweight materials, sideboards would be a reasonable consideration.

SCRAPER PRODUCTION CYCLE

The production *cycle* for a scraper consists of six operations—(1) loading, (2) haul travel, (3) dumping and spreading, (4) turning, (5) return travel, and (6) turning and positioning to pick up another load (Eq. [7.1]):

$$T_s = \text{load}_t + \text{haul}_t + \text{dump}_t + \text{turn}_t + \text{return}_t + \text{turn}_t \qquad \textbf{[7.1]}$$

Loading time is fairly consistent regardless of the scraper size. Even though large scrapers carry larger loads, they load just as fast as smaller machines. This is attributable to the fact that the larger scraper has more horsepower, and it will be matched with a larger push tractor. The average load time for push-loaded scrapers in common earth is 0.85 min. Tandem-powered scrapers will load slightly faster. Equipment manufacturers will supply load times for their machines based on the use of a specific push tractor [1]. The economical load time for a self-loading elevating scraper is usually around 1 min. Both the haul and return times depend on the distance traveled and scraper speed. Hauling and returning are usually at different speed ranges. Therefore, it is necessary to determine the time for each separately. If the haul road has multiple grade or rolling resistance conditions, a speed should be calculated for each segment of the route. In the calculations, it is appropriate to use a short distance at a slower speed when coming out of the pit, approaching the dump area, leaving the dump area, and again when entering the pit, to account for acceleration/deceleration time. A distance on the order of 200 ft at a speed of about 5 mph in each case is usually appropriate. Extremely steep downhill (favorable) grades can result in longer travel times than those calculated. This is caused by the fact that operators tend to downshift to keep speeds from becoming excessive. When using the charts, always consider the human element; do not blindly accept chart speeds.

SCRAPER PRODUCTION ESTIMATING

GENERAL INFORMATION

The most profitable methods and equipment to be used on any job can only be determined by a careful project investigation. The basic approach is a systematic analysis of the scraper cycle with a determination of the cost per cubic yard under the conditions existing or anticipated on the project. Such an analysis is helpful when making decisions for estimating purposes and for job control.

Next we set forth a format that can be used to analyze scraper production. The calculations are based on using a CAT 631E scraper for which the specification information is given in Table 7.1 and the performance charts are shown in Figs. 7.8 and 7.9. The total length of haul when moving from the cut to the fill is 4,000 ft in segments:

1,200 ft	+4% grade
1,400 ft	+2% grade
1,400 ft	−2% grade

The soil being moved is a clay having a unit weight of 3,000 lb per bcy. The earth haul road will be well maintained, therefore the assumed rolling resistance is 80 lb per ton or 4%. Assume an average load time of 0.85 min. If it is also assumed that the shape of load-growth curve (Fig. 7.10) is valid for this example, the expected load will be 96% of heaped capacity. A 50 min per hour efficiency factor will be used.

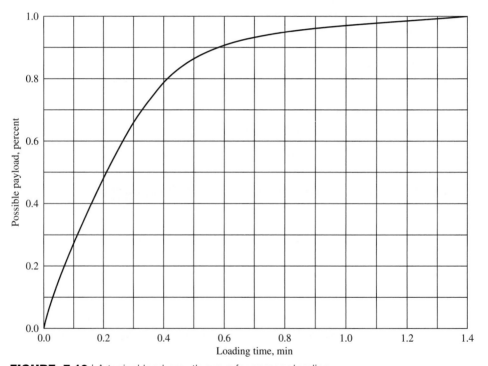

FIGURE 7.10 I A typical load-growth curve for scraper loading.

STEP 1: WEIGHT

The first step in calculating scraper production is to determine the

■ Empty vehicle weight (EVW),

■ Load weight, and

■ Gross vehicle weight (GVW).

Empty Vehicle Weight

To determine the *EVW,* reference the manufacturer's data for the specific make and model of scraper being considered. When referring to the manufacturer's data be careful about notes concerning what is included in the stated weight. Typically, empty operating weight includes a full fuel tank, coolants, and lubricants, a rollover protective system (ROPS) canopy, and the operator.

Load Weight

The *load weight* is a function of scraper load volume and the unit weight of the material being hauled. The load volume is a loose material volume, so the unit weight needs to be an lcy unit weight or the loose volume can be converted to match the units of the unit weight.

Gross Vehicle Weight

The GVW is the sum of the EVW and the load weight.

Example Data and Calculations

Empty weight (EVW)	Table 7.1	96,880 lb
Load volume: (based on load time, Fig. 7.10 for this example)	0.96 × 31 cy = 29.8 lcy	
From Table 4.3:	swell factor clay = 0.74	
Load volume bank measure:	29.8 lcy × 0.74 × 1.1 = 24.3 bcy	
Weight of load:	24.3 bcy × 3,000 lb per bcy =	72,900 lb
	Gross weight (GVW)	169,780 lb

STEP 2: ROLLING RESISTANCE

Rolling resistance (RR) is the result of a conscious management decision as to how much effort will be expended maintaining the haul road. As the effort (and money) expended for maintenance of haul roads increases, production will also increase. The reverse is also true—if no effort is made to improve, the haul-road production suffers. Graders and sometimes dozers are necessary to eliminate ruts and rough (washboard) surfaces. Water trucks will be required to provide moisture for compacting the haul road and for dust control. Visibility is improved by keeping roads moist. This lessens the chance of accidents. Dust control also helps to alleviate additional mechanical wear.

Rolling Resistance and Scraper Production

Haul-road rolling resistance is a job condition that is sometimes neglected. The effect of haul-road conditions on rolling resistance, and thereby on scraper production and the cost of hauling earth should always be carefully analyzed. A well-maintained haul road permits faster travel speeds and reduces the costs of scraper maintenance and repair. Figure 7.11 is from a field study of scraper haul times.

The shaded area represents the range of average travel times on numerous projects. The lower boundary indicates times on projects having well-maintained haul roads. The upper boundary indicates poor haul roads. Considering a haul distance of 4,000 ft, the total time for the haul and return combined on a well-maintained haul road was 5.20 min. Under poor conditions, this could be as high as 7.55 min. A difference of 2.35 min in cycle time is a 4.7% production loss in a 50-min hour.

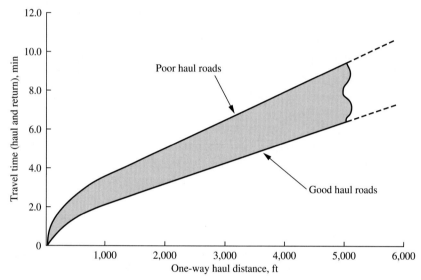

FIGURE 7.11 | Average travel times single-powered axle scrapers, capacity < 25 cy, negligible grade.

Source: U.S. Department of Transportation, FHWA.

Representative rolling resistance values are given in Table 5.1. Rolling resistance can be expressed as either a percentage or as pounds per ton of vehicle weight.

For this example it is anticipated that the earth haul road will be well maintained, and have a rolling resistance of 80 lb per ton, or 4%.

STEP 3: GRADE RESISTANCE/ASSISTANCE

Grade resistance (GR) or assistance (GA) is usually a function of the project topography. From where the material must be excavated to where it must be hauled is a physical condition imposed by the project requirements. Plan the job to avoid adverse grades that could drastically reduce production. There can sometimes be a selection of haul-route placement that allows flatter grades, but those possibilities must usually be considered in terms of increased haul-route length. Such alternatives may mean that several estimates of scraper production will have to be produced.

Grade resistance can be expressed as either a percentage or as pounds per ton of vehicle weight.

For this example the haul-route length and grades from the cut to the fill are given as

1,200 ft	80 lb per ton	+4% grade
1,400 ft	40 lb per ton	+2% grade
1,400 ft	−40 lb per ton	−2% grade

STEP 4: TOTAL RESISTANCE/ASSISTANCE

The total resistance (TR) or assistance (TA) for each segment of the haul and return must be determined. This is the sum of the rolling resistance and the grade resistance/assistance for each segment. It is recommended practice to use a table format when developing an estimate of scraper production. Such a format is illustrated here. A separate row can be used for each segment of both the haul and return routes.

To account for acceleration, 200 ft at the beginning of the first segment on both the haul and return routes will be at a reduced speed. Likewise, to account for deceleration, 200 ft at the end of the last segment on both the haul and return routes will be at a reduced speed. These do not represent additional distance, they are part of their respective segments; but instead of using a speed appropriate to the weight of the scraper and the total resistance, a reduced speed will be used in calculating travel time for these segments. This is illustrated in the table.

	Distance (ft)	Step 2: RR (%)	Step 3: GR (%)	Step 4: TR (lb per ton)	Step 4: TR (%)
Haul (169,780 lb or 84.89 tons)	200 (acc.)	4	4	160	8
	1,000	4	4	160	8
	1,400	4	2	120	6
	1,200	4	−2	40	2
	200 (dec.)	4	−2	40	2
Return (96,880 lb or 48.44 tons)	200 (acc.)	4	2	120	6
	1,200	4	2	120	6
	1,400	4	−2	40	2
	1,000	4	−4	0	0
	200 (dec.)	4	−4	0	0

Total resistance can be calculated either as lb per ton or as a percent. Both are included in the chart to illustrate this fact. However, it is not necessary to use both when making a production calculation, one or the other will suffice. The choice does affect how the performance charts are used but either method will yield the same speed.

STEP 5: TRAVEL SPEED

Steady-state travel speeds based on total vehicle weight and total resistance can be determined from the manufacturer's performance or retarder charts for the specific scrapers under consideration. If the total resistance is a positive number, the performance chart should be used and if total resistance is a negative number, the retarder chart should be used. The speeds in the chart represent that which *can be achieved* in a steady-state condition. They do necessarily represent a safe operating speed for specific job-site conditions. Before selecting a speed for estimating purposes, visualize the project conditions and adjust chart speeds to the expected conditions. An example of this is hauling downhill with a full load. The chart may indicate that the maximum vehicle speed can be achieved, but many times operators are not comfortable traveling that fast downhill (they realize it is an unsafe condition) and will shift into a lower gear.

	Distance (ft)	Step 2: RR (%)	Step 3: GR (%)	Step 4: TR (lb rimpull)	Step 4: TR (%)	Step 5: speed (mph)
Haul (169,780 lb or 84.89 tons)	200* (acc.)	4	4	13,582	8	5†
	1,000	4	4	13,582	8	10
	1,400	4	2	10,187	6	14
	1,200	4	−2	3,396	2	32
	200* (dec.)	4	−2	3,396	2	16†
Return (96,880 lb or 48.44 tons)	200* (acc.)	4	2	5,813	6	11†
	1,200	4	2	5,813	6	23
	1,400	4	−2	1,938	2	33
	1,000	4	−4	0	0	33
	200* (dec.)	4	−4	0	0	16†

*Assumed acceleration and deceleration distances.
†Reduced speed to account for acceleration or deceleration.

STEP 6: TRAVEL TIME

Travel time is the sum of the time it takes the scraper to traverse each segment of the haul and return routes. Based on the speeds determined from the performance or retarder charts, or the speeds assumed because of job conditions, travel time can be calculated using Eq. [7.2].

$$\text{Travel time per segment (min)} = \frac{\text{Segment distance (ft)}}{88 \times \text{travel speed (mph)}} \qquad \textbf{[7.2]}$$

	Distance (ft)	Step 2: RR (%)	Step 3: GR (%)	Step 4: TR (%)	Step 5: speed (mph)	Step 6: time (min)
Haul (169,780 lb)	200* (acc.)	4	4	8	5†	0.45
	1,000	4	4	8	10	1.14
	1,400	4	2	6	14	1.14
	1,200	4	−2	2	32	0.43
	200* (dec.)	4	−2	2	16†	0.14
Return (96,880 lb)	200* (acc.)	4	2	6	11†	0.21
	1,200	4	2	6	23	0.56
	1,400	4	−2	2	33	0.48
	1,000	4	−4	0	33	0.34
	200* (dec.)	4	−4	0	16†	0.14
					Total travel time	5.03

*Assumed acceleration and deceleration distances.
†Reduced speed to account for acceleration or deceleration.

STEP 7: LOAD TIME

Load time is a management decision that should be made after a careful evaluation of the production and cost effects. The tendency in the field is to take too long in loading scrapers [5].

Scraper Load-Growth Curve

Without a critical evaluation of available information, it may appear that the lowest cost for moving earth with scrapers is to load every scraper to its maximum capacity before it leaves the cut. However, numerous studies of loading practices have revealed that loading scrapers to their maximum capacities will usually reduce, rather than increase, production (see Table 7.2).

TABLE 7.2 | Variations in the rates of production of scrapers with loading times* (a 2500-ft one-way haul distance)

Loading time (min)	Other time (min)	Cycle time* (min)	Number of trips per hour	Payload†		Production per hour	
				cy	(cu m)	cy	(cu m)
0.5	5.7	6.2	8.07	17.4	(13.3)	140	(107)
0.6	5.7	6.3	7.93	18.3	(14.0)	145	(111)
0.7	5.7	6.4	7.81	18.9	(14.5)	147	(112)
0.8	5.7	6.5	7.70	19.2	(14.7)	148‡	(113)
0.9	5.7	6.6	7.57	19.5	(14.9)	147	(112)
1.0	5.7	6.7	7.46	19.6	(15.0)	146	(112)
1.1	5.7	6.8	7.35	19.7	(15.1)	145	(111)
1.2	5.7	6.9	7.25	19.8	(15.2)	143	(109)
1.3	5.7	7.0	7.15	19.9	(15.2)	142	(109)
1.4	5.7	7.1	7.05	20.0	(15.3)	141	(108)

*For a 50-min hour.
†Determined from measured performance.
‡The economical loading time is 0.8 min.

   When a scraper starts loading, the earth flows into it rapidly and easily, but as the quantity of earth in the bowl increases, the incoming earth encounters greater resistance and the rate of loading decreases quite rapidly, as illustrated in Fig. 7.10. This figure is a load-growth curve for a specific scraper and pusher combination, and material; it shows the relation between the load in the scraper and the loading time. An examination of the curve reveals that during the first 0.5 min, the scraper loads about 85% of its maximum possible payload. During the next 0.5 min it loads an additional 12%, and if loading is continued to 1.4 min, the gain in volume during the last 0.4 min is only 3%—an instructive example of the law of diminishing returns.
   Economical loading time is a function of haul distance. As haul distance increases, economical load time will increase. Figure 7.12 illustrates the relationship between loading time, production, and haul distance; it is presented only to demonstrate graphically those interrelations. The economical haul distance for

scrapers is usually much less than a mile. This can be recognized from the rapid rate at which production falls with increasing haul distance.

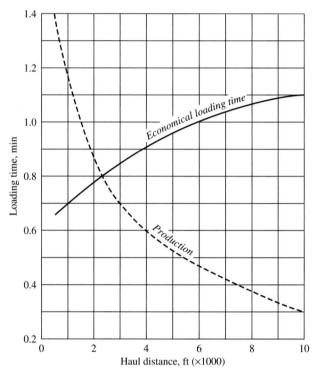

FIGURE 7.12 | The effect of haul distance on the economical loading time of a scraper.

For this example, an average load time 0.85 min has been assumed.

Step 6: Travel time   5.03 min
Step 7: Load time   0.85 min.

STEP 8: DUMP TIME

Dump times vary with scraper size, but project conditions will affect the duration. Average values for dump time are presented in Table 7.3. Physical constraints in the dump area may dictate scraper-dumping techniques. The most common method is for the scraper to dump before turning. This uses the haul-speed momentum to carry the scraper through the dump. It reduces the possibility of the scraper becoming stuck in the newly dumped material and yields a fairly even spreading of material, thereby reducing spreading and compaction equipment effort.

Sometimes it is necessary to negotiate the turn before dumping. This generally increases dump time and the chances of the scraper becoming stuck while dumping. The last method is to dump during the turning maneuver. This definitely increases dumping time, results in uneven spreading, and can cause

TABLE 7.3 | Scraper dump cycle times

Scraper size (cy) heaped	Scraper type	
	Single engine (min)	Tandem-powered (min)
<25	0.30	—
25–34	0.37	0.26
35–44	0.44	0.28

Source: U.S. Department of Transportation, FHWA.

scrapers to become bogged down. When access is limited at bridge abutments, or when backfilling culverts, a turning/dumping maneuver may be necessary. Wet material is difficult to eject and will increase the dump time.

Considering the condition of the example, a 631 scraper, that is, a single-engine machine, with a rated heaped capacity of 31 cy (Table 7.1); the dump time will be 0.37 min (Table 7.3).

STEP 9: TURNING TIMES

Turning time does not seem to be significantly affected by either the type or size of scraper. Based on FHWA studies, the average turn time in the cut is 0.30 min, and on the fill, the average turn time is 0.21 min [6]. The slightly slower turn in the cut is primarily caused by congestion in the area and the necessity to spot the scraper for the pusher.

STEP 10: TOTAL CYCLE TIME

Total scraper cycle time is the sum of the times for the six operations that have been examined in steps 1 to 9—haul travel and return travel, which have been combined; loading; dumping and spreading; turning on the fill; and turning and positioning to pick up another load at the cut.

Step 6:	Travel time	5.03 min
Step 7:	Load time	0.85 min
Step 8:	Dump time	0.37 min
Step 9:	Turn time fill	0.21 min
	Turn time cut	0.30 min
Step 10:	Total cycle time scraper	6.76 min

STEP 11: PUSHER CYCLE TIME

If push-loaded tractor scrapers are to attain their volumetric capacities, they need the assistance of a push tractor during the loading operation. Such assistance will reduce the loading time duration and thereby reduce total cycle time. When using push tractors, the number of pushers must be matched with the number of scrapers available at a given time. If either the pusher or scraper must wait for the other, operating efficiency is lowered and production costs are increased.

The pusher cycle time includes the time required to push-load the scraper (the contact time) and the time required to move into position to load the next scraper. The cycle time for a push tractor will vary with the conditions in the loading area, the relative size of the tractor and the scraper, and the loading method. Figure 7.13 shows three loading methods. *Back-track loading* is the

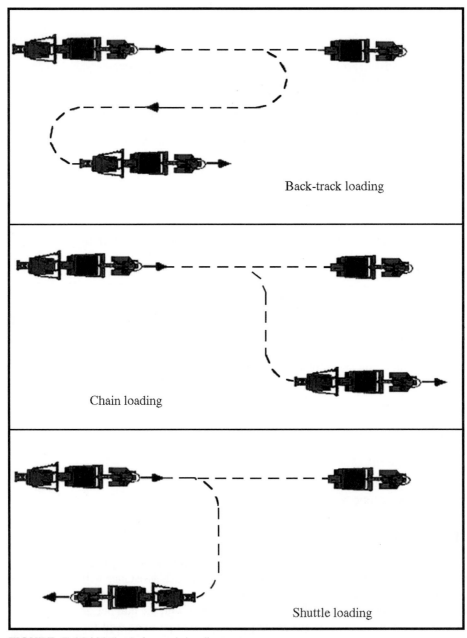

Back-track loading

Chain loading

Shuttle loading

FIGURE 7.13 | Methods for push-loading scrapers.

most common method employed. It offers the advantage of always being able to load in the direction of the haul. *Chain loading* can be used when the excavation is conducted in a long cut. *Shuttle loading* is used infrequently. However, if one pusher can serve scrapers hauling in opposite directions from the cut, it is a viable method. In several past publications, Caterpillar recommends calculating back-track push tractor cycle time, T_p, by the formula

$$T_p = 1.4L_t + 0.25 \qquad [7.3]$$

where L_t is the scraper load time (pusher contact time).

The formula is based on the concept that pusher cycle time is a function of four components:

1. Load time of the scraper.
2. Boost time, time assisting scraper out of the cut, 0.15 min.
3. Maneuver time, 40% of load time.
4. Positioning for contact time, 0.10 min.

Pusher cycle time will be less when using the chain or shuttle methods.

The load time in this example is 0.85 min (this is sometimes referred to as the contact time), therefore applying Eq. [7.3]:

$$T_p = 1.4\,(0.85) + 0.25 \Rightarrow 1.44 \text{ min}$$

STEP 12: BALANCED FLEET

The number of scrapers a push tractor can serve is simply the ratio of the scraper cycle time to the pusher cycle time:

$$N = \frac{T_s}{T_p} \qquad [7.4]$$

where N is the number of scrapers per one pusher.

Rarely, if ever, will the value of N be an integer. This means that either the pusher or a scraper will be idle some of the time.

Favorable loading conditions will reduce loading time and increase the number of scrapers that a pusher can serve. A large pit or cut, ripping hard soil prior to loading, loading downgrade, and using a push tractor whose power is matched with the size of the scraper are all factors that can improve loading times. Likewise, tight soils with no prior ripping, very large scrapers, and loading rock are factors that may create a situation where multiple pushers are required to load a scraper effectively.

In this example, the total cycle time for the scraper was 6.76 min and 1.44 min for the pusher, therefore applying Eq. [7.4]:

$$N = \frac{6.76 \text{ min}}{1.44 \text{ min}} \Rightarrow 4.7$$

Consequently the economics of using either four or five scrapers should be investigated.

STEP 13: EFFICIENCY

The term *efficiency* or *operating efficiency* is used to account for the actual productive operations in terms of an average number of minutes per hour that the machine will operate. A 50-min hour average would yield a 0.83 (50/60) efficiency factor. The 50-min hour is a reasonable starting point if no company-and/or equipment-specific efficiency data are available. The estimator should always try to visualize the work site and how the work will likely be performed in the field before applying a factor. If the pit will not be congested and if the dump area is wide open, a 55-min hour may be appropriate. But if the cut involves a tight area, such as a ditch, or if the embankment area is a narrow bridge header, the estimator should consider a 45-min hour.

For the example, a 50-min hour efficiency factor has been assumed.

STEP 14: PRODUCTION

If the number of scrapers placed on the job is less than the balance number from Eq. [7.4], the scrapers will control production.

$$\text{Production (scrapers controlling)} = \frac{\text{efficiency, min/hr}}{\text{total cyc. time scraper, min}}$$
$$\times \text{ No. of scrapers} \times \text{volume per load} \quad \textbf{[7.5]}$$

If the number of scrapers placed on the job is greater than the balance from Eq. [7.4], the pusher will control production.

$$\text{Production (pusher controlling)} = \frac{\text{efficiency, min/hr}}{\text{total cyc. time pusher, min}} \times \text{volume per load}$$
$$\textbf{[7.6]}$$

In this example if only four scrapers on the job, production would be

$$\text{Production (scrapers controlling)} = \frac{50 \text{ min/hr}}{6.76 \text{ min}} \times 4 \times 24.3 \text{ bcy} \Rightarrow 719 \text{ bcy/hr}$$

If five scrapers were used on the job, production would be

$$\text{Production (pusher controlling)} = \frac{50 \text{ min/hr}}{1.44 \text{ min}} \times 24.3 \text{ bcy} \Rightarrow 844 \text{ bcy/hr}$$

STEP 15: COST

If production is the important factor, it is obvious in this example that five scrapers should be used on the job because of the increased production when compared to using four scrapers. But usually the decision concerning the number of scrapers to employ is a question of unit production cost. Let us assume that each of these scrapers has an O&O cost of $89 per hour and that the O&O cost for the push tractor is \$105 per hour. Furthermore, assume that the scraper operators are paid \$12 per hour and the pusher operator \$20 per hour. With this cost information available it is possible to determine the unit cost for moving the material.

4 scrapers @ \$89/hr	+ 1 pusher @ \$105/hour	= \$461/hr
4 operators @ \$12/hr	+ 1 operator @ \$20/hr	= 68/hr
	Cost for a four-scraper spread	\$529/hr

5 scrapers @ \$89/hr	+ 1 pusher @ \$105/hour	= \$550/hr
5 operators @ \$12/hr	+ 1 operator @ \$20/hr	= 80/hr
	Cost for a five-scraper spread	\$630/hr

Unit cost to move the material using a four-scraper spread:

$$\frac{\$529/hr}{719 \text{ bcy/hr}} = \$0.736/\text{bcy}$$

Unit cost to move the material using a five-scraper spread:

$$\frac{\$630/hr}{844 \text{ bcy/hr}} = \$0.746/\text{bcy}$$

The unit costs are very close. If this were for a project having a large quantity of material to move it would most likely be best to use a five-scraper spread of equipment. However, if it is for a project having a very limited amount of material to move, the mobilization cost of the fifth scraper might amount to more of an expense than the extra unit cost.

The final decision concerning the size of spread to use must include consideration of the costs to mobilize the equipment and the costs of daily overhead for the project. When daily overhead expenses are great, higher production is usually justified.

INCREASING SCRAPER PRODUCTION

Scrapers are best suited for medium-haul distance earthmoving operations. Distances greater than 500 ft but less that 3,000 are typical, although with larger units the maximum distance can approach a mile. The selection of a particular type of scraper for a project should take into consideration the type of material being loaded and transported. Figure 7.14 provides a summary of applicable scraper type based on project material type.

To obtain a higher profit from earthwork, a contractor must organize and operate the spread in a manner that will ensure maximum production at the lowest cost. There are several methods whereby this objective can be attained.

Ripping

Most types of tight soils will load faster if they are ripped ahead of the scraper. Additionally, delays pertaining to equipment repairs will be reduced substantially as the scraper will not be operated under as much strain. If the value of the

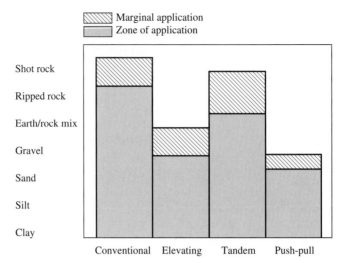

FIGURE 7.14 | Zones of application for different type scrapers.

increased production resulting from ripping exceeds the ripping cost, the material should be ripped.

When rock is ripped for scraper loading, the depth ripped should always exceed the depth to be excavated. This is done so that there will always be a loose layer of material under the tires to provide better traction and to reduce the wear on the tracks and tires.

Prewetting the Soil

Some soils will load more easily if they are reasonably moist. Prewetting can be performed in conjunction with ripping or ahead of loading, to permit a uniform penetration of the moisture into the soil.

Prewetting the soil in the cut can reduce or eliminate the use of water trucks on the fill, thereby reducing the congestion of equipment on the fill. The elimination of excess moisture on the surface of the fill may facilitate the movement of the scrapers on the fill.

Loading Downgrade

When it is practicable to do so, scrapers should be loaded downgrade and in the direction of haul. Downgrade loading results in faster loading times, whereas loading in the direction of haul both shortens the length of haul and eliminates the need to turn in the cut with a loaded scraper. Each 1% of favorable grade is the equivalent of increasing the loading force by 20 lb per ton of gross weight of the push tractor and scraper unit.

Supervision

Full-time supervisory control should be provided in the cut. A more efficient operation will result through the elimination of confusion and traffic congestion.

A spotter should always control the fill operations, being responsible for coordinating the scrapers with the spreading and compacting equipment. The spotter's job is to maintain the scrapers' dumping pattern. Typically, the spotter will direct each scraper operator to dump the load at the end of the preceding spread until the end of the fill is reached. Then the next spread is started parallel to the first. This enables compaction equipment to work the freshly dumped material without interfering with the scrapers.

SUMMARY

Profitable scraper employment on any job can only be determined by a careful project investigation. Equipment manufacturers provide scraper performance charts that are used to analyze the performance of a scraper under various operating conditions. The first part of the analysis is to assume a scraper load time and volume of load. With the load information and haul-route data generated from a study of the project layout, and possibly a mass diagram, a scraper cycle time can be calculated. Spread production is calculated after deciding on the number of scrapers to employ by examining the ratio of the pusher cycle time and the scraper cycle time. Critical learning objectives would include:

- An ability to properly use scraper performance charts.
- An ability to calculate the time required to complete each of the six operations of the scraper production cycle.
- An ability to calculate pusher cycle time.
- An ability to calculate scraper spread production.

These objectives are the basis for the problems that follow.

PROBLEMS

7.1 What is the travel time for a scraper moving at 18 mph over a distance of 2,300 ft? (1.45 min)

7.2 What is the travel time for a scraper moving at 23 mph over a distance of 1,300 ft?

7.3 A wheel-tractor scraper is operating on a level grade. Assume no power derating is required for equipment condition, altitude, temperature, and so on. Use the scraper specifications in Table 7.1 and the "Performance Charts" in Figs. 7.8 and 7.9.

    a. Disregarding traction limitations, what is the maximum value of rolling resistance (in pounds per ton) over which the fully loaded unit can maintain a speed of 15 mph?

    b. What minimum value of coefficient of traction between the tractor wheels and the traveling surface is needed to satisfy the requirements of part a?

7.4 Determine the cycle time for a single-engine scraper rated at 20 cy heaped, which is used to haul material from a pit to a fill 1,700 ft away under severe conditions. The average haul speed will be 10 mph and the average return speed will be 16 mph. Assuming 200 ft is required to both accelerate and decelerate at an average speed of 5 mph. The operating efficiency will be equal to a 50-min hour. It will take 0.85 min. to load this scraper.

7.5 Based on the scraper specifications in Table 7.1 and on the "Performance Charts" in Figs. 7.8 and 7.9, and for haul conditions as stated here, analyze the probable scraper production. How many scrapers should be used and what will be the production in bcy per hour? The material to be hauled is a sandy clay (dry earth), 2,800 lb per bcy. The expected rolling resistance for the well-maintained haul road is 40 lb per ton. Assume using a 0.85-min load time that an average load of 91% heaped capacity will be obtained and that 200 ft is required to both accelerate and decelerate at an average speed of 5 mph (miles per hour). Use a 50-min hour efficiency factor. The total length of haul is 3,200 ft and has the following individual segments when moving from cut to fill:

$$\begin{array}{ll} 600 \text{ ft} & +3\% \text{ grade} \\ 2,200 \text{ ft} & 0\% \text{ grade} \\ 400 \text{ ft} & +4\% \text{ grade} \end{array}$$

7.6 Based on the scraper specifications in Table 7.1 and the "Performance Charts" in Figs. 7.8 and 7.9, and for haul conditions as stated here, analyze the probable scraper production. How many scrapers should be used and what will be the production in bcy per hour? The material to be hauled is cohesive. It has a swell factor of 0.76 and a unit weight of 2,900 lb per bcy. The expected rolling resistance for the well-maintained haul road is +3%. Assume using a 0.80-min load time that an average load of 90% heaped capacity will be obtained and that 200 ft is required to both accelerate and decelerate at an average speed of 5 mph. Use a 50-min hour efficiency factor. The total length of haul is 2,600 ft and has the following individual segments when moving from cut to fill:

$$\begin{array}{ll} 600 \text{ ft} & +5\% \text{ grade} \\ 1,800 \text{ ft} & -2\% \text{ grade} \\ 200 \text{ ft} & -4\% \text{ grade} \end{array}$$

7.7 Determine the maximum hauling production given the following conditions. As many scrapers as required can be used, but only one push tractor will be available. The material is a sandy clay (dry earth), 2,900 lb per bcy. The expected haul-road rolling resistance is 80 lb per ton. The average scraper load will be 28.2 lcy. There are three segments to the haul route: 600 ft at a grade of +3%; 2,200 ft at 0% grade; and 400 ft at +4% grade (moving from the cut to the fill). To account for acceleration and deceleration use an average speed of 4 mph for 200 ft at each end of the haul and return. Use the scraper specifications in Table 7.1 and the "Performance Charts" in Figs. 7.8 and 7.9. Assume a 50-min hour efficiency.

REFERENCES

1. *Caterpillar Performance Handbook,* Caterpillar Inc., Peoria, Ill., issued annually.
2. *Fundamentals of Earthmoving,* Caterpillar Tractor Co., Peoria, Ill., May 1979.
3. *Making the Most of Scraper Potential,* Caterpillar Inc., Peoria, Ill., 1993.
4. "Danish Quarry Uses Auger Scrapers in Lime Production," *Rock Products,* July 1993.
5. *Optimum Scraper Load Time,* Caterpillar Inc., Peoria, Ill., 1994.
6. *Production Efficiency Study on Rubber-Tired Scrapers,* FHWA DP-PC-920, Federal Highway Administration, Arlington, Va., April 1977.

8

Excavators

Hydraulic power is the key to the utility of many excavators. Hydraulic front shovels are used predominantly for hard digging above track level and for loading haul units. Hydraulic hoe excavators are used primarily to excavate below the natural surface of the ground on which the machine rests. The loader is a versatile piece of equipment designed to excavate at or above wheel/track level. Unlike a shovel or hoe, to position the bucket to dump, a loader must maneuver and travel with the load. Additionally, there are a variety of excavators available for specialty applications.

HYDRAULIC EXCAVATORS

There are many variations in hydraulic excavators. They may be either crawler or rubber-tire-carrier-mounted, and there are many different operating attachments. With the options in types, attachments, and sizes of machines, there are differences in appropriate applications and therefore variations in economical advantages. This chapter takes into account the important operating features and calls attention to cost consequences of specific machine applications.

Hydraulic power is the key to the advantages offered by these machines. The hydraulic control of machine components provides

- Faster cycle times.
- Outstanding control of attachments.
- High overall efficiency.
- Smoothness and ease of operation.
- Positive control that offers greater accuracy and precision.

Hydraulic excavators are classified by the digging motion of the hydraulically controlled boom and stick to which the bucket is attached (see Fig. 8.1). A downward arc unit is classified as a "hoe." It develops excavation breakout force by pulling the bucket toward the machine and curling the bucket inward. An upward motion unit is known as a "front shovel." A shovel develops breakout

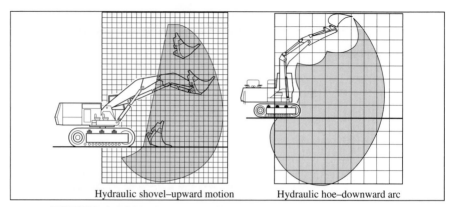

Hydraulic shovel–upward motion   Hydraulic hoe–downward arc

FIGURE 8.1 | Digging motion of hydraulic excavators.

force by crowding material away from the machine. The downward swing of a hoe dictates usage for excavating below the running gear. The boom of a shovel swings upward to load; therefore, the machine requires a material face above the running gear to work against.

If an excavator is considered as an independent unit (a one-link system) its production rate can be estimated using the following steps.

Step 1. Obtain the heaped bucket load volume from the manufacturers' data sheet. This would be a loose volume (lcy) value.

Step 2. Apply a bucket fill factor based on the type of machine and the class of material being excavated.

Step 3. Estimate a peak cycle time. This is a function of machine type and job conditions to include angle of swing, depth or height of cut, and, in the case of loaders, travel distance.

Step 4. Apply an efficiency factor.

Step 5. Conform the production units to the desired volume or weight (lcy to bcy or tons).

Step 6. Calculate the production rate.

The basic production formula is: Material carried per load × cycles per hour. In the case of excavators, this formula can be refined and written as

$$\text{Production} = \frac{3{,}600 \text{ sec} \times Q \times F \times (\text{AS:D})}{t}$$

$$\times \frac{E}{60\text{-min hr}} \times \frac{1}{\text{volume correction}} \qquad \textbf{[8.1]}$$

where

Q = heaped bucket capacity (lcy)

F = bucket fill factor

AS:D = angle of swing and depth (height) of cut correction

t = cycle time in seconds

E = efficiency (min per hour)

volume correction for loose volume to bank volume, $\dfrac{1}{1 + \text{swell factor}}$;

for loose volume to tons, $\dfrac{\text{loose unit weight, lb}}{2{,}000 \text{ lb/ton}}$.

FRONT SHOVELS

GENERAL INFORMATION

Front shovels are used predominantly for hard digging above track level and for loading haul units. Loading of shot rock would be a typical application (see Fig. 8.2). Shovels are capable of developing high breakout force with their buckets but the material being excavated should be such that it will stand as a vertical bank, i.e., a wall of material that stands perpendicular to the ground. Most shovels are crawler mounted and have very slow travel speeds, less than 3 mph. The parts of the shovel are designed for machine balance; each element of the front-end attachment is designed for anticipated load. The front-end attachment weighs about one-third as much as the superstructure with its power parts and cab.

Size Rating of Front Shovels

The size of a shovel is indicated by the size of the bucket, expressed in cubic yards. There are three different bucket-rating standards, Power Crane and Shovel Association (PCSA) Standard No. 3, Society of Automotive Engineers

FIGURE 8.2 | A hydraulic-operated front shovel loading shot rock into a truck.

(SAE) Standard J-296, and the Committee on European Construction Equipment (CECE) method. All of these methods are based only on the physical dimensions of the bucket and do not address the "bucket loading motion" of a specific machine. For buckets greater than 3-cy capacity, ratings are in $\frac{1}{4}$-cy intervals and in $\frac{1}{8}$-cy intervals for buckets less than 3 cy in size.

> *Struck capacity.* The volume actually enclosed by the bucket with no allowance for bucket teeth is the struck capacity.
>
> *Heaped capacity.* Both PCSA and SAE use a 1:1 angle of repose for evaluating heaped capacity. CECE specifies a 2:1 angle of repose.
>
> *Fill factors.* The amount of material actually in a bucket compared to its volume is a vital factor in determining shovel production. Materials that are easy to dig and that can be described as flowing (sand, gravel, or loose earth) should easily fill the bucket to capacity with a minimum of void space. At the other extreme are the hard, rocky materials that will have lots of void spaces. If the material dug has a significant amount of oversize chunks or is extremely sticky, the average bucket load will be reduced.

Rated-heaped capacities represent a net section bucket volume; therefore they must be corrected to average bucket payload based on the characteristics of the material being handled. Manufacturers usually suggest factors, commonly called "fill factors," for making such corrections. Fill factors are percentages that, when multiplied by a rated-heaped capacity, adjust the volume by accounting for how the specific material will load into the bucket (see Table 8.1). It is best, when possible, to conduct field tests based on the weight of material per bucket load to validate fill factors.

TABLE 8.1 | Fill factors for front shovel buckets

Material	Fill factor* (%)
Bank clay; earth	100–110
Rock-earth mixture	105–115
Rock—poorly blasted	85–100
Rock—well blasted	100–110
Shale; sandstone—standing bank	85–100

*Percentage of heaped bucket capacity
Reprinted courtesy of Caterpillar Inc.

Basic Parts and Operation

The basic parts of a front shovel include the mounting, cab, boom, stick, and bucket (see Fig. 8.3). With a shovel in the correct position, near the face of the material to be excavated, the bucket is lowered to the floor of the pit, with the teeth pointing into the face. A crowding force is applied by hydraulic pressure to the stick cylinder at the same time the bucket cylinder rotates the bucket through the face.

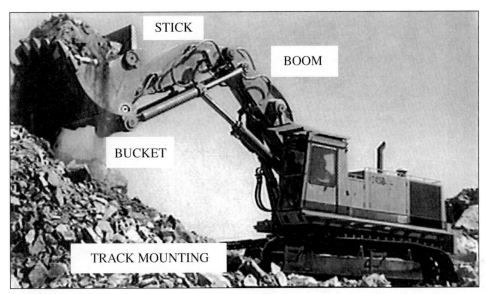

FIGURE 8.3 | Basic parts of a hydraulic front shovel.

SELECTING A FRONT SHOVEL

In selecting a shovel, the two fundamental factors that should be taken into account are (1) the cost per cubic yard of material excavated and (2) the job conditions under which the shovel will operate.

In estimating the cost per cubic yard, one should consider the following factors:

1. The size of the job; a job involving a large quantity of material may justify the higher cost of a larger shovel.
2. The cost of transporting the machine; a large shovel will involve more cost than a small one.
3. The combined cost of drilling, blasting, and excavating; for a large shovel, these costs may be less than for a small shovel, as a large machine will handle more massive rocks than a small one. Large shovels may permit savings in drilling and blasting cost.

These job conditions should be considered in selecting the size of a shovel:

1. If the material is hard to excavate, the bucket of the large shovel that exerts higher digging pressures will handle the material more easily.
2. If blasted rock is to be excavated, the large-size bucket will handle larger individual pieces.
3. The size of available hauling units should be considered in selecting the size of a shovel (see Fig. 8.4). If small hauling units must be used, the size of the shovel should be small, whereas if large hauling units are available, a large shovel should be used. The haul unit capacity should be approximately five times excavator bucket size. This ratio provides for an

FIGURE 8.4 I Size of haul truck matched to hydraulic front shovel bucket size.

efficient balance between sizes. Planning for this will eliminate the wasted cycle time of dealing with partially loaded buckets.

Manufacturers' specifications should always be consulted for exact values of machine clearances. The maximum bucket dumping height is especially important when the shovel is loading haul units.

SHOVEL PRODUCTION

There are four elements in the production cycle of a shovel:

1. Load bucket.
2. Swing with load.
3. Dump load.
4. Return swing.

It should be noted that a shovel does not travel during the digging and loading cycle. Travel is limited to moving into or along the face as the excavation progresses. One study of shovel travel found that on the average it was necessary to move after about 20 bucket loads. This movement into the excavation took an average of 36 sec.

Typical cycle element times under average conditions, for 3-to 5-cy-size shovels, are

1. Load bucket   7–9 sec.
2. Swing with load   4–6 sec.

3. Dump load   2–4 sec.

4. Return swing   4–5 sec.

The actual production of a shovel is affected by numerous factors, including the:

1. Class of material.

2. Height of cut.

3. Angle of swing.

4. Operator skill.

5. Condition of the shovel.

6. Haul-unit exchange.

7. Size of hauling units.

8. Handling of oversize material.

9. Cleanup of loading area.

Haul-unit exchange refers to the total time required for a loaded truck to clear its loading position under the excavator and for the next empty truck to be positioned for loading.

When handling shot rock, carefully evaluate the amount of oversize material to be moved. A machine with a bucket whose bite width and pocket are satisfactory for the average-size pieces may spend too much time handling individual oversize pieces. A larger bucket, or a larger machine, or changing the blasting pattern should be considered when there is a large percentage of oversize material.

The use of auxiliary equipment in the loading area, such as a dozer, can reduce cleanup delays. Control of haul units and operator breaks are within the control of field management.

The capacity of a bucket is based on its heaped volume. This would be loose cubic yards (lcy). To obtain the bank-measure volume of a bucket when considering a particular material, the average loose volume should be divided by 1 plus the material's swell. For example, if a 2-cy bucket, excavating material whose swell is 25%, will handle an average loose volume of 2.25 lcy, the bank-measure volume will be $\frac{2.25}{1.25} = 1.8$ bcy. If this shovel can make 2.5 cycles per min that includes no allowance for lost time, the output will be $2.5 \times 1.8 = 4.5$ bcy per min, or 270 bcy per hour. This is an ideal production. Ideal production is based on digging at optimum height with a 90° swing and no delays.

HEIGHT OF CUT EFFECT ON SHOVEL PRODUCTION

A loose, flowing material will fill a digging bucket in a shorter sweep up the embankment than will a chunky material. If the height of the face from which a shovel is excavating material is too shallow, it will be difficult or impossible to fill the bucket in one pass up the face. The operator will have a choice of making more than one pass to fill the bucket, which will increase the time per cycle, or with each cycle, the operator may carry a partly filled bucket to the hauling unit. In either case, the effect will be to reduce the production of the shovel.

If the height of the face is greater than the minimum required for filling the bucket, the operator is presented with three options. The depth of bucket penetration into the face may be reduced in order to fill the bucket in one full stroke. This will increase the time for a cycle. The operator may maneuver the bucket so as to begin digging above the base of the face, and then remove the lower portion of the face later. Or the bucket may be run up the full height of the face and the excess earth is allowed to spill down to the bottom. This spillage will have to be picked up later. The choice of any one of the procedures will result in lost time, based on the time required to fill the bucket when digging at optimum height.

The PCSA has published findings on the optimum height of cut based on data from studies of small cable-operated shovels (see Table 8.2). In the table, the percent of optimum height of cut is obtained by dividing the actual height of cut by the optimum height for the given material and bucket, then multiplying the result by 100. Thus, if the actual height of cut is 6 ft and the optimum height is 10 ft, the percentage of optimum height of cut is $\frac{6}{10} \times 100 = 60\%$. In most cases, other types of excavators, track or rubber tire loaders, have replaced the small shovels of the PCSA studies. But some general guidelines can still be gleaned from the data.

TABLE 8.2 | Factors for height of cut and angle of swing effect on shovel production

Percentage of	Angle of swing (degrees)						
optimum depth	45	60	75	90	120	150	180
40	0.93	0.89	0.85	0.80	0.72	0.65	0.59
60	1.10	1.03	0.96	0.91	0.81	0.73	0.66
80	1.22	1.12	1.04	0.98	0.86	0.77	0.69
100	1.26	1.16	1.07	1.00	0.88	0.79	0.71
120	1.20	1.11	1.03	0.97	0.86	0.77	0.70
140	1.12	1.04	0.97	0.91	0.81	0.73	0.66
160	1.03	0.96	0.90	0.85	0.75	0.67	0.62

Optimum height of cut ranges from 30 to 50% of maximum digging height, with the lower percentage being representative of easy-to-load materials, such as loam, sand, or gravel. Hard-to-load materials, sticky clay or blasted rock, necessitate a greater optimum height, in the range of 50% of the maximum digging height value. Common earth would require slightly less than 40% of the maximum digging height.

ANGLE OF SWING EFFECT ON SHOVEL PRODUCTION

The angle of swing of a shovel is the horizontal angle, expressed in degrees, between the position of the bucket when it is excavating and the position where it discharges the load. The total time in a cycle includes digging, swinging to the dumping position, dumping, and returning to the digging position. If the angle of swing is increased, the time for a cycle will be increased, whereas if the angle of swing is decreased, the time for a cycle will be decreased. Ideal production of a shovel is based on operating at a 90° swing and optimum height of cut. The effect of the angle of swing on the production of a shovel is illustrated in Table 8.2. The ideal production should be multiplied by the proper conversion factor to correct the production for any given height and swing angle.

Proper planning can control the necessary angle of swing for the shovel's excavation cycle. For example, if a shovel that is digging at optimum depth has the angle of swing reduced from 90 to 60°, the production will be increased by 16%.

EXAMPLE 8.1

A shovel with a 5-cy heaped capacity bucket is loading poorly blasted rock, a situation similar to that shown in Fig. 8.2. It is working a 12-ft-high face. The shovel has a maximum rated digging height of 34 ft. The haul units can be positioned so the swing angle is only 60°. What is a conservative ideal loose cubic yard production if the ideal cycle time is 21 sec?

Step 1. Size of bucket, 5 cy

Step 2. Bucket fill factor (Table 8.1) for poorly blasted rock: 85 to 100%, use 85%, conservative

Step 3. Cycle time given 21 sec

Average height of excavation 12 ft

Optimum height for this machine and material (poorly blasted rock):

$$0.50 \times 34 \text{ ft (max. height)} \approx 17 \text{ ft}$$

Percentage optimum height, $\dfrac{12 \text{ ft}}{17 \text{ ft}} = 0.71$

Correcting for height and swing from Table 8.2, by interpolation, 1.08

Step 4. Efficiency factor—ideal production, 60-min hour

Step 5. Production will be in lcy

Step 6. Ideal production per 60-min hour

$$\frac{3600 \text{ sec/hr} \times 5 \text{ cy} \times 0.85 \text{ (fill factor)} \times 1.08 \text{ (height-swing factor)}}{21 \text{ sec/cycle}} = 787 \text{ lcy/hr}$$

Although the information given in the text and in Tables 8.1 and 8.2 is based on extensive field studies, the reader is cautioned against using it too literally without adjusting for conditions that will probably exist on a particular project.

PRODUCTION EFFICIENCY FACTOR

As every contractor knows, no two projects are alike. There are certain conditions at every job over which the contractor has no control. These conditions must be considered in estimating the probable production of any piece of equipment including shovels.

A shovel may operate in a large, open quarry situation with a firm well-drained floor, where trucks can be spotted on either side of the machine to eliminate lost time waiting for haul units. The terrain of the natural ground may be uniformly level so that the height of cut will always be close to optimum. The haul road is not affected by climatic conditions, such as rains. These would be excellent work conditions.

Another shovel may be used to excavate material for a highway cut through a hill. The height of cut may vary from zero to considerably more than the optimum height. The sides of the cut must be carefully sloped. The cut may be so narrow that a loaded truck must be moved out before an empty truck can be backed into the loading position. As the truck must be spotted behind the shovel, the angle of swing will approximate 180°. The floor of the cut may be muddy, which will delay the movement of the trucks. On a project of this type, the shovel will have a significantly reduced productivity.

Besides project-specific factors, there are contractor or management factors that will suppress or enhance production efficiency. The advanced planning and foresight of the contractor in controlling the work and organizing the job will affect production. Factors to always be considered and addressed are

1. Maintenance of equipment.
2. Haul-road condition.
3. Loading area layout.
4. Haul-unit sizing and number.
5. Competency of field management.

The estimator must consider all of these factors and decide upon an efficiency factor with which to adjust peak production. Experience and good judgment are essential to selecting the appropriate efficiency factor. Transportation Research Board (TRB) studies have shown that actual production times for shovels used in highway construction excavation operations are 50 to 75% of available working time. Therefore, production efficiency is only 30 to 45 min per hour. The best estimating method is to develop specific historical data by type of machine and project factors. But information such as the TRB study that presents data from thousands of shovel cycles provides a good benchmark for selecting an efficiency factor.

EXAMPLE 8.2

A shovel with a 3-cy heaped capacity bucket is loading well-blasted rock on a highway project. The average face height is expected to be 22 ft. The shovel has a maximum rated digging height of 30 ft. Most of the cut will require a 140° swing of the shovel to load the haul units. What is a conservative production estimate in bank cubic yards?

Step 1. Size of bucket, 3 cy

Step 2. Bucket fill factor (Table 8.1) for well-blasted rock 100 to 110%, use 100%, conservative estimate

Step 3. Cycle element times

Load   9 sec (because of material, rock)

Swing loaded   4 sec (small machine, 3 cy)

Dump   4 sec (into haul units)

Swing empty   4 sec (small machine, 3 cy)

Total time   21 sec

Average height of excavation 22 ft

Optimum height: 50% of max.: 0.5 × 30 ft = 15 ft

Percent optimum height: $\dfrac{22\ \text{ft}}{15\ \text{ft}} \times 100 = 147\%$

Height and swing factor: From Table 8.2, for 147%, by interpolation, 0.73

Step 4. Efficiency factor: If the TRB information were used, the efficiency would be 30 to 45 working minutes per hour. Assume 30 min for a conservative estimate.

Step 5. Class of material, well-blasted rock, swell 60% (Table 4.3)

Step 6. Production:

$$\frac{3600\ \text{sec/hr} \times 3\ \text{cy} \times 1.0 \times 0.73}{21\ \text{sec/cycle}} \times \frac{30\ \text{min}}{60\ \text{min}} \times \frac{1}{(1 + 0.6)} = 117\ \text{bcy/hr}$$

HOES

GENERAL INFORMATION

Hoes are used primarily to excavate below the natural surface of the ground on which the machine rests. A hoe is sometimes referred to by other names, such as backhoe or back shovel. Hoes are adept at excavating trenches and pits for basements, and the smaller machines can handle general grading work. Because of their positive bucket control they are superior to draglines in operating on close-range work and loading into haul units (see Fig. 8.5).

Wheel-mounted hydraulic hoes (see Fig. 8.6) are available with buckets up to $1\frac{1}{2}$ cy. Maximum digging depth for the larger machines is about 25 ft. With all four outriggers down, the large machines can handle 10,000-lb loads at a 20-ft radius. These are not production excavation machines. They are designed for mobility and general-purpose work.

Basic Parts and Operation of a Hoe

The basic parts and operating ranges of a hoe are illustrated in Fig. 8.7. Table 8.3 gives representative dimensions and clearances for hydraulic crawler-mounted hoes. Buckets are available in varying widths to suit the job requirements.

FIGURE 8.5 | Crawler-mounted hydraulic hoe loading an off-highway truck.

FIGURE 8.6 | Wheel-mounted hydraulic hoe.

Penetration force into the material being excavated is achieved by the stick cylinder and the bucket cylinder. Maximum crowd force is developed when the stick cylinder operates perpendicular to the stick (see Fig. 8.8). The ability to break material loose is best at the bottom of the arc because of the geometry of the boom, stick, and bucket and the fact that at that point, the hydraulic cylinders exert the maximum force drawing the stick in and curling the bucket.

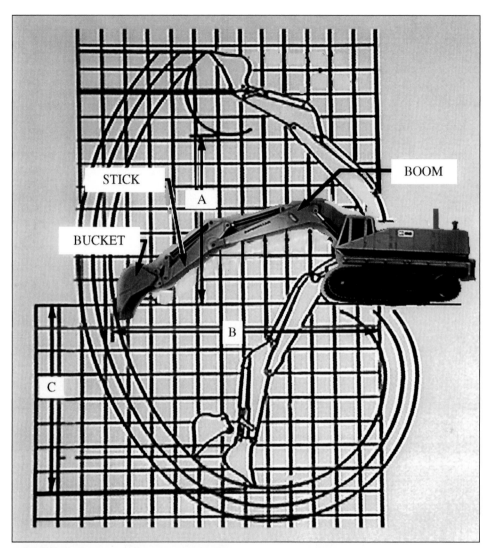

FIGURE 8.7 I Basic parts and operating ranges of a hydraulic hoe: A, dumping height; B, digging reach; C, maximum digging depth.

BUCKET RATING FOR HYDRAULIC HOES

Hoe buckets are rated like shovel buckets by PCSA and SAE standards using a 1:1 angle of repose for evaluating heaped capacity (see Fig. 8.9). Buckets should be selected based on the material being excavated. The hoe can develop high penetration forces. By matching bucket width and bucket tip radius to the resistance of the material, full advantage can be taken of the hoe's potential. For easily excavated materials, wide buckets should be used. When excavating rocky

TABLE 8.3 | Representative dimensions, loading clearance, and lifting capacity hydraulic crawler hoes

Size bucket (cy)	Stick length (ft)	Maximum reach at ground level (ft)	Maximum digging depth (ft)	Maximum loading height (ft)	Lifting capacity at 15 ft			
					Short stick		Long stick	
					Front (lb)	Side (lb)	Front (lb)	Side (lb)
$\frac{3}{8}$	5–7	19–22	12–15	14–16	2,900	2,600	2,900	2,600
$\frac{3}{4}$	6–9	24–27	16–18	17–19	7,100	5,300	7,200	5,300
1	5–13	26–33	16–23	17–25	12,800	9,000	9,300	9,200
$1\frac{1}{2}$	6–13	27–35	17–21	18–23	17,100	10,100	17,700	11,100
2	7–14	29–38	18–27	19–24	21,400	14,500	21,600	14,200
$2\frac{1}{2}$	7–16	32–40	20–29	20–26	32,600	21,400	31,500	24,400
3	10–11	38–42	25–30	24–26	32,900*	24,600*	30,700*	26,200*
$3\frac{1}{2}$	8–12	36–39	23–27	21–22	33,200*	21,900*	32,400*	22,000*
4	11	44	29	27	47,900*	33,500*		
5	8–15	40–46	26–32	25–26	34,100†	27,500†	31,600†	27,600†

*Lifting capacity at 20 ft.
†Lifting capacity at 25 ft.

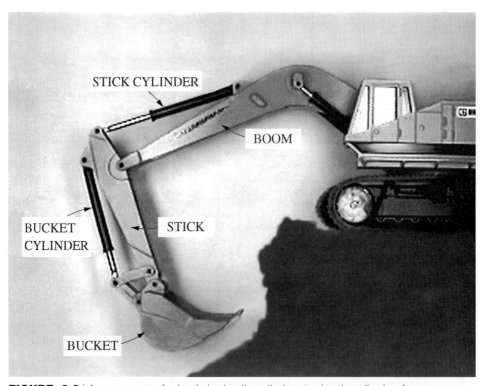

FIGURE 8.8 | Arrangement of a hoe's hydraulic cylinders to develop digging forces.

material or blasted rock, a narrow bucket with a short tip radius is best. In utility work, the width of the required trench may be the critical consideration. Fill factors for hydraulic hoe buckets are presented in Table 8.4.

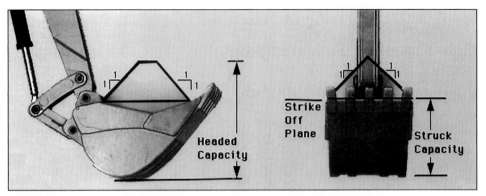

FIGURE 8.9 I Hydraulic hoe bucket capacity rating dimensions.

TABLE 8.4 I Fill factors for hydraulic hoe buckets

Material	Fill factor* (%)
Moist loam/sandy clay	100–110
Sand and gravel	95–110
Rock—poorly blasted	40–50
Rock—well blasted	60–75
Hard, tough clay	80–90

*Percentage of heaped bucket capacity
Reprinted courtesy of Caterpillar Inc.

SELECTING A HOE

In the selection of a hoe for use on a project these factors must be considered:

1. Maximum excavation depth required.
2. Maximum working radius required for digging and dumping.
3. Maximum dumping height required.
4. Hoisting capability required (where applicable, i.e., handling pipe).

Multipurpose Tool Platform

It must also be recognized that the hydraulic hoe has evolved from a single-purpose excavating machine into a versatile, multipurpose tool (see Fig. 8.10). It is a platform designed for literally hundreds of applications.

A quick-coupler enables the hoe to change attachments and perform a variety of tasks in rapid succession. There are rock drills (see Fig. 12.12), earth augers, grapples for land clearing, impact hammers and demolition jaws, and vibratory plate compactors, all of which can easily be attached to the stick in place of the normal excavation bucket. Additionally, there is a broad range of special purpose buckets such as trapezoidal buckets for digging and cleaning irrigation ditches, round-bottom buckets for cast-in-place pipe operations, and

(a) Wheel-mounted hydraulic hoe with clamshell bucket   **(b) Hydraulic hoe with land-clearing grapple**

(c) Special round-bottom hoe bucket   **(c) Demolition jaws attachment**

FIGURE 8.10 | The hydraulic hoe as a multipurpose tool platform.

clamshell buckets for vertical excavation of footings that enhance this machine versatility.

Rated Hoist Load

In storm drain and utility work, the hoe can perform the trench excavation and handle the pipe, eliminating a second machine (see Fig. 8.11). Manufacturers provide machine-lifting capacities (rated hoist load) based on reach from the

FIGURE 8.11 | Crawler-mounted hydraulic hoe handling reinforced concrete pipe.

center of gravity of the bucket load to either the front or side of the tract rails. Some typical data are provided in Table 8.3.

Rated hoist load is typically established based on these guidelines:

1. Rated hoist load shall not exceed 75% of the tipping load.
2. Rated hoist load shall not exceed 87% of the excavator's hydraulic capacity.
3. Rated hoist load shall not exceed the machine's structural capabilities.

HOE PRODUCTION

The same elements that affect shovel production are applicable to hoe excavation operations. Hoe cycle times are approximately 20% longer in duration than those of a similar-size shovel because the hoisting distance is greater as the boom and stick must be fully extended to dump the bucket.

Optimum depth of cut for a hoe will depend on the type of material being excavated and bucket size and type. As a rule, the optimum depth of cut for a hoe is usually in the range of 30 to 60% of the machine's maximum digging depth. Table 8.5 presents cycle times for hydraulic track hoes based on bucket size and average conditions. No one has developed tables to relate average hoe cycle time to variations in depth of cut and horizontal swing. Therefore, when using Table 8.5, consideration must be given to those two factors when deciding upon a load bucket time and the two swing times.

TABLE 8.5 I Excavation cycle times for hydraulic crawler hoes under average conditions.*

Bucket size (cy)	Load bucket (sec)	Swing loaded (sec)	Dump bucket (sec)	Swing empty (sec)	Total cycle (sec)
<1	5	4	2	3	14
$1-1\frac{1}{2}$	6	4	2	3	15
$2-2\frac{1}{2}$	6	4	3	4	17
3	7	5	4	4	20
$3\frac{1}{2}$	7	6	4	5	22
4	7	6	4	5	22
5	7	7	4	6	24

*Depth of cut 40 to 60% of maximum digging depth; swing angle 30 to 60°; loading haul units on the same level as the excavator.

The basic production formula for a hoe used as an excavator is

$$\text{Hoe (excavation) production} = \frac{3600 \text{ sec} \times Q \times F}{t}$$

$$\times \frac{E}{60\text{-min hour}} \times \frac{1}{\text{volume correction}} \qquad [8.2]$$

where

Q = heaped bucket capacity in lcy
F = bucket fill factor for hoe buckets
t = cycle time in seconds
E = efficiency (in minutes per hour)

volume correction for loose volume to bank volume, $\dfrac{1}{1 + \text{swell factor}}$; for

loose volume to tons, $\dfrac{\text{loose unit weight, lb}}{2{,}000 \text{ lb/ton}}$

EXAMPLE 8.3

A crawler hoe having a $3\frac{1}{2}$-cy bucket is being considered for use on a project to excavate very hard clay from a borrow pit. The clay will be loaded into trucks having a loading height of 9 ft 9 in. Soil-boring information indicates that below 8 ft, the material changes to an unacceptable silt material. What is the estimated production of the hoe in cubic yards bank measure, if the efficiency factor is equal to a 50-min hour?

Step 1.   Size of bucket, $3\frac{1}{2}$ cy

Step 2.   Bucket fill factor (Table 8.4), hard clay 80 to 90%; use average 85%

Step 3. Typical cycle element times

Optimum depth of cut is 30 to 60% of maximum digging depth. From Table 8.3 for a $3\frac{1}{2}$-cy-size hoe maximum digging depth is 23 to 27 ft

Depth of excavation, 8 ft

$$\frac{8\text{ ft}}{23\text{ ft}} \times 100 = 34\% \geq 30\%; \text{ okay}$$

$$\frac{8\text{ ft}}{27\text{ ft}} \times 100 = 30\% \geq 30\%; \text{ okay}$$

Therefore, under average conditions and for a $3\frac{1}{2}$-cy-size hoe cycle times from Table 8.5 would be

1. Load bucket	7 sec	very hard clay
2. Swing with load	6 sec	load trucks
3. Dump load	4 sec	load trucks
4. Return swing	5 sec	
Cycle time	22 sec	

Step 4. Efficiency factor, 50-min hour

Step 5. Class of material, hard clay, swell 35% (Table 4.3)

Step 6. Probable production:

$$\frac{3600\text{ sec/hr} \times 3\frac{1}{2}\text{ cy} \times 0.85}{22\text{ sec/cycle}} \times \frac{50\text{ min}}{60\text{ min}} \times \frac{1}{(1+0.35)} = 300\text{ bcy/hr}$$

Check maximum loading height to ensure the hoe can service the trucks, from Table 8.3, 21 to 22 ft

$$9\text{ ft }9\text{ in.} < 21\text{ okay}$$

As stated, hoe cycle times are usually of greater duration than shovel times. Part of the reason for this increase in cycle time is that after making the cut, the hoe bucket must be raised above the ground level to load a haul unit or to get above a spoil pile. If the trucks can be spotted on the floor of the pit, the bucket will be above the truck when the cut is completed (see Fig. 8.5). Then it would not be necessary to raise the bucket any higher before swinging and dumping the load. Every movement of the bucket equals increased cycle time. The spotting of haul units below the level of the hoe will increase production. A study by the second author found a 12.6% total cycle-time savings between loading at the same level and working the hoe from a bench above the haul units [3].

In trenching operations, the volume of material moved is usually not the question. The primary concern is to match the hoe's ability to excavate linear feet of trench per unit of time with the pipe-laying production.

LOADERS

GENERAL INFORMATION

Loaders are used extensively in construction work to handle and transport bulk material, such as earth and rock; to load trucks; to excavate earth; and to charge aggregate bins at asphalt and concrete plants. The loader is a versatile piece of equipment designed to excavate at or above wheel/track level. The hydraulic-activated lifting system exerts maximum breakout force with an upward motion of the bucket. It does not require other equipment to level, smooth, or clean up the area in which it has been working.

Types and Sizes

Classified on the basis of running gear, there are two types of loaders: (1) the crawler-tractor-mounted type (see Fig. 8.12) and (2) the wheel-tractor-mounted type (see Fig. 8.13). They may be further grouped by the capacities of their buckets or the weights that the buckets can lift. Wheel loaders may be steered by the rear wheels, or they may be articulated to permit steering, as indicated in Fig. 8.14. To increase stability during load lifting, the tracks of crawler loaders are usually longer and wider than those found on tractors. Figure 8.14 gives important dimensional specifications.

LOADER BUCKETS/ATTACHMENTS

The most common loader attachments are a shovel-type bucket and a forklift. The loader's hydraulic system provides the control necessary for operating these attachments.

FIGURE 8.12 | Track-type loader.

FIGURE 8.13 | Wheel-tractor loader.

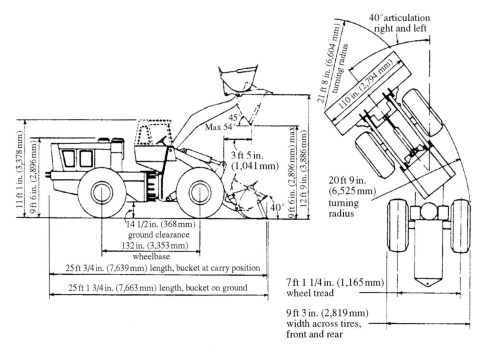

FIGURE 8.14 | Articulated wheel loader.

Source: International Harvester Company.

Buckets

The most common buckets are the one-piece conventional type, *general-purpose;* the hinged-jaw, *multipurpose;* and the heavy-duty, *rock bucket.* The selected bucket is attached to the tractor by a push frame and lift arms.

> *General purpose.* The general-purpose (one-piece) bucket is made of heavy-duty, all-welded steel. The replaceable teeth or cutting edge are bolted onto the bucket proper. Bolt-on-type replaceable teeth are provided for excavation of medium-type materials.
>
> *Multipurpose.* The segmented (two-piece) hinged-jawed bucket is made of heavy-duty, all-welded steel. It has bolted replaceable cutting edges. Bolt-on-type replaceable teeth are provided for excavation of medium-type materials. The two-piece bucket provides capabilities not available with a single-piece bucket. It enables the loader to also be used as a dozer, and to grab material.
>
> *Rock.* The rock bucket is of one piece, heavy-duty construction, having a protruding V-shaped cutting edge (see Fig. 8.15). This protruding edge can be used for prying up and loosening shot rock.
>
> *Forklift.* The forklift is attached to the loader in place of a bucket. Designed for material handling, it is made of steel with two movable tines.
>
> *Other.* Other buckets and attachments available include side dump buckets for use in cramped quarters; demolition buckets; plow blades for snow removal; brush rakes for clearing applications; heavy-duty sweeper brooms; and front booms, designed for lifting and moving sling loads.

Fill Factors for Loaders

The heaped capacity of a loader bucket is based on SAE standards. That standard specifies a 2:1 angle of repose for the material above the struck load. This repose angle (2:1) is different from that specified by both SAE and PCSA for shovel and hoe buckets, (1:1). The fill factor correction for a loader bucket (see Table 8.6) adjusts heaped capacity based on the type of material being handled and the type of loader, wheel or track. Mainly because of the relationship between traction and developed breakout force, the bucket fill factors for the two types of loaders are different.

Operating Loads

Once the bucket volumetric load is determined, a check must be made of payload weight. Unlike a shovel or hoe, to position the bucket to dump, a loader must maneuver and travel with the load. A shovel or hoe simply swings about its center pin and does not require travel movement when moving the bucket from loading to dump position. SAE has established operating load weight limits for loaders. A wheel loader is limited to an operating load, by weight, that is less than 50% of rated full-turn static tipping load considering the combined weight of the bucket and the load, measured from the center of gravity of the extended bucket

FIGURE 8.15 | Wheel-tractor loader with V-shaped rock bucket.

TABLE 8.6 | Bucket fill factors for wheel and track loaders

Material	Wheel loader fill factor (%)	Track loader fill factor (%)
Loose material		
Mixed moist aggregates	95–100	95–100
Uniform aggregates:		
up to $\frac{1}{8}$ in.	95–100	95–110
$\frac{1}{8} - \frac{3}{8}$ in.	90–95	90–110
$\frac{1}{2} - \frac{3}{4}$ in.	85–90	90–110
1 in. and over	85–90	90–110
Blasted rock		
Well blasted	80–95	80–95
Average	75–90	75–90
Poor	60–75	60–75
Other		
Rock-dirt mixtures	100–120	100–120
Moist loam	100–110	100–120
Soil	80–100	80–100
Cemented materials	85–95	85–100

Reprinted courtesy of Caterpillar Inc.

at its maximum reach, with standard counterweights and nonballasted tires. In the case of track loaders, the operating load is limited to less than 35% of static tipping load. The term "operating capacity" is sometimes used interchangeably for operating load. Most buckets are sized based on 3,000 lb per lcy material.

OPERATING SPECIFICATIONS

Representative operating specifications for a wheel loader furnish information such as that listed below:

> Engine flywheel hp 119 at 2,300 rpm
>> Speeds, forward and reverse:
>
> Low   0–3.9 mph
>
> Intermediate   0–11.1 mph
>
> High   0–29.5 mph
>
> | | |
> |---|---|
> | Operating load (SAE) | 6,800 lb |
> | Tipping load, straight ahead | 17,400 lb |
> | Tipping load, full turn | 16,800 lb |
> | Lifting capacity | 18,600 lb |
> | Breakout force, maximum | 30,000 lb |

Tables 8.7 and 8.8 present the operating specifications across the ranges of commonly available wheel loaders and track loaders.

PRODUCTION RATES FOR WHEEL LOADERS

Two critical factors to be considered in choosing a loader are (1) the type of material and (2) the volume of material to be handled. Wheel loaders are excellent machines for soft to medium-hard material. However, wheel loader production rates decrease rapidly when used in medium to hard material. Another factor to consider is the height that the material must be lifted. To be of value in loading trucks, the loader must be able to reach over the side of the truck's dump bed. A wheel loader attains its highest production rate when working on flat, smooth-surfaced areas with enough space to maneuver. In poor underfoot conditions or when there is a lack of space to maneuver efficiently, other equipment may be more effective.

Wheel loaders work in repetitive cycles, constantly reversing direction, loading, turning, and dumping. The production rate for a wheel loader will depend on the

1. Fixed time required to load the bucket, maneuver with four reversals of direction, and dump the load.
2. Time required to travel from the loading to the dumping position.
3. Time required to return to the loading position.
4. The actual volume of material hauled each trip.

Table 8.9 gives fixed cycle times for both wheel and track loaders. Figure 8.16 illustrates a typical loading situation. Because wheel loaders are more maneuverable and can travel faster on smooth haul surfaces, their production rates should be higher than those of track units under favorable conditions requiring longer maneuver distances.

TABLE 8.7 I Representative specifications for wheel loaders

Size, heaped bucket capacity (cy)	Bucket dump clearance (ft)	Static tipping load, at full turn (lb)	Maximum forward speed				Maximum reverse speed				Raise/ dump/ lower cycle (sec)
			First (mph)	Second (mph)	Third (mph)	Fourth (mph)	First (mph)	Second (mph)	Third (mph)	Fourth (mph)	
1.25	8.4	9,600	4.1	7.7	13.9	21	4.1	7.7	13.9	—	9.8
2.00	8.7	12,700	4.2	8.1	15.4	—	4.2	8.3	15.5	—	10.7
2.25	9.0	13,000	4.1	7.5	13.3	21	4.4	8.1	14.3	23	11.3
3.00	9.3	17,000	5.0	9.0	15.7	26	5.6	10.0	17.4	29	11.6
3.75	9.3	21,000	4.6	8.3	14.4	24	5.0	9.0	15.8	26	11.8
4.00	9.6	25,000	4.3	7.7	13.3	21	4.9	8.6	14.9	24	11.6
4.75	9.7	27,000	4.4	7.8	13.6	23	5.0	8.9	15.4	26	11.5
5.50	10.7	37,000	4.0	7.1	12.4	21	4.6	8.1	14.2	24	12.7
7.00	10.4	50,000	4.0	7.1	12.7	22	4.6	8.2	14.5	25	16.9
14.00	13.6	98,000	4.3	7.6	13.0	—	4.7	8.3	14.2	—	18.5
23.00	19.1	222,000	4.3	7.9	13.8	—	4.8	8.7	15.2	—	20.1

TABLE 8.8 | Representative specifications for track loaders

Size, heaped bucket capacity (cy)	Bucket dump clearance (ft)	Static tipping load (lb)	Maximum forward speed (mph)	Maximum reverse speed (mph)	Raise/dump/ lower cycle (sec)
1.00	8.5	10,500	6.5	6.9	11.8
1.30	8.5	12,700	6.5	6.9	11.8
1.50	8.6	17,000	5.9*	5.9*	11.0
2.00	9.5	19,000	6.4*	6.4*	11.9
2.60	10.2	26,000	6.0*	6.0*	9.8
3.75	10.9	36,000	6.4*	6.4*	11.4

*Hydrostatic drive.

TABLE 8.9 | Fixed cycle times for loaders

Loader size, heaped bucket capacity (cy)	Wheel loader cycle time* (sec)	Track loader cycle time* (sec)
1.00–3.75	27–30	15–21
4.00–5.50	30–33	—
6.00–7.00	33–36	—
14.00–23.00	36–42	—

*Includes load, maneuver with four reversals of direction (minimum travel), and dump.

FIGURE 8.16 | Loading travel cycle for a loader.

When travel distance is more than minimum, it will be necessary to add a travel time to the fixed cycle time. For travel distances of less than 100 feet, a wheel loader should be able to travel, with a loaded bucket, at about 80% of its maximum speed in low gear and return empty at about 60% of its maximum speed in second gear. In the case of distances over 100 ft, return travel should be at about 80% of its maximum speed in second gear. If the haul surface is not well maintained, or is rough, these speeds should be reduced accordingly.

Consider a wheel loader with a $2\frac{1}{2}$-cy-heaped capacity bucket, handling well-blasted rock weighing 2,700 lb per lcy, for which the swell is 25%. This unit, equipped with a torque converter and a power-shift transmission, has these speed ranges, forward and reverse:

Low	0–3.9 mph
Intermediate	0–11.1 mph
High	0–29.5 mph

The average speeds [in feet per minute (fpm)] should therefore be about

Hauling, all distances	0.8×3.9 mph $\times$ 88 fpm per mph = 274 fpm
Returning, 0–100 ft	0.6×11.1 mph $\times$ 88 fpm per mph = 586 fpm
Returning over, 100 ft	0.8×11.1 mph $\times$ 88 fpm per mph = 781 fpm

The effect of haul-increased haul distance on production is shown by the following calculations provided below:

Cycle Time					
Haul distance (ft)	**25**	**50**	**100**	**150**	**200**
Fixed time	0.45	0.45	0.45	0.45	0.45
Haul time	0.09	0.18	0.36	0.55	0.73
Return time	0.04	0.09	0.13	0.19	0.26
Cycle time (min)	0.58	0.72	0.94	1.19	1.44
Trips per 50-min hour	86.2	69.4	53.2	42.0	34.7
Production (tons)*	262	210	161	127	105

*0.9 bucket fill factor.

The following example demonstrates the process for estimating loader production.

EXAMPLE 8.4

A 4-cy wheel loader will be used to load trucks from a quarry stockpile of processed aggregate having a maximum size of $1\frac{1}{4}$ in. The haul distance will be negligible. The aggregate has a loose unit weight of 3,100 lb per cy. Estimate the loader production in tons based on a 50-min hour efficiency factor. Use a conservative fill factor.

Step 1. Size of bucket, 4 cy

Step 2. Bucket fill factor (Table 8.6), aggregate over 1 in., 85 to 90%; use 85% conservative. Check tipping:

Load weight:

$$4 \text{ cy} \times 0.85 = 3.4 \text{ lcy}$$

3.4 lcy $\times$ 3,100 lb/lcy (loose unit weight of material) = 10,540 lb

From Table 8.7: 4-cy machine static tipping load at full turn is 25,000 lb

Therefore, operating load (50% static tipping at full turn) is

$$0.5 \times 25,000 \text{ lb} = 12,500 \text{ lb}$$

10,540 lb actual load < 12,500 lb operating load; therefore okay.

Step 3. Typical fixed cycle time (Table 8.9) 4-cy wheel loader, 30 to 33 sec; use 30 sec.

Step 4. Efficiency factor, 50-min hour.

Step 5. Class of material, aggregate 3,100 lb per lcy.

Step 6. Probable production:

$$\frac{3600 \text{ sec/hr} \times 4 \text{ cy} \times 0.85}{30 \text{ sec/cycle}} \times \frac{50 \text{ min}}{60 \text{ min}} \times \frac{3,100 \text{ lb/lcy}}{2,000 \text{ lb/ton}} = 527 \text{ ton/hr}$$

EXAMPLE 8.5

The loader in Example 8.4 will also be used to charge the aggregate bins of an asphalt plant that is located at the quarry. The one-way haul distance from the $1\frac{1}{4}$-in. aggregate stockpile to the cold bins of the plant is 220 ft. The asphalt plant uses 105 tons per hour of $1\frac{1}{4}$-in. aggregate. Can the loader meet this requirement?

Step 3. Typical fixed cycle time (Table 8.9) 4-cy wheel loader, 30 to 33 sec.; use 30 sec.

From Table 8.7: Travel speeds forward.

First, 4.3 mph; second, 7.7 mph; third, 13.3 mph

Travel speeds reverse

First, 4.9 mph; second, 8.6 mph; third, 14.9 mph

Travel loaded: 220 ft, because of short distance and time required to accelerate and brake, use 80% of first gear maximum speed.

$$\frac{4.3 \text{ mph} \times 80\% \times 88 \text{ fpm/mph}}{60 \text{ sec/min}} = 5.04 \text{ ft per sec}$$

Return empty: 220 ft; because of short distance and time required to accelerate and brake, use 80% of second gear maximum speed.

$$\frac{7.7 \text{ mph} \times 80\% \times 88 \text{ fpm/mph}}{60 \text{ sec/min}} = 9.03 \text{ ft per sec}$$

1.	Fixed time	30 sec	4-cy wheel loader
2.	Travel with load	44 sec	220 ft, 80% first gear
3.	Return travel	24 sec	220 ft, 80% second gear
	Cycle time	98 sec	

Step 6. Probable production:

$$\frac{3600 \text{ sec/hr} \times 4 \text{ cy} \times 0.85}{98 \text{ sec/cycle}} \times \frac{50 \text{ min}}{60 \text{ min}} \times \frac{3{,}100 \text{ lb/lcy}}{2{,}000 \text{ lb/ton}} = 161 \text{ ton/hr}$$

161 tons per hour > 105 tons per hour required

The loader will meet the requirement.

PRODUCTION RATES FOR TRACK LOADERS

The production rates for track loaders are determined in the same manner as for wheel loaders.

EXAMPLE 8.6

A track loader having the following specifications is used to load trucks from a bank of moist loam. This operation will require that the loader travel 30 ft for both the haul and return. Estimate the loader production in bcy based on a 50-min-hour efficiency factor. Use a conservative fixed cycle time.

Bucket capacity, heaped, 2-cy travel speed by gear		
Gear	**mph**	**fpm**
Forward		
First	1.9	167
Second	2.9	255
Third	4.0	352
Reverse		
First	2.3	202
Second	3.6	317
Third	5.0	440

Assume that the loader will travel at an average of 80% of the specified speeds in second gear, forward and reverse. The fixed time should be based on time studies for the particular equipment and job; Table 8.9 provides average times by loader size.

Step 1. Size of bucket, 2 cy

Step 2. Bucket fill factor (Table 8.6), moist loam, 100 to 120%; use average 110%. Check tipping:

Load weight:

$$2 \text{ cy} \times 1.10 = 2.2 \text{ lcy}$$

Unit weight moist loam (earth, wet) (Table 4.3) 2,580 lb lcy

$$2.2 \text{ lcy} \times 2{,}580 \text{ lb/lcy} = 5{,}676 \text{ lb}$$

From Table 8.8: 2-cy track machine static tipping load is 19,000 lb

Therefore, operating load (35% static tipping) is

$$0.35 \times 19{,}000 \text{ lb} = 6{,}650 \text{ lb}$$

5,676 lb actual load < 6,650 lb operating load, therefore okay.

Step 3. Typical fixed cycle time (Table 8.9) 2-cy track loader, 15 to 21 sec; use 21 sec to be conservative.

Travel loaded: 30 ft, use 80% of first gear maximum speed.

$$\frac{1.9 \text{ mph} \times 80\% \times 88 \text{ fpm/mph}}{60 \text{ sec/min}} = 2.229 \text{ ft per sec}$$

Return empty: 30 ft, use 60% (less than 100 ft) of second gear maximum speed.

$$\frac{2.9 \text{ mph} \times 60\% \times 88 \text{ fpm/mph}}{60 \text{ sec/min}} = 2.552 \text{ ft per sec}$$

1. Fixed time	30 sec	2-cy track loader
2. Travel with load	13 sec	30 ft, 80% first gear
3. Return travel	12 sec	30 ft, 60% second gear
Cycle time	55 sec	

Step 4. Efficiency factor, 50-min hour

Step 5. Class of material, moist loam, swell 25 (Table 4.3)

Step 6. Probable production:

$$\frac{3600 \text{ sec/hr} \times 2 \text{ cy} \times 1.1}{55 \text{ sec/cycle}} \times \frac{50 \text{ min}}{60 \text{ min}} \times \frac{1}{1.25} = 96 \text{ bcy/hr}$$

LOADER SAFETY

Wheel loaders must be operated carefully, since they are easy to overturn. With either type of loader, do not extend any portion of the body between the cab and the raised bucket arms. Loader operators should be aware of these unsafe practices:

1. Not wearing the safety belt.
2. Personnel working between wheels and frame while the engine is running.
3. Operating too close to the edge of a trench when backfilling.
4. Digging into banks or stockpiles creating overhangs, then working under the overhangs.
5. Riding with the bucket above axle height when traveling.
6. Forgetting to ground the bucket or set the parking brake before leaving the machine.

SPECIALTY EXCAVATORS

A variety of excavators are available for specialty applications. There are machines designed strictly for one work application, such as trenchers, but there are also general utility machines such as backhoe-loaders. If there is a requirement to excavate larger quantities of material from an extensive borrow area, the Holland Loader is designed for such mass excavation tasks. However, if the excavation is in a busy city street, the contractor should consider a vac excavator.

TRENCHING MACHINES

Trenching or ditching machines are designed for excavating trenches or ditches of considerable length and having a variety of widths and depths. The term "trenching machine," as used in this book, applies to the wheel-and-ladder-type machines. Most work is no wider than 2 ft or deeper than 7 ft, so 95% of the machines are in the 40- to 150-hp range. These machines are satisfactory for digging utility trenches for water, gas, and shoulder drains on highways. The larger machines dig wide and deep trenches necessary to lay large-diameter cross-country pipelines or water and sewage system main-distribution lines. The larger machines can have gross horsepower ratings of over 340.

There are fully hydrostatic trenchers having self-leveling tracks. A vertical trench can be maintained on uneven terrain having up to an 18.5% slope.

Trenchers provide relatively fast digging, with positive controls of depths and widths of trenches, reducing expensive finishing. Most large trenchers are crawler-mounted to increase their stability and to distribute the weight over a greater area.

WHEEL-TYPE TRENCHING MACHINES

Figure 8.17 illustrates a wheel-type trenching machine. These machines are available with maximum cutting depths exceeding 8 ft, with trench widths from 12 in. to approximately 60 in. Many are available with 25 or more digging speeds to permit the selection of the most suitable speed for almost any job condition.

The excavating part of the machine consists of a power-driven wheel, on which are mounted a number of removable buckets, equipped with cutter teeth. Buckets are available in varying widths to which side cutters may be attached when it is necessary to increase the width of a trench. The machine is operated by lowering the rotating wheel to the desired depth, while the unit moves forward slowly. The earth is picked up by the buckets and deposited onto an endless belt conveyor that can be adjusted to discharge the earth on either side of the trench.

Table 8.10 gives representative specifications for wheel-type trenching machines. These specifications do not necessarily include all machines that are available. The various trench widths for a given machine are obtained by using different bucket widths and installing side cutters.

FIGURE 8.17 | Wheel-type trenching machine.

TABLE 8.10 | Representative specifications for wheel-type trenching machines

Max. trench depth [ft (m)]	Trench width [in. (mm)]	Engine power [hp (kW)]	Wheel speed [fpm (m/sec)]	Travel speed [mph (km/hr)]	Digging speed [fpm (m/min)]
5.5 (1.67)	15–18–21 (380–450–532) 20–23–26 (507–583–660)	55 (41)	36–266 (0.18–1.35)	0.5–2.7 (0.8–4.3)	0.2–10 (0.06–0.30)
6.0 (1.82)	16–18–20 (405–457–517) 20–22–24 (507–559–610) 24–26–28 (610–660–710) 28–30 (710–760)	67 (50)	153–410 (0.78–2.08)	0.16–4.6 (0.26–7.4)	2.8–57.5 (0.08–17.4)
8.5 (2.58)	38–40 (965–1,015) 40–51 (1,015–1,290)	110 (82)	243 (1.23)	1.9 (3.1)	1.3–35.0 (0.42–10.8)

Wheel-type machines are especially suited to excavating trenches for water, gas, and oil pipelines and pipe drains that are placed in relatively shallow trenches. They may be used to excavate trenches for sewer pipes up to the maximum digging depths.

LADDER-TYPE TRENCHING MACHINES

Figure 8.18 illustrates a ladder-type trenching machine. By installing extensions to the ladders or booms, and by adding more buckets and chain links, it is possible to dig trenches in excess of 30 ft deep with large machines. Trench widths in excess of 12 ft may be dug. Most of these machines have booms whose lengths may be varied, thereby permitting a single machine to be used on trenches varying considerably in depth. This eliminates the need of owning a different machine for each depth range. A machine may have 30 or more digging speeds to suit the needs of a given job.

The excavating part of the machine consists of two endless chains that travel along the boom, to which there are attached cutter buckets equipped with teeth. In addition, shaft-mounted side cutters may be installed on each side of the boom to increase the width of a trench. As the buckets travel up the underside of the boom, they bring out earth and deposit it on a belt conveyor that discharges it along either side of the trench. As a machine moves over uneven ground it is possible to vary the depth of cut by adjusting the position, but not the length of the boom.

Table 8.11 gives representative specifications for ladder-type trenching machines. The various trench widths for a given machine are obtained by using different bucket widths and installing side cutters.

As can be seen from Table 8.11, ladder-type trenching machines have considerable flexibility with regard to trench depths and widths. However, the machines are not suitable for excavating trenches in rock or where large quantities of groundwater, combined with unstable soil, prevent the walls of a trench from

FIGURE 8.18 | Ladder-type trenching machine.

TABLE 8.11 | Representative specifications for ladder-type trenching machines

Max. trench depth [ft (m)]	Trench width [in. (mm)]	Engine power [hp (kW)]	Bucket speed [fpm (m/sec)]	Travel speed [mph (km/hr)]	Digging speed [fpm (m/min)]
4.5	6–8	47	245–538	0.7–3.4	2.2–21.8
(1.37)	(152–203)	(35)	(1.24–2.72)	(1.1–5.5)	(0.67–6.6)
8.5	16–36	55	96–225	1.4–3.2	0.5–13.8
(2.58)	(407–920)	(41)	(0.48–1.14)	(2.2–5.1)	(0.15–4.2)
12.5	16–42	74	135–542	1.4–3.2	0.3–9.7
(3.81)	(407–1,070)	(55)	(0.68–2.74)	(2.2–5.1)	(0.09–2.95)
15.0	18–54	90	103–168	1.7	0.7–15.5
(4.57)	(457–1,370)	(67)	(0.52–0.85)	(2.7)	(0.21–4.75)

remaining in place. If the soil, such as loose sand or mud, tends to flow into the trench, it may be desirable to adopt some other method of excavating the trench.

SELECTING SUITABLE EQUIPMENT FOR EXCAVATING TRENCHES

The choice of equipment to be used in excavating a trench will depend on

1. The job conditions.
2. The depth and width of the trench.
3. The class of soil.
4. The extent to which groundwater is present.
5. The width of the right-of-way for disposal of excavated earth.

If a relatively shallow and narrow trench is to be excavated in firm soil, the wheel-type machine is probably the most suitable. However, if the soil is rock that requires blasting, the most suitable excavator will be a hoe. If the soil is an unstable, water-saturated material, it may be necessary to use a hoe or a clamshell and let the walls establish a stable slope. If it is necessary to install solid sheeting to hold the walls in place, either a hoe or a clamshell that can excavate between the trench braces that hold the sheeting in place, will probably be the best equipment for the job.

EXAMPLE 8.7

Consider the selection of a machine to excavate a trench 24 ft deep and 10 ft wide in soil that is sufficiently firm to require only shoring to hold the walls in place. A trench of this size can be excavated with a ladder-type machine, provided that the length and height of the conveyor belt are adequate to dispose of the earth along one side of the trench. The cross-sectional area of the trench will be 240 sq ft (sf). If the loose earth has a 30% swell, the cross-section area of the spoil pile will be

$$240 \text{ sf} \times 1.3 = 312 \text{ sf}$$

If the excavated earth will repose with 1:1 side slopes, the pile will have a height of 17.6 ft and a base width of 35.2 ft. If a minimum of 4 ft of clear berm is required along the side of the trench, the end of the conveyor must have a height clearance of 17.6 ft and a length of approximately 27 ft, measured from the center of the trench. The casting effect on the earth, as it leaves the end of the conveyor belt, may permit the use of a slightly shorter conveyor. Unless the machine under consideration satisfies these clearances, it is probable that difficulties will be experienced in disposing of the earth.

TRENCHING MACHINE PRODUCTION

Many factors influence the production rates of trenching machines. These include the class of soil; depth and width of the trench; extent of shoring required; topography; climatic conditions; extent of vegetation such as trees, stumps, and roots; physical obstructions such as buried pipes, sidewalks, paved streets, buildings; and the speed with which the pipe can be placed in the trench. Any factor that may affect the progress of the work should be considered in estimating the probable digging speed of a trenching machine.

In laying oil and gas pipelines through open, level country, with no physical obstructions to interfere with the progress, it is possible to install in excess of 6,000 ft of pipe in an 8-hr day. This is equivalent to approximately 800 ft per hr, which is not excessive for a wheel-type machine. However, if a trench must be excavated into rock over rough terrain covered with heavy timber, it may not be possible to excavate more than a few hundred feet per day.

If a trench is dug for the installation of sewer pipe under favorable conditions, it is possible that the machine could dig 300 ft of trench per hour. However, an experienced pipe-laying crew may not be able to lay more than 25 joints of small-diameter pipe, 3 ft long, in an hour. Thus the speed of the machine will be limited to about 75 ft per hr regardless of its ability to excavate more trench. In estimating the probable rate of digging a trench, one must apply an appropriate operating factor to the speed at which the machine could dig if there were no interferences.

EXAMPLE 8.8

Estimate the probable average production rate, in feet per hour, in excavating a trench 36 in. wide, with a maximum depth of 12 ft, in hard, tough clay. The trench will be dug for the installation of a 21-in.-diameter sewer pipe that can be laid at a rate of approximately 30 ft per hr. An examination of the site along the trench reveals that obstructions will reduce the digging speed to approximately 60% of the theoretically possible speeds. This will require the application of an operating factor of 0.6 to the speed of the machine.

Table 8.11 presents information on a ladder-type machine with a maximum digging depth of 12.5 ft. Considering the class of soil and the depth and width of the trench, one finds that the maximum possible digging speed should be about 1 fpm, or 60 ft per hour. The application of the operating factor will reduce the average speed to 36 ft per hour. However, since only 30 ft of pipe can be laid per hour, this will be the controlling speed.

TRENCH SAFETY

Time and again noncompliance with trench safety guidelines and common sense results in lost-time injuries and loss of life from cave-ins and entrapments. The death rate for trench-related accidents is nearly double that for any other type of construction accident. The first line of defense against cave-ins is a basic knowledge of soil mechanics and of the terrain in question. A critical issue often neglected is previous disturbance of the material being excavated. OSHA regulations provide good guidance for eliminating these tragedies [2].

Any trench measuring 5 ft or more in depth must be sloped, shored, or shielded (see Fig. 8.19). *Sloping* is the most common method employed to protect workers in trenches. The trench walls are excavated in a V-shaped manner so that the angle of repose prevents the cave-in. Required slope angles vary depending on the specific soil type and moisture. *Benching* is a subsidiary class of sloping that involves the formation of "steplike" horizontal levels. Both sloping and benching require ample right of way. When the area is restricted, shoring or shielding is necessary. *Shoring* is a system that applies pressure against the excavation's wall to prevent collapse. *Shields* or trench boxes are designed to protect the workers, not the excavation from collapse.

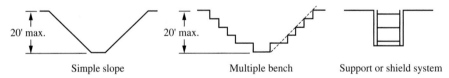

| Simple slope | Multiple bench | Support or shield system |

FIGURE 8.19 | Methods for protecting workers in trenches.

In trenches 4 ft or more deep, "A stairway, ladder, ramp or other safe means of egress shall be located . . . so as to require no more than 25 ft of lateral travel for employees"[2]. Spoil piles, tools, equipment, and materials must be kept at least 2 ft from the excavation's edge. A registered professional engineer must design the excavation protection in the case of trenches 20 ft or more in depth.

BACKHOE-LOADERS

The success of the backhoe-loader (see Fig. 8.20) comes from the number of tasks it can perform. This is not a high-production machine for any one task, but it provides flexibility to accomplish a variety of work tasks.

Loader functions. A backhoe-loader can function as a dozer for backfilling. It can also load trucks.

Backhoe functions. A variety of backhoe bucket options allow bucket-to-job matching. It can excavate the trench and handle placement of the pipe.

FIGURE 8.20 I Backhoe-loader using the rear hoe to excavate.

The hoe bucket can be replaced with a breaker or hammer turning the unit into a demolition machine.

Four-wheel drive. Because of its four-wheel-drive capability the backhoe-loader can work in unstable ground conditions.

Production. The backhoe-loader is an excellent excavator for digging loosely packed moist clay or sandy clay. It is average over a broad range of soil conditions from sand to hard clay. It is not really suitable for continuous high-impact diggings such as with hard clay or caliche.

HOLLAND LOADERS

A Holland Loader is an excavating unit mounted between two crawler tractors that operate in tandem from a single operator in the lead tractor. In continuous passes through excavation areas the loader carves material from the ground and belt loads it into hauling units (see Fig. 8.21). There are both vertical- and horizontal-cut Holland Loaders. Working with large-bottom dump trailers these loaders have reported production rates between 2,000 and 3,300 bcy per hour.

VAC EXCAVATORS

There are vac excavation units that make use of powerful vacuum systems to safely excavate around buried utilities (Fig. 8.22). There are similar hydrovac units that excavate using a combination of high-pressure water jets and vacuum

FIGURE 8.21 | Side-cut Holland Loader loading a bottom dump trailer.

FIGURE 8.22 | Truck-mounted vacuum excavate system.

systems. These units are excellent tools when it is necessary to work between previously buried electrical and communication conducts and utility pipelines. They are very good for city work where there is no space to dispose of excavated material. The larger units have debris body tanks capable of holding up to 15 cy. The excavated material can then be easily transported away for disposal without the need for rehandling.

SUMMARY

The same elements govern the production of all types of excavators. The first issue is how much material is actually loaded into the bucket. This is a function of the bucket volume and the type of material being excavated; a bucket fill factor is applied to the rated-heaped capacity of the bucket. The second issue is cycle time. In the case of shovels and hoes, cycle time is a function of height or depth of cut and swing angle. When travel distance is more than minimum for a loader, the travel cycle time is influenced by travel speed. Critical learning objectives include:

■ An ability to adjust bucket volume based on appropriate fill factors.
■ An ability to ascertain bucket cycle time as a function of machine type and job conditions to include angle of swing, depth or height of cut, and in the case of loaders, travel distance.
■ An ability to select an efficiency factor to use in the production calculation by considering both project-specific physical conditions and contractor management ability.

These objectives are the basis for the problems that follow.

PROBLEMS

8.1 What is a good efficiency factor for a front shovel working on highway construction projects?

8.2 A contractor has both a 3-cy and a 5-cy shovel in the equipment fleet. Select the minimum size shovel that will excavate 400,000 bcy of common earth in a minimum of 130 working days of 8 hr each. The average height of excavation will be 15 ft, and the average angle of swing will be 120°. The 3-cy shovel has a maximum digging height of 30 ft and the 5-cy machine's maximum digging height is 34 ft. The efficiency factor will be a 45-min hour. Appropriate-size haul units can be used with either shovel. How many days will it require to complete the work? (5-cy machine, 121 days to complete)

8.3 For each of the stated conditions determine the probable production expressed in cubic yards per hour bank measure for a shovel equipped with a 3-cy bucket. The shovel has a maximum digging height of 32 ft. Use a 30-min hour efficiency factor.

	Class of material			
Condition	**Common earth**	**Common earth**	**Rock-earth/ earth-gravel**	**Shale**
Height of excavation (ft)	12	8	12	19
Angle of swing (degrees)	90	120	60	130
Loading haul units	No	Yes	No	No

8.4 A shovel having a 5-cy bucket whose cost per hour, including the wages to an operator, is $96, will excavate well-blasted rock and load haul units under each of the stated conditions. The maximum digging height of the machine is 35 ft. Determine the cost per bank cubic yard for each condition.

Condition	(1)	(2)	(3)	(4)
Height of excavation (ft)	10	18	24	28
Angle of swing (degrees)	60	90	120	150
Efficiency factor (min-hr)	45	40	30	45

8.5   A crawler hoe having a $2\frac{1}{4}$-cy bucket and whose cost per hour, including the wages to an operator, is $67, will excavate and load haul units under each of the stated conditions. The maximum digging height of the machine is 20 ft. Determine the cost per bank cubic yard for each condition. (Condition 1: 300 bcy/hr, $0.223/bcy; Condition 2: 319 bcy/hr, $0.210/bcy; Condition 3: 128 bcy/hr, $0.523/bcy)

Condition	(1)	(2)	(3)
Material	Moist loam, earth	Sand and gravel	Rock, well blasted
Depth of excavation (ft)	12	16	14
Angle of swing (degrees)	60	80	120
Percent swell		14	
Efficiency factor (min-hr)	45	50	45

8.6   A crawler hoe having a 3-cy bucket and whose cost per hour, including the wages to an operator, is $86, will excavate and load haul units under each of the stated conditions. The maximum digging height of the machine is 25 ft. Determine the cost per bank cubic yard for each condition.

Condition	(1)	(2)	(3)	(4)
Material	Sandy clay	Hard clay	Rock, well blasted	Sand and gravel
Depth of excavation (ft)	10	18	15	20
Angle of swing (degrees)	60	90	90	150
Percent swell	20			13
Efficiency factor (min-hr)	45	50	50	55

8.7   A 3-cy wheel loader will be used to load trucks from a quarry stockpile of processed aggregates having a maximum size of $\frac{1}{4}$ in. The haul distance will be negligible. The aggregate has a loose unit weight of 2,950 lb per cy. Estimate the loader production in tons based on a 50-min hour efficiency factor. Use an aggressive cycle time and fill factor. (311 ton/hr)

8.8   A 7-cy wheel loader will be used to load a crusher from a quarry stockpile of blasted rock (average breakage) 180 ft away. The rock has a loose unit weight of 2,700 lb per cy. Estimate the loader production in tons based on a 50-min hour efficiency factor.

REFERENCES

1.   *Caterpillar Performance Handbook,* Caterpillar Inc., Peoria, Ill. (published annually) www.cat.com.

2. *Construction Standards for Excavations,* AGC publication No. 126, promulgated by the Occupational Safety and Health Administration, Associated General Contractors of America, Washington, D.C.

3. Lewis, Chris R., and Cliff J. Schexnayder, "Production Analysis of the CAT 245 Hydraulic Hoe," in *Proceedings Earthmoving and Heavy Equipment Specialty Conference,* American Society of Civil Engineers, February 1986, pp. 88–94.

4. Nichols, Herbert L., Jr., and David A. Day, *Moving the Earth, the Workbook of Excavation,* 4th ed., McGraw-Hill, New York, 1998.

5. O&K Inc. 8055 Troon Circle, Suite A, Austell, GA 30168-7849, www.oandk.com/.

6. O'Brien, James J., John A. Havers, and Frank W. Stubbs, Jr., *Standard Handbook of Heavy Construction,* 3rd ed., McGraw-Hill, New York, 1996.

7. Schexnayder, Cliff, Sandra L. Weber, and Brentwood T. Brooks, "Effect of Truck Payload Weight on Production," *Journal of Construction Engineering and Management, ASCE,* Vol. 125, No. 1, January–February, 1999, pp. 1–7.

8. Terex Corporation, 500 Post Road East, Suite 320, Westport, Conn., 06880, www.terex.com/.

9

Finishing Equipment

Finishing, finish grading, and fine grading are all terms used in reference to the process of shaping materials to the required line and grade. Graders are multipurpose machines used for finishing and shaping. The gradall is a utility machine that combines the operating features of the hoe, dragline, and motor grader. It is designed as a versatile machine for both excavation and finishing work. There are a variety of highly specialized trimming machines. These automatic trimmers use an automatic control system. A force-feed loader is a self-propelled machine designed for loading windrows of loose material.

INTRODUCTION

Finishing, finish grading, and fine grading are all terms used in reference to the process of shaping materials to the required line and grade specified in the contract documents. Finishing operations follow closely behind excavation (rough grade) operations or compaction of embankments. This includes finishing to prescribed grade those sections supporting structural members, and the smoothing and shaping of slopes. On many projects, graders are used as the finishing machine. In the case of long linear projects, such as roads and airfields, there are special trimming machines to accomplish the finishing under the pavement sections.

GRADERS

GENERAL INFORMATION

Graders (see Fig. 9.1) are multipurpose machines used for finishing, shaping, bank sloping, and ditching. They are also used for mixing, spreading, side casting, leveling and crowning, light stripping operations, general construction, and dirt road maintenance. A grader's primary purpose is cutting and moving material with the moldboard. They are restricted to making shallow cuts in medium-hard materials;

FIGURE 9.1 I Grader with rear ripper leveling a building site.

they should not be used for heavy excavation. A grader can move small amounts of material but cannot perform dozer-type work because of the structural strength and location of its moldboard.

Graders are capable of working on slopes as steep as 3:1. However, it is not advisable to use graders to construct ditches running parallel on such a slope because they have a comparatively high center of gravity, and the right pressure at a critical point on the moldboard could cause the machine to roll over. Graders are capable of progressively cutting ditches to a depth of 3 ft. It is more economical to use other types of equipment to cut ditches deeper than 3 ft.

The components of the grader that actually do the work are the moldboard (blade) and the scarifier (see Fig. 9.2). Graders may also be equipped with lightweight rear-mounted rippers (see Figs. 4.8 and 9.1).

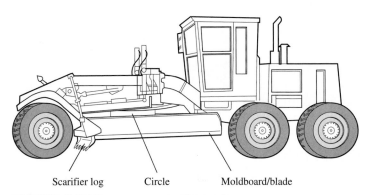

Scarifier log   Circle   Moldboard/blade

FIGURE 9.2 I The components of a grade.

Moldboard

The moldboard, commonly referred to as the blade, is the working member of the grader. A rotating circle carries the moldboard. Through sophisticated hydraulics, the moldboard can be placed into many positions, either under the grader or to the side. It can be side-shifted horizontally for increased reach outside the tires.

The moldboard is used to side cast the material it encounters. The ends of the moldboard can be raised or lowered together or independently of one another. By convention, the toe of the moldboard is the foremost end of the moldboard in the direction of travel and the heel is the discharge end.

Moldboard Angle

The moldboard can be angled (positioned) at almost any angle to the line of travel, parallel to the direction of travel, shifted to either side, or raised into vertical position (see Figs. 9.3 and 9.4).

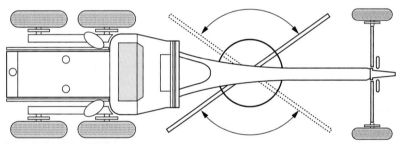

FIGURE 9.3 | Moldboard rotation.

FIGURE 9.4 | Moldboard positions.

Moldboard Pitch

For normal work, the moldboard is kept near the center of the pitch adjustment, which keeps the top of the moldboard directly over the cutting edge. However, the top of the moldboard can be pitched (leaned) forward or backward (see Fig. 9.5).

When leaned forward, the cutting ability of the moldboard is decreased, and it has more dragging action. It will tend to ride over material rather than cut and push, and it has less chance of catching on solid obstructions. A forward pitch is used to make light, rapid cuts and to blend materials. When leaned to the rear, the moldboard cuts readily but tends to let material spill back over the moldboard.

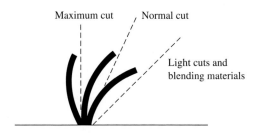

Maximum cut Normal cut

Light cuts and blending materials

FIGURE 9.5 | Grader moldboard pitch.

Scarifier

Material too hard to cut with the moldboard should be broken up with the scarifier. A scarifier is an attachment hung between the front axle and the moldboard. It is composed of a scarifier log with removable teeth. The teeth can be adjusted to cut to a depth of 12 in. When operating in hard material, it may be necessary to remove some of the teeth from the scarifier log. A maximum of five teeth may be removed from the log. If more than five teeth are removed, the force against the remaining teeth could shear them off. When removing teeth, take the center one out first, and then alternately remove the other four teeth. This balances the scarifier and distributes the load evenly. With the top of the scarifier pitched to the rear, the teeth lift and tear the material being loosened. This position is also used for breaking up asphalt pavement. The pitch of the scarifier log can be adjusted for the particular material being manipulated.

GRADER OPERATIONS

When the moldboard is set at an angle, the load being pushed tends to drift off to the trailing end of the moldboard. Rolling action caused by the moldboard curve assists this side movement. As the moldboard is angled more sharply, the speed of the side drift increases, so that material is not carried forward as far and deeper cuts can be made. To shape and maintain most roads, set the moldboard at a 25° to 30° angle. The angle should be decreased for spreading windrows and increased for hard cuts and ditching.

Planing Surfaces

Set the moldboard at an angle to plane off irregular surfaces and use the resulting material to fill low spots. Cut enough material to keep some in front of the moldboard. Move the loosened material forward and sideward to distribute it evenly. On the next pass, pick up the windrow left at the trailing edge of the moldboard. On the final pass, make a lighter cut and lift the trailing edge of the moldboard enough to allow the surplus material to go under rather than around it. This will avoid leaving a ridge. Do not pile windrows in front of the ridge. Also, do not pile them in front of the rear wheels because they will adversely affect traction and grader accuracy.

Road and embankment finish work and shallow ditch construction are basic grader operations. These operations are normally performed as follows.

Ditch Cuts

Normally, ditching cuts are done in second gear at full throttle. For better grader control and straighter ditches, make a 3- to 4-in.-deep marking cut at the outer edge of the bank slope (usually identified by slope stakes) on the first pass. The toe of the moldboard should be in line with the outside edge of the lead tire. This marking cut provides a guide for subsequent operations. Make each ditch cut as deep as possible without stalling or losing control of the grader. Each successive cut is brought in from the edge of the bank slope so the toe of the moldboard will be in line with the ditch bottom on the final cut.

Creating a Bank Slope

Sloping the bank on a road cut prevents slope failure. It also prevents excessive erosion of the bank that could fill the roadside ditch. Make cuts as close to the design slope as possible to prevent having to make additional passes. The material cut from the outer slope is initially cast into the bottom of the ditch and must be removed later.

Cleaning Ditches

To remove unwanted material that was pushed into the ditch during the bank slope operation, place the moldboard in the same position as used for the ditching cuts. This casts the material onto the shoulder.

Moving Windrows

Side draft is the force at the moldboard that tends to pull the front of the grader to one side. When a grader makes a cut, a windrow will form between the heel of the moldboard and the left rear wheel. This windrow will impart a side draft force. The front wheels of a grader can be leaned both left and right. Lean the wheels against the direction of side draft and the windrow can then be shifted across the grade with successive cuts. Sometimes cuts produce more material than needed for the roadbed and shoulders. This excess material can be used as fill at other locations throughout the project. In this case, the excess material is drifted into a windrow and picked up by an elevating scraper. The scraper can haul the material to the appropriate location on the project.

Haul-Road Maintenance

Keep haul roads in good condition. This will increase the efficiency of scrapers or trucks on large earthmoving operations. Graders are the best machines for maintaining haul roads. The most efficient method of road maintenance is to use sufficient graders to complete one side of a road with one pass of each grader (tandem operation). In this method, one side of the road is complete while the other side is left open to traffic.

Ordinarily leveling and maintaining a surface is done by working the material across the road from one side to the other. However, to maintain a satisfactory surface in dry weather, work traffic-eroded material from the edges and shoulders of the road toward the center. The surface is easier to work if it is damp; therefore after a rain is a good time to perform surface maintenance. A water truck may be necessary to dampen the material that is too dry.

Smoothing Pitted Surfaces   When binder (fine material) is present and moisture content is appropriate, rough or badly pitted surfaces may be cut smooth. The cut surface material is then respread over the smooth base. Again, the best time to reshape earth and gravel roads is after a rain. Dry roads should be watered using a water distributor. This ensures that the material will have sufficient moisture content to recompact readily.

Correcting Corrugated Roads   When correcting corrugated roads, be careful not to make the situation worse. Deep cuts on a washboard surface will set up moldboard *chatter,* which emphasizes rather than corrects corrugations. Scarifying may be required if the surface is too badly corrugated. With proper moisture content, the surface can be leveled by cutting across the corrugations. Alternate the moldboard so the cutting edge will not follow the rough surface. Cut the surface to the bottom of the corrugations. Then reshape the road surface by spreading the windrows in an even layer across the road. Rolling after shaping gives better, longer-lasting results.

Spreading

Graders are often used to spread and mix dumped loads. Because of their mechanical structure and operating characteristics, graders can only be effective spreading and mixing free-flowing materials. A general formula for figuring grader spreading and mixing production is

$$\text{Production (bcy) per hr} = 3.0 \times \text{hp} \qquad \textbf{[9.1]}$$

where hp is the engine flywheel-brake horsepower of the grader and efficiency is assumed to be a 50-min working hour.

Proper Working Speeds

Always operate as fast as the skill of the operator and the condition of the road permit. Operate at full throttle in each gear. If less speed is required, use a lower gear, rather than run at less than full throttle. Correct gear ranges for various grader operations, under normal conditions, are listed in Table 9.1.

TABLE 9.1 | Proper gear ranges for grader operations

Operation	Gear
Road maintenance	Second to third
Spreading	Third to fourth
Mixing	Fourth to sixth
Bank sloping	First
Ditching	First to second
Finishing	Second to fourth

Turns

When making a number of passes over a short distance (less than 1,000 ft), backing the grader to the starting point is normally more efficient than turning it around and continuing the work from the far end. Never make turns on newly laid bituminous road or runway surfaces.

Number of Passes

Grader efficiency is in direct proportion to the number of passes made. Operator skill, coupled with planning, is most important in eliminating unnecessary passes. For example, if four passes will complete a job, every additional pass increases the time and cost of the job.

Tire Inflation

Overinflated tires cause less contact between the tires and the road surface, resulting in a loss of traction. Air pressure differences in the rear tires cause wheel slippage and grader bucking. It is necessary to always keep tires properly inflated to get the best results.

TIME ESTIMATES

The following formula may be used to prepare estimates of the total time (in hours or minutes) required to complete a grader operation:

$$\text{Total time} = \frac{P \times D}{S \times E} \qquad [9.2]$$

where

P = number of passes required

D = distance traveled in each pass, in miles or feet

S = speed of grader, in mph or feet per minute (fpm) (use 88 fpm to change mph to fpm)

E = grader efficiency factor

Factors in Formula

The number of passes depends on project requirements and is estimated before construction begins. (For example, five passes may be needed to clean out a ditch and reshape a road.) Travel distance per pass is determined before construction begins.

Speed   Speed is the most difficult factor in the formula to estimate accurately. As work progresses, conditions may require that speed estimates be increased or decreased. Work output should be computed for each rate of speed used in an operation. The speed depends largely on the skill of the operator and the type of material.

Efficiency Factor   A reasonable efficiency factor for grader operations is 60%.

EXAMPLE 9.1

Mile Formula

Maintenance of 5 miles of haul road requires cleaning the ditches and leveling and reshaping the roadway. Use a grader with an efficiency factor of 0.60. Cleaning the ditches requires two passes in first gear (2.3 mph); leveling the road requires two passes in second gear (3.7 mph); and final shaping of the road requires three passes in fourth gear (9.7 mph).

$$\text{Total time} = \frac{2 \times 5 \text{ miles}}{2.3 \text{ mph} \times 0.60} + \frac{2 \times 5 \text{ miles}}{3.7 \text{ mph} \times 0.60} + \frac{3 \times 5 \text{ miles}}{9.7 \text{ mph} \times 0.60}$$

$$= 7.3 \text{ hr} + 4.5 \text{ hr} + 2.6 \text{ hr} \Rightarrow 14.4 \text{ hr}$$

EXAMPLE 9.2

Feet Formula

A haul road of 1,500 ft requires leveling and reshaping. Use a grader with an efficiency factor of 0.60. The work requires two passes in second gear (3.7 mph) and three passes in third gear (5.9 mph).

$$\text{Total time} = \frac{2 \times 1,500 \text{ ft}}{88 \text{ fpm/mph} \times 3.7 \text{ mph} \times 0.60} + \frac{3 \times 1,500 \text{ ft}}{88 \text{ fpm/mph} \times 5.9 \text{ mph} \times 0.60}$$

$$= 15.4 \text{ min} + 14.4 \text{ min} \Rightarrow 29.8 \text{ min}$$

GRADER SAFETY

Listed here are some specific safety rules for grader operations.

- When operating a grader slowly on a highway or roadway, display a red flag or flashing light on a staff at least 6 ft above the left rear wheel.
- Never allow other personnel to ride on the tandem, moldboard, or rear of grader.
- Always keep the grader in low gear when going down steep slopes.

- When working on hillsides, take extra care to drive slowly and watch out for holes or ditches.
- Never use graders to pull stumps or other heavy loads.
- Keep the moldboard angled well under the machine when not in use.

GRADALLS

GENERAL INFORMATION

The *gradall* is a utility machine that combines the operating features of the hoe, dragline, and motor grader (see Fig. 9.6). The full revolving superstructure of the unit can be mounted on either crawler tracks or wheels. The unit is designed as a versatile machine for both excavation and finishing work. Being designed as a multiuse machine affects production efficiency in respect to individual applications, when compared to a unit designed specifically for a particular application. The gradall will have lower production capability than those single-purpose units.

The bucket of a gradall can be rotated (that is, the gradall's arm can rotate) 90° or more, allowing it to be effective in reaching restricted working areas and where special shaping of slopes is required. The three-part telescoping boom can

FIGURE 9.6 | Gradall shaping the side slopes of a ditch.

be hydraulically extended or retracted to vary digging or shaping reach. It can exert breakout force either above or below ground level.

When used in a hoe application to excavate below the running gear, its production rate will be less than that of a hoe equipped with an equal-size bucket. Similarly, it can perform dragline-type tasks, but it does have limited reach compared to a dragline. Because the machine provides the operator with positive hydraulic control of the bucket, it can be used as a finishing tool for fine-grading slopes and confined areas, tasks that would normally be grader work if there were no space constraints.

TRIMMERS

GENERAL INFORMATION

In the case of linear projects, there are a variety of large, highly specialized but extremely versatile machines to do trimming. The result is better accuracy and greater production compared to fine-grading with a grader. It has been reported that the production from one dual-lane trimmer is equal to that achievable with four to six graders. The automatic trimmers also enable grade control to closer tolerances.

OPERATION

There are large, four-track multilane trimmers and smaller three-track machines (see Fig. 9.7) for bridge approach work and parking lot applications. Both the large and small machines work using the same general control and mechanical systems.

Grade Control

The automatic trimmer centers around the control system that is either a sensor arm riding on a grade wire or a ski riding on an established grade. A tightly stretched wire set at a known elevation above the specified project grade and having a known offset distance from the alignment establishes a reference for the trimmer. A horizontal movable arm, mounted on the trimming machine and spring-tensioned, rides on the wire. This arm is connected to an electric switch and activates control signals for vertical movement of the trimmer. A similar vertical arm travels along the wire, steering the trimmer.

On projects where it is necessary to match an existing grade, a ski-type runner travels over the existing surface and controls the vertical motion of the trimmer. In this case, the alignment is usually controlled manually by the operator.

The frames of four-track multilane trimmers can provide a trimming width of 40 ft or more. These machines typically have four crawler-tracks, one at each corner. Each crawler mount has a vertical member that is hydraulically adjustable either manually or by the automatic control system.

FIGURE 9.7 I Small trimmer working off a wire grade line and loading a bottom-dump truck.

Trimmer Assembly

The actual cutting of the grade is accomplished by a series of cutting teeth mounted on a full-width *cutter mandrel.* The cutter assembly is usually fixed at the ends, but it can break, allowing for the programming of a crown in the grade. The cutter mandrel is followed by an adjustable *moldboard assembly* that strikes off the material. An *auger* for directing the spoil follows the moldboard. Typically, the spoil can be cast to either or both sides of the machine. On many large machines, the right and left augers are independently driven. There are typically *wastegates* located at each end of the auger and adjacent to the centerline of the machine. These are for depositing excess material in a windrow on the grade.

A gathering and discharge conveyor system can be added to many trimmers. These attach to the rear for removing and reclaiming material from the finished grade. The material can be directly loaded into haul units with these systems or deposited on the shoulder of the finished grade.

Production

The large, full-width trimmers have operating speeds of about 30 fpm but this is dependent on the amount (depth of cut) of material being handled. In the case of the smaller, single-lane trimmers, operating speed increases significantly. Some of these machines are rated at 128 fpm, but again speed is controlled by the amount of material being cut. As operating speed is increased, there is usually a decrease in quality.

FORCE-FEED LOADERS

GENERAL INFORMATION

A force-feed loader (Fig. 9.8) is a self-propelled machine designed for loading windrows of loose material. A set of plowshare wings rests on the ground and funnels material to the center of the paddles as the loader moves forward into the windrow. The paddles lift the material onto a conveyor for discharge into hauling units. The conveyor can be positioned directly behind the loader for loading trucks operating in the same lane or it can be swiveled to the side for a parallel operation. These machines can handle about 10 lcy per minute.

SUMMARY

A reasonable production estimate for a grader operation can be made by considering the specific operation and the proper gear range for that operational task. The number of passes, the anticipated speed, the distance of the operation, and an efficiency factor are the input variables for a production calculation. The speed depends largely on the skill of the operator and the type of material.

If a gradall is used as an excavator, production can be estimated by using the general excavator production Eq. [8.1]. Trimmer production is a function of forward operating speed. Critical learning objectives include:

- An ability to estimate grader speed based on operational task.
- An ability to calculate grader production.

These objectives are the basis for the problems that follow.

FIGURE 9.8 I A force-feed loader.

PROBLEMS

9.1 A grader will be used for maintenance of 3 miles of haul road. It is estimated that this work will require two passes in first gear; leveling the road requires two passes in second gear; and final shaping of the road requires three passes in fourth gear. Use an efficiency factor of 60%. Grader speeds are given in the chart. What is the time requirement for this task?

Forward gears	Maximum travel speed (mph)
First	2.2
Second	3.0
Third	4.3
Fourth	6.0

9.2 A grader will be used for maintenance of 1,800 ft of haul road. It is estimated that this work will require two passes in second gear and three passes in third gear. Use an efficiency factor of 60%. Grader speeds are given in the Problem 9.1 chart. What is the time requirement for this task?

9.3 A grader will be used for maintenance of 2.5 miles of haul road. It is estimated that this work will require three passes in first gear; leveling the road requires two passes in second gear; and final shaping of the road requires two passes in fourth gear. Use an efficiency factor of 60%. Grader speeds are given in the chart. What is the time requirement for this task?

Forward gears	Maximum travel speed (mph)
First	2.3
Second	3.2
Third	4.4
Fourth	6.4

9.4 A grader will be used for maintenance of 1,400 ft of haul road. It is estimated that this work will require three passes in second gear and four passes in third gear. Use an efficiency factor of 60%. Grader speeds are given in the Problem 9.3 chart. What is the time requirement for this task?

REFERENCES

1. Athey Products Corporation, 1839 S. Main Street, Wake Forest, NC 27587. www.athey.com.
2. Caterpillar Inc., Peoria, Ill. www.cat.com.
3. CMI Corporation, P.O. Box 1985, Oklahoma City, OK 73101-1985. www.cmicorp.com.
4. Deere & Company, One John Deere Place, Moline, IL (USA) 61265. www.deere.com.
5. Gradall, 406 Mill Ave. SW, New Philadelphia, OH 44663. www.gradall.com.

C H A P T E R

10

Trucks and Hauling Equipment

Trucks are hauling units that provide relatively low hauling costs because of their high travel speeds. The weight capacity of a truck may limit the volume of the load that a unit may haul. The productive capacity of a truck depends on the size of its load and the number of trips it can make in an hour. The number of trips completed per hour is a function of cycle time. Truck cycle time has four components: (1) load time, (2) haul time, (3) dump time, and (4) return time. Tires for trucks and all other haul units should be suitably matched to the job requirements.

TRUCKS

In transporting excavated material, processed aggregates, and construction materials, and for moving other pieces of construction equipment (see Fig. 10.1), trucks serve one purpose: they are hauling units that, because of their high travel speeds, provide relatively low hauling costs. The use of trucks as the primary hauling unit provides a high degree of flexibility, as the number in service can usually be increased or decreased easily to permit modifications in the total hauling capacity of a fleet. Most trucks may be operated over any haul road for which the surface is sufficiently firm and smooth, and on which the grades are not excessively steep. Some units are designated as off-highway trucks because their size and weight are greater than that permitted on public highways (see Fig. 10.2). Off-highway trucks are used for hauling materials in quarries and on large projects involving the movement of substantial amounts of earth and rock. On such projects, the size and costs of these large trucks is easily justified because of the increased production capability they provide.

Trucks can be classified by many factors, including

1. The method of dumping the load—rear-dump, bottom-dump, or side-dump.
2. The type of frame—rigid-frame or articulated.
3. The size and type of engine—gasoline, diesel, butane, or propane.

FIGURE 10.1 | Truck tractor unit towing a low profile trailer hauling an off-highway truck.

FIGURE 10.2 | Off-highway truck.

4. The kind of drive—two-wheel, four-wheel, or six-wheel.
5. The number of wheels and axles, and the arrangement of driving wheels.
6. The class of material hauled—earth, rock, coal, or ore.
7. The capacity—gravimetric (tons) or volumetric (cubic yards).

If trucks are to be purchased for general material hauling, the purchaser should select units adaptable to the multipurposes for which they will be used. However, if trucks are to be used on a given project for a single purpose, they should be selected specifically to fit the requirements of the project.

FIGURE 10.3 | Highway rigid-frame rear-dump truck.

RIGID-FRAME REAR-DUMP TRUCKS

Rigid-frame rear-dump trucks are suitable for use in hauling many types of materials (see Fig. 10.3). The shape of the body, such as the extent of sharp angles and corners, and the contour of the rear, through which the materials must flow during dumping, will affect the ease or difficulty of loading and dumping. The bodies of trucks that will be used to haul wet clay and similar materials should be free of sharp angles and corners. Dry sand and gravel will flow easily from almost any shape of body. When hauling rock, the impact loading on the truck body is extremely severe. Continuous use under such conditions will require a heavy-duty rock body made of high-tensile steel. Even with the special body, the loader operator must use care in placing a load in the truck.

Off-highway dump trucks do not have tailgates, therefore the body floor slopes forward at a slight angle, typically less than 15°. The floor shape perpendicular to the length of the body for some models is flat while other models utilize a "V" bottom to reduce the shock of loading and to help center the load. Low sides and longer, wider bodies are a better target for the excavator bucket. The result of such a configuration is quicker loading cycles.

ARTICULATED REAR-DUMP TRUCKS

The articulated dump truck (ADT) is specifically designed to operate over rough or soft ground, and in confined working locations where a rigid-frame truck would have problems (see Fig. 10.4). An articulated joint and oscillating ring between the tractor and dump body permit all wheels to maintain contact with

FIGURE 10.4 | An articulated dump truck.

FIGURE 10.5 | An articulated dump truck moving through soft ground.

the ground at all times. The articulation, all-wheel drive, high clearance, and low-pressure radial tires combine to produce a truck capable of moving through soft or sticky ground (see Fig. 10.5).

When haul-route grades are an operating factor, articulated trucks can typically climb steeper grades than rigid-frame trucks. Articulated trucks can operate on grades up to about 35%, whereas rigid-frame trucks can navigate only grades of 20% for short distances, and for continuous grades, 8 to 10% is a more reasonable limit.

The most common ADTs are the 4 × 4 models, but there are larger 6 × 6 models. Articulated dump trucks usually have high hydraulic dumping pressures, which means they hoist their beds faster. The bed also achieves a steeper dump angle. One model can attain a 72° dump angle in 15 sec. The combination of these two attributes, hoist speed and a steep angle, translates into quick discharge times. To solve the problem of unloading sticky materials, one manufacturer is equipping its truck bed with an ejector.

Rear-dumps, be they rigid-frame or articulated, should be considered when

1. The material to be hauled is free flowing or has bulky components.
2. The hauling unit must dump into restricted locations or over the edge of a bank or fill.
3. Maximum maneuverability in the loading or dumping area is required.

TRACTORS WITH BOTTOM-DUMP TRAILERS

Tractors towing bottom-dump trailers are economic haulers when the material to be moved is free flowing, such as sand, gravel, reasonably dry earth, and coal. The use of bottom-dump trailers will reduce the time required to unload the material. To take full advantage of this time saving, there must be a large clear dumping area where the load can be spread into windrows. Bottom-dump units are also very good for unloading into a drive over hopper. The rapid rate of discharging the load gives the bottom-dump wagons a time advantage over rear-dump trucks. There are both large off-highway units (see Fig. 10.6) and highway-sized units (see Fig. 10.7). With either the large off-highway or the highway units, relatively flat haul roads are required if maximum travel speed is to be obtained.

FIGURE 10.6 | Off-highway tractor towing a loaded bottom-dump trailer.

FIGURE 10.7 I Highway bottom-dump.

The clamshell doors through which these units discharge their loads have a limited opening width. Difficulties may be experienced in discharging materials such as wet, sticky clay, especially if the material is in large lumps.

Tractors towing bottom-dump trailers are often used as hauling units on projects where large quantities of materials are to be transported and haul roads can be kept in reasonable condition. The bottom-dump trailer can have a single axle, tandem axles, or even triaxles. Hydraulic excavators, loaders, draglines, or belt loaders may be used to load these units.

Bottom-dumps should be considered when

1. The material to be hauled is free flowing.
2. There are unrestricted loading and dump sites.
3. The haul-route grades are less than about 5%. Because of its unfavorable power-weight ratio and the fact that there is less weight on the drive wheels of the tractor unit, thereby limiting traction, bottom-dump units have limited ability to pull steep grades.

CAPACITIES OF TRUCKS AND HAULING EQUIPMENT

There are at least three methods of rating the capacities of trucks and wagons:

1. *Gravimetric*—the load that it will carry, expressed as a weight.
2. *Struck volume*—the volumetric amount it will carry, if the load was water level in the body.
3. *Heaped volume*—the volumetric amount it will carry, if the load was heaped on a 2:1 slope above the body.

The gravimetric rating is usually expressed in pounds or kilograms and the latter two ratings in cubic yards or cubic meters.

The struck capacity of a truck is the volume of material that it will haul when it is filled level to the top of the body sides (see Fig. 10.8). The heaped capacity is the volume of material that it will haul when the load is heaped above the sides. The standard for rated-heaped capacity is an assumed 2:1 slope (see Fig. 10.8). The actual heaped capacity will vary with the material that is being hauled. Wet earth or sandy clay may be hauled with a slope of about 1:1, while dry sand or gravel may not permit a slope greater than about 3:1. To determine the actual heaped capacity of a unit, it is necessary to know the struck capacity, the length and width of the body, and the slope at which the material will remain stable while the unit is moving. Smooth haul roads will permit a larger heaped capacity than rough haul roads.

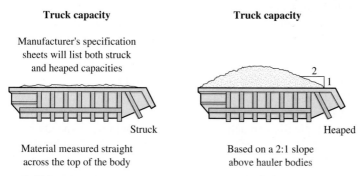

FIGURE 10.8 | Measurement of volumetric capacity.

The weight capacity may limit the volume of the load when a unit is used to haul a material having a high unit weight, such as iron or even wet sand. However, when the unit weight of the materials is such that the safe load is not exceeded, a unit may be filled to its heaped capacity. Always check to ensure that the volumetric load does not cause a condition where the load weight exceeds the gravimetric capacity of the truck or trailer. Overloading will cause the unit's tires to flex too much, and flexing produces excessive internal tire temperature. Such a condition will cause permanent tire damage.

In some instances, it is possible to add sideboards to increase the depth of the truck or wagon body, thereby permitting it to haul a larger load. When this is done, the weight of the new volume must be checked against the vehicle's load capacity. If the weight is greater than the rated gravimetric capacity, the practice will probably increase the hourly cost of operating a unit because of higher fuel consumption, reduced tire life, more frequent failure of parts (such as axles, gears, brakes, and clutches), and higher maintenance costs. If the value of the extra material hauled is greater than the total increase in the cost of operating the vehicle, overloading is justified. In considering the option of sideboarding and hauling larger volumes of materials, the maximum safe loads on the tires must be checked to prevent excessive loading, which might result in considerable lost time due to tire failures. Tires are about 35% of a truck's operating cost. Overloading a truck abuses the tires. The Rubber Manufacturers Association publication *Care and Service of Off-the-Highway Tires* addresses overloading and provides load and inflation tables.

EFFECTS OF TRUCK SIZE

The productive capacity of a truck depends on the size of its load and the number of trips it can make in an hour. The number of trips completed per hour is a function of cycle time. Truck cycle time has four components: (1) load time, (2) haul time, (3) dump time, and (4) return time. Examining a match between truck body size and excavator bucket size yields the size of the load and the load time. The haul and return cycle times will depend on the weight of the vehicle, the horsepower of the engine, the haul and return distance, and the condition of the roads traversed. Dump time is a function of the type of equipment and conditions in the dump area.

When an excavator is used to load earth into trucks, the size of the trucks may introduce several factors, which will affect the production rate and the cost of handling earth.

1. Advantages of small compared with large trucks:
   a. They are more flexible in maneuvering, which may be an advantage on restricted work sites.
   b. They typically can achieve higher haul and return speeds.
   c. The loss in production is less when one truck in a fleet breaks down.
   d. It is easier to balance the number of trucks with the output of the excavator, which will reduce the time lost by the trucks or the excavator.

2. Disadvantages of small compared with large trucks:
   a. A small truck is more difficult for the excavator to load owing to the small target for depositing the bucket load.
   b. More total spotting time is lost in positioning the trucks because of the larger number required.
   c. More drivers are required to haul a given output of material.
   d. The greater number of trucks increases the danger of units bunching at the pit, along the haul road, or at the dump.

3. Advantages of large compared with small trucks:
   a. Fewer trucks are required, which may reduce the total investment in hauling units and the cost of maintenance and repairs.
   b. Fewer drivers are required.
   c. The smaller number of trucks facilitates synchronizing the equipment and reduces the danger of bunching by the trucks. This is especially true for long hauls.
   d. There is a larger target for the excavator during loading.
   e. The frequency of spotting trucks under the excavator is reduced.
   f. There are fewer trucks to maintain and repair, and fewer parts to stock.

4. Disadvantages of large compared with small trucks:
   a. The cost of truck time at loading is greater, especially with small excavators.
   b. The heavier loads may cause more damage to the haul roads, thus increasing the cost of mechanical maintenance to the trucks and requiring more support equipment for maintenance of the haul road.

c.   It is more difficult to balance the number of trucks with the output of
the excavator.

d.   The largest sizes may not be permitted to haul on highways.

Balancing the capacities of hauling units with the excavator bucket size is
important. When loading with excavators such as hydraulic hoes or shovels,
draglines, or loaders, it is desirable to use haul units whose load body volume is
balanced with the bucket size of the excavator. If this is not done, operating difficulties will develop and the combined cost of excavating and hauling material
will be higher than when balanced units are used.

A practical rule-of-thumb frequently used in selecting the size of trucks is
to use trucks with a minimum capacity of four to five times the capacity of the
excavator bucket. The dependability of this practice is discussed in the following analysis.

EXAMPLE 10.1

Consider a 3-cy shovel excavating good common earth with a 90° swing, with no
delays waiting for hauling units, and with a 20-sec-cycle time. Assume for this example that if the bucket and the trucks are operated at their heaped capacities, the
swelling effect of the earth will permit each truck to carry its rated struck capacity,
expressed in cubic yards bank measure (bcy). Assume that the number of buckets
required to fill a truck will equal the capacity of the truck divided by the size of the
bucket, both expressed in cubic yards. Assume further that the time for the travel
and dump cycle, *excluding the time for loading,* will be the same for the several
sizes of trucks considered. The time for a travel cycle, which includes traveling to
the dump, dumping, and returning to the shovel, will be 6 min.

If **12-cy trucks** are used, it will require four buckets (12 ÷ 3) to fill a truck. The
time that is required to load a truck would be 80 sec, or 1.33 min. The minimum
round-trip cycle for a truck will be 7.33 min. The minimum number of trucks required
to keep the shovel busy will be 7.33 ÷ 1.33 = 5.51.

Five Trucks: The time required to load five trucks will be 5 × 1.33 = 6.65 min.
Thus the shovel will lose 7.33 − 6.65 = 0.68 min when only five trucks are used. The
percentage of time lost will be (0.68 ÷ 6.65) × 100 = 10.2%.

Six Trucks: If six trucks are used, the total loading time required will be 6 × 1.33
= 7.98 min. As this will increase the total round-trip cycle of each truck from 7.33 to
7.98 min, the lost time per truck cycle will be 0.65 min per truck. This will result in a
loss of

$$\frac{0.65}{7.98} \times 100 = 8.2\%, \text{ for each truck}$$

which is equivalent to an operating factor of 91.8% for the trucks.

If **24-cy trucks** are used, it will require eight buckets to fill a truck. The time that
is required to load a truck would be 160 sec, or 2.66 min. The minimum round-trip
cycle for a truck will be 8.66 min. The minimum number of trucks required to keep
the shovel busy will be 8.66 ÷ 2.66 = 3.26.

Four Trucks: Using four trucks, the time required to load will be 4 × 2.66 = 10.64 min, the lost time per truck cycle will be 10.64 − 8.66 = 1.98 min per truck. This will produce an operating factor of $\frac{8.66}{10.64}$ × 100 = 81.4% for the trucks.

In Example 10.1, note that the production of the shovel is based on a 60-min hour. This policy should be followed when balancing a servicing (loading) unit with the units being served (hauling) because at times both types of units will operate at maximum capacity if the number of units is properly balanced. However, the average production of a unit, excavator or truck, for a sustained period of time, should be based on applying an appropriate efficiency factor to the maximum productive capacity. Attention must be called to the fact that in this example we have chosen truck sizes that exactly match the loader, i.e., to load a 12-cy truck with a 3-cy shovel results in an integer number of bucket swings. In practice, this is not always the case, but physically only an integer number of swings can be used in loading the truck.

CALCULATING TRUCK PRODUCTION

The following is a format that can be used to calculate truck production.

Step 1. Number of Bucket Loads

The first step in analyzing truck production is to determine the number of excavator bucket loads it takes to load the truck.

$$\text{Balanced number of bucket loads} = \frac{\text{Truck capacity (lcy)}}{\text{Bucket capacity (lcy)}} \qquad \textbf{[10.1]}$$

Step 2. Load Time

The actual number of bucket loads placed on the truck should be an integer number. It is possible to light load a bucket to match the bucket volume to the truck volume, but that practice is usually inefficient as it results in wasted loading time.

If one less bucket load is placed on the truck, the loading time will be reduced; but the truckload is also reduced. Sometimes job conditions will dictate that a fewer number of bucket loads be placed on the truck, i.e., the load size is adjusted if haul roads are in poor condition or if the trucks must traverse steep grades. The truckload in such cases will equal the bucket volume multiplied by the number of bucket swings.

For the case where the number of bucket loads is *rounded down to an integer lower* than the balance number of swings or reduced because of job conditions:

$$\text{Load time} = \text{Number of bucket swings} \times \text{bucket cycle time} \qquad \textbf{[10.2]}$$

$$\text{Truckload (volumetric)} = \text{Number of bucket swings} \times \text{volume of the bucket}$$

$$\textbf{[10.3]}$$

If the division of truck body volume by the bucket volume is *rounded to the next higher integer* and that higher number of bucket swings is used to load the truck, excess material will spill off the truck. In such a case, the loading duration equals the bucket cycle time multiplied by the number of bucket swings. But the volume of the load on the truck equals the truck capacity, not the number of bucket swings multiplied by the bucket volume.

Number of bucket loads rounded up to *next higher* integer above the balance number of swings:

$$\text{Load time} = \text{Number of bucket swings} \times \text{bucket cycle time} \qquad \textbf{[10.2]}$$

$$\text{Truckload (volumetric)} = \text{Volumetric capacity of the truck} \qquad \textbf{[10.4]}$$

Always check the load weight against the gravimetric capacity of the truck.

$$\text{Truckload (gravimetric)} = \text{Volumetric (lcy)} \times \text{unit weight (loose vol. lb/lcy)}$$

$$\textbf{[10.5]}$$

$$\text{Truckload gravimetric} < \text{Rated gravimetric payload?} \qquad \textbf{[10.6]}$$

Step 3. Haul Time

Hauling should be at the highest safe speed and in the proper gear. To increase efficiency, use one-way traffic patterns.

$$\text{Haul time (min)} = \frac{\text{Haul distance (ft)}}{88 \text{ fpm/mph} \times \text{Haul speed (mph)}} \qquad \textbf{[10.7]}$$

Based on the gross weight of the truck with the load, and considering the rolling and grade resistance from the loading area to the dump point, haul speeds can be determined using the truck manufacturer's performance chart (see Figure 10.9).

EXAMPLE 10.2

The specifications for the truck are as follows:

Engine, 239 fwhp (flywheel horsepower)

Capacity
  Struck, 14.7 cu yd
  Heaped, 2:1, 18.3 cu yd

Net weight empty   = 36,860 lb
Payload   = 44,000 lb
Gross vehicle weight = 80,860 lb

Determine the maximum speed for the truck when it is hauling a load of 22 tons up a 6% grade on a haul road having a rolling resistance of 60 lb per ton, equivalent to a 3% adverse grade. Because the chart is based on zero rolling resistance,

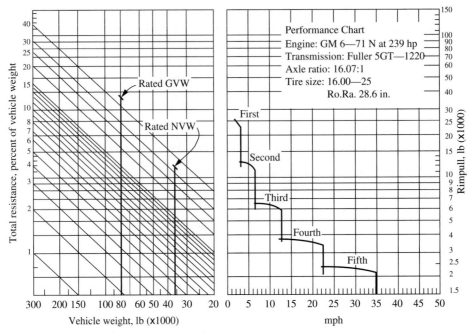

Vehicle weight, lb (x1000)   mph

FIGURE 10.9 | Performance chart for a 22-ton rear-dump truck.

it is necessary to combine the grade and rolling resistance, which gives an equivalent total resistance equal to 6 + 3 = 9% of the vehicle weight.

The procedure for using Fig. 10.9 is

1. Find the vehicle weight on the lower left horizontal scale.
2. Read up the vehicle weight line to the intersection with the slanted total resistance line.
3. From this intersection, read horizontally to the right to the intersection with the performance curve.
4. From this intersection, read down to find the vehicle speed.

Following these four steps it is found that the truck will operate in the second gear range, and that its maximum speed will be 6.5 mph.

The chart should be used to determine the maximum speed for each section of a haul road having a significant difference in grade or rolling resistance.

While a performance chart indicates the maximum speed at which a vehicle can travel, the vehicle will not necessarily travel at this speed. A performance chart makes no allowance for acceleration or deceleration. Additionally, other roadway conditions and safety (see Fig. 10.10) can control travel speed. Before using a performance chart speed in an analysis, always consider such factors as congestion, narrow roads, or traffic signals, when hauling on public roads, because these can limit the speed to less than the value given in the chart. The anticipated effective speed is what should be used in calculating travel time.

FIGURE 10.10 I The driver of an off-highway truck has a limited field of vision.

Step 4. Return Time

Based on the empty vehicle weight, and the rolling and grade resistance from the dump point to the loading area, return speeds can be determined using the truck manufacturer's performance chart.

$$\text{Return time (min)} = \frac{\text{Return distance (ft)}}{88 \text{ fpm/mph} \times \text{Return speed (mph)}} \qquad \textbf{[10.8]}$$

Step 5. Dump Time

Dump time will depend on the type of hauling unit and congestion in the dump area. Consider that the dumping area is usually crowded with support equipment. Dozers are spreading the dumped material and multiple pieces of compaction equipment may be working in the area. Rear-dumps must be spotted before dumping. This usually means that the truck must come to a complete stop and then back up some distance. Total dumping time in such cases can exceed 2 min. Bottom-dumps will customarily dump while moving. After dumping, the truck normally turns and returns to the loading area. Under favorable conditions, a rear-dump can dump and turn in 0.7 min but an average unfavorable time is about 1.5 min. Bottom-dumps can dump in 0.3 min under favorable conditions, but they too may average 1.5 min when conditions are unfavorable. Always try to visualize the conditions in the dump area when estimating dump time.

Step 6. Truck Cycle Time

The cycle time of a truck is the sum of the load time, the haul time, the dump time, and the return time:

$$\text{Truck cycle time} = \text{Load}_{\text{time}} + \text{Haul}_{\text{time}} + \text{Dump}_{\text{time}} + \text{Return}_{\text{time}} \qquad \textbf{[10.9]}$$

Step 7. Number of Trucks Required

The number of trucks required to keep the loading equipment working at capacity:

$$\text{Number of trucks} = \frac{\text{Truck cycle time (min)}}{\text{Loader cycle time (min)}} \qquad \textbf{[10.10]}$$

Step 8. Production

The number of trucks must be an integer number, so if an integer number of trucks *lower* than the result of Eq. [10.10] is chosen then the trucks will control production.

$$\text{Production (lcy/hr)} = \text{Truck load (lcy)} \times \text{Number trucks}$$
$$\times \frac{60 \text{ min}}{\text{Truck cycle time (min)}} \qquad \textbf{[10.11]}$$

When an integer number of trucks greater than the result of Eq. [10.10] is selected, production is controlled by the loading equipment.

$$\text{Production (lcy/hr)} = \text{Truck load (lcy)} \times \frac{60 \text{ min}}{\text{Loader cycle time (min)}} \qquad \textbf{[10.12]}$$

As a rule, it is better to never keep the loading equipment waiting. If there is not a sufficient number of haul trucks there will be a loss in production. Truck bunching or queuing will reduce production 10 to 20%, even when there is a perfect match between loader capability and the number of trucks. If there are extra haul units, this queuing effect is reduced. Therefore, it is usually best to have more trucks; i.e., with Eq. [10.10] round up to the next integer.

Step 9. Efficiency

The production calculated with either Eq. [10.11] or [10.12] is based on a 60 min working hour. That production should be adjusted by an efficiency factor. Longer hauling distances usually result in better driver efficiency. Driver efficiency increases as haul distances increase out to about 8,000 ft, after which efficiency remains constant. Other critical elements effecting efficiency are bunching, discussed above, and equipment condition.

$$\text{Adjusted production (lcy/hr)} = \text{Production (lcy)}$$
$$\times \frac{\text{Working time (min/hr)}}{60 \text{ min}} \qquad \textbf{[10.13]}$$

CONTROL OF HAULING COST

Economical hauling is the result of good job planning and good field management.

■   It involves selection of material sources only after a careful investigation of the hauling cost.

- It requires management attention to haul-route grades and condition.
- It requires a complete analysis of production options.

Material Source Selection

In constructing a fill it is frequently possible to obtain the necessary material from a borrow pit located either above or below the fill. If the borrow pit is above the fill, the effect of the favorable grade on the loaded truck is to reduce the required rimpull by 20 lb per gross ton for each 1% of grade. If the borrow pit is below the fill, the effect of the adverse grade on the loaded truck is to increase the required rimpull by 20 lb per gross ton for each 1% grade. Obviously the grade of the haul road will affect the hauling capacity of a truck, its performance, and the cost of hauling material. It may be more economical to obtain material from a borrow pit above, instead of below, the fill, even though the haul distance from the higher pit is greater than from the lower pit. This is an item that should be given consideration in locating borrow pits.

If material is hauled downhill, it may be possible to add sideboards to the vehicle to increase the hauling capacity, up to the maximum load, which the truck frame, dumping system, and the tires can carry. In some instances, it will be desirable to use larger tires to permit the trucks to haul even greater loads. If the material is hauled uphill, it may be necessary to reduce the size of the load or the travel speed of the truck, either of which will increase the cost of hauling.

Haul Route

An important factor that affects the production capacity of a truck or a tractor-pulled wagon is the rolling resistance of the haul road. Rolling resistance is determined primarily by two factors, the physical condition of the road and the tires used on the hauling unit. Rolling resistance can be greatly reduced by properly maintaining the road and by properly selecting tire sizes and keeping them inflated to the correct pressure. Money spent for these purposes will return dividends, through reduced hauling costs, far in excess of the expenditures. This is one field where the application of engineering knowledge will yield excellent returns.

A haul road that is given little or no maintenance will soon become rough, loose, and soft, and may develop a rolling resistance of 150 lb per ton or more, depending on the type of material and weather conditions. If a road is properly maintained with a grader, sprinkled with water, and compacted as required, it may be possible to reduce the rolling resistance to 50 lb per ton or less. Sprinkling the road will reduce the damage to hauling equipment by eliminating dust. It will reduce the danger of vehicular collision by improving visibility and it will prolong the life of tires because of the cooling effect that the moisture has on the tires.

Production Options

To lower the cost per unit of material moved, high production must be obtained from the hauling fleet. Idle time, such as occurs while loading, should be kept to a minimum. When possible, loading with overhead hoppers should be considered because of the faster loading time obtained when this arrangement is practical.

Generally, three to six passes of an excavator bucket to fill a haul unit represents a good balance.

TIRES

Tires for trucks and all other haul units should be suitably matched to the job requirements. The selection of proper tire sizes and the practice of maintaining correct air pressure in the tires will reduce that portion of the rolling resistance due to the tires. A tire supports its load by deforming where it contacts the road surface until the area in contact with the road will produce a total force on the road equal to the load on the tire. Neglecting any supporting resistance furnished by the side walls of the tire, if the load on a tire is 5,000 lb and the air pressure is 50 psi, the area of contact will be 100 sq in. If, for the same tire, the air pressure is permitted to drop to 40 psi, the contact area will be increased to 125 sq in. The additional area of contact will be produced by additional deformation of the tire. This will increase the rolling resistance because the tire will be continually climbing a steeper grade as it rotates.

The tire size selected and the inflated pressure should be based on the resistance, which the surface of the road offers to penetration, by the tire. For rigid road surfaces, such as concrete, small-diameter high-pressure tires will give lower rolling resistance, while, for soft road surfaces, large-diameter low-pressure tires will give lower rolling resistance because the larger areas of contact will reduce the tire penetration depth.

Many tire failures can be traced to constant overload, excessive speed, incorrect tire selection, and poorly maintained haul roads. Underinflating tires can cause radial sidewall cracking and ply separation. Overinflation subjects the tire to excessive wear in the center of the tread. Mismatched duals will cause unequal weight distribution, overloading the larger tire.

Tires generate heat as they roll and flex. As a tire's operating temperature increases, the rubber and textiles within significantly lose strength. Earthmover tire manufacturers will provide a ton-miles-per-hour (TMPH) limit for their tires. It is good practice to calculate a job TMPH rate.

$$\text{TMPH job rate} = \text{Average tire load} \times \text{Average speed during a day's operation}$$

$$[10.14]$$

$$\text{Average tire load (tons)} =$$

$$\frac{\text{"Empty" tire load (tons)} + \text{"Loaded" tire load (tons)}}{2} \quad [10.15]$$

$$\text{Average speed (mph)} = \frac{\text{Round trip distance (miles)} \times \text{number of trips}}{\text{Total hours worked}}$$

$$[10.16]$$

When calculating the job TMPH rate always select the tire that carries the highest average load.

EXAMPLE 10.3

An off-highway truck weighs 70,000 lb empty and 150,000 lb when loaded. The weight distribution empty is 50% front and 50% rear. The weight distribution loaded is 33% front and 67% rear. The truck has two front tires and four rear tires. The truck works an 8-hr shift hauling blast rock to a crusher. The one-way haul distance is 5.5 miles. The truck can make 14 trips per day.

	Front, two tires	Rear, four tires
Empty weight 70,000 lb 50% front and 50% rear	35,000 lb	35,000 lb
Loaded weight 150,000 lb 33.3% front and 66.6% rear	50,000 lb	100,000 lb

	Weight per tire	Weight per tire
Empty	17,500 lb	8,750 lb
Loaded	25,000 lb	25,000 lb
Average tire load	10.6 ton	8.4 ton

The front tire carries the highest average load.

$$\text{Average speed} = \frac{(2 \times 5.5 \text{ mi}) \times 14 \text{ trips}}{8 \text{ hr}}$$

$$\text{Average speed} = 19.25 \text{ mph}$$

$$\text{TMPH job rate} = 10.6 \text{ tons} \times 19.25 \text{ mph}$$

$$\text{TMPH job rate} = 204$$

This means that a tire with a TMPH rating of 204 or higher must be used under these job conditions. If the tires actually used have a TMPH rating of less than 204, either the speed or the load, or both, must be reduced.

TRUCK PERFORMANCE CALCULATION

The following example analyzes the performance of a fleet of 22-ton rear-dump trucks being loaded by hydraulic hoe having a 3-cy bucket.

EXAMPLE 10.4

Rear-dump trucks with specifications as follows are used to haul sandy clay. The performance chart shown in Fig. 10.9 is valid for these trucks.

Capacity
  Struck, 14.7 cy
  Heaped, 2:1, 18.3 cy
Net weight empty   = 36,860 lb
Payload   = 44,000 lb
Gross vehicle weight = 80,860 lb

The trucks will be loaded by a hydraulic hoe having a 3-cy bucket. The haul road from the borrow site to the fill is a 3-mile downhill grade of 1%. It is earth, poorly maintained. Dump time will average 2 min because of expected congestion on the fill. The hoe should be able to cycle in 20 sec. The sandy clay has a loose unit weight of 2,150 lb/cy. A realistic efficiency estimate for this work is a 50-min hour.

Step 1. Number of bucket loads. The bucket fill factor for the hoe handling sandy clay has been determined to be 110%. The hoe bucket volume will therefore be 3.3 lcy (3 × 1.1). The heaped capacity of the truck is 18.3 lcy.

$$\text{Balanced number of bucket loads } \frac{18.3 \text{ lcy}}{3.3 \text{ lcy}} = 5.5$$

The actual number of buckets should be an integer number; therefore, two cases, placing either five or six bucket loads on the truck, should be investigated.

Step 2. Load time. Check production based on both possible situations, 5 or 6 bucket loads to fill the truck.

Load time (five buckets) $5 \times \dfrac{20 \text{ sec}}{60 \text{ sec per min}} = 1.66 \text{ min}$

Load volume (five buckets)   5 × 3.3 lcy/bucket load = 16.5 lcy

Check load weight   16.5 lcy × 2,150 lb per lcy = 35,045 lb

     Okay   35,045 lb < 44,000 lb rated payload

Load time (six buckets) $6 \times \dfrac{20 \text{ sec}}{60 \text{ sec per min}} = 2.00 \text{ min}$

Load volume (six buckets) equals truck capacity 18.3 lcy, excess spills off.

Check load weight   18.3 lcy × 2,150 lb per lcy = 39,345 lb

     Okay   39,345 lb < 44,000 lb rated payload

Step 3. Haul time.

Rolling resistance   Table 5.1; earth, poorly maintained, 100 to 140 lb per ton

                       Using an average value 120 lb per ton or 6.0%

Grade resistance   −1%

Total resistance   5.0%   [6.0% + (−1%)]

	Five buckets	**Six buckets**
Empty truck net weight	36,860 lb	36,860 lb
Load weight	35,045 lb	39,345 lb
Gross weight	71,905 lb	76,205 lb
Speed (Fig. 10.9)	16 mph	13 mph

In this particular case, the effect of total resistance and the difference in gross weight for the two load scenarios results in a speed difference for the conditions. If the total resistance had been only 4% the achievable speed would have been 22 mph for both load conditions.

$$\text{Haul time (five buckets)} \frac{3 \text{ miles} \times 5{,}280 \text{ ft/mile}}{88 \text{ fpm/mph} \times 16 \text{ mph}} = 11.25 \text{ min}$$

$$\text{Haul time (six buckets)} \frac{3 \text{ miles} \times 5{,}280 \text{ ft/mile}}{88 \text{ fpm/mph} \times 13 \text{ mph}} = 13.85 \text{ min}$$

Step 4. Return time.

Rolling resistance	Table 5.1; earth, poorly maintained, 100 to 140 lb per ton
	Using an average value 120 lb per ton or 6.0%
Grade resistance	1%
Total resistance	7.0% [6.0% + (+1%)]
Empty truck weight	36,860 lb
Speed (Fig. 10.9)	22 mph

$$\text{Return time} \quad \frac{3 \text{ miles} \times 5{,}280 \text{ ft/mile}}{88 \text{ fpm/mph} \times 22 \text{ mph}} = 8.18 \text{ min}$$

Step 5. Dump time. Congestion on the fill, dump time expected to be 2 min.

Step 6. Truck cycle time.

	Five bucket loads on the truck (min)	Six bucket loads on the truck (min)
Load time	1.66	2.00
Haul time	11.25	13.85
Dump time	2.00	2.00
Return time	8.18	8.18
Truck cycle time	23.09	26.03

Step 7. Number of trucks required.

	Five bucket loads on the truck	Six bucket loads on the truck
Truck cycle time	23.09 min	26.03 min
Loader cycle time	1.66 min	2.00 min
Number of trucks	13.9	13.0

Step 8. Production. The number of trucks must be an integer number. For the case of five buckets to load the truck, consider using 13 or 14 trucks. If 13 trucks are used, the loader will have to wait for trucks; therefore, the truck cycle will control production.

Production (five buckets and 13 trucks) 16.5 lcy × 13 trucks

$$\times \frac{60 \text{ min}}{23.09 \text{ min}} = 557 \text{ lcy per hr}$$

If 14 trucks are used, loader will control production and the trucks will sometimes wait to be loaded.

Production (five buckets and 14 trucks) 16.5 lcy

$$\times \frac{60 \text{ min}}{1.66 \text{ min}} = 596 \text{ lcy per hr}$$

Considering the case of six bucket loads and using 13 trucks

Production (six buckets and 13 trucks) 18.3 lcy × 13 trucks

$$\times \frac{60 \text{ min}}{26.03 \text{ min}} = 548 \text{ lcy per hr}$$

In this particular case, considering only production, it is best to load light and use 14 trucks for an hourly production of 596 lcy. If the effort were made to better maintain the haul road, the rolling resistance would decrease, and the travel speed advantage to the lighter load would disappear.

For the case of an earth, compacted, and maintained haul road the rolling resistance would be between 40 and 70 lb per ton considering rubber high-pressure tires. Using an average condition, the total resistance for the haul would be just over 4% and the haul speed for both loading conditions (five or six buckets) would be 22 mph. This would reduce the number of trucks required because of the faster cycle time. Note in Table 10.1 that if the haul road is improved the same 596 lcy per hr production can be achieved with one less truck. The same production with fewer trucks will cause production cost to be reduced. The analysis also illustrates the fact that there is a mismatch between the size of the hoe bucket and the size of the truck. It is very inefficient to be loading so light or to be wasting so much spillage with the last bucket.

TABLE 10.1 | Production comparison with various numbers of bucket loads to fill the truck and different numbers of trucks

Total resistance ≈ 4%	Five bucket loads on the truck	Six bucket loads on the truck
Truck cycle time	20.02 min	20.36 min
Load time	1.66 min	2.00 min
Balance no. of trucks	12.06	10.18
Production 10 trucks	495 lcy/hr	539 lcy/hr
Production 11 trucks	544 lcy/hr	549 lcy/hr
Production 12 trucks	593 lcy/hr	549 lcy/hr
Production 13 trucks	596 lcy/hr	549 lcy/hr

Step 9. Efficiency.

Adjusted production   596 lcy per hr $\times \dfrac{50 \text{ min}}{60 \text{ min}} = 497 \text{ lcy per hr}$

Finally, this production can be converted as necessary into bank cubic yards or tons by using material property information specific to the job or average values as found in Table 4.3.

Adjusted production   497 lcy per hr $\times \dfrac{2{,}150 \text{ lb per lcy}}{2{,}000 \text{ lb per ton}} = 534 \text{ ton per hr}$

Adjusted production   497 lcy per hr × 0.74 = 367 bcy per hr

Because the cost of the excavation equipment is usually greater than the cost of a haul truck it is common practice to use a greater number of trucks than the balance number derived from the ratio of the loader and truck cycle times. When considering this decision the mechanical condition of the trucks should be considered. Another consideration is the availability of standby trucks. These are not necessarily idle units but could be trucks assigned to lower priority tasks from which they can easily be diverted.

After the job has started, the number of trucks required may vary because of changes in haul-road conditions, reductions or increases in the length of hauls, or changes in conditions at either the loading or dumping areas. Management should always continue to monitor hauling operations for changes in assumed conditions.

SUMMARY

The use of trucks as the primary hauling unit provides a high degree of flexibility, as the number in service can usually be increased or decreased easily to permit modifications in the total hauling capacity. When estimating what a truck will carry both the rated *gravimetric* load and the rated-heaped *volume* must be examined. The heaped capacity is the volume of material that the truck will haul when the load is heaped above the sides. The actual heaped capacity will vary with the material that is being hauled. Critical learning objectives include:

■ An understanding of the necessity to achieve balance between excavator bucket volume and truck load volume.

■ An ability to use performance charts to calculate truck speed.

■ An understanding of the job site constraints that affect dump times.

■ An ability to calculate the number of trucks required to keep the excavating equipment working at capacity.

These objectives are the basis for the problems that follow.

PROBLEMS

10.1 A truck for which the information in Fig. 10.9 applies operates over a haul road with a +3% slope and a rolling resistance of 140 lb per ton. If the gross vehicle weight is 90,000 lb, determine the maximum speed of the truck.
(rimpull 9,000 lb, speed 6.5 mph)

10.2 A truck for which the information in Fig. 10.9 applies operates over a haul road with a +4% slope and a rolling resistance of 90 lb per ton. If the gross vehicle weight is 70,000 lb, determine the maximum speed of the truck.

10.3 A truck for which the information in Fig. 10.9 applies operates over a haul road with a −4% slope and a rolling resistance of 200 lb per ton. If the gross vehicle weight is 80,000 lb, determine the maximum speed of the truck.
(rimpull 4,800 lb, speed 12 mph)

10.4 The truck of Problem 10.2 operates on a haul road having a −4% slope; determine the maximum speed.

10.5 An articulate truck weighs 46,300 lb empty and 96,300 lb when loaded. This truck has one front axle and two rear axles. The weight distribution empty is

58% front, 21% center, and 21% rear. The weight distribution loaded is 32% front, 34% center, and 34% rear. The truck has two tires on each axle. The truck works a 10-hr shift hauling sand. The one-way haul distance is 3.5 mi. The truck can make 24 trips per day. The truck is equipped with tires having a 110 TMPH rating. Are these satisfactory under these job conditions?

10.6   Prepare a table similar to Table 10.1, using a $3\frac{1}{2}$-cy shovel loading poorly blasted rock (2,600 lb per lcy) and 15- and 20-cy-size trucks. Assume that both trucks can handle the gravimetric load and the shovel bucket swing cycle time is 26 sec. Use a bucket fill factor of 100%. Using the information in Fig. 10.9, the net empty weight of the 15-cy truck is 44,000 lb and that of the 20-cy truck is 50,000 lb. Dump time will be 1.5 min. The haul distance is 4 mi from the excavation area to the fill up a 2% grade. The rolling resistance of the haul road will be maintained at 4%.

10.7   Evaluate cost and production using a 3-cy hoe to load wet gravel into 14-cy-size trucks. Assume a bucket fill factor of 105% and that the hoe bucket swing cycle time is 24 sec. Using the information in Fig. 10.9, the net empty wt. of the 14-cy truck is 36,860 lb and the rated payload is 40,000 lb. Dump time will be 1.3 min. The haul distance is 2.5 mi from the pit to the plant up a 3% grade. The rolling resistance of the haul road will be maintained at 3%. The hoe and operator cost $97 per hour and trucks cost $49 per hour. What is the most cost-effective mix of bucket loads to use in loading the trucks and number of trucks to place on this job?

REFERENCES

1.   *Care and Service of Off-the-Highway Tires,* Rubber Manufacturers Association, Washington, D.C.

2.   Gove, D., and W. Morgan, "Optimizing Truck-Loader Matching," *Mining Engineering,* Vol. 46, October 1994.

3.   Hull, Paul E., "Moving Materials," *World Highways/Routes Du Monde,* November–December 1999.

4.   Ingle, John H., "Good Tire Management Cuts Cost, Downtime," *Engineering and Mining Journal,* 16AA–16EE, February 1992.

5.   Karshenas, S., "Truck Capacity Selection for Earthmoving," *Journal of Construction Engineering and Management,* ASCE, Vol. 115, No. 2, pp. 212–227, June 1989.

6.   Monroe, D. M., "Optimizing Truck and Haul Road Economics," *Mining Engineering,* pp. 311–314, April 1992.

7.   Morgan, W. C., and L. L. Peterson, "Determining Shovel-Truck Productivity," *Mining Engineering,* pp. 76–80, December 1988.

8.   Schexnayder, Cliff, Sandra L. Weber, and Brentwood T. Brooks, "Effect of Truck Payload Weight on Production," *Journal of Construction Engineering and Management,* ASCE, Vol. 125, No. 1, pp. 1–7, January–February 1999.

9.   Zaburunov, Steven A., "Trends in Surface Mining Equipment," *Engineering and Mining Journal,* pp. 22–27, October 1990.

11

Compressed Air

Compressed air is used extensively on construction projects for powering rock drills, pile driving air hammers, air motors, hand tools, and pumps, and for cleaning. When air is compressed, it receives energy from the compressor. As air is a gas, it obeys the fundamental laws that apply to gases. The capacity of an air compressor is determined by the amount of free air that it can compress to a specified pressure in 1 min, under standard conditions. The loss in pressure due to friction as air flows through a pipe or a hose is a factor that must be considered in selecting the size of a pipe or hose to transport air to tools or equipment on the construction site.

GENERAL INFORMATION

Compressed air is used extensively on construction projects for powering rock drills, pile driving air hammers (see Fig. 11.1), air motors, hand tools, and pumps, and for cleaning. In many instances, the energy supplied by compressed air is the most convenient method of operating equipment and tools. A compressed-air system consists of one or more compressors together with a distribution system to carry the air to the points of use.

When air is compressed, it receives energy from the compressor. This energy is transmitted through a pipe or hose to the operating equipment, where a portion of the energy is converted into mechanical work. The operations of compressing, transmitting, and using air will always result in a loss of energy that will give an overall efficiency less than 100%, sometimes considerably less.

As air is a gas, it obeys the fundamental laws that apply to gases. The laws are related to the pressure, volume, temperature, and transmission of air.

GLOSSARY OF GAS-LAW TERMS

The following glossary defines the important terms used in developing and applying the laws that relate to compressed air.

FIGURE 11.1 | Air compressor mounted on the back of a crane for powering a pile hammer.

Absolute pressure. The total pressure measured from absolute zero. It is equal to the sum of the gauge and the atmospheric pressure, corresponding to the barometric reading. *The absolute pressure is used in dealing with the gas laws.*

Absolute temperature. The temperature of a gas measured above absolute zero. It equals degrees Fahrenheit plus 459.6 or, as more commonly used, 460.

Atmospheric pressure. The pressure exerted by Earth's atmosphere at any given position. Also referred to as barometric pressure.

Celsius temperature. The temperature indicated by a thermometer calibrated according to the Celsius scale. For this scale, pure water freezes at 0°C and boils at 100°C, at a pressure of 14.7 psi.

Density of air. The weight of a unit volume of air, usually expressed as pounds per cubic foot (pcf). Density varies with the pressure and temperature of the air. The weight of air at 60°F and 14.7 psi absolute pressure is 0.07658 pcf. The volume per pound is 13.059 cubic feet (cf).

Fahrenheit temperature. The temperature indicated by a measuring device calibrated according to the Fahrenheit scale. For this thermometer, pure water freezes at 32°F and boils at 212°F, at a pressure of 14.7 psi. Thus, the number of degrees between the freezing and boiling point of water is 180.

Gauge pressure. The pressure exerted by the air in excess of atmospheric pressure. It is usually expressed in psi or inches of mercury and is measured by a pressure gauge or a mercury manometer.

psf. The abbreviation for pounds per square foot of pressure.

psi. The abbreviation for pounds per square inch of pressure.

Relation between Fahrenheit and Celsius temperatures. A difference of 180° on the Fahrenheit scale equals 100° on the Celsius scale; 1°C equals 1.8°F. A Fahrenheit thermometer will read 32° when a Celsius thermometer reads 0°.

Let T_F = Fahrenheit temperature and T_C = Celsius temperature. For any given temperature, the thermometer readings are expressed by

$$T_F = 32 + 1.8T_C \qquad\qquad \text{[11.1]}$$

Standard conditions. Because of the variations in the volume of air with pressure and temperature, it is necessary to express the volume at standard conditions if it is to have a definite meaning. Standard conditions are an absolute pressure of 14.696 psi (14.7 psi is commonly used in practice) and a temperature of 60°F.

Temperature. A measure of the amount of heat contained by a unit quantity of gas (or other material). It is measured with a thermometer or some other suitable temperature-measuring device.

Vacuum. A measure of the extent to which pressure is less than atmospheric pressure. For example, a vacuum of 5 psi is equivalent to an absolute pressure of 14.7 − 5 = 9.7 psi.

TYPES OF COMPRESSION

Isothermal compression. When a gas undergoes a change in volume without any change in temperature, this is referred to as *isothermal expansion* or *compression*.

Adiabatic compression. When a gas undergoes a change in volume without gaining or losing heat, this is referred to as *adiabatic expansion* or *compression*.

Boyle's law states that when a gas is subjected to a change in volume due to a change in pressure, at a constant temperature, the product of the pressure times the volume will remain constant. This relation is expressed by the equation

$$P_1V_1 = P_2V_2 = K \qquad\qquad \text{[11.2]}$$

where

P_1 = initial absolute pressure
V_1 = initial volume
P_2 = final absolute pressure
V_2 = final volume
K = a constant

EXAMPLE 11.1

A compressor on a project takes in 1,000 cf of air and increases the gauge pressure from 20 to 120 psi, with no change in temperature. The barometer indicates an atmospheric pressure of 14.7 psi. Determine the final volume of the air.

$$P_1 = 20 + 14.7 = 34.7 \text{ psi}$$

$$P_2 = 120 + 14.7 = 134.7 \text{ psi}$$

$$V_1 = 1,000 \text{ cf}$$

From Eq. [11.2]

$$V_2 = \frac{P_1 V_1}{P_2} = \frac{34.7 \text{ psi} \times 1,000 \text{ cf}}{134.7 \text{ psi}} = 257.6 \text{ cf}$$

BOYLE'S AND CHARLES' LAW

When a gas undergoes a change in volume or pressure with a change in temperature, Boyle's law will not apply. Charles' law determines the effect of absolute temperature on the volume of a gas when the pressure is maintained constant. It states that the volume of a given weight of gas at constant pressure varies in direct proportion to its absolute temperature. It may be expressed mathematically by

$$\frac{V_1}{T_1} = \frac{V_2}{T_2} = C \tag{11.3}$$

where

V_1 = initial volume
T_1 = initial absolute temperature
V_2 = final volume
T_2 = final absolute temperature (A)
C = a constant

The laws of Boyle and Charles may be combined to give the equation

$$\frac{P_1 V_1}{T_1} = \frac{P_2 V_2}{T_2} = \text{a constant} \tag{11.4}$$

Equation [11.4] may be used to express the relations between pressure, volume, and temperature for any given gas, such as air. This is illustrated by Example 11.2.

EXAMPLE 11.2

One thousand cubic feet of air, at an initial gauge pressure of 40 psi and temperature of 50°F, is compressed to a volume of 200 cf at a final temperature of 110°F. Determine the final gauge pressure. The atmospheric pressure is 14.46 psi.

$$P_1 = 40 \text{ psi} + 14.46 \text{ psi} = 54.46 \text{ psi}$$

$$V_1 = 1,000 \text{ cf}$$

$$T_1 = 460° + 50°F = 510°A$$

$$V_2 = 200 \text{ cf}$$

$$T_2 = 460° + 110°F = 570°A$$

Rewriting Eq. [11.4] and substituting these values, we get

$$P_2 = \frac{P_1 V_1}{T_1} \times \frac{T_2}{V_2} \Rightarrow \frac{54.46 \text{ psi} \times 1,000 \text{ cf}}{510°A} \times \frac{570°A}{200 \text{ cf}} = 304.34 \text{ psi}$$

Final gauge: 304.34 psi − 14.46 psi = 289.88 psi or 290 psi

ENERGY REQUIRED TO COMPRESS AIR

Equation [11.2] can be expressed as $PV = K$, where K is a constant so long as the temperature remains constant. However, in actual practice, the temperature usually will not remain constant, and the equation must be modified to provide for the effect of changes in temperature. The effect of temperature may be provided for by introducing an exponent n to V. Thus, Eq. [11.2] can be rewritten as

$$P_1 V_1^n = P_2 V_2^n = K \qquad [11.5]$$

For air, the values of n will vary from 1.0 for isothermal (no change in temperature) compression to 1.4 for adiabatic (without gaining or losing heat) compression. The actual value for any compression condition can be determined experimentally from an indicator card obtained for a given compressor.

When the pressure of a given volume of air is increased by an air compressor, it is necessary to furnish energy to the air. Consider a single compression cycle for an air compressor, as indicated in Fig. 11.2. Air is drawn into the cylinder at pressure P_1 and is discharged at pressure P_2. P_1 does not need to be atmospheric pressure. The initial volume is V_1. As the piston compresses the air, the pressure-volume will follow the curve CD. At D, when the pressure is P_2, the discharge valve will open and the pressure will remain constant while the volume decreases to V_2, as indicated by line DE. Point E represents the end of the piston stroke. At point E, the discharge valve will close, and as the piston begins its return stroke, the pressure will decrease along line EB to a value of P_1, when the intake valve will open and allow additional air to enter the cylinder. This will establish line BC.

The work done along the line CD may be obtained by integrating the equation $dW = VdP$.

Equation [11.5] can be rearranged so that $V^n = K/P$. If both sides of the equation are raised to the $1/n$ power, the equation will be

$$V = \left(\frac{K}{P} \right)^{1/n}$$

Substituting this value of V gives

$$dW = \left(\frac{K}{P} \right)^{1/n} dP$$

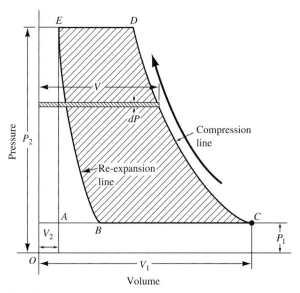

FIGURE 11.2 | Cycle for isothermal compression of air.

Integrating gives

$$W = K^{1/n} \int_1^2 \frac{dP}{P^{1/n}} \qquad \text{[11.6]}$$

For isothermal compression, $n = 1$. Substituting this value in Eq. [11.6] gives

$$W = K \int_1^2 \frac{dP}{P} = -K \log_e \frac{P_2}{P_1} + C$$

When $P_2 = P_1$ and no work is done, the constant of integration is equal to zero. The minus sign may be disregarded. Thus, for *isothermal compression of air*, the equation may be written as

$$W = K \log_e \frac{P_2}{P_1} \qquad \text{[11.7]}$$

If it is desired to convert from natural to common logarithms, $\log_e \frac{P_2}{P_1}$ can be replaced by $2.3026 \log \frac{P_2}{P_1}$. For the given compression conditions, $n = 1$ and $K = P_1 V_1$. If the compression begins with air at standard conditions, P_1 will be 14.7 psi at 60°F. Since work is commonly expressed in foot-pounds (ft-lb), it is necessary to express P_1 in psf. This is done by multiplying P_1 by 144. After making these substitutions and using exact values for e and P_1, one can write Eq. [11.7] as

$$W = 14.696 \text{ psi} \times \frac{144 \text{ in.}^2}{\text{ft}^2} V_1 \times 2.3026 \log \frac{P_2}{P_1}$$

$$= 4{,}873 \text{ psf } V_1 \log \frac{P_2}{P_1} \qquad \text{[11.8]}$$

The value of W will be in foot-pounds per cycle when V_1 is expressed as cubic feet. One horsepower is equivalent to 33,000 ft-lb per min. If V_1 in Eq. [11.8] is replaced by V, the volume of free air per minute at standard conditions, the horsepower required to compress V cf of air from an absolute pressure of P_1 to P_2 psi will be

$$\text{hp} = \frac{4{,}873V \log(P_2/P_1)}{33{,}000}$$

$$= 0.1477V \log \frac{P_2}{P_1} \qquad \textbf{[11.9]}$$

EXAMPLE 11.3

Determine the theoretical horsepower required to compress under isothermal conditions 100 cf of free air per minute, measured at standard conditions, from atmospheric pressure to 100 psi gauge pressure. Substituting in Eq. [11.9], the result is:

$$\text{hp} = 0.1477 \times 100 \text{ cf} \times \log \frac{114.7 \text{ psi}}{14.7 \text{ psi}}$$

$$= 14.77 \times \log 7.8027$$

$$= 14.77 \times 0.8922$$

$$= 13.2 \text{ hp}$$

If air is compressed under other than isothermal conditions, the equation for the required horsepower may be derived in a similar manner. However, since n will not equal 1, it must appear as an exponent in the equation. Equation [11.10] gives the *horsepower for nonisothermal conditions*:

$$\text{hp} = \frac{n}{n-1} 0.0642V \left[\left(\frac{P_2}{P_1} \right)^{(n-1)/n} - 1 \right] \qquad \textbf{[11.10]}$$

where the terms are the same as those used in Eq. [11.9].

EXAMPLE 11.4

Determine the theoretical horsepower required to compress 100 cf of free air per minute, measured at standard conditions, from atmospheric pressure 100 psi gauge pressure, under adiabatic conditions. The value n will be 1.4 for adiabatic compression of air. Substituting in Eq. [11.10], we get

$$\text{hp} = \frac{1.4}{1.4 - 1} 0.0642 \times 100(7.8027^{0.4/1.4} - 1)$$

$$= 22.47(7.8027^{0.2857} - 1)$$

$$= 22.47 \times 0.7986$$

$$= 17.9 \text{ hp}$$

For air compressors used on construction projects, the compression will be performed under conditions between isothermal and adiabatic. Thus, the theoretical horsepower will be between 13.2 and 17.9, the actual value depending on the extent to which the compressor is cooled during operation. The difference in the horsepower required (35% in this example) illustrates the importance of operating an air compressor at the lowest practical temperature.

EFFECT OF ALTITUDE ON THE POWER REQUIRED TO COMPRESS AIR

When a given volume of air, measured as free air prior to its entering a compressor, is compressed, the original pressure will average 14.7 psi absolute pressure at sea level. If the same volume of free air is compressed to the same gauge pressure at a higher altitude, the volume of the air after being compressed will be less than the volume compressed at sea level. The reason for this difference is that the density of a cubic foot of free air at higher altitudes is less than at sea level. Thus, while a compressor may compress air to the same discharge pressure at a higher altitude, the volume supplied in a given time interval will be less at the higher altitude. The use of Eq. [11.2] and information on pressure variation with altitude demonstrate this fact.

Because a compressor of a specified capacity actually supplies a smaller volume of air at a given discharge pressure at a higher altitude, it requires less power to operate a compressor at a higher altitude, as illustrated in Table 11.1.

GLOSSARY OF AIR-COMPRESSOR TERMS

The following glossary defines the important terms that are pertinent to air compressors.

Aftercooler. A heat exchanger that cools the air after it is discharged from a compressor.

Air compressor. A machine used to increase the pressure of air by reducing its volume.

Brake horsepower. The actual horsepower input required by a compressor.

Centrifugal compressor. A machine in which the compression is effected by a rotating vane or impeller that imparts velocity to the flowing air to give it the desired pressure.

cfm. The abbreviation for cubic feet per minute.

Compressor efficiency. The ratio of the theoretical horsepower to the brake horsepower.

Compression ratio. The ratio of the absolute discharge pressure to the absolute inlet pressure.

Discharge pressure. The absolute pressure of the air at the outlet from a compressor.

Diversity factor. The ratio of the actual quantity of air required for all uses to the sum of the individual quantities required for each use.

TABLE 11.1 | Theoretical horsepower required to compress 100 cf of free air per minute at different altitudes

| | Isothermal compression—single- and two-stage gauge pressure (psi) | | | | Adiabatic compression | | | | | | |
| | | | | | Single-stage gauge pressure (psi) | | | Two-stage gauge pressure (psi) | | | |
Altitude (ft)	60	80	100	125	60	80	100	60	80	100	125
0	10.4	11.9	13.2	14.4	13.4	15.9	18.1	11.8	13.7	15.4	17.1
1,000	10.2	11.7	12.9	14.1	13.2	15.6	17.8	11.6	13.5	15.1	16.8
2,000	10.0	11.4	12.6	13.8	13.0	15.4	17.5	11.4	13.2	14.8	16.4
3,000	9.8	11.2	12.3	13.5	12.8	15.2	17.2	11.2	13.0	14.5	16.1
4,000	9.6	11.0	12.1	13.2	12.6	14.9	16.9	11.0	12.7	14.2	15.7
5,000	9.4	10.7	11.8	12.8	12.4	14.7	16.5	10.8	12.5	13.9	15.4
6,000	9.2	10.5	11.5	12.5	12.2	14.4	16.2	10.6	12.2	13.6	15.1
7,000	9.0	10.3	11.2	12.2	12.0	14.2	16.0	10.4	12.0	13.4	14.8
8,000	8.9	10.0	11.0	11.9	11.8	14.0	15.7	10.2	11.8	13.1	14.5
9,000	8.7	9.8	10.7	11.6	11.6	13.7	15.4	10.0	11.6	12.8	14.1
10,000	8.5	9.6	10.4	11.4	11.5	13.5	15.1	9.8	11.3	12.6	13.8

Source: Compressed Air and Gas Institute.

Double-acting compressor. A machine that compresses air in both ends of a cylinder.

Free air. Air as it exists under atmospheric conditions at any given location.

Inlet pressure. The absolute pressure of the air at the inlet to a compressor.

Intercooler. A heat exchanger that is placed between two compression stages to remove the heat of compression from the air.

Load factor. The ratio of the average load during a given period of time to the maximum rated load of a compressor.

Multistage compressor. A compressor that produces the desired final pressure through two or more stages.

Reciprocating compressor. A machine that compresses air by means of a piston reciprocating in a cylinder.

Rotary compressor. A machine in which the compression is effected by the action of rotating elements.

Single-acting compressor. A machine that compresses air in only one end of a cylinder.

Single-stage compressor. A machine that compresses air from atmospheric pressure to the desired discharge pressure in a single operation.

Theoretical horsepower. The horsepower required to adiabatically compress the air delivered by a compressor through the specified pressure range, without any provision for lost energy.

Two-stage compressor. A machine that compresses air in two separate operations. The first operation compresses the air to an intermediate pressure, whereas the second operation further compresses it to the desired final pressure.

Volumetric efficiency. The ratio of the capacity of a compressor to the piston displacement of the compressor.

COMPRESSORS

The capacity of an air compressor is determined by the amount of free air that it can compress to a specified pressure in 1 min, under standard conditions (absolute pressure of 14.7 psi at 60°F). The number of pneumatic tools that can be operated from one air compressor depends on the air requirements of the specific tools. On construction projects, it is important to always locate air compressors upwind from the work to keep foreign material out of the air intake.

Stationary Compressors

Stationary compressors are generally used for installations where there will be a requirement for compressed air over a long duration of time at fixed locations. The compressors may be reciprocating or rotary types and single-stage or multistage. One or more compressors may supply the total quantity of air. The installed cost of a single compressor will usually be less than for several compressors having the same capacity. However, several compressors provide better flexibility for varying load demands, and in the event of a shutdown for repairs the entire plant does not need to be stopped.

Steam, electric motors, or internal-combustion engines can drive stationary compressors.

Portable Compressors

Portable compressors are more commonly used on construction sites where it is necessary to meet frequently changing job demands, typically at a number of locations on the job site. The compressors may be mounted on rubber tires (see Fig. 11.3), steel wheels, or skids. They are usually powered by gasoline or diesel engines and most of those used in construction work are of the rotary-screw type.

Reciprocating Compressors

A reciprocating compressor depends on a piston that moves back and forth in a cylinder, for the compressing action. The piston may compress air while moving in one or both directions. For the former, it is defined as single acting, while for the latter it is defined as double acting. A compressor may have one or more cylinders.

Rotary-Screw Compressors

These machines offer several advantages compared with reciprocating compressors, such as compactness, light weight, uniform flow, variable output, maintenance-free operation, and long life (see Fig. 11.4). The working parts of a screw compressor are two helical rotors. The male rotor has four lobes and rotates 50% faster than the female rotor, which has six flutes with which the male motor meshes. As the air enters and flows through the compressor, it is compressed in the space between the lobes and the flutes. The inlet and outlet ports are

FIGURE 11.3 I Portable air compressor on a construction project.

FIGURE 11.4 I Portable rotary-screw air compressor, 185 cfm.

Source: Sullair Corp.

automatically covered and uncovered by the shaped ends of the rotors as they turn. There are also monoscrew compressors that operate in a similar manner.

Rotary-screw compressors are available in a relatively wide range of capacities, with single-stage or multistage compression and with rotors that operate under oil-lubricated conditions or with no oil, the latter to produce oil-free air.

Rotary-screw compressors offer several advantages when compared with other types of compressors, including but not limited to

1. Quiet operation, with little or no loss in output, to satisfy a wide range of legal requirements limiting permissible noise
2. Few moving parts, with minimum mechanical wear and few maintenance requirements
3. Automatic controls actuated by the output pressure that regulate the speed of the driving unit and the compressor to limit the output to only the demand required
4. Little or no pulsation in the flow of air, and hence reduced vibrations

Table 11.2 lists information applicable for representative screw compressors operated under absolute inlet air pressure equal to 1 bar (14.5 psi or 1.02 kp/sq cm) and inlet air temperature and inlet coolant temperature equal to 60°F (15°C).

TABLE 11.2 | Representative specifications for rotary screw air compressors

Free air delivered		Rated operating pressure	
cfm	L/sec	psi gauge	bar
100	47	100	6.9
130	61	100	6.9
160	75	100	6.9
175	82	100	6.9
185	87	100	6.9
185	87	125	8.6
250	118	100	6.9
300	142	150	10.3
375	177	100	6.9
400	189	200	13.8
450	212	150	10.3
525	248	125	8.6
600	283	100	6.9

COMPRESSOR CAPACITY

Air compressors are rated by the piston displacement in cfm. However, the capacity of a compressor will be less than the piston displacement because of valve and piston leakage and the air left in the end-clearance spaces of the cylinders.

The capacity of a compressor is the actual volume of free air drawn into a compressor in a minute. It is expressed in cubic feet. For a reciprocating compressor in good mechanical condition, the actual capacity should be 80 to 90% of the piston displacement. This is illustrated by an analysis of a 315-cfm two-stage portable compressor. The manufacturer's specifications give this information:

Number of low-pressure cylinders, 4
Number of high-pressure cylinders, 2
Diameter of low-pressure cylinders, 7 in.

Diameter of high-pressure cylinders, $5\frac{3}{4}$ in.

Length of stroke, 5 in.

rpm, 870

Consider only the piston displacement of the low-pressure cylinders as they determine the capacity of the unit.

Area of cylinder, $\dfrac{\pi \times (7 \text{ in.})^2}{4 \text{ cylinders} \times 144 \text{ in.}^2/\text{sf}} = 0.2673 \text{ sf}$

Displacement per cylinder per stroke, $0.2673 \text{ sf} \times \dfrac{5 \text{ in.}}{12 \text{ in./ft}} = 0.1114 \text{ cf}$

Displacement per minute, 4 cylinders $\times$ 0.1114 cf $\times$ 870 rpm = 388 cfm

Specified capacity, 315 cfm

Volumetric efficiency, $\dfrac{315 \text{ cfm}}{388 \text{ cfm}} \times 100 = 81\%$

EFFECT OF ALTITUDE ON COMPRESSOR CAPACITY

The capacity of an air compressor is rated on the basis of its performance at sea level, where the normal absolute barometric pressure is about 14.7 psi. If a compressor is operated at a higher altitude, such as 5,000 ft above sea level, the absolute barometric pressure will be about 12.2 psi. Thus at the higher altitude, air density is less and the weight of air in a cubic foot of free volume is less than at sea level. If the air is discharged by the compressor at a given pressure, the compression ratio will be increased, and the capacity of the compressor will be reduced. This can be demonstrated by applying Eq. [11.2].

Assume the 100 cf of free air at sea level is compressed to 100-psi gauge with no change in temperature. Applying Eq. [11.2], we obtain

$$V_2 = \frac{P_1 V_1}{P_2}$$

where

$V_1 = 100 \text{ cf}$

$P_1 = 14.7 \text{ psi absolute}$

$P_2 = 114.7 \text{ psi absolute}$

$V_2 = \dfrac{14.7 \text{ psi} \times 100 \text{ cf}}{114.7 \text{ psi}} = 12.82 \text{ cf}$

At 5,000 ft above sea level,

$V_1 = 100 \text{ cf}$

$P_1 = 12.2 \text{ psi absolute}$

$P_2 = 112.2 \text{ psi absolute}$

$V_2 = \dfrac{12.2 \text{ psi} \times 100 \text{ cf}}{112.2 \text{ psi}} = 10.87 \text{ cf}$

Table 11.3 shows the percentage of volumetric efficiency for various types of compressors at different altitudes. The table is developed based on the compressor delivering the air at 100-psi gauge.

TABLE 11.3 | Efficiency of air compressors at various altitudes based on a 100-psi gauge output pressure

Altitude above sea level		Single-stage reciprocating compressor (percent efficiency)	Two-stage reciprocating compressor (percent efficiency)	Rotary compressor (percent efficiency)
ft	m			
2,000	610	98.7	99.4	100.0
5,000	1,525	92.5	98.5	100.0
7,000	2,135	—	—	100.0
8,000	2,440	87.3	97.6	99.9
10,000	3,050	84.0	97.0	—
12,000	3,660	—	—	98.6

INTERCOOLERS

Frequently intercoolers are installed between the stages of a compressor to reduce the temperature of the air and to remove moisture from the air. The reduction in temperature prior to additional compression can reduce the total power required by as much as 15%. Unless an intercooler is installed, the power required by a two-stage compressor will be the same as for a single-stage compressor.

An intercooler requires a continuous supply of circulating cool water to remove the heat from the air. It will require 1.0 to 1.5 gal of water per minute for each 100 cfm of air compressed, the actual amount depending on the temperature of the water.

AFTERCOOLERS

Aftercoolers are sometimes installed at the discharge side of a compressor to cool the air to the desired temperature and to remove moisture from the air. It is highly desirable to remove excess moisture from the air, as it tends to freeze during expansion in air tools, and it washes the lubricating oil out of tools, thereby reducing the lubricating efficiency.

RECEIVERS

An air receiver should be installed on the discharge side of a compressor to equalize the compressor pulsations and to serve as a condensing chamber for the removal of water and oil vapors. A receiver should have a drain cock at its bottom to permit the removal of the condensate. Its volume should be one-tenth to one-sixth of the capacity of the compressor. A blowoff valve to limit the maximum pressure is desirable.

COMPRESSED-AIR DISTRIBUTION SYSTEM

The objective of installing a compressed-air distribution system is to provide a sufficient volume of air to the work site at pressures adequate for efficient tool operation. Any drop in pressure between the compressor and the point of use is an irretrievable loss. Therefore, the distribution system is an important element in the total air supply scheme. These are general rules for designing a compressed-air distribution system:

- Pipe sizes should be large enough so that the pressure drop between the compressor and the point of use does not exceed 10% of the initial pressure.
- Each header or main line should be provided with outlets as close as possible to the point of use. This permits shorter hose lengths and avoids large pressure drops through the hose.
- Condensate drains should be located at appropriate low points along the header or main lines.

Air Manifolds

Many construction projects require more compressed air per minute than any single job-site compressor will produce. An air manifold is a large-diameter pipe used to transport compressed air from one or more air compressors without a detrimental friction-line loss.

Manifolds can be constructed of any durable pipe. Compressors are connected to the manifold with flexible hoses. A one-way check valve must be installed between the compressor and the manifold. This valve keeps manifold back pressure from possibly forcing air back into an individual compressor's receiver tank. The compressors grouped to supply an air manifold may be of different capacities, but the final discharge pressure of each should be coordinated at a specified pressure. In the case of construction, this is usually 100 psi. Compressors of different types should not be used on the same manifold. The difference in the pressure control systems of a rotary and a reciprocating compressor could cause one compressor to become overloaded while the other compressor idles.

Loss of Air Pressure in Pipe

The loss in pressure due to friction as air flows through a pipe or a hose is a factor that must be considered in selecting the size of a pipe or hose. Failure to use a sufficiently large line may cause the air pressure to drop so low that it will not satisfactorily operate the tool to which it is providing power.

Pipe size selection for an air line is an economics problem. The efficiency of most equipment operated by compressed air drops off rapidly as the pressure of the air is reduced. When the cost of lost efficiency exceeds the cost of providing a larger line, it is cost effective to install a larger line. The manufacturers of pneumatic equipment generally specify the minimum air pressure at which the equipment will operate satisfactorily. However, these values should be considered as minimum and not desirable operating pressures. The actual pressure should be higher than the specified minimum.

EXAMPLE 11.5

The cost of lost efficiency on a project resulting from the operation of pneumatic equipment at reduced pressure is estimated to be $1,000. Installing a larger pipe at an additional cost of $600 can eliminate the lost efficiency. In this instance the contractor will save $400 by installing the larger pipe. Thus, it is cost effective to spend $600 to eliminate an operating loss of $1,000.

Several formulas are used to determine the loss of pressure in a pipe due to friction. This equation has been used extensively:

$$f = \frac{CL}{r} \times \frac{Q^2}{d^5} \qquad [11.11]$$

where

f = pressure drop in psi
L = length of pipe, in feet
Q = cubic feet of free air per second
r = ration of compression
d = actual inside diameter (ID) of pipe in inches
C = experimental coefficient

For ordinary steel pipe, the value of C has been found to equal $0.1025/d^{0.31}$. If this value is substituted in Eq. [11.11], the result is

$$f = \frac{0.1025L}{r} \times \frac{Q^2}{d^{5.31}} \qquad [11.12]$$

A chart for determining the loss in pressure in a pipe is given in Fig. 11.5.

EXAMPLE 11.6

This example illustrates the use of the chart in Fig. 11.5. Determine the pressure loss per 100 ft of pipe resulting from transmitting 1,000 cfm of free air, at 100 psi gauge pressure, through a 4-in. standard-weight steel pipe. Enter the chart at the top at 100 psi and proceed vertically downward to a point opposite 1,000 cfm. From the 1,000 cfm intersect point, proceed parallel to the sloping guide lines to a point opposite the 4-in. pipe; then proceed vertically downward to the bottom of the chart, where the pressure drop is indicated to be 0.225 psi per 100 ft of pipe.

Table 11.4 gives the loss of air pressure in 1,000 ft of standard-weight pipe due to friction. For longer or shorter lengths of pipe, the friction loss will be in proportion to the length.

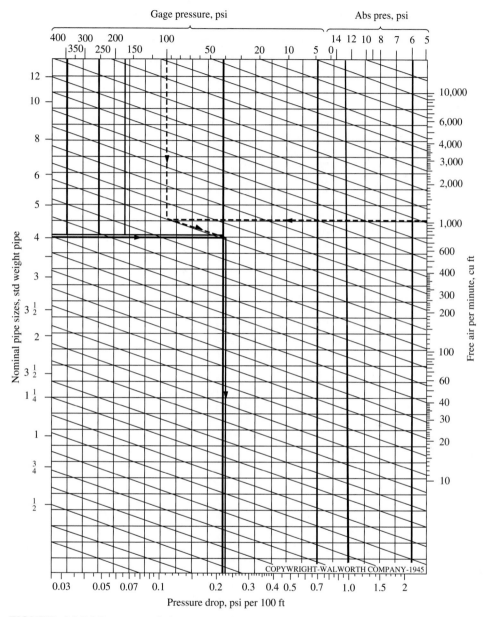

FIGURE 11.5 I Compressed-air pressure drop chart.

The losses given in the table are for an initial gauge pressure of 100 psi. If the initial pressure is other than 100 psi, the corresponding losses may be obtained by multiplying the values in Table 11.4 by a suitable factor. Reference to Eq. [11.12] reveals that for a given rate of flow through a given size pipe the only variable is r, that is, the ratio of compression, based on absolute pressures.

TABLE 11.4 | Loss of pressure in psi in 1,000 ft of standard-weight pipe due to friction for an initial gauge pressure of 100 psi

Free air per min (cf)	Nominal diameter (in.)												
	$\frac{1}{2}$	$\frac{3}{4}$	1	$1\frac{1}{4}$	$1\frac{1}{2}$	2	$2\frac{1}{2}$	3	$3\frac{1}{2}$	4	$4\frac{1}{2}$	5	6
10	6.50	0.99	0.28										
20	25.90	3.90	1.11	0.25	0.11								
30	68.50	9.01	2.51	0.57	0.26								
40	—	16.00	4.45	1.03	0.46								
50	—	25.10	6.96	1.61	0.71	0.19							
60	—	36.20	10.00	2.32	1.02	0.28							
70	—	49.30	13.70	3.16	1.40	0.37							
80	—	64.50	17.80	4.14	1.83	0.49	0.19						
90	—	82.80	22.60	5.23	2.32	0.62	0.24						
100	—	—	27.90	6.47	2.86	0.77	0.30						
125	—	—	48.60	10.20	4.49	1.19	0.46						
150	—	—	62.80	14.60	6.43	1.72	0.66	0.21					
175	—	—	—	19.80	8.72	2.36	0.91	0.28					
200	—	—	—	25.90	11.40	3.06	1.19	0.37	0.17				
250	—	—	—	40.40	17.90	4.78	1.85	0.58	0.27				
300	—	—	—	58.20	25.80	6.85	2.67	0.84	0.39	0.20			
350	—	—	—	—	35.10	9.36	3.64	1.14	0.53	0.27			
400	—	—	—	—	45.80	12.10	4.75	1.50	0.69	0.35	0.19		
450	—	—	—	—	58.00	15.40	5.98	1.89	0.88	0.46	0.25		
500	—	—	—	—	71.60	19.20	7.42	2.34	1.09	0.55	0.30		
600	—	—	—	—	—	27.60	10.70	3.36	1.56	0.79	0.44		
700	—	—	—	—	—	37.70	14.50	4.55	2.13	1.09	0.59		
800	—	—	—	—	—	49.00	19.00	5.89	2.77	1.42	0.78		
900	—	—	—	—	—	62.30	24.10	7.60	3.51	1.80	0.99		
1,000	—	—	—	—	—	76.90	29.80	9.30	4.35	2.21	1.22		
1,500	—	—	—	—	—	—	67.00	21.00	9.80	4.90	2.73	1.51	0.57
2,000	—	—	—	—	—	—	—	37.40	17.30	8.80	4.90	2.73	0.99
2,500	—	—	—	—	—	—	—	58.40	27.20	13.80	8.30	4.20	1.57
3,000	—	—	—	—	—	—	—	84.10	39.10	20.00	10.90	6.00	2.26
3,500	—	—	—	—	—	—	—	—	58.20	27.20	14.70	8.20	3.04
4,000	—	—	—	—	—	—	—	—	69.40	35.50	19.40	10.70	4.01
4,500	—	—	—	—	—	—	—	—	—	45.00	24.50	13.50	5.10
5,000	—	—	—	—	—	—	—	—	—	55.60	30.20	16.80	6.30
6,000	—	—	—	—	—	—	—	—	—	80.00	43.70	24.10	9.10
7,000	—	—	—	—	—	—	—	—	—	—	59.50	32.80	12.20
8,000	—	—	—	—	—	—	—	—	—	—	77.50	42.90	16.10
9,000	—	—	—	—	—	—	—	—	—	—	—	54.30	20.40
10,000	—	—	—	—	—	—	—	—	—	—	—	—	25.10
11,000	—	—	—	—	—	—	—	—	—	—	—	—	30.40
12,000	—	—	—	—	—	—	—	—	—	—	—	—	36.20
13,000	—	—	—	—	—	—	—	—	—	—	—	—	42.60
14,000	—	—	—	—	—	—	—	—	—	—	—	—	49.20
15,000	—	—	—	—	—	—	—	—	—	—	—	—	56.60

For a gauge pressure of 100 psi, $r = \frac{114.7}{14.7} = 7.80$, whereas for a gauge pressure of 80 psi, $r = \frac{94.7}{14.7} = 6.44$. The ratio of these values of $r = \frac{7.80}{6.44} = 1.211$. Thus, the loss for an initial pressure of 80 psi will be 1.211 times the loss for an initial pressure of 100 psi. For other initial pressures the factors are given below:

Gauge pressure (psi)	Factor
80	1.211
90	1.095
100	1.000
110	0.912
120	0.853
125	0.822

To calculate the loss of pressure resulting from the flow of air through fittings, it is common practice to convert a fitting to equivalent lengths of pipe having the same nominal diameter. This equivalent length should be added to the actual length of the pipe in determining pressure loss. Table 11.5 gives the equivalent length of standard-weight pipe for computing pressure losses.

TABLE 11.5 | Equivalent length in feet of standard-weight pipe having the same pressure losses as screwed fittings

Nominal pipe size (in.)	Gate valve	Globe valve	Angle valve	Long-radius ell or on run of standard tee	Standard ell or on run of tee	Tee through side outlet
$\frac{1}{2}$	0.4	17.3	8.6	0.6	1.6	3.1
$\frac{3}{4}$	0.5	22.9	11.4	0.8	2.1	4.1
1	0.6	29.1	14.6	1.1	2.6	5.2
$1\frac{1}{4}$	0.8	38.3	19.1	1.4	3.5	6.9
$1\frac{1}{2}$	0.9	44.7	22.4	1.6	4.0	8.0
2	1.2	57.4	28.7	2.1	5.2	10.3
$2\frac{1}{2}$	1.4	68.5	34.3	2.5	6.2	12.3
3	1.8	85.2	42.6	3.1	6.2	15.3
4	2.4	112.0	56.0	4.0	7.7	20.2
5	2.9	140.0	70.0	5.0	10.1	25.2
6	3.5	168.0	84.1	6.1	15.2	30.4
8	4.7	222.0	111.0	8.0	20.0	40.0
10	5.9	278.0	139.0	10.0	25.0	50.0
12	7.0	332.0	166.0	11.0	29.8	59.6

Recommended Sizes of Pipe for Transmitting Compressed Air

In transmitting air from a compressor to pneumatic equipment, it is necessary to limit the pressure drop along the line. If this precaution is not taken, the pressure may drop below that for which the equipment was designed and production will suffer.

At least two factors should be considered in determining the minimum size pipe:

1. The necessity of supplying air at the required pressure.
2. The desirability of supplying energy (compressed air) at the lowest total cost, taking into account the cost of the pipe and the cost of production obtained from the equipment being powered by the air.

Considering the first factor, a smaller pipe may be used for a short run than for a long run. While this is possible, it may not be economical. For the latter factor, economy may dictate the use of a pipe larger than the minimum possible size. The cost of installing large pipe will be more fully justified for an installation that will be used for a long period of time than for one that will be used for a short period of time.

No book, table, or fixed data can give the correct pipe size for all installations. The correct method of determining the size pipe for a given installation is to make a complete engineering analysis of the particular operations required.

Table 11.6 gives recommended sizes of pipe for transmitting compressed air for various lengths of run. This information is useful as a guide in selecting pipe sizes.

TABLE 11.6 | Recommended pipe sizes for transmitting compressed air at 80 to 125 psi gauge

Volume of air (cfm)	Length of pipe (ft)				
	50–200	200–500	500–1,000	1,000–2,500	2,500–5,000
	Nominal size pipe (in.)				
30–60	1	1	$1\frac{1}{4}$	$1\frac{1}{2}$	$1\frac{1}{2}$
60–100	1	$1\frac{1}{4}$	$1\frac{1}{4}$	2	2
100–200	$1\frac{1}{4}$	$1\frac{1}{2}$	2	$2\frac{1}{2}$	$2\frac{1}{2}$
200–500	2	$2\frac{1}{2}$	3	$3\frac{1}{2}$	$3\frac{1}{2}$
500–1,000	$2\frac{1}{2}$	3	$3\frac{1}{2}$	4	$4\frac{1}{2}$
1,000–2,000	$2\frac{1}{2}$	4	$4\frac{1}{2}$	5	6
2,000–4,000	$3\frac{1}{2}$	5	6	8	8
4,000–8,000	6	8	8	10	10

Loss of Air Pressure in Hose

Air-line hose is a rubber-covered, pressure-type hose used for transmitting compressed air. Hose is usually furnished equipped with quick-acting fittings for attaching a tool, a compressor, or another hose. Hose size is based on the amount of air that must be delivered to the tool. When transmitting compressed air at 80 to 125 psi gauge the guidance described here is valid for short hose lengths (25 ft or less). As the required hose length increases, the nominal size must also be increased. The guidelines in selecting air-line hose are

- Nominal $\frac{1}{4}$-in. size is recommended for tools requiring up to 15 cfm of air: small drills and air hammers.
- Nominal $\frac{3}{8}$-in. size is recommended for tools requiring up to 40 cfm of air: impact wrenches, grinders, and chipping hammers.
- Nominal $\frac{1}{2}$-in. size is recommended for tools requiring up to 80 cfm of air: heavy chipping and rivet hammers.
- Nominal $\frac{3}{4}$-in. size is recommended for tools requiring up to 100 cfm of air: 35- to 55-lb rock drills, large concrete vibrators, and sump pumps.
- Nominal 1-in. size is recommended for tools requiring 100 to 200 cfm of air: 75-lb rock drills, and drifters.

The loss of pressure resulting from the flow of air through hose is given in Table 11.7.

TABLE 11.7 I Loss of pressure, in psi, in 50 ft of hose and ending couplings

Size of hose (in.)	Gauge pressure at line (psi)	Volume of free air through hose (cfm)													
		20	30	40	50	60	70	80	90	100	110	120	130	140	150
$\frac{1}{2}$	50	1.8	5.0	10.1	18.1										
	60	1.3	4.0	8.4	14.8	23.5									
	70	1.0	3.4	7.0	12.4	20.0	28.4								
	80	0.9	2.8	6.0	10.8	17.4	25.2	34.6							
	90	0.8	2.4	5.4	9.5	14.8	22.0	30.5	41.0						
	100	0.7	2.3	4.8	8.4	13.3	19.3	27.2	36.6						
	110	0.6	2.0	4.3	7.6	12.0	17.6	24.6	33.3	44.5					
$\frac{3}{4}$	50	0.4	0.8	1.5	2.4	3.5	4.4	6.5	8.5	11.4	14.2				
	60	0.3	0.6	1.2	1.9	2.8	3.8	5.2	6.8	8.6	11.2				
	70	0.2	0.5	0.9	1.5	2.3	3.2	4.2	5.5	7.0	8.8	11.0			
	80	0.2	0.5	0.8	1.3	1.9	2.8	3.6	4.7	5.8	7.2	8.8	10.6		
	90	0.2	0.4	0.7	1.1	1.6	2.3	3.1	4.0	5.0	6.2	7.5	9.0		
	100	0.2	0.4	0.6	1.0	1.4	2.0	2.7	3.5	4.4	5.4	6.6	7.9	9.4	11.1
	110	0.1	0.3	0.5	0.9	1.3	1.8	2.4	3.1	3.9	4.9	5.9	7.1	8.4	9.9
1	50	0.1	0.2	0.3	0.5	0.8	1.1	1.5	2.0	2.6	3.5	4.8	7.0		
	60	0.1	0.2	0.3	0.4	0.6	0.8	1.2	1.5	2.0	2.6	3.3	4.2	5.5	7.2
	70	—	0.1	0.2	0.4	0.5	0.7	1.0	1.3	1.6	2.0	2.5	3.1	3.8	4.7
	80	—	0.1	0.2	0.3	0.5	0.7	0.8	1.1	1.4	1.7	2.0	2.4	2.7	3.5
	90	—	0.1	0.2	0.3	0.4	0.6	0.7	0.9	1.2	1.4	1.7	2.0	2.4	2.8
	100	—	0.1	0.2	0.2	0.4	0.5	0.6	0.8	1.0	1.2	1.5	1.8	2.1	2.4
	110	—	0.1	0.2	0.2	0.3	0.4	0.6	0.7	0.9	1.1	1.3	1.5	1.8	2.1
$1\frac{1}{4}$	50	—	—	0.2	0.2	0.2	0.3	0.4	0.5	0.7	1.1				
	60	—	—	—	0.1	0.2	0.3	0.3	0.5	0.6	0.8	1.0	1.2	1.5	
	70	—	—	—	0.1	0.2	0.2	0.3	0.4	0.4	0.5	0.7	0.8	1.0	1.3
	80	—	—	—	—	0.1	0.2	0.2	0.3	0.4	0.5	0.6	0.7	0.8	1.0
	90	—	—	—	—	0.1	0.2	0.2	0.3	0.3	0.4	0.5	0.6	0.7	0.8
	100	—	—	—	—	—	0.1	0.2	0.2	0.3	0.4	0.4	0.5	0.6	0.7
	110	—	—	—	—	—	0.1	0.2	0.2	0.3	0.3	0.4	0.5	0.5	0.6
$1\frac{1}{2}$	50	—	—	—	—	—	0.1	0.2	0.2	0.2	0.3	0.3	0.4	0.5	0.6
	60	—	—	—	—	—	—	0.1	0.2	0.2	0.2	0.3	0.3	0.4	0.5
	70	—	—	—	—	—	—	0.1	0.2	0.2	0.2	0.3	0.3	0.4	
	80	—	—	—	—	—	—	—	0.1	0.2	0.2	0.2	0.2	0.3	0.4
	90	—	—	—	—	—	—	—	—	0.1	0.2	0.2	0.2	0.2	0.2
	100	—	—	—	—	—	—	—	—	—	0.1	0.2	0.2	0.2	
	110	—	—	—	—	—	—	—	—	—	0.1	0.2	0.2	0.2	

DIVERSITY FACTOR

It is necessary to provide as much compressed air as required. However, to supply the needs of all operating equipment and the many different tools that can be connected to one system (see Fig. 11.6) would require more air capacity than is actually needed. It is probable that all equipment nominally used on a project

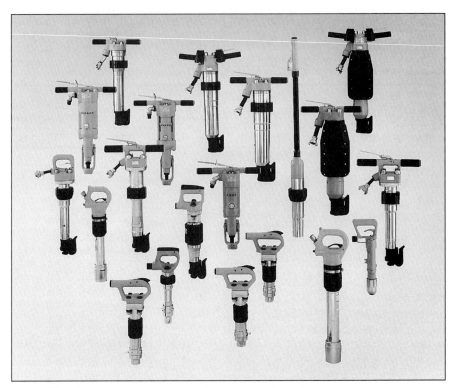

FIGURE 11.6 | Air-operated tools that can be connected to an air system.
Source: Sullair Corp.

will not be in operation at the exact same time. An analysis of the job should be made to determine the probable maximum *actual* need, prior to designing the compressed-air system.

As an example, if 10 jackhammers are on a job drilling, normally not more than 5 or 6 will be consuming air at a given time. The others will be out of use temporarily for changes in bits or drill steel or moving to new locations. Thus, the *actual* amount of air demand should be based on 5 or 6 drills instead of 10. The same condition will apply to other pneumatic tools.

The *diversity factor* is the ratio of the actual load to the maximum calculated load that would exist if all tools were operating at the same time. For example, if a jackhammer required 90 cfm of air, 10 hammers would require a total of 900 cfm if they were all operated at the same time. However, with only 5 hammers operating at one time, the demand for air would be 450 cfm. Thus, the diversity factor would be 450 ÷ 900 = 0.5.

The approximate quantities of compressed air required by pneumatic equipment and tools are given in Table 11.8. The quantities are based on continuous operation at a pressure of 90 psi gauge.

TABLE 11.8 | Quantities of compressed air required by pneumatic equipment and tools*

| Equipment or tools | Capacity or size | | Air consumption (cfm) |
	Weight (lb)	Depth of hole (ft)	
Jackhammers	10	0–2	15–25
	15	0–2	20–35
	25	2–8	30–50
	35	8–12	55–75
	45	12–16	80–100
	55	16–24	90–110
	75	8–24	150–175
Paving breakers	35	—	30–35
	60	—	40–45
	80	—	50–50

Equipment or tools	Capacity or size	Air consumption (cfm)
Chipping hammers	Light	15–25
	Heavy	25–30
Clay diggers	Light, 20 lb	20–25
	Medium, 25 lb	25–30
	Heavy, 35 lb	30–35
Concrete vibrators	$2\frac{1}{2}$-in.-tube diameter	20–30
	3-in.-tube diameter	40–50
	4-in.-tube diameter	45–55
	5-in.-tube diameter	75–85
Drills or borers	1-in. diameter	35–40
	2-in. diameter	50–75
	4-in. diameter	50–75
Hoist	Single-drum, 2,000-lb pull	200–220
	Double-drum, 2,400-lb pull	250–260
Impact wrenches	$\frac{5}{8}$-in. bolt	15–20
	$\frac{3}{4}$-in. bolt	30–40
	$1\frac{1}{4}$-in. bolt	60–70
	$1\frac{1}{2}$-in. bolt	70–80
	$1\frac{3}{4}$-in. bolt	80–90
Saws		
Circular	12-in. blade	40–60
Chain	18–30-in. blade	85–95
	36-in. blade	135–150
	48-in. blade	150–160
Reciprocating	20-in.	45–50
Spray guns	Light duty	2–3
	Medium duty	8–15
	Heavy duty	14–30
Sump pump	Single-stage, 10–40-ft head	80–90
	Single-stage, 100–150-ft head	150–170
	Two-stage, 100–150-ft head	160–180
Trampers, earth	35 lb	30–35
	60 lb	40–45
	80 lb	50–60
Wagon drills—drifters	3-in. piston	150–175
	$3\frac{1}{2}$-in. piston	180–210
	4-in. piston	225–275

*Air pressure at 90 psi gauge.

EFFECTS OF ALTITUDE ON THE CONSUMPTION OF AIR BY ROCK DRILLS

As already explained in this chapter, the capacity of an air compressor is the volume of free air that enters the compressor during a stated time, usually expressed in cfm. Because of the lower atmospheric pressure at higher altitudes, the quantity of air supplied by a compressor at a given gauge pressure will be less than at sea level. It is necessary to provide more compressor capacity at higher altitudes to assure an adequate supply of air at the specified pressure to rock drills.

Table 11.9 gives representative factors to be applied to specified compressor capacities to determine the required capacities at different altitudes. For example, if a single drill requires a capacity of 600 cfm of air at sea level, it will require a capacity of 600 × 1.2 = 720 cfm at an altitude of 5,000 ft, and 600 × 1.3 = 780 cfm at an altitude of 10,000 ft.

TABLE 11.9 | Factors to be used in determining the capacities of compressed air required by rock drills at different altitudes

	Number of drills									
	1	2	3	4	5	6	7	8	9	10
Altitude (ft)	Factor									
0	1.0	1.8	2.7	3.4	4.1	4.8	5.4	6.0	6.5	7.1
1,000	1.0	1.9	2.8	3.5	4.2	4.9	5.6	6.2	6.7	7.3
2,000	1.1	1.9	2.9	3.6	4.4	5.1	5.8	6.4	7.0	7.6
3,000	1.1	2.0	3.0	3.7	4.5	5.3	5.9	6.6	7.2	7.8
4,000	1.1	2.1	3.1	3.9	4.7	5.5	6.1	6.8	7.4	8.1
5,000	1.2	2.1	3.2	4.0	4.8	5.6	6.3	7.0	7.6	8.3
6,000	1.2	2.2	3.2	4.1	4.9	5.8	6.5	7.2	7.8	8.5
7,000	1.2	2.2	3.3	4.2	5.0	5.9	6.6	7.4	8.0	8.7
8,000	1.3	2.3	3.4	4.3	5.2	6.1	6.8	7.6	8.2	9.0
9,000	1.3	2.3	3.5	4.4	5.3	6.2	7.0	7.7	8.4	9.2
10,000	1.3	2.4	3.6	4.5	5.4	6.3	7.1	7.9	8.6	9.4
12,000	1.4	2.5	3.7	4.6	5.6	6.6	7.4	8.2	8.9	9.7
15,000	1.4	2.6	3.9	4.7	5.9	6.9	7.7	8.6	9.3	10.2

Source: Compressed Air and Gas Institute.

The values of the factors are adjusted to reflect representative diversity factors for the use of multidrills. Because these factors will not necessarily apply to all drills and projects, they should be used only as a guide.

THE COST OF COMPRESSED AIR

The cost of compressed air may be determined at the compressor or at the point of use. The former will include the cost of compressing plus transmitting, including line losses.

The cost of compressing should include the total cost of the compressor, both ownership and operation. The cost is usually based on 1,000 cf of free air.

EXAMPLE 11.7

Determine the cost of compressing 1,000 cf of free air to a gauge pressure of 100 psi by using a 600-cfm two-stage portable compressor driven by a 180-hp diesel engine.

This information will apply:

Annual ownership cost = $19,686

Based on a 5-yr life at 1,400 hr per yr

Fuel consumed per hour, 0.04 × 180 = 7.2 gal

Lubricating oil consumed per hour, 0.125 gal

Hourly costs:

Fixed cost,	$19,686 ÷ 1,400 hour =	$14.06
Fuel,	7.2 gal @ $1.00 =	7.20
Lubricating oil,	0.125 gal @ $3.20 =	0.40
Operator,	$\frac{1}{2}$ time @ $16.00 per hour =	8.00*
	Total cost per hour =	$29.66

*This cost will vary with location and company benefit packages.

Volume of air compressed per 50-min hour, 50 × 600 = 30,000 cf.

Cost per 1,000 cf, $29.66 ÷ 30 = $0.989 or $0.99

The cost per 1,000 cf of air for a compressor operating under various load factors might be

	Load factor (%)		
	100	**75**	**50**
Hours costs:			
Fixed cost*	$14.06	$14.06	$14.06
Fuel	7.20	5.38	4.28
Lubricating oil	0.40	0.30	0.24
Operator, $\frac{1}{2}$ time†	8.00	8.00	8.00
Total cost per hour	$29.66	$27.74	$26.58
Volume of air per hour (cf)	30,000	22,500	15,000
Cost per 1,000 cf	$0.99	$1.23	$1.77

*This cost may vary slightly with the load factor.
†This cost may vary with the locations.

THE COST OF AIR LEAKS

The loss of air through leakage in a transmission line can be surprisingly large and costly. Leakage results from poor pipe connections, loose valve stems, deteriorated hose, and loose hose connections. If the cost of such leaks were more fully known, most of them would be eliminated. The rate of leakage through an opening of known size can be determined by applying a formula for the flow of air through an orifice.

Table 11.10 illustrates the cost of air leakage for various sizes of openings and costs per 1,000 cf of air.

TABLE 11.10 | Cost per month for air leakage

Size of opening (in.)	Cubic feet of air lost per month at 100 psi	For indicated cost per 1,000 cf			
		$0.45	**$0.60**	**$0.75**	**$1.20**
$\frac{1}{32}$	45,500	$ 19.14	$ 27.30	$ 34.14	$ 45.50
$\frac{1}{16}$	182,300	82.20	109.50	136.95	182.30
$\frac{1}{8}$	740,200	333.00	444.00	555.00	740.20
$\frac{1}{4}$	2,920,000	1,314.00	1,782.00	2,190.00	2,920.00

THE COST OF USING LOW AIR PRESSURE

The effect on the cost of production of operating pneumatic equipment at less than the recommended air pressure can be demonstrated by analyzing the performance equipment under different pressures. Tests conducted on drilling equipment indicate that the drills operating at lower pressure, namely, 75.2 psi, will have an efficiency equal to about 80% of those operating at 89.4 psi. Thus, the increase in the depth of hole drilled at the higher pressure will be $100 - 80$, equaling 20%.

SAFETY

Extreme care should be exercised when working with compressed air. At close range, it is capable of putting out eyes, bursting eardrums, causing serious skin blisters, or even killing an individual.

Air Compressors

- Be sure the intake air is cool and free from flammable gases or vapors.
- Make sure that all the pressure gauges are in good working order.
- Check the safety valves and regulators to determine if they are working properly before starting the air compressor.

Pneumatic Tools

- Wear protective clothing and equipment (such as goggles, gloves, and respirators) appropriate for the particular tool being operated.
- Maintain a good footing and proper balance at all times while operating pneumatic tools.
- Turn off the air and disconnect the tool when repairs or adjustments are being made or the tool is not in use.
- Inspect the hose to ensure it is in good condition and free from obstructions before connecting a pneumatic tool.
- Remove leaking or defective hoses from service. The air hose must be able to withstand the pressure required for the tool.

■   Do not kink a hose to stop the air flow.

■   Never point an air hose directly at other personnel.

SUMMARY

In many instances, the energy supplied by compressed air is the most convenient method of operating construction equipment and tools. A compressed-air system consists of one or more compressors together with a distribution system to carry the air to the points of use. Portable compressors are more commonly used on construction sites where it is necessary to meet frequently changing job demands, typically at a number of locations on the job site. Pipe and hose size selection for an air line is an economics problem. Critical learning objectives include:

■   An understanding of the difference between isothermal and adiabatic compression.

■   An ability to apply the gas laws as appropriate.

■   An ability to calculate the pressure loss as air flows through a pipe or hose.

■   An understanding of the diversity factor in calculating actual load on an air system.

These objectives are the basis for the problems that follow.

PROBLEMS

11.1   An air compressor draws in 1,000 cf of air at a gauge pressure of 0 psi and a temperature of 70°F. The air is compressed to a gauge pressure of 100 psi at a temperature of 140°F. The atmospheric pressure is 14.0 psi. Determine the volume of air after it is compressed. (139 cf)

11.2   An air compressor draws in 1,000 cf of free air at a gauge pressure of 0 psi and a temperature of 60°F. The air is compressed to a gauge pressure of 100 psi at a temperature of 130°F. The atmospheric pressure is 12.30 psi. Determine the volume of air after it is compressed.

11.3   Determine the theoretical horsepower required to compress 500 cfm of free air, measured at standard conditions, from atmospheric pressure to 125-psi gauge pressure when the compression is performed under isothermal conditions. (72.2 hp)

11.4   Determine the theoretical horsepower required to compress 600 cfm of free air, measured at standard conditions, from atmospheric pressure to 100-psi gauge pressure when the compression is performed under isothermal conditions.

11.5   Solve Problem 11.4 if the air is compressed under adiabatic conditions.

11.6   Determine the difference in horsepower required to compress 600 cfm of free air under adiabatic conditions for altitudes of 3,000 and 8,000 ft. The air will be compressed in a single stage to 100-psi gauge pressure at each altitude.

11.7   A rotary compressor has a capacity of 500 cfm of free air at 100-psi gauge pressure at sea level. If the compressor is operating at an altitude of 12,000 ft, determine the capacity when the air is compressed to 100-psi gauge pressure, with no change in temperature. (493 cfm)

11.8   A single-stage reciprocating compressor has a capacity of 600 cfm of free air at 100-psi gauge pressure at zero altitude. If the compressor is operating at an altitude of 8,000 ft, determine the capacity when the air is compressed to 100-psi gauge pressure, with no change in temperature.

11.9 Using the Fig. 11.5 chart determine the pressure loss per 100 ft of pipe resulting from transmitting 200 cfm of free air, at 100-psi gauge pressure, through a 2-in. standard-weight steel pipe. (0.27 psi per 100 ft)

11.10 A two-stage reciprocating compressor operating at sea level will supply enough air to operate nine drills. If the compressors and drills are operated at an altitude of 10,000 ft, how many drills can the compressor serve?

11.11 What will be the pressure at the air tool end of a 1-in.-diameter hose 85 ft long if the tool requires 120 cfm. The pressure entering the hose is 90 psi. (87.11 psi)

11.12 A 3-in. pipe with screwed fittings is used to transmit 1,000 cfm of free air at an initial pressure of 100-psi gauge pressure. The pipeline includes these items:

900 ft of pipe

Three gate valves

Eight standard on-run tees

Six standard ells

Determine the total loss of pressure in the pipeline.

11.13 If the air from the end of the pipeline of Problem 11.12 is delivered through 50 ft of 1-in. hose to a rock drill that requires 140 cfm of air, determine the pressure at the drill.

REFERENCES

1. Atlas Copco, Inc., 70 Demarest Drive, Wayne, NJ 07470. www.atlascopco.com/.

2. Chicago Pneumatic Tool Company, 2200 Bleecker Street, Utica, NY 13501. www.chicagopneumatic.com/.

3. Ingersoll-Rand Company, Portable Compressor Division, 501 Sanford Ave., Mocksville, NC 27028. www.ingersoll-rand.com/.

4. "Selecting a Compressor," *Equipment Guide News*, May 1986, pp. 26–27.

5. Sullair Corporation, 3700 East Michigan Boulevard, Michigan City, IN 46360. www.sullair.com/.

CHAPTER

12

Drilling Rock and Earth

Because the purposes for which drilling is performed vary a great deal from general applications to highly specialized, it is desirable to select the equipment and methods best suited to the specific service. The rates of drilling rock will vary with a number of factors such as the type of drill and bit size, hardness of the rock, depth of holes, drilling pattern, terrain, and time lost because of sequencing other operations. The first step in estimating drilling production is to make an assumption about the type of equipment that will be used and then to account for the total depth to be drilled, penetration rate, time for changing steel and bits, and time to clean the hole.

INTRODUCTION

The manner of achieving a hole in hard materials did not change from ancient times until about the middle of the 19th century. Across that long span of time, drilling was accomplished by manpower—a man swinging a hammer against a pointed drill. In 1861 the first practical mechanical drilling machine was employed in the Alps on the Mount Cenis tunnel work. The first pneumatic drill employed in the United States was at the eastern header of the Hoosac tunnel in western Massachusetts in June 1866.

This chapter deals with the equipment and methods used by the construction and mining industries to drill holes in both rock and earth. Drilling may be performed to explore the types of materials to be encountered on a project (exploratory drilling), or it may involve production work such as drilling holes for explosive charges. Other purposes would include holes for grouting or rock bolt stabilization work and boring for the placement of utility lines. Some jobs also require seep holes for drainage to reduce hydrostatic pressure. Rock (see Fig. 12.1) and earth drilling (see Fig. 12.2) will be treated separately in this chapter, although in some instances the same or similar equipment may be used for drilling both materials.

Because the purposes for which drilling is performed vary a great deal from general to highly specialized applications, it is desirable to select the equipment

FIGURE 12.1 I Drilling rock to load explosives and for the placement of rock bolts in a tunnel.

FIGURE 12.2 I Drilling earth with an auger for a bridge pier.

and methods best suited to the specific service. A contractor engaged in highway construction must usually drill rock under varying conditions; therefore, equipment that is suitable for various services would be selected. However, if equipment is needed to drill rock in a quarry where the material and conditions will not vary, specialized equipment should be considered. In some instances custom-made equipment designed for use on a single project may be justified.

GLOSSARY OF TERMS

The following glossary defines the important terms used in describing drilling equipment and procedures.

Bit. The portion of a drill that contacts the rock and disintegrates it. Many types are used.

> *Carbide-insert bit.* A detachable bit whose cutting edges consist of tungsten carbide embedded in a softer steel base.
>
> *Detachable bit.* A bit that can be attached to or removed from the drill steel or drill stem.
>
> *Diamond bit.* A detachable bit whose cutting elements consist of diamonds embedded in a metal matrix.
>
> *Downhole bit or downhole drill.* Instead of driving the bit by a mechanical mechanism located on the drill platform, this is a combination bit and power unit, usually air driven, that is suspended in the hole.
>
> *Forged bit.* A bit that is forged on the drill steel.

Burden. The horizontal distance from a rock face to the first row of drill holes or the distance between rows of drill holes.

Coupling. A short, hollow steel pipe having interior threads. The coupling is used to hold pieces of drill steel together or to the shank. The percussion energy is transferred through the steel, not the coupling; therefore, the coupling *must* allow the two pieces of drill steel to butt together.

Cuttings. The disintegrated rock particles removed from a drill hole.

Depth per bit. The depth of hole that can be drilled by a bit before it is replaced.

Drifter. An air-operated percussion-type drill, similar to a jackhammer; it is so large, however, that it requires a mechanical mounting.

Drills

> *Abrasion.* A drill that grinds rock into small particles through the abrasive effect of a bit that rotates in the hole.
>
> *Blast-hole.* A rotary drill consisting of a steel-pipe drill stem on the bottom of which is a roller bit. As the bit rotates it crumbles the rock. A stream of compressed air removes the cuttings.

Churn. A percussion-type drill consisting of a long steel bit that is mechanically lifted and dropped to disintegrate the rock. It is used to drill deep holes, usually 6 in. in diameter or larger.

Core. A drill designed for obtaining samples of rock from a hole, usually for exploratory purposes. Diamond and shot drills are used for core drilling.

Diamond. A rotary abrasive-type drill whose bit consists of a metal matrix in which are embedded a large number of diamonds. As the drill rotates, the diamonds disintegrate the rock. This drill is used extensively to obtain core samples from hard rock.

Downhole drill or downhole bit. The bit and the power system providing rotation and percussion are one unit suspended at the bottom of the drill steel.

Dry. A drill that uses compressed air to remove the cuttings from a hole.

Percussion. A drill that breaks rock into small particles by the impact from repeated blows. Compressed air or hydraulic fluids can power percussion drills.

Shot. A rotary abrasive-type drill whose bit consists of a section of steel pipe with a roughened surface at the bottom. As the bit is rotated under pressure, chilled-steel shot are supplied under the bit to accomplish the disintegration of the rock. The cuttings are removed by water.

Wet. A drill that uses water to remove the cuttings from a hole.

Drilling pattern. The regular distance between the drill holes in both directions usually stated as "burden distance × spacing." The burden distance denotes the distance between the rock face and a row of holes or between adjacent rows of holes, and spacing refers to the distance between adjacent holes in the same row (see Fig. 13.1).

Drilling rate. The number of feet of hole drilled per hour per drill.

Drill steel, or rods. Steel rods that transmit the blow energy and drill rotation from the shank to the bit. Drill steel is threaded on both ends and usually comes in standard 10-, 12-, and 20-ft. lengths.

Face. The approximately vertical surface extending upward from the floor of a pit to the level at which drilling is accomplished.

Jackhammer, or sinker. An air-operated percussion-type drill that is small enough to be handled by one worker.

Stoper. An air-operated percussion-type drill, similar to a drifter, that is used for overhead drilling, as in a tunnel.

Shank, or striker bar. A short piece of steel that attaches to the percussion drill piston for receiving the blow and transferring the energy to the drill steel.

Subdrilling. The depth to which a blasthole will be drilled below the proposed final grade is commonly referred to as subdrilling depth. This extra depth is necessary to ensure that rock breakage will occur completely to the required elevation.

The dimension terminology frequently used for drilling is illustrated in Fig. 13.1.

BITS

The *bit* is the essential part of a drill, as it is the part that must engage and disintegrate the rock. The success of a drilling operation depends on the ability of the bit to remain sharp under the impact of the drill. Many types and sizes are available. Most bits are units that screw to the drill steel. They are easily replaced or some can be resharpened. Bits are available in various sizes, shapes, and hardnesses.

Steel bits for jackhammers and drifters are available in sizes from 1 to $4\frac{1}{2}$ in., the gauge size varying in steps of $\frac{1}{8}$ in. These bits can be resharpened two to six times.

The depth of the hole that can be drilled with a steel bit will vary from a few inches to 40 ft or more, depending on the type of rock.

Carbide-Insert Bits

Some types of rock are so abrasive that steel bits must be replaced after they have drilled only a few inches of hole. The cost of the bits and lost time to production in changing bits is so great that it will usually be economical to use carbide-insert bits. This bit is illustrated in Fig. 12.3. As can be noted in the figure, the actual drilling edges consist of a very hard metal, tungsten carbide, that is embedded in steel. Although these bits are considerably more expensive than steel bits, the increased drilling rate and depth of hole obtained per bit provide an overall economy in drilling hard rock. Typical sizes for carbide-insert bits are from $1\frac{3}{8}$ to 5 in. in diameter.

Carbide-insert bits are available in four grades, in order of increasing hardness:

Grade	Abrasion resistance
Shock	Fair
Intermediate	Good
Wear	Excellent
Extra wear	Outstanding

Susceptibility to breakage increases with hardness. However, abrasion resistance also increases. If excessive bit breakage occurs using a specific grade, a softer grade should be tried.

A contractor on a highway project in Pennsylvania found that when drilling diabase rock the depth per steel bit was $\frac{1}{2}$ to 2 in. When he changed to carbide bits an average depth per bit of 1,992 ft was obtained.

Button Bits

Figure 12.4 illustrates button bits that are available in numerous sizes. These bits, which are available in different cutting-face designs with a choice of insert grades, require no regrinding or sharpening. Button bits can yield faster penetration rates

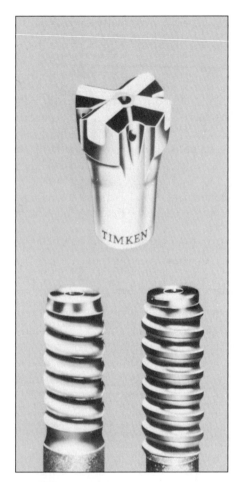

FIGURE 12.3 I Carbide-insert rock bit.

Source: The Timken Company.

in a wide range of drilling applications. Most button bits are run to destruction and never reconditioned.

JACKHAMMERS

Jackhammers are handheld air-operated percussion-type drills that are used primarily for drilling holes in a downward direction. For this reason, they are frequently called "sinkers." They are classified according to their weight, such as 45 or 55 lb. A complete drilling unit consists of a hammer, drill steel, and bit. As the compressed air flows through a hammer, it causes a piston to reciprocate at a speed up to 2,200 blows per minute that produces the hammer effect. The energy of this piston is transmitted to a bit through the drill steel. Air flows through a hole in the drill steel and the bit to remove the cuttings from the hole

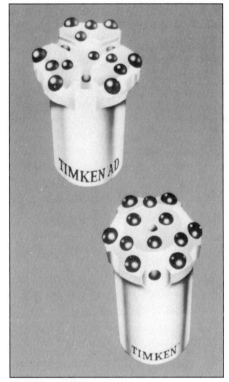

FIGURE 12.4 | Button bits.

Source: The Timken Company.

and to cool the bit. For wet drilling, water is used instead of air to remove the cuttings. The drill steel is rotated slightly following each blow so that the points of the bit will not strike at the same place each time.

Although jackhammers may be used to drill holes in excess of 20 ft deep, they are seldom used for holes exceeding 10 ft in depth. The heavier hammers will drill holes up to $2\frac{1}{2}$ in. in diameter. Drill steel usually is supplied in 2-, 4-, 6-, and 8-ft lengths.

DRIFTERS

Drifter drills are similar to jackhammers in operation, but they are larger and are used as mounted tools for drilling downward, horizontal, or upward holes. They vary in weight from 75 to 260 lb and are capable of drilling holes up to $4\frac{1}{2}$ in. in diameter. These tools are used extensively in rock excavation, mining, and tunneling. Either air or water can be used to remove the cuttings.

The drifter's weight is usually sufficient to supply the necessary feed pressure for down drilling. But when used for horizontal or up drilling, a hand-operated screw, or either a pneumatic or hydraulic piston (see Fig. 12.1, lower left), supplies the feed pressure. Steel changes can be obtained in lengths of 24, 30, 36, 45, and 60 in.

ROTARY DRILLING

With rotary drilling, the rock is ground away by applying a down pressure on the drill steel and bit, and at the same time continuously rotating the bit in the hole. Additionally, compressed air is constantly forced down through the drill pipe and the bit to remove the rock cuttings and cool the bit (see Fig. 12.5). Rotary or blasthole drills are self-propelled drills that can be mounted on a truck or on crawler tracks (see Fig. 12.6). Rigs are available to drill holes to different diameters and to depths up to approximately 300 ft. These drills are suitable for drilling soft to medium rock, such as hard dolomite and limestone, but are not suitable for drilling the harder igneous rocks.

In drilling dolomite for the Ontario Hydro-canal, heavyweight drills were used to drill 25- and 50-ft-deep holes. The 25-ft holes were drilled on 10- × 10- and 12- × 12-ft patterns, using $6\frac{1}{4}$- and $6\frac{3}{4}$-in. bits. The average drilling speed was approximately 30 ft per hr, including moving and positioning the drill rig. The average life of the bits was 958 ft for the $6\frac{1}{4}$ in. and 1,374 ft for the $6\frac{3}{4}$-in. bits.

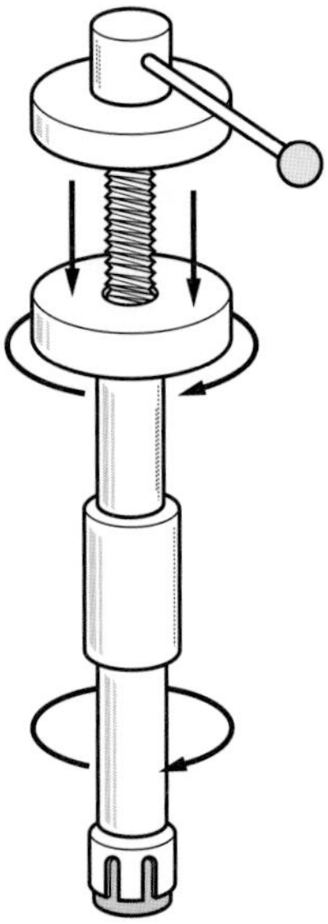

FIGURE 12.5 |
Rotary drilling
mechanism.

Source: Ingersoll-Rand Co.

FIGURE 12.6 | Two rotary (blasthole) drills working
on a highway project.

On other projects, drilling speeds have varied from $1\frac{1}{2}$ ft per hr in dense, hard dolomite to 50 ft per hr in limestone. The speed of drilling is regulated by pressure delivered through a hydraulic feed.

ROTARY-PERCUSSION DRILLING

Rotary-percussion drilling combines the hard-hitting reciprocal action of the percussion drill with the turning-under-pressure action of the rotary drill (Fig. 12.7). Whereas the percussion drill only has a rotary action to reposition the bit's cutting edges, the rotation of this combination drill, with the bit under constant pressure, has demonstrated its ability to drill much faster than the regular percussion drill. On one project, rotary-percussion drills are reported to have drilled blastholes three times as fast as regular percussion drills. Rotary-percussion drills require special carbide bits, with the carbide inserts set at a different angle than those used with standard carbide bits.

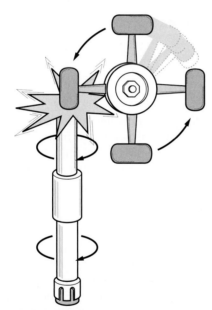

FIGURE 12.7 | Rotary-percussion drilling mechanism.

Source: Ingersoll-Rand Co.

DOWNHOLE DRILLING

When drilling deep holes, it can often be more efficient to place the driving mechanism in the hole with the bit (see Fig. 12.8). This eliminates having to transmit the rotation and percussion forces through the drill steel. Typically

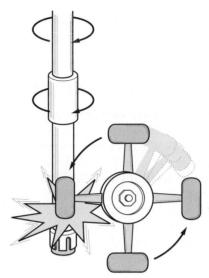

FIGURE 12.8 | Downhole drilling mechanism.

Source: Ingersoll-Rand Co.

these units are air-operated hammers (see Fig 12.9). They can be operated under water by maintaining a higher air pressure in the hammer than the pressure outside. Air or water can be used to clear cuttings from the hole. Standard sizes are available from 4 in. (102 mm) to 30 in. (762 mm). Special drill rigs with 34-in. (864 mm) downhole bits were used to construct a cutoff wall in dike 1 at Beaver Dam in Arkansas.

TRACK-MOUNTED (AIR-TRACK) DRILLS

The track-mounted drills illustrated in Figs. 12.10 and 12.11 are the work-horse drills on construction projects. These are often referred to as "air-track" drills. They are very productive tools because of their ability to move quickly between locations and because the hydraulically operated boom enables easy positioning of the drill. Holes can be drilled at any angle from under 15° back from the vertical to above the horizontal, ahead, or on either side of the unit. All operation, including tramming (travel), can be powered by compressed air. There are also hydraulic-powered models, but compressed air is still used for hole cleaning.

DRILLS MOUNTED ON EXCAVATORS

Figure 12.12 illustrates a drill mounted on a small hydraulic excavator boom. These drills are similar in sizes and capacities to the track-mounted drills. However, these drills require a nearly level ground surface to operate effectively.

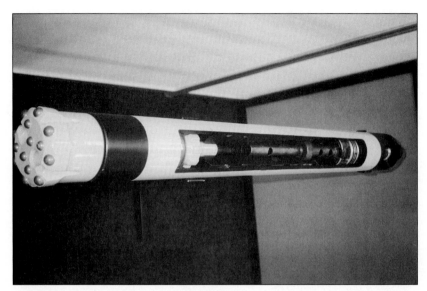

FIGURE 12.9 | Cutaway of a downhole drill.

FIGURE 12.10 | Track-mounted drill with an air
compressor mounted on the drill.

FIGURE 12.11 | Air-track drill towing the air compressor.

SHOT DRILLS

A shot drill is a tool that depends on the abrasive effect of chilled steel shot to penetrate the rock. The essential parts include a shot bit, core barrel, sludge barrel, drill rod, water pump, and power-driven rotation unit. The bit consists of a section of steel pipe with a serrated lower end extending through the drill rod. The cutting medium, chilled steel shot, is fed with wash water. It lodges around and is partially embedded in a coring bit of slot steel. To be effective, the shot must be crushed during the coring. Precrushed shot called Calyxite is often used in coring relatively soft rock. Water that is supplied through the drill rod forces the rock cuttings up around the outside of the drill, where they settle in a sludge barrel, to be removed when the entire unit is pulled from the hole. The flow of water must be carefully regulated so that it removes the cuttings but not the chilled shot. Periodically, it is necessary to break the core off and remove it from the hole so that drilling may proceed.

Standard shot drills are capable of drilling holes up to 600 ft in depth, with diameters varying from $2\frac{1}{2}$ to 20 in. Special equipment has been used to drill up to 6 ft in diameter with depths in excess of 1,000 ft. Shot drilling can be used in nearly all types of rock, but it is most effective in medium-hard and uniform rock. In soft rock, the shot instead of grinding may become embedded in the

FIGURE 12.12 | Air-operated drill mounted on a hydraulic excavator boom.

rock. In seamy or cavernous rock, the shot is often lost, and in hard rock the rate of progress is slow. The method can be used only for downward boring and is best suited for vertical holes, but it can be used in holes with an angle of inclination up to 30°.

A primary purpose of small hole-shot drilling is to provide continuous cores for examination for structural information, as rock of any hardness can be drilled.

The rate of drilling with a shot drill is relatively slow, sometimes less than 1 ft per hr, depending on the size of the drill and the hardness of the rock.

On a project for an electric utility company near Oak Park, Ohio, the contractor used a shot drill to drill 100 large-diameter footings for columns in hard limestone rock. The holes varied from 30 in. (76 cm) to 60 in. (152 cm) in diameter and averaged about 12 ft (3.7 m) in depth. Holes 42 in. (107 cm) in diameter were drilled at an average rate of 1 to $1\frac{1}{2}$ ft (0.3 to 0.5 m) per hour, while holes 60 in. (152 cm) in diameter were drilled at an average rate of about 1 ft (0.3 m) per hour [8].

DIAMOND DRILLS

Diamond drills are used primarily for exploration drilling, where cores are desired for the purpose of studying the rock structure. The Diamond Core Drill Manufacturers' Association lists four sizes as standard—$1\frac{1}{2}$, $1\frac{7}{8}$, $2\frac{3}{8}$, and 3 in. Larger sizes are available, but the investment in diamonds increases so rapidly with an increase in size that shot drills may be more economical for larger-diameter holes.

A drilling rig consists of a diamond bit, a core barrel, a jointed driving tube, and a rotary head to supply the driving torque. Water is pumped through the driving tube to remove the cuttings. The pressure on the bit is regulated through a screw or hydraulic-feed swivel head. Core barrels are available in lengths varying from 5 to 15 ft. When the bit advances to a depth equal to the length of the core barrel, the core is broken off and the drill is removed from the hole. Diamond drills can drill in any desired direction from vertically downward to upward.

The selection of the size of diamonds depends on the nature of the formation to be drilled. Large stones are preferred for the softer formations and small stones for fine-grained solid formations.

Diamond drills are capable of drilling to depths in excess of 1,000 ft. Bit speeds can be varied between 200 and 1,200 rpm. The drilling rate will vary from less than a foot to several feet per hour, depending on the type of rock.

MANUFACTURERS' REPORT ON DRILLING EQUIPMENT AND TECHNIQUES [4]

A magazine devoted primarily to construction methods and equipment published an article presenting the views of representatives of drilling equipment manufacturers, with suggestions listed for selecting and using drilling equipment to achieve increased production at reduced costs. Some of those ideas are presented here.

The manufacturers' representatives who assessed drilling productivity included as important factors: maintainability, mobility, operator expertise, the operability of equipment, the use of auxiliary attachments, the method of drilling employed, and the interrelationship between the drill bit and the drilling rig. Each user or prospective user of drilling equipment should examine these ideas and adopt the ones that apply to the particular operations being considered.

The user of drills to be operated in a quarry should select equipment that will maximize productivity within the limits of the loading, hauling, and crushing equipment. The equipment selected must work under the most severe and grueling conditions of the given project. Before any equipment is selected, it should be tested for performance.

It was suggested that when drilling holes deeper than about 50 ft, downhole drilling, i.e., the process of moving the cylinder and its percussive impact into the hole, should be considered. This move can increase the effective energy by as much as 10% by eliminating the loss of energy in extended lengths of drill rods.

Another suggestion was that more tests be conducted to determine the best sizes of blast holes, ratios of burden to spacing, density of explosives, and delay patterns. Such tests can lead to an increase in the yield of blasted rock per foot of hole, and that, in turn, can reduce the cost of production. In determining the cost of production, the cost of the changes in drilling, the cost of explosives, and the cost of excavation and hauling must all be considered.

The drill operators can exercise greater care as they add sections of steel rods to produce deeper holes. The *abutting ends* of drill rods must *bear tightly* against each other to transmit the drilling forces to the bits. If the couplings transmit the drilling forces, these forces may cause excessive splitting of the couplings, with possible loss of bits and drill rods down the hole.

The use of button bits as a means of more evenly distributing the forces of the drills on the rock may be desirable on some projects. The superior geometric pattern of the buttons can help increase the penetration by as much as 20%, smooth the drilling operation, and reduce the stress on the drilling equipment when compared with conventional blade bits. However, button bits are not recommended for drilling long, close, parallel holes, as in presplitting. Holes drilled with button bits may drift or even cross each other. Moreover, their small contact surface may cause greater wear in drilling hard abrasive rock.

In some drilling operations, increasing the pressure and volume of air supplied to the drill has increased productivity. This maintains a cleaner blasthole. At the same time, increasing the down pressure on the bit has improved the cutting action of the bit.

Maintenance and repair services for equipment can be improved, with less downtime for repairs and replacement of parts, if the equipment is standardized to the extent permitted by the operations for which it is used. With fewer models of equipment to be serviced, an inventory of spare parts that fail most frequently can be maintained on the job.

The availability of extendable booms on track drills increases the service ranges and performance of drills. The availability of drills with dust control systems permits the use of drills in locations where environmental restrictions would otherwise preclude their operation.

Because of the high cost, complexity, and sophistication of drilling equipment, some manufacturers have developed training programs to assist the users and operators of their equipment in increasing its efficiency.

SELECTING THE DRILLING METHOD AND EQUIPMENT

Holes are drilled for various purposes, such as to receive charges of explosives, for exploration, or for ground modification by the injection of grout. Within practical limits, the equipment that will produce the greatest overall economy for the particular project is the most satisfactory. Many factors affect the selection of equipment. Among these are

1. The nature of the terrain. Rough surfaces may dictate jackhammers regardless of other factors.
2. The required depth of holes.
3. The hardness of the rock.

4. The extent to which the formation is broken or fractured.
5. The size of the project.
6. The extent to which the rock must be broken for handling or crushing.
7. The availability of water for drilling purposes. Lack of water favors dry drilling.
8. The purpose of the holes, such as blasting, exploration, or grout injection.
9. The size cores required for exploration. Small cores permit the use of diamond drills, whereas large cores suggest shot drills.

For small-diameter, shallow blastholes, especially on rough surfaces where larger drills cannot operate, it is usually necessary to use jackhammers or track-mounted drills even though the production rates will be low and the costs higher.

For blastholes up to about 6 in. in diameter and up to about 50 ft deep, where machines can operate, the choice may be between track-mounted, rotary-percussion, or piston drills.

For drilling holes from 6 to 12 in. in diameter, from 50 to 300 ft deep, the rotary or blasthole drill is usually the best choice, but the type of rock affects the choice.

If cores up to 3 in. are desired, the diamond coring drill is the most satisfactory.

If intermediate-size cores (3- to 8-in. outside diameter) are desired, the choice will be between a diamond drill and a shot drill. A diamond drill will usually drill faster than a shot drill. Also, a diamond drill can drill holes in any direction, whereas a shot drill is limited to holes vertically down, or nearly so.

SELECTING THE DRILLING PATTERN

The pattern selected for drilling holes to be loaded with explosives will vary with the type and size drill used, the depth of the holes, the type of rock, the maximum rock breakage size permissible, and other factors.

Drilling operations for rock excavation where the material will be used in an embankment fill must consider the project specifications concerning the maximum physical size of individual pieces placed in the fill. The drilling pattern should be planned to produce rock sizes small enough to permit most of the blasted material to be handled by the excavator, such as a loader or shovel, or to pass into the crusher opening without secondary blasting. While meeting either condition is possible, the cost of excess drilling and greater amounts of explosives to produce such material may be so high that the production of some oversize rocks will be cost effective. The oversize rocks will still have to be handled on an individual basis, possibly with a headache ball.

If small-diameter holes are spaced close together, the better distribution of the explosives will result in a more uniform rock breakage. However, if the added cost of drilling more holes (i.e., more drilling footage) exceeds the value of the benefits resulting from better breakage, the close spacing is not justified.

Large-diameter holes permit greater explosive loading per hole, making it possible to increase the spacing between holes, and thereby reduce the number of holes and the cost of drilling.

In analyzing a job for drilling and blasting operations, four factors should be considered.

1.   The cubic yards of rock produced per linear foot of hole.
2.   The number of pounds of explosive required per cubic yard of rock.
3.   The number of pounds of explosive used per linear foot of hole.
4.   Will the resulting breakage meet the job requirements?

The value of each of the first three factors may be estimated in advance of drilling and blasting operations, but after experimental drilling operations are conducted, it probably will be desirable to modify the values to achieve better results. The fourth factor is more subjective but the relationship between hole size and spacing gives some indication of expected results.

The relationships between the first three factors are illustrated in Table 12.1. The volumes of rock per linear foot of hole are based on the net depth of holes and do not include subdrilling that will usually be necessary. The pounds of explosive per linear foot of hole are based on filling the holes completely with 60% dynamite. The pounds of explosive per cubic yard or rock are based on filling each hole to 100, 75, and 50% of its total capacity with explosive. When a hole is not filled completely with dynamite, the surplus volume is filled with stemming (see Chapter 13, "Blasting Rock").

RATES OF DRILLING ROCK

The rates of drilling rock will vary with a number of factors such as the type of drill and bit size, hardness of the rock, depth of holes, drilling pattern, terrain, and time lost because of sequencing other operations. If pneumatic drills are used, the rate of drilling will vary with the pressure of the air, as demonstrated in Chapter 11 on compressed air.

The rate at which a drill will penetrate rock is a function of the rock, the drilling method, and the size and type of bit. The four critical rock properties that affect penetration rate are

■   Hardness.
■   Texture.
■   Breaking characteristic.
■   Formation.

Hardness is the resistance of a smooth surface to abrasion. It is measured by the MOH scale. The scale is from 1 to 10 with talc as 1, the softest, and diamond rated as 10, the hardest (Table 12.2). The effect of rock hardness on drilling speed is presented in Table 12.3.

The term *texture* in relation to rock refers to the grain structure. A loose-grained structured rock (porous, cavities) drills fast. If the grains are large enough to be seen individually (granite), the rock will drill medium. Fine-grained rocks drill slow.

The impact of rock *breaking characteristics* on drilling rates is shown in Table 12.4.

TABLE **12.1** | Drilling and blasting data

Hole size (in.)	Hole pattern (ft)	Area per hole (sf)	Volume of rock per linear foot hole† (cy)	Pounds of explosive per linear foot of hole*	Pounds of explosive per cubic yard of rock* (% of hole filled)		
					100	75	50
$1\frac{1}{2}$	4 × 4	16	0.59	0.9	1.52	1.14	0.76
	5 × 5	25	0.93	0.9	0.97	0.73	0.48
	6 × 6	36	1.33	0.9	0.68	0.51	0.34
	7 × 7	49	1.81	0.9	0.50	0.38	0.25
2	5 × 5	25	0.93	1.7	1.83	1.37	0.92
	6 × 6	36	1.33	1.7	1.28	0.96	0.64
	7 × 7	49	1.81	1.7	0.94	0.71	0.47
	8 × 8	64	2.37	1.7	0.72	0.54	0.36
3	7 × 7	49	1.81	3.9	2.15	1.61	1.08
	8 × 8	64	2.37	3.9	1.65	1.24	0.83
	9 × 9	81	3.00	3.9	1.30	0.97	0.65
	10 × 10	100	3.70	3.9	1.05	0.79	0.53
	11 × 11	121	4.48	3.9	0.87	0.65	0.44
4	8 × 8	64	2.37	7.5	3.16	2.37	1.58
	10 × 10	100	3.70	7.5	2.03	1.52	1.02
	12 × 12	144	5.30	7.5	1.42	1.06	0.71
	14 × 14	196	7.25	7.5	1.03	0.77	0.52
	16 × 16	256	9.50	7.5	0.79	0.59	0.40
5	12 × 12	144	5.30	10.9	2.05	1.54	1.02
	14 × 14	196	7.25	10.9	1.50	1.13	0.75
	16 × 16	256	9.50	10.9	1.15	0.86	0.58
	18 × 18	324	12.00	10.9	0.91	0.68	0.46
	20 × 20	400	14.85	10.9	0.73	0.55	0.37
6	12 × 12	144	5.30	15.6	2.94	2.20	1.47
	14 × 14	196	7.25	15.6	2.05	1.54	1.02
	16 × 16	256	9.50	15.6	1.64	1.23	0.82
	18 × 18	324	12.00	15.6	1.30	0.97	0.65
	20 × 20	400	14.85	15.6	1.05	0.79	0.53
	24 × 24	576	21.35	15.6	0.73	0.55	0.37
9	20 × 20	400	14.85	35.0	2.36	1.77	1.18
	24 × 24	576	21.35	35.0	1.64	1.23	0.82
	28 × 28	784	29.00	35.0	1.21	0.91	0.61
	30 × 30	900	33.30	35.0	1.05	0.79	0.53
	32 × 32	1,024	37.90	35.0	0.92	0.69	0.46

*Based on using dynamite weighing 80 lb per cf.
†Does not account for subdrilling.

TABLE **12.2** | MOH scale for rock hardness

Rock	MOH number	Scratch test
Diamond	10.0	
Schist	5.0	Knife
Granite	4.0	Knife
Limestone	3.0	Copper coin
Potash	2.0	Fingernail
Gypsum	1.5	Fingernail

TABLE 12.3 | Hardness effect on drilling speed

Hardness	Drilling speed
1–2	Fast
3–4	Fast to medium
5	Medium
6–7	Slow to medium
8–9	Slow

TABLE 12.4 | Effect of rock breaking characteristics on drilling speed

Breaking characteristics	Drilling speed
Shatters	Fast
Brittle	Fast to medium
Shaving	Medium
Strong	Slow to medium
Malleable	Slow

Formation describes how the rock mass is structured and structure affects drilling speed. Solid rock masses tend to drill fast. If there are horizontal strata (layers), the rock should drill between medium and fast. Dipping planes drill slow to medium. Dipping planes also make it difficult to maintain the drill hole alignment. All of these factors should be carefully considered when estimating drilling penetration rate without the benefit of actual field tests.

Another item that influences the rate of drilling is the availability factor. Because of the nature of the work that they do, drills are subjected to severe vibration and wear that may result in frequent failures of critical parts, or a deterioration of the whole unit, entailing mechanical delays. The portion of time that a drill is operative is defined as the *availability factor,* which is usually expressed as a percentage of the total time the drill is expected to be working.

Historical drill penetration rates based on very broad rock type classification are shown in Table 12.5. These rates should be used only as an order-or-magnitude guide. Actual project estimates need to be based on the results of drilling tests on the specific rock that will be encountered.

Drill bits, steel, and couplings are high-wear items, and the time required to replace or change each affects drilling production. Table 12.6 gives the average life of these high-wear items based on drill footage and type on rock.

The Effect of Air Pressure on the Rate of Drilling Rock

The cost of energy furnished by compressed air is high when compared with the cost of energy supplied by electricity or diesel fuel. The ratio of costs may be as high as 6:1. For this reason, it is essential that every reasonable effort be made to increase the efficiency of the compressed air system and the equipment that

TABLE 12.5 | Order-of-magnitude drilling production rates

Bit size	Drill type—compressed air	Direct penetration rate		Estimated* production rate good conditions	
		Granite (ft/hr)	Dolomite (ft/hr)	Granite (ft/hr)	Dolomite (ft/hr)
	Rotary-percussion				
$3\frac{1}{2}$	750 cfm at 100 psi	65	125	35	55
$3\frac{1}{2}$	900 cfm at 100 psi	85	175	40	65
	Downhole drill				
$4\frac{1}{2}$	600 cfm at 250 psi	70	110	45	75
$6\frac{1}{2}$	900 cfm at 350 psi	100	185	65	90
	Rotary				
$6\frac{1}{4}$	30,000 pulldown	NR	100	NR	65
$6\frac{3}{4}$	40,000 pulldown	75	120	30	75
$7\frac{7}{8}$	50,000 pulldown	95	150	45	85

NR—Not recommended.

*Estimated productions are for ideal conditions, but they do account for all delays including blasting.

TABLE 12.6a | Igneous rock: Average life for drill bits and steel in feet

	Drill bits		Igneous rock				
in.		Type	High silica LA < 20 (Rhyolite) (ft)	High silica 20 < LA < 50 (Granite) (ft)	Medium silica LA < 50 (Granite) (ft)	Low silica LA < 20 (Basalt) (ft)	Low silica LA > 20 (Diabase) (ft)
3		B	250	500	750	750	1,000
3		STD	NR	NR	NR	NR	750
$3\frac{1}{2}$		STD	NR	NR	NR	750	1,500
$3\frac{1}{2}$		HD	200	575	1,000	1,400	2,000
$3\frac{1}{2}$		B	550	1,200	2,500	2,700	3,200
4		B	750	1,500	2,800	3,000	3,500
Rotary bits							
5		ST	NR	NR	NR	NR	NR
$5\frac{7}{8}$		ST	NR	NR	NR	NR	NR
$6\frac{1}{4}$		ST	NR	NR	NR	NR	NR
$6\frac{3}{4}$		ST	NR	NR	NR	NR	800
$6\frac{3}{4}$		CB	NR	NR	1,500	2,000	4,000
$7\frac{7}{8}$		CB	NR	1,700	2,400	3,500	6,000
Downhole bits							
$6\frac{1}{2}$		B	500	1,000	1,800	2,200	3,000
Drill steel							
Shanks			2,500	4,500	5,800	5,850	6,000
Couplings			700	700	800	950	1,100
Steel	10 ft		1,450	1,500	1,600	1,650	2,200
Steel	12 ft		2,200	2,600	3,000	3,500	5,000
5 in.	20 ft		25,000	52,000	60,000	75,000	100,000

B = button, CB = carbide button, HD = heavy duty, ST = steel tooth, STD = standard, NR = not recommended.

uses compressed air as a source of energy. One method of increasing the efficiency of pneumatic drills is to be certain that the specified, at the drill, air pressure is available.

It has been shown that the energy of a rock drill can be represented by the equation [5.7]

$$E \propto \frac{P^{1.5}A^{1.5}S^{0.5}}{W^{0.5}}$$
[12.1]

where

E = energy per blow
P = air pressure
A = area of piston stroke
S = length of piston stroke
W = weight of piston

TABLE 12.6b | Metamorphic rock: Average life for drill bits and steel in feet

		Metamorphic rock				
Drill bits in.	Type	High silica LA < 35 (Quartzite) (ft)	Medium silica low mica (Schist; Gneiss) (ft)	Medium silica high mica (Schist; Gneiss) (ft)	Medium silica LA < 25 (Metalatite) (ft)	Low silica LA > 45 (Marble) (ft)
3	B	200	1,200	1,500	800	1,300
3	STD	NR	800	900	400	850
3½	STD	NR	1,300	1,700	850	1,600
3½	HD	NR	1,800	2,200	1,200	2,100
3½	B	450	3,000	3,500	2,000	3,300
4	B	600	3,300	3,800	2,300	3,700
Rotary bits						
5	ST	NR	NR	NR	NR	NR
5⅞	ST	NR	NR	NR	NR	1,200
6¼	ST	NR	NR	NR	NR	2,000
6¾	ST	NR	NR	750	NR	4,500
6¾	CB	NR	3,700	4,200	1,200	9,000
7⅞	CB	NR	5,500	6,500	2,200	13,000
Downhole bits						
6½	B	500	2,700	3,200	1,500	4,500
Drill steel						
Shanks		5,000	5,700	6,200	5,550	5,800
Couplings		900	1,000	1,200	750	800
Steel	10 ft	1,700	2,100	2,300	1,500	1,600
Steel	12 ft	3,000	3,300	3,800	2,800	3,000
5 in.	20 ft	50,000	90,000	100,000	85,000	175,000

B = button, CB = carbide button, HD = heavy duty, ST = steel tooth, STD = standard, NR = not recommended.

TABLE 12.6c | Sedimentary rock: Average life for drill bits and steel in feet

in. Drill bits Type		High silica fine grain (Sandstone) (ft)	Medium silica coarse grain (Sandstone) (ft)	Low silica fine grain (Dolomite) (ft)	Low silica fine to medium grain (Shale) (ft)	Low silica coarse grain (Conglomerate) (ft)
				Sedimentary		
3	B	800	1,200	1,300	2,000	1,800
3	STD	NR	850	900	1,500	1,200
$3\frac{1}{2}$	STD	NR	1,500	1,800	3,000	2,500
$3\frac{1}{2}$	HD	850	2,000	2,200	3,500	3,000
$3\frac{1}{2}$	B	2,000	3,100	3,500	4,500	4,000
4	B	2,500	3,500	2,000	5,000	4,800
Rotary bits						
5	ST	NR	1,000	NR	8,000	6,000
$5\frac{7}{8}$	ST	NR	2,500	NR	15,000	13,000
$6\frac{1}{4}$	ST	NR	4,000	4,000	18,000	14,000
$6\frac{3}{4}$	ST	500	6,000	8,000	20,000	15,000
$6\frac{3}{4}$	CB	2,000	8,000	10,000	25,000	20,000
$7\frac{7}{8}$	CB	3,000	10,000	15,000	25,000	20,000
Downhole bits						
$6\frac{1}{2}$	B	2,500	3,500	5,500	7,500	6,000
Drill steel						
Shanks		5,000	5,500	6,000	7,000	6,500
Couplings		1,000	1,200	1,500	2,000	1,750
Steel	10 ft	2,000	2,300	2,500	4,000	3,500
Steel	12 ft	4,500	5,000	6,000	7,500	7,000
5 in.	20 ft	65,000	250,000	200,000	300,000	250,000

B = button, CB = carbide button, HD = heavy duty, ST = steel tooth, STD = standard, NR = not recommended.

With all the factors constant in a given drill except the pressure of the air, this equation indicates that the energy delivered per blow varies with the 1.5 power of the pressure.

In the past, factors discouraging the use of higher pressures to increase the rate of drilling have been the limitations imposed by the design of the drills and reduced life spans of the drill steel and bits. To a large extent, these limitations have been overcome today.

The potential increase in production of a drill when operating at a higher pressure should not be the sole factor in a decision to use higher pressure. The value of the increased production should be compared with the probable increase in the cost of air and maintenance and repairs for the drill, including drill steel and bits. Any increase in maintenance and repairs may reduce the availability factor for the drill. The optimum pressure is the pressure that will result in the minimum cost of drilling a foot of hole depth, taking into account all factors related to the drilling including, but not limited to,

1. The value of the increased production.
2. The increased cost of providing air at a higher pressure.
3. The cost of increased line leakage.
4. The increased cost of maintenance and repairs for the drill.
5. The adverse effect, if any, of increased noise in some instances, such as tunneling.
6. The effect that a higher pressure may have on the availability factor for the drill or air compressor.

On most construction projects it is not practical to conduct studies to evaluate each of these factors. However, a limited number of studies have been made under conditions that did permit evaluation of the effects of varying the pressure of the air.

Determining the Optimum Air Pressure for Drilling Rock

This is a report on the results of tests that were conducted in a mine in Ontario, Canada [4]. The walls of the test station were marked off in panels, so that by drilling in each of the panels at every stage of testing, variations in the drillability of the rock were minimized.

Prior to starting the tests, seven new jackleg drills were obtained from five manufacturers and divided into two groups. Holes were drilled at pressures varying from 90 to 140 psi. For each pressure increment, a new carbide-insert bit was used on both 6-ft and 12-ft steel. All bits were $1\frac{1}{4}$ in. in diameter.

Figure 12.13 shows the relationship between the average rate of penetration and the operating pressure for each group of drills. Figure 12.14 is a nomogram based on the information appearing in Fig. 12.13, which indicates the percentage increase in penetration resulting from an increase in air pressure. For example, if the pressure is increased from 90 to 100 psi, the increase in penetration will be 38%.

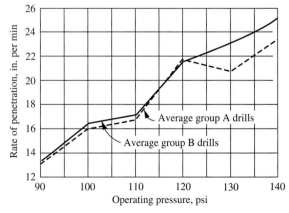

FIGURE 12.13 | Variations in the rate of penetration with air pressure.

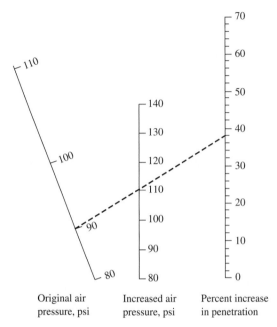

Original air   Increased air   Percent increase
pressure, psi   pressure, psi   in penetration

FIGURE 12.14 | Variations in the rates of
penetration with air pressure.

DETERMINING THE INCREASE IN PRODUCTION RESULTING FROM AN INCREASE IN AIR PRESSURE

If a drill is presently operating at a given air pressure, such as 90 psi, Fig. 12.14 indicates that if the pressure is increased to 110 psi, the rate of penetration of the drill will be increased 38%. This will not result in an increase of 38% in the production on the project. The increased rate of penetration is effective only during the time the drill is *actually* producing hole or drilling. Thus, the increase does not apply to the time that the drill is not, in fact, drilling. This nonproductive time will generally remain the same regardless of the rate of penetration.

A relationship between an increase in production resulting from an increase in the rate of penetration can be derived. These symbols will be used:

T = elapsed time that the drill is on the job in hours

T_1 = time actually devoted to drilling in hours

D = drilling factor, the portion of the elapsed time devoted to

  drilling $= \dfrac{T_1}{T}$

Q_1 = total depth of hole drilled during T hours in feet

R_1 = average rate of drilling during T hours $= \dfrac{Q_1}{T}$ in feet per hour

P = increase in rate of penetration resulting from increase in pressure, expressed as a fraction

R_2 = average rate of drilling resulting from increase in pressure

$\quad = R_1(1 + P)$

T_2 = time required to drill Q_1 ft of hole at increased rate R_2,

$\quad$ hours $= \dfrac{T_1}{1 + P}$

T_s = time saved by increased rate of drilling

$$T_s = T_1 - T_2 = T_1 - \frac{T_1}{1 + P}$$

$$= \frac{T_1(1 + P) - T_1}{1 + P} = \frac{T_1 + T_1 P - T_1}{1 + P} = \frac{T_1 P}{1 + P}$$

T_1 equals TD and, thus, the result is

$$T_s = \frac{TDP}{1 + P}$$

Let Q_2 = the increased depth of hole drilled at the increased rate of penetration during time T. Thus,

$$Q_2 = T_s R_2$$

$$= \frac{TDP}{1 + P} \times R_1(1 + P)$$

$$= TDPR_1$$

But $Q_1 = TR_1$. Thus,

$$Q_2 = Q_1 DP$$

and

$$\frac{Q_2}{Q_1} = DP \qquad\qquad [12.2]$$

which is the ratio of the increased production divided by the original production, expressed as fraction.

EXAMPLE 12.1

Consider a 1,000-hr elapsed time for a drill on a project. During this time the drill actually penetrates rock 300 hr for a drilling factor of 0.3, for a depth of hole equal to 10,000 ft. The initial operating air pressure at the drill is 90 psi. If the pressure is increased to 110 psi, what is the probable total depth of hole drilled in 1,000 hr, based on the information appearing in Fig. 12.14? Reference to this figure indicates an increased rate of penetration equal to 38%. Thus $P = 0.38$.

Applying Eq. [12.2],

$$\frac{Q_2}{Q_1} = DP = 0.3 \times 0.38 = 0.114$$

$$Q_2 = 0.114Q_1 = 0.114 \times 10,000$$

$$= 1,140 \text{ ft additional depth of hole}$$

The total depth of hole will be 10,000 + 1,140 or 11,140 ft. This should result in an increase of 11.4% in production if the increased depth of hole is reflected in increased production.

It should be emphasized that the information appearing in Fig. 12.14 does not necessarily apply to all drilling conditions. For other projects, the increased rate of penetration may be more or less than the values obtained from this figure.

ESTIMATING DRILLING PRODUCTION

The first step in estimating drilling production is to make an assumption about the type of equipment that will be used. The type of rock to be drilled will guide that first assumption. Both Tables 12.5 and 12.6 give information useful in making such a decision. However, it must again be emphasized, the final decision on type of equipment should be made only after test drilling the specific formation. The drilling test should yield data on penetration rate based on bit size and type. Once a drill type and bit is selected, the format given in Fig. 12.15 can be used to estimate the hourly production.

```
(1)  Depth of hole:        (a) _____ ft pull, (b) _____ ft drill
(2)  Penetration rate:         _____ ft/min

(3)  Drilling time:            _____ min     (1b)/(2)
(4)  Change steel:             _____ min
(5)  Blow hole:                _____ min
(6)  Move to next hole:        _____ min
(7)  Align steel:              _____ min
(8)  Change bit:               _____ min
                           _____
(9)      Total time:           _____ min


(10) Operating rate:           _____ ft/min   (1b)/(9)
(11) Production efficiency:     _____ min/hr
(12) Hourly production:         _____ ft/hr    (11)×(10)
```

FIGURE 12.15 | Format for estimating drilling production.

Subdrilling

Usually when drilling for loading explosives and blasting, it is necessary to subdrill below the desired final bottom or floor elevation. This is because when explosives are fired in blastholes breakage is not normally achieved to the full depth of the hole. This extra depth is dependent on the blasting design. Factors that impact blasthole drilling include hole diameter, hole spacing, pounds of explosive per cubic yard, and firing sequence. Normally 2 or 3 ft of extra depth is required. Therefore, if the depth to finish grade is 25 ft, (1a) pull depth, it may be necessary to actually drill 28 ft, (1b) drilling depth.

Drilling Time

Penetration rate will be an average rate developed from the test-drilling program and based on a specific bit size and type. Knowing hole depth and drilling rate allows the calculation of drilling time.

Changing Steel

If the drilling depth is greater than the steel length, it will be necessary to add steel during the drilling and to remove steel when coming out of the hole. For track-mounted rotary-percussion drills, standard steel lengths are 10 to 12 ft. Average weights for these steel lengths are given in Table 12.7. They require about 0.5 min or less to add or remove a length.

TABLE 12.7 | Average weights for drill steel

Size (in.)	Length (ft)	Weight (lb)
1.50	10	53
1.50	12	64
1.75	10	60
1.75	12	71

The single-pass capability of rotary drills varies considerably, in the range of 20 to 60 ft. A limited study by the second author found it took an average of 1.1 min to add a 20-ft length of steel and 1.5 min to remove the same length steel.

Blow Hole

After the actual drilling is completed, it is good practice to blow out the hole to ensure all cuttings are removed. However, some drillers prefer to simply drill an extra foot and pull the drill out without blowing the hole clean.

Move

The time required to move between drill hole locations is a function of the distance (blasting pattern) and terrain. Track-mounted rotary-percussion drills can move from 1 to 3 mph. Track-mounted rotary drills with their high masts can move at a maximum speed of about 2 mph. It should be remembered that hole spacing is often less than 20 ft and the operator is maneuvering to place the drill over an exact spot, so travel speed is slow. There are some drill rigs equipped with global positioning system (GPS) technology that enables the operator to accurately place holes. If a high-mast drill must traverse unlevel ground, it may be necessary to lower the mast before moving and then raise it again when the move is accomplished. This will significantly lengthen the movement time.

Align

Once over the drilling location, the mast or steel must be aligned. In the case of rotary drills, the entire machine is leveled by use of hydraulic jacks. This usually takes about 1 min.

Change Bit

A time allowance must be made for changing bits, shanks, couplings, and steel. Table 12.6 provides information for determining the frequency of such changes.

Efficiency

Finally, as with all production estimating, the effect of job and management factors must be taken into account. With experienced drillers working on a large project, a 50-min production hour should be achievable. If the situation is sporadic drilling with knowledgeable people, a 40-min production hour might be more appropriate. The estimator needs to consider the specific project requirements and the skill of the available labor pool before deciding on the appropriate production efficiency.

EXAMPLE 12.2

A project utilizing experienced drillers will require drilling and blasting of high silica, fine-grained sandstone rock. From field drilling tests, it was determined that a direct drilling rate of 120 ft per hour could be achieved with a $3\frac{1}{2}$ HD bit on a rotary-percussion drill at 100 psi. The drills to be used take 10-ft steel. The blasting pattern will be a 10- $\times$ 10-ft grid with 2 ft of subdrilling required. On the average, the specified finish grade is 16 ft below the existing ground surface. Determine the drilling production based on a 50-min hour.

Using the Fig. 12.15 format,

(1) Depth of hole	(a) 16 ft pull	(b) 18 ft drill (16 ft + 2 ft)
(2) Penetration:	2.00 ft/min	(120 ft ÷ 60 min)
(3) Drilling time:	9.00 min	(18 ft ÷ 2 ft/min)
(4) Change steel:	1.00 min	(1 add and 1 remove at 0.5 each)
(5) Blow hole:	0.18 min	(about 0.1 min per 10 ft of hole)
(6) Move 10 ft:	0.45 min	(10 ft at $\frac{1}{4}$ mph)
(7) Align steel:	0.50 min	(not a high-mast drill)
(8) Change bit:	0.09 min	$\left(4 \text{ min} \times \dfrac{18 \text{ ft per hole}}{850 \text{ ft life (Table 12.6c)}} \right)$
(9) Total time:	11.22 min	

(10) Operating rate:	1.60 ft/min	(18 ft ÷ 11.22 min)	
(11) Production efficiency:	50 min/hr		
(12) Hourly production:	80.2 ft/hr	[(50min/hr)×(1.60ft/min)]	

High-Wear Items

From Table 12.6c, the expected life of the high-wear items, the bit, shank, couplings, and steel (see Fig 12.16), of the drill can be found.

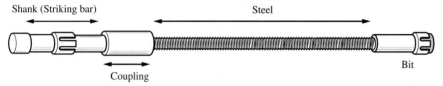

FIGURE 12.16 | High-wear drilling items: shank, coupling, steel, and bit.

EXAMPLE 12.3

Assuming project conditions and drilling equipment are as described in Example 12.2, what is the expected life, in holes that can be completed, for each of the high-wear items?

For an average hole depth of 18 ft, the following number of holes can be completed per each replacement.

$3\frac{1}{2}$ HD bit	850-ft life	47 holes
Shank	5,000-ft life	278 holes
Couplings	1,000-ft life	56 holes
Steel	2,000-ft life	111 holes

Rock Production

Drilling is only one part of the process of excavating rock, so when considering production it is good to consider the cost and output in terms of cubic yards of rock. With the 10- × 10-ft pattern, the 18 ft of drilling, used in Example 12.2, yields 16 ft of excavation. Therefore, each foot of *drill* hole produces 3.3 cy of bank measure (bcy) rock.

$$\frac{10 \text{ ft} \times 10 \text{ ft} \times 16 \text{ ft}}{27 \text{ cf/cy}} \times \frac{1}{18 \text{ ft}} = 3.3 \text{ cy/ft}$$

If the hourly drilling production is 80.2 ft, then the rock production is 80.2 × 3.3, equaling 265 cy. This should be matched to the blasting production and to the loading and hauling production. For example, if the loading and hauling capability is 500 cy per hr, it will be necessary to employ two drills.

In calculating cost, it is good practice to make the analysis in terms of both feet of hole drilled and cubic yard of rock produced. Considering only the high-wear items of Examples 12.2 and 12.3, if bits cost $200 each, shanks $105, couplings $50, and a 10-ft steel $210, what is the cost per cu yd of rock produced?

Bits	$200 ÷ 850 ft = $0.235/ft
Shank	$105 ÷ 5000 ft = $0.021/ft
Couplings	(2 × $50) ÷ 1000 ft = $0.100/ft
Steel	(2 × $210) ÷ 2000 ft = $0.210/ft
	$0.566/ft

$$\text{or} \quad \frac{\$0.566 \text{ per ft}}{3.3 \text{ cy per ft}} = \$0.172 \text{ per cy}$$

GPS and Computer Monitoring Systems

Current technology for monitoring drilling and blasting operations is progressing from using external and somewhat subjective measurement methods of performance, i.e., time sheets, foreman- or operator-generated drilling footage reports, to onboard data acquisition and management systems. There are now available drills that use GPS systems to locate the position of the boreholes. These systems reduce the survey work required for laying out drilling patterns. The drill pattern design, with hole attributes of depth and angle, is loaded into onboard computer systems that display positions on an LCD (liquid crystal device) screen for the operator. There are electronic sensor systems that provide a production-monitoring capability to include the recording and displaying of penetration rate, rotary torque and speed, bailing air pressure and volume, pulldown pressure, and hole depth. The bailing air pressure and volume are important to productivity and bit life as pressure and volume are the critical elements for cooling the bit and ensuring that the cuttings are cleared from the face of the bit to prevent regrinding of material. The systems can provide detailed knowledge of the relative hardness of the strata drilled. That strata information can be imported to computer blast design programs to improve explosive loading calculations.

Most of these computer-based monitoring systems are being developed and used in the mining industry. But that technology will soon find its way into the construction industry. In the mining industry, there have been reports that this on-line drill monitoring capability, which furnishes detailed information on rock quality and fracture zones, can provide a 20 to 25% saving in blasting cost. Accurate rock property data obtained from drill monitoring allows a computer-based explosives truck to generate loading designs on the bench. The onboard computer in the explosives truck can formulate and control the loading of the blasthole with the exact quality and type explosive within minutes after the drill has moved to the next blasthole.

DRILLING EARTH

This section discusses various types of equipment used to drill holes in earth, as distinguished from rock. Some equipment, such as that used for exploratory purposes in securing core samples and similar operations, can be used for drilling rock or earth.

Purposes for Drilling Holes in Earth

In the construction and mining industries, holes are drilled into earth for many purposes, including, but not limited to,

1. Obtaining samples of soil for test purposes (see Fig. 12.17).
2. Locating and evaluating deposits of aggregate suitable for mining.
3. Locating and evaluating deposits of minerals.
4. Installing cast-in-place piles or shafts for structural support.
5. Enabling the driving of load-bearing piles into hard and tough formations.
6. Providing wells for supplies of water or for deep drainage purposes.
7. Providing shafts for ventilating mines, tunnels, and other underground facilities.
8. Providing horizontal holes through embankments, such as those for the installation of utility conduits.

Sizes and Depths of Holes Drilled into Earth

Most holes are drilled by rotating bits or heads attached to the lower end of a shaft called a "kelly bar." An external motor or engine rotates this bar that is supported by a truck, a tractor, a skid mount, or a crane (see Figs. 12.2, and 12.18).

The sizes of holes drilled may vary from a few inches to more than 12 ft (3.7 m). Drills may be equipped with a device attached to the lower end of the drill shaft, known as an *underreamer,* that will permit a gradual increase in the diameter of the hole at the bottom. This enlargement permits a substantial increase in the bearing area under a shaft-type concrete footing. Underream diameters as great as 144 in. have been drilled. Under favorable conditions, it is possible to drill holes as deep as 200 ft for underreamed foundations.

A truck-mounted rig, whose essential parts include a power unit, cable drum, boom, rotary table, drill stem, and drill, drills the holes. The shaft is drilled first with a large earth auger or a bucket drill, equipped with cutting blades at the bottom. After the shaft is completed, the bottom portion of the hole is enlarged with the underreamer.

When drilling holes through unstable soils, such as mud, sand, or gravel, containing water, it may be necessary to use a casing. Sometimes the shaft is drilled entirely through the unstable soil and then a temporary steel casing is installed in the hole to eliminate groundwater and caving. An alternative method is to add sections to the casing as drilling progresses until the hole is completed through the full depth of unstable soil. The remainder of the hole can often then be completed without additional casing. When the hole is filled with concrete, the casing is pulled before the concrete sets.

This type of foundation has been used extensively in areas whose soils are subject to changes in moisture content to considerable depth. By placing the footings below the zone of moisture change, the effects of soil movements due to changes in moisture are minimized.

FIGURE 12.17 I Tractor-mounted auger-type earth drill.

FIGURE 12.18 | Crane-mounted auger-type drill.

Among the advantages of drilled and underreamed foundations, compared with piles and conventional spread footings, are

1. They are less expensive for some soils and projects.
2. They make it easy to adjust the depth when there are varying soil conditions.
3. They permit inspection of soil prior to establishing depth or placing concrete.
4. They eliminate damage to adjacent structures due to vibration of the pile hammer.
5. They eliminate the use of forms for concrete.

Removal of Cuttings

Several methods are used to remove the cuttings from the holes. One method of removing the cuttings is to attach an auger to the drill head. The drill head is the actual cutting tool at the bottom of the drill stem. The auger extends from the drill head to above the surface of the ground (see Fig. 12.17). As the drill shaft and the auger rotate, the earth is forced to the top of the hole, where it is removed and wasted. However, the depth of a hole for which this method may be used is limited by the diameter of the hole, the class of soil, and the moisture content of the soil.

Another method of removing the cuttings is to attach the drill head to only a section of the auger. When the auger section is filled with cuttings, it is raised above the surface of the ground and reverse rotated rapidly to free it of the cuttings (see Figs. 12.2 and 12.18).

A third method of removing the cuttings is to use a combination of a drill head with a cylindrical bucket, whose diameter is the same as the diameter of the hole. As the bucket is rotated, steel cutting blades attached to the bottom of the bucket force the cuttings up and into the bucket. When the bucket is filled, it is raised to the surface of the ground and emptied.

A fourth method of removing the cuttings is to force air and water through the hollow kelly bar and drill shaft to the bottom of the hole and then upward around the drill shaft. The air or water carries the cuttings to the surface of the ground for disposal.

Directional Drilling

The drilling technique of directional boring is not new technology. The technology was perfected in the oil industry and that industry has used it with vertical drilling for decades. During the last 10 yr, directional-boring technology has been adapted to the construction industry for horizontal work such as installing utilities. The technology enables the installation of underground utilities without the necessity of disrupting surface facilities that are already in place and being used. Directional boring and horizontal boring are similar in job function but differ in procedure and technology. Directional boring converts the old oil industry technology and uses it for utility work. But there can be problems with alignment accuracy in soils containing rock. Another system, horizontal boring, is extremely accurate; however, there is a requirement for boring pits, shoring boxes, and special track installations. Horizontal boring is also a more labor-intensive activity than directional boring. The cost of directional boring is typically less than that of horizontal boring. In the case of a 24-in. casing the price difference can be as much as 25% per linear foot [9].

Project soil conditions are a critical factor in directional boring work, with the most important consideration being the presence of rock. Directional boring equipment capable of boring through rock is available. However, if rock is expected anywhere along the bore path, even for only a few feet in a bore of hundreds of feet, the whole bore must be treated as a rock bore. This is because it is not practical to bore part of the distance using standard equipment and to then pull that equipment out of the bore and replace it with a special rock-boring head. Additionally, having to change the boring head a second time once the rock portion of the bore is completed would further complicate the procedure.

Standard directional boring equipment cannot effectively bore through rock. When standard boring equipment is used and boulders or cobble are encountered, the procedure for advancing the bore is to direct the boring head to follow a path either above or below the obstruction. Such a procedure results in a bore that does not conform to a uniform grade, a standard requirement for sewer work.

Directional boring, unlike horizontal boring, is a relatively new field. The process has two stages. In the first stage a small pilot hole is drilled. To drive the pilot bore, a flexible drill steel is used to push a steerable drill head through the ground. The steering control is done manually with steering rods housed within the drill steel and lasers are used to verify line and grade. A transmitter located behind the drill head enables the head to be tracked by a walkover system on the ground surface. With this system, precise and accurate pilot holes can be achieved when soil conditions are favorable.

Typically with directional boring, the diameter of the casing to be placed is greater than the diameter of the pilot bore; in which case, a second stage is necessary. This second stage is a reaming process. The pilot hole can be reamed simultaneously with pulling the casing. The casing is attached to a puller and swivel, which, in turn, is attached behind the reamer. This arrangement enables the pipe to follow directly behind the reamer as it cuts the soil. A thin pressure grout is pumped into the pilot hole through the reamer. When dry, this grout forms something similar to soil cement due to the mixing with the dirt outside the casing. This grout ensures that there are zero voids outside the casing once it is in place. The size of the reamer can vary considerably, and there are different types of reamers available for different soil conditions, including rock.

Horizontal Boring

A rigid horizontal drilling process where the boring machine sits on guide rails in a boring pit is known as horizontal boring. The pit is excavated to a depth so that the drill, when placed on its tracks, is aimed on the alignment required of the pipe. As rams push the drill, sections of the sleeve (casing) pipe are pushed into the bore directly behind the drill head. The sleeve pipe should follow no more than 1 in. behind the drill head. After each section of sleeve is pushed into the bore, the pusher rams are retracted and a new sleeve section is joined to those already in the bore. This process is repeated until the bore is completed. Pressure grout is fed through the drill bit to encase the sleeve pipe. Dirt tailing is augered out of the sleeve and back to the pit where it is removed.

As the boring advances, the machine automatically maintains a forward thrust on the casing and the auger. These machines can be powered hydraulically, by air, or by electric motors. Figure 12.19 illustrates a horizontal boring machine cutter head for use in soft rock.

In the case of either directional boring or horizontal boring, alignment of the utility line in the sleeve is controlled with skates. They serve as spacers that hold the utility pipe at the specified invert. The skates are usually pressure-treated redwood and are located in groups of three, spaced at equal distances around the circumference of the utility pipe. They are bound to the pipe with metal straps. Their thickness is dictated by the required final invert of the utility pipe. Spacing of the skates is typically on 15 or 20 ft centers along the utility pipe. Once the

FIGURE 12.19 | Cutter head for an auger-type boring machine used in soft rock.

utility pipe is braced in place by the skates, sand is blown into the void between the sleeve and the pipe.

With water lines, bends are not a significant problem as the water flows under pressure. As long as a connection can be made to the existing lines, the service can be acceptable. With sewer lines, however, adherence to grade and invert specifications is critical as sewers usually work by gravity flow. Therefore, understanding the soil properties of a project site and the limitations of the two drilling methods is crucial to successful placement of utility lines.

SUMMARY

Drilling may be performed to explore the types of materials to be encountered on a project (exploratory drilling), or it may involve production work such as drilling holes for explosive charges. Other purposes include holes for grouting or rock bolt stabilization work and boring for the placement of utility lines. The rate at which a drill will penetrate rock is a function of the rock, the drilling method, and the size and type of bit. The four critical rock properties that affect penetration rate are (1) hardness, (2) texture, (3) breaking characteristic, and (4) formation.

Most holes drilled in earth use rotating bits or heads attached to the lower end of a shaft called a "kelly bar." During the last 10 years, directional-boring technology has been adapted to the construction industry for horizontal work such as installing utilities. The technology allows the installation of underground utilities without the necessity of disrupting surface facilities that are already in place and being used. Critical learning objectives include:

- An understanding of the different types of drilling equipment.
- An appreciation of the factors that influence the rates of drilling rock.
- An ability to prepare a drilling production estimate.
- A basic understanding of the new directional drilling technology that is available.

These objectives are the basis for the problems that follow.

PROBLEMS

12.1 A project utilizing experienced drillers will require drilling and blasting of medium silica, granite. No field drilling tests were conducted. It is proposed to use a $6\frac{1}{2}$ B downhole drill at 350 psi. The drills to be used have a single-pass capability of 20 ft and take steel in 20-ft lengths. The blasting pattern will be a 10- $\times$ 12-ft grid with 3 ft of subdrilling required. On the average, the specified finish grade is 30 ft below the existing ground surface. Determine the drilling production, assuming it takes 30 min to change a $6\frac{1}{2}$-in. downhole hammer. (66.7 ft/hr)

12.2 A project in medium silica, sandstone is being investigated. Field drilling tests proved that air-track drills with $3\frac{1}{2}$-in. HD bits can achieve a penetration rate of 2.5 ft per min. The drills use 12-ft steel. The blasting pattern will be 6 $\times$ 8 ft with 2 ft of subdrilling. The average depth to finish grade is 8 ft. The drilling production must match that of the loading and hauling which is 210 bcy per hour. Assuming a 50-min. production hour many drill units will be required?

12.3 A highway cut through shale is being constructed. Drilling tests indicate that air-track drills with $3\frac{1}{2}$-in. STD bits can penetrate 3.5 ft per min. The drills use 10-ft steel. The blasting pattern will be 8 $\times$ 10 ft with 2 ft of subdrilling. The average depth to finish grade is 15 ft. The project labor scale is $16 per hour for drillers. The air-track drill and compressor cost $46 per hour; bits are $185 each; shanks, $115; couplings, $54; and a 10 ft of steel, $200. Because of very difficult job conditions use a 45-min hour to calculate production. What is the hourly drilling production in both feet and bcy? What is the cost of the high-wear items? What is the bcy cost including equipment and labor, and the high-wear items? (109.2 ft/hr; 284 bcy/hr; $0.232/ft; $0.307/bcy)

12.4 The president of "Low Bid" construction company has just visited one of his jobs. At the project he found most of the rock trucks parked. When he questioned the foreman as to why this new equipment with the capability of hauling 600 cy per hr was not working, the foreman said the drilling could not keep up. The president has told you to solve the problem immediately.

The project is in a dolomite formation. Company drilling experience in this area indicates that compressed air rotary-percussion drills with $3\frac{1}{2}$-in. button bits can average 1.5 ft of penetration per minute. Drilling efficiency is typically a 50-min hour. These drills use 12-ft drill steel. The best blasting pattern is 10 $\times$ 10 ft with 2 ft of subdrilling. The average excavation depth is 8 ft. How many drills will you recommend for use on the job?

12.5 Buffet Inc. has determined that there is an untapped party market at the North Pole. Intensive marketing studies indicate that a "Margaritaville Bar & Grille" would be very successful during the long dark arctic winters. Joe Cool, now living in Key West, has been employed to manage the construction of this project and has employed you to analyze the foundation-drilling program.

The plan is to construct the establishment 20 ft below the surface of the ice. This will require an extra 2 ft of drilling to ensure that the final floor can be excavated to the required depth after blasting. From field testing with a 90-psi rotary-percussion drill, it has been determined that a direct drilling rate, through the ice, of 175 ft per hr can be achieved when using a $3\frac{1}{2}$ HD bit. This drill uses 12-ft drill steel. Studies have indicated that ice acts similar to low silica, LA > 20, diabase. The blasting pattern will be on a 12- × 12-ft grid. The cool climate will affect production.

Because of the high freight cost to the site, $3\frac{1}{2}$-bits cost $500 each; shanks, $400; couplings, $150; and steel, $600. All prices are FOB (free on board) at the North Pole. What drilling production can be expected and what will be the cost on a per cubic yard basis for the drill's high-wear items that you should tell Mr. Cool to use in the project's construction cost estimate?

12.6 A project in medium silica, granite is being investigated. Field drilling tests proved that air-track drills with $3\frac{1}{2}$-in. B bits can achieve a penetration rate of 1.1 ft per min. The drills use 10-ft steel. The blasting pattern will be 8 × 8 ft with 2 ft of subdrilling. The average depth to finish grade is 16 ft. The drilling production must match that of the loading and hauling, which is 300 bcy per hr. Assuming a 40-min production hour drilling, how many drill units will be required?

REFERENCES

1. Ariaratnam, Samuel T., and Erez N. Allouche, "Suggested Practices for Installations Using Horizontal Directional Drilling," *Practice Periodical on Structural Design and Construction,* ASCE, Vol. 5, No. 4, pp. 142–149, November 2000.

2. Bruce, Donald A., and Dugnani, Giovanni, "Pile Wall Cuts Off Seepage," *Civil Engineering,* Vol. 66, No. 7, pp. 8A–11A, July 1996.

3. Dick, Richard A., Larry R. Fletcher, and Dennis W. D'Andrea, "Explosives and Blasting Procedures," GPO No. 024-004-02115-6, Government Printing Office, Washington, D.C.

4. Higgins, Lindley R., "Drills Play Dramatic Role in Profit Production," *Construction Methods and Equipment,* Vol. 58, pp. 54–61, September 1976.

5. Nelmark, Jack D., "Large Diameter Blast Hole Drills," *Journal of the Mining Congress,* August 1980.

6. Pasieka, A. R., and J. C. Wilson, "The Importance of High-Pressure Compressed Air to Mining Operations," *Canadian Mining and Metallurgical Bulletin,* Vol. 59, pp. 1093–1102, September 1966.

7. "Shot-Drill Cuts Hard Rock Sockets for Footings," *Construction Methods and Equipment,* Vol. 53, pp. 84–85, May 1971.

8. Tulloss, Michael D., and Cliff J. Schexnayder, "Horizontal Directional Boring at Lewis Prison Complex," *Practice Periodical on Structural Design and Construction,* ASCE, Vol. 4, No. 3, pp. 119, 120, August 1999.

DRILLING EQUIPMENT WEBSITES

1. www.atlascopco.se/rde. Atlas Copco is a manufacturer of state-of-the-art rock drills and drill rigs. The company offers drifting and tunneling equipment, production drilling equipment as well as bolting equipment.

2. www.mining-technology.com/contractors/drilling/archway/index.html. Archway Engineering is a manufacturer of equipment for core drilling, soil sampling, core barrels, drill rods, casings, and drill bits, with applications for geotechnical investigation and environmental assessment of contaminated ground.

3. www.tamrock.com/. Tamrock is a supplier of mechanized underground and surface drilling equipment and tools.

4. www.halcodrilling.com/. Halco Group manufactures a comprehensive range of downhole hammers and drill bits for drilling hole diameters from $2\frac{3}{4}$ in. (70 mm) to in excess of 12 in. (30 mm).

5. www.rocktek.com/. RockTek develops, manufactures, and markets innovative new systems for controlled rock breaking in the mining, quarrying, and civil construction industries. RockTek USA Ltd. is a member of the Brandrill Limited Group with corporate headquarters in Western Australia.

CHAPTER

13

Blasting Rock

When there is a requirement to remove hard rock, blasting should be considered as it is almost always more economical than mechanical excavation. The major mechanisms of rock breakage result from the sustained gas pressure buildup in the borehole by the explosion. Every blast must be designed to meet the existing conditions of the rock formation and overburden, and to produce the desired final result. The economics of handling the fractured rock is a factor that should be considered in blast design. The blast pattern will affect such considerations as type of equipment and the bucket fill factor of the excavators.

BLASTING

The operation referred to as "blasting" is performed to break rock so that it can be quarried for processing in an aggregate production operation, or to excavate a right-of-way. Blasting is accomplished by discharging an explosive that has been placed in either an unconfined manner, such as mud capping boulders, or is confined as in a borehole. There are two forms of energy released when high explosives are detonated: (1) shock and (2) gas. An unconfined charge works by shock energy, whereas a confined charge has a high gas energy output.

When there is a requirement to remove hard rock, blasting should be considered as it is almost always more economical than mechanical excavation. One reported comparison of mechanical excavation costs versus removal by blasting identified a difference of \$4.05 per cy [9].

There are many types of explosives and methods of using them. A full treatment of each explosive and method is too comprehensive for inclusion in this book. A complete discussion of explosives can be found in handbooks on blasting published by manufacturers of explosives [1, 2].

GLOSSARY OF TERMS

The following glossary defines the important terms that are used in describing blasting operations.

ANFO. A blasting agent that is produced by mixing prilled industrial grade ammonium nitrate and fuel oil.

Back break. Rock broken beyond the last row of holes.

Bench height. The vertical distance between the floor or base of an excavation and the berm or ledge, above where the blastholes are drilled and shot.

Blasthole (borehole). A hole drilled into rock to permit the placing of an explosive.

Blasting (shot). The detonation of an explosive to fracture the rock.

Blasting agent. The classification of a particular type of explosive compound from the standpoint of storage and transportation. It is a material or mixture intended for blasting, consisting of an oxidizer and a fuel. It is less sensitive to initiation and cannot be detonated with a No. 8 blasting cap when unconfined. Therefore, blasting agents are covered by different handling regulations than high explosives.

Blasting cap. A device consisting of a small hollow metal tube that is filled with a high explosive. The blasting cap is detonated within or adjacent to a primer as a means of detonating the blasting agent. There are two types of blasting caps: electric and nonelectric.

Blasting machine. A machine used to generate the electric current that detonates an electric blasting cap.

Booster. A chemical agent used to intensify an explosive reaction. A booster does not include an initiating device.

Burden. The distance from the explosive charge to the nearest free or open face is referred to as burden or burden distance. There can be an apparent burden and a true burden. True burden is in the direction that the displacement of broken rock will move following the firing of the charge.

Charge. The total explosive loaded in a blasthole, to include the blasting agent, booster, and primer.

Crimping. An act of reducing the diameter of a blasting cap near its open end for the purpose of securely holding the fuse in the cap.

Cutoff. The breaking of a fuse or electric circuit to a blasting cap in a primer. The result of such an occurrence is that a portion of a column of explosives will fail to detonate.

Decking (deck stemming). The operation of placing inert material in a blasthole at spacings to separate explosive charges in the hole.

Delay. The term used to describe noninstantaneous firing of a charge or group of charges usually accomplished by blasting caps having predetermined built-in time delays. Can also be accomplished using a sequential blasting machine.

Density. An explosive's density is its specific weight expressed as grams per cubic centimeter (g/cc). When comparing granular explosives with other granular explosives, density and energy are correlated.

Detonation rate. The velocity at which a detonation progresses through an explosive.

Downline. Primacord lines extending into the blasthole from a trunkline on the surface at the top of the hole. At the bottom end it is attached to the primer.

Electric blasting cap (EBC). A small metal tube loaded with a powder charge. It is detonated by the heat produced from an electric current flowing through a wire bridge inside the cap.

Explosive. A chemical compound or mixture that reacts at high speed, liberating gas and heat, thus causing very high pressure. High explosives contain at least one high explosive ingredient. Low explosives contain no ingredients that by themselves can explode. Both high and low explosives can be initiated by a single No. 8 blasting cap.

Flammability. The characteristic of an explosive that describes its ease of initiation from spark, fire, or flame.

Floor. The horizontal, or nearly so, bottom plane of an excavation.

Flyrock. Rock that is ejected into the air by an explosion. It is an indication of wasted energy.

Fragmentation. The degree to which mass rock is broken into pieces by blasting.

Fumes. The amount of toxic gases produced by an explosive in the detonation process.

Initiation. The act of detonating a high explosive.

Lead wires. The wires that conduct the electric current from its source to the leg wires of the electric blasting caps.

Leg wires. The wires that conduct the electric current from the lead wires to an electric blasting cap.

Magazine. A building or chamber for the storage of explosives.

Mat (blasting mat). A blanket, usually of woven wire rope, used to restrict or contain flyrock.

MS delay cap (millisecond). A cap that has a built-in delay time element. These caps are commonly available in $\frac{25}{1,000}$-sec increments.

Mud capping (plaster shot). An operation where an explosive is placed in contact with the surface of a boulder or rock and covered with mud or earth for firing.

Nitroglycerin. A colorless explosive liquid obtained by treating glycerol with a mixture of nitric and sulfuric acids.

Overbreak. Rock that is fractured outside the desired excavation limits. (Many construction contracts stipulate nonpayment for rock fracture or removed beyond a specified overbreak line.)

Overburden. The depth of material lying above the rock that is to be shot.

PETN. The abbreviation for the chemical (pentaerythritol tetranitrate) of a high explosive having a very high rate of detonation that is used in detonating cord.

Powder factor. The quantity of explosive used to fracture a specific volume of rock, e.g., pounds of explosive per cubic yard of rock.

Premature. A charge that detonates before it is intended to explode.

Presplitting. A stress relief line of small diameter drill holes spaced at close intervals, which are lightly loaded and detonated to rupture the web of rock remaining between the holes. The main charge is fired after firing the presplit holes.

Prill. In the United States most ammonium nitrate, both agricultural and blasting grade, is produced by the prill tower method. Ammonium nitrate liquor is released as a spray at the top of a prilling tower. Prills of ammonium nitrate congeal in the upcoming steam and air of the tower. The moisture driven out by the dropping process leaves voids within the prills.

Propagation. The movement of a detonation wave, either in the hole or from hole to hole.

Relative bulk strength. The relative strength of an explosive compared to the strength of standard ANFO. Standard ANFO is assigned a relative bulk strength rating of 100.

Rounds. A term that includes all the blastholes that are drilled, loaded, and fired at one time.

Safety fuse. A fuse containing a low explosive enclosed in a suitable covering. When the fuse is ignited, it will burn at a predetermined speed. It is used to initiate explosions under certain conditions.

Secondary blasting. An operation performed, after the primary explosion, to reduce remaining oversize material to a desirable size.

Sensitiveness. A measure of an explosive's cartridge-to-cartridge propagating ability.

Sensitizer. An ingredient used in explosive compounds to promote greater ease in initiation or propagation of the reactions.

Stemming. The adding of inert material, such as drill cuttings, on top of the explosive in a blasthole to confine the energy of the explosion.

Tamping. The process of compacting the stemming or explosive in a blasthole.

Throw/heave. The displacement of rock as a result of a detonation and the expansion of gases.

TNT. A high explosive whose chemical content is trinitrotoluene, or trinitrotoluol.

Toe. The horizontal distance measured at floor level from the free face to the blasthole. If the blasthole is vertical this is equal to the burden.

Trunkline. The main line of a detonating cord on the surface, in the case of a nonelectric system, extending from the ignition point to the blastholes containing explosives to be detonated. Secondary lines (downlines) of detonating cord attached to the trunkline are used to detonate the blasting caps in the primers.

COMMERCIAL EXPLOSIVES

Commercial explosives are compounds that detonate upon introduction of a suitable initiation stimulus. The ingredients of an explosive compound are converted into high-pressure, high-temperature gases upon detonation. Pressures just behind the detonation front are in the order of 150,000 to 3,980,000 psi, while temperature can range from 3,000°F to 7,000°F.

There are four main categories of commercial high explosives: (1) dynamite, (2) slurries, (3) ANFO, and (4) two-component explosives. To be a high explosive, the material must be cap sensitive, react at a speed faster than the speed of sound, and the reaction must be accompanied by a shock wave. The first three categories—(1) dynamite, (2) slurries, and (3) ANFO—are the principle explosives used for blasthole charges. Two-component or binary explosives are normally not classified as an explosive until mixed. Therefore, they offer advantages in shipping and storage that make them attractive alternatives on small jobs. But their unit price is significantly greater than that of other explosives.

Dynamite

This nitroglycerin-based product is the most sensitive of all the generic classes of explosives in use today. It is available in many grades and sizes to meet the requirements of a particular job. Straight dynamite is not appropriate for construction applications because it is very sensitive to shock. With straight dynamite, sympathetic detonation can result from adjacent holes that are fired on an earlier delay.

The most widely used product is known as "high-density extra dynamite," but individual explosive manufacturers have their own trade names for dynamite products. In this product, which is less sensitive to shock than straight dynamite, some of the nitroglycerin has been replaced with ammonium nitrate.

The approximate strength of a dynamite is specified as a percentage that is an indication of the ratio of the weight of nitroglycerin to the total weight of a cartridge. Individual cartridges vary in size from approximately 1 to 8 in. in diameter and 8 to 24 in. long.

Dynamite is used extensively for charging blastholes, especially for the smaller sizes. For best results, dynamite cartridges, except for the prime cartridge, should be expanded to fill the entire diameter of the hole. As cartridges are placed in a hole, they are tamped sharply with a wooden pole, expanding the cartridges to fill the hole. For this purpose, it may be desirable to split the sides of a cartridge, or use cartridges with perforated shells. A blasting cap or a Primacord fuse may be used to fire the dynamite. If a blasting cap is used, one of the cartridges serves as a primer. The cap is placed within a hole made in this cartridge.

Slurries

This is a generic term for both water gels and emulsions. Water gels are water-resistant explosive mixtures of oxidizing salts, fuels, and sensitizers. The primary sensitizing methods are the introduction of air throughout the mixture, the addition of aluminum particles, or the addition of nitrocellulose. Emulsions are mixtures of oxidizing salts and fuels but contain no chemical sensitizers. In comparison to ANFO (see next section), slurries have a higher cost per pound and they have less energy. However, in wet conditions they are very competitive with ANFO, because the ANFO is water-sensitive and must be protected in lined holes or a bagged product has to be used. Both of these measures add to the total cost of the ANFO. Emulsions will have a somewhat higher detonation velocity than water gels.

An advantage of slurries over dynamite is that the separate ingredients can be hauled to the project in bulk and mixed immediately before loading the blastholes. The mixture can be poured directly into the hole. Some emulsions tend to be wet and will adhere to the blasthole, causing bulk loading problems. Slurries may be packaged in plastic bags for placement in the holes. Because they are denser than water, they will sink to the bottom of holes containing water.

Slurries are detonated by special primers, such as dynamite or PETN, using electric blasting caps or Primacord.

ANFO

This explosive is used extensively on construction projects and represents about 80% of all explosives used in the United States. "ANFO," an ammonium nitrate and fuel oil mixture, is synonymous with dry blasting agents. This explosive is the cheapest source of explosive energy. Because it must be detonated by special primers, it is much safer than dynamite.

Standard ANFO is a mixture of prilled industrial grade ammonium nitrate and 5.7% No. 2 diesel fuel oil. This is the optimum mixture. The detonation efficiency is controlled by the amount of fuel oil. It is less detrimental to have a fuel deficiency but both extremes affect the blast. With too little fuel, the explosive will not perform properly. With too much fuel, maximum energy output is reduced. The prills should not be confused with ammonium nitrate fertilizer prills. A blasting prill is porous to better distribute the fuel oil.

Because the mixture is free flowing, it can be either blown or augered from bulk trucks directly into the blastholes. It is detonated by primers consisting of charges of explosive placed at the bottoms of the holes and sometimes at intermediate depths. Electric blasting caps or Primacord may be used to detonate the primer.

ANFO is not water resistant. Detonation will be marginal if ANFO is placed in water and shot, even if the interval between loading and shooting is very short. If it is to be used in wet holes, there is a *densified* ANFO cartridge. This product has a density greater than water, so it will sink to the bottom of a wet hole. Standard ANFO has a product density of 0.84 g/cm^3. The poured density of ANFO is generally between 0.78 and 0.85 g/cm^3. Cartridges or bulk product sealed in plastic bags will not sink. Another method to permit use in wet conditions is to preline the holes with plastic tubing, closed at the bottom, to exclude the water. The tubes, whose diameters should be slightly smaller than the holes, are installed in the holes by placing rocks or other weights in their bottoms.

PRIMERS

A primer is a cap-sensitive explosive loaded with a firing device that initiates the explosion. The resulting detonation is then transmitted to an equal or less sensitive explosive. Primers produce a high detonation pressure for initiation of ANFO products and certain water gels. Good priming will give improved overall explosive performance. It improves fragmentation, increases productivity, and lowers overall cost.

To obtain optimum performance of the ANFO, the size of the primer must match the borehole diameter. As the mismatch in the size of the primer to the borehole diameter increases, the detonation velocity of the ANFO will decrease. The primer should never be tamped.

The primer, which is the point of initiation of an explosive column, should always be at the point of maximum confinement. This is usually at the bottom of the blasthole. There are exceptions and care should be exercised when there are soft seams or fractures in the rock. Contact between the primer and the explosive column is extremely important. A bottom primer should be just off the bottom of the hole. When using bulk ANFO, 1 to 2 ft should be placed in the hole before loading the primer. This will ensure good contact between the ANFO column and the primer.

INITIATING AND DELAY DEVICES

It is common practice to fire several holes or rows of holes at one time. Fragmentation, backbreak, vibration, and violence of a blast are all controlled by the firing sequence of the individual blastholes. The order and timing of the detonation of individual holes are controlled by the initiation system. Electric and nonelectric initiation systems are available. When selecting the proper system, one should consider both blast design and safety. Electrical systems are more sensitive to lightning than nonelectric systems, but both are susceptible.

Electric Blasting Caps

Blasting caps are used for initiating high explosives. An electric cap passing an electric current through a wire bridge, similar to an electric lightbulb filament, causes an explosion. The current, approximately 1.5 amps, heats the bridge to incandescence and ignites a heat-sensitive flash compound. The ignition sets off a primer that in turn fires a base charge in the cap. This charge detonates with sufficient violence to fire a charge of explosive.

Electric caps are supplied with two leg wires in lengths varying from 2 to 100 ft. These wires are connected with the wires from other holes to form a closed electric circuit for firing purposes. The leg wires of electric blasting caps are made of either iron or copper. For ease in wiring a blast, each leg wire on an electric cap is a different color. Regular electric blasting caps are made to fire within a few milliseconds after current is applied.

When two or more electric blasting caps are connected in the same circuit, they must be products of the same manufacturer. This is essential to prevent misfires, because blasting caps of different manufacturers do not have the same electrical characteristics. Blasting caps are extremely sensitive. They must be protected from shock and extreme heat. They are never to be stored or transported with other explosives.

Delay Blasting Systems

Delay blasting caps are used to obtain a specified firing sequence. Such caps are available for delay intervals varying from a small fraction of a second to about 7 sec. When explosive charges in two or more rows of holes parallel to a face are fired in one shot, it is desirable to fire the charges in the holes nearest the face a short time ahead of those in the second row. This procedure will reduce the apparent burden for the holes in the second row, and thereby will permit the explosive in the second row to break the rock more effectively. In the case of more than two rows, this same delayed firing sequence would be followed for each successive row.

In construction and surface blasting applications, millisecond delay electric blasting caps are frequently used. They have individual intervals ranging from 25 to about 650 milliseconds (msec). Long-period-delay electric blasting caps have individual intervals ranging from about $\frac{2}{10}$ sec to over 7 sec. They are primarily used in tunneling and underground mining.

Detonating Cord *(Primacord)*

This is a nonelectric initiation system consisting of a flexible cord having a center core of high explosive, usually PETN. It is used to detonate dynamite and other cap-sensitive explosives. The explosive core is encased within textile fibers and covered with a waterproof sheath for protection. Detonating cord is insensitive to ordinary shock or friction. The explosive has a detonation rate of approximately 21,000 ft per sec. Delays can be achieved by attaching in-line delay devices. When several blastholes are fired in a round, the cord is laid on the surface between the holes as a trunkline. At each hole, one end of a detonating cord downline is attached to the trunkline, whereas the other end of the downline extends into the blasthole. If it is necessary to use a blasting cap and/or a primer to initiate the blast in the hole, the bottom end of the downline may be cut square and securely inserted into the cap.

Sequential Blasting Machine

There are condenser-discharge-blasting machines for firing electric blasting caps. Special machines have sequential timers, permitting precisely timed firing intervals for blasting circuits. This provides the blaster the option of many delays within a blast. Since many delays are available, the pounds of explosive fired per delay can be reduced to better control noise and vibration.

ROCK BREAKAGE

The major mechanisms of rock breakage result from the sustained gas pressure buildup in the borehole by the explosion. First, this pressure will cause radial cracking. Such cracking is similar to what happens in the case of frozen water pipes—a longitudinal split occurs parallel to the axis of the pipe. A borehole is analogous to the frozen pipe in that it is a cylindrical pressure vessel. But there is a difference in the rate of loading. A blasthole is pressurized instantaneously. Failure, therefore, instead of being at the one weakest seam, is in *many* seams parallel with the borehole. Burden distance and direction to the free face will control the course and extent of the radial crack pattern.

When the radial cracking takes place, the rock mass is transformed into individual rock wedges. If relief is available perpendicular to the axis of the blasthole, the gas pressure pushes against these wedges, putting the opposite sides into tension and compression. The exact distribution of such stresses is affected by the location of the charge in the blasthole. In this second breakage mechanism, flexural rupture is controlled by the burden distance and bench height. The ratio of the bench height divided by the burden distance is known as the "stiffness ratio." This is the same mechanism a structural engineer is concerned with when analyzing the length of a beam or column in relation to its thickness.

There is a greater degree of difficulty in breaking rock when the burden distance is equal to the bench height. As bench height increases compared to burden distance, the rock is more easily broken. If the blast is not designed properly and the burden distance is too great, relief will not be available and the hole will either crater or the stemming will blow out. Overcharging with explosives will result in backbreak and/or excessive throw rock. Undercharging will result in large boulders at the face.

BLAST DESIGN

Every blast must be designed to meet the existing conditions of the rock formation and overburden, and to produce the desired final result. There is no single solution to this problem. Rock is not a homogeneous material. There are fracture planes, seams, and changes in burden to be considered. Wave propagation is faster in hard rock than in soft rock. Initial blast designs use idealized assumptions. The engineer does this realizing that discontinuities exist in the field. Because of these facts, it must always be understood that the theoretical design is only the starting point for blasting operations in the field. A trial blast should always be performed. It will either validate the initial assumptions or provide the information needed for final blast design.

Blast design is not an exact science. Empirical formulas provide an estimate of the work that can be accomplished by a given explosive. The application of these formulas that are given in the following sections results in a series of blasting dimensions (burden, blasthole size, stemming, subdrilling, and spacing) suitable for trial shots. Adjustments made from investigating the product of the trial shots should result in the optimum blast dimensions.

Burden

Burden distance is the most critical and important dimension in blasting. This is the shortest distance to stress relief at the time a blasthole detonates. It is normally the distance to the free face in an excavation, be it a quarry situation or a highway cut (see Fig. 13.1). Internal faces can be created by blastholes fired on an earlier delay within a shot. When the burden distance is insufficient, rock will be thrown for excessive distances from the face, fragmentation may be excessively fine and air blast levels will be high.

An empirical formula for approximating a burden distance to be used on a first trial shot is

$$B = \left(\frac{2SG_e}{SG_r} + 1.5 \right) D_e \qquad \text{[13.1]}$$

FIGURE 13.1 | Blasthole dimensional terminology.

where

B = burden in feet

SG_e = specific gravity of the explosive

SG_r = specific gravity of the rock

D_e = diameter of explosive in inches

The actual explosive diameter will depend on the manufacturer's packaging container thickness. If the specific explosive product is known, the exact specific gravity information should be used; but in the case of developing a design before settling on a specific product, an allowance should be made. Rock density is an indicator of strength, which, in turn, governs the amount of energy required to cause breakage. The approximate specific gravities of rocks are given in Table 13.1.

TABLE 13.1 | Density by nominal rock classifications

Rock classification	Specific gravity	Density broken (ton/cy)
Basalt	2.8–3.0	2.36–2.53
Diabase	2.6–3.0	2.19–2.53
Diorite	2.8–3.0	2.36–2.53
Dolomite	2.8–2.9	2.36–2.44
Gneiss	2.6–2.9	2.19–2.44
Granite	2.6–2.9	2.19–2.28
Gypsum	2.3–2.8	1.94–2.36
Hematite	4.5–5.3	3.79–4.47
Limestone	2.4–2.9	1.94–2.28
Marble	2.1–2.9	2.02–2.28
Quartzite	2.0–2.8	2.19–2.36
Sandstone	2.0–2.8	1.85–2.36
Shale	2.4–2.8	2.02–2.36
Slate	2.5–2.8	2.28–2.36
Trap rock	2.6–3.0	2.36–2.53

EXAMPLE 13.1

A contractor plans to use bulk ANFO, specific gravity 0.8, to open an excavation in granite rock. The drilling equipment available will drill a 3-in. blasthole. What is the recommended burden distance for the first trial shot?

From Table 13.1, specific gravity of granite 2.6 to 2.9, use an average value 2.75:

$$B = \left(\frac{2 \times 0.8}{2.75} + 1.5 \right) \times 3 = 6.2 \text{ ft}$$

EXAMPLE 13.2

A contractor plans to use a packaged explosive having a specific gravity of 1.3, to open an excavation in granite rock. The drilling equipment available will drill a 3-in. blasthole. The explosive comes packaged in $2\frac{1}{2}$-in-diameter sticks. What is the recommended burden distance for the first trial shot?

From Table 13.1, specific gravity of granite 2.6 to 2.9, use an average value 2.75:

$$B = \left(\frac{2 \times 1.3}{2.75} + 1.5\right) \times 2.5 = 6.1 \text{ ft}$$

Explosive density is used in Eq. [13.1] because of the proportional relationship between explosive density and strength. There are, however, some explosive emulsions that exhibit differing strengths at equal densities. In such a case, Eq. [13.1] will not be valid. An equation based on relative bulk strength instead of density can be used in such situations.

Relative bulk strength is the strength ratio for a constant volume compared to the strength of standard ANFO. Standard ANFO is assigned a relative bulk strength rating of 100.

The relative bulk strength rating of an explosive should be based on test data under specified conditions, but sometimes the rating is based on calculations. Manufacturers will supply specific values for their individual products. Again a caution is important: These design equations only provide a starting point for blasting activities in the field. The relative energy equation for burden distance is

$$B = 0.67D_e \sqrt[3]{\frac{St_v}{SG_r}} \qquad [13.2]$$

where St_v = relative bulk strength compared to ANFO = 100

When one or two rows of blastholes are used the burden distance between rows would be equal. If more than two rows are to be fired in a single shot, either the burden distance of the rear holes must be adjusted or delay devices must be used to allow the face rock from the front rows to move before the back rows are fired.

The burden distance may have to be adjusted because of geological variations. Rock is not the homogeneous material as assumed in all the formulas; therefore, it is often necessary to employ correction factors for specific geological conditions. Table 13.2 provides burden distance correction factors for rock deposition, K_d (see Fig. 13.2) and rock structure, K_s.

$$B_{\text{corrected}} = B \times K_d \times K_s \qquad [13.3]$$

TABLE 13.2 | Burden distance correction factors

Rock deposition	K_d
Bedding steeply dipping into cut	1.18
Bedding steeply dipping into face	0.95
Other cases of deposition	1.00

Rock structure	K_s
Heavily cracked, frequent weak joints, weakly cemented layers	1.30
Thin, well-cemented layers with tight joints	1.10
Massive intact rock	0.95

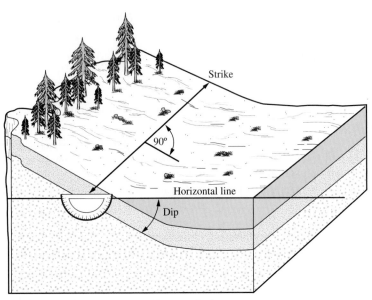

FIGURE 13.2 I Rock deposition terminology.

EXAMPLE 13.3

A new quarry is being opened in a limestone (SG, 2.6) formation having horizontal bedding with numerous weak joints. From a borehole test-drilling program, it is believed that the limestone is highly laminated with many weakly cemented layers. Because of possible wet conditions a cartridged slurry (relative bulk strength of 140) will be used as the explosive. The 6.5-in. blastholes will be loaded with 5-in-diameter cartridges. What is the calculated burden distance to be used initially?

$$B = 0.67 \times 5 \times \sqrt[3]{\frac{140}{2.6}} = 12.65 \text{ ft}$$

Correction factors from Table 13.2, $K_d = 1$, horizontal bedding, and $K_s = 1.3$, numerous weakly cemented layers.

$$B_{\text{corrected}} = 12.65 \times 1 \times 1.3 = 16.4 \text{ ft}$$

Blasthole Size

The size (diameter) of the blasthole will affect blast considerations concerning fragmentation, air blast, flyrock, and ground vibration. The economics of drilling is a consideration in determining blasthole size. Larger holes are usually more economical to drill but they introduce possible problems concerning blast product. Once again, the second mechanism of rupture and the stiffness ratio need to be considered. The *stiffness ratio* for blasting purposes is the bench height divided by the burden distance. In some situations, as in a quarry, the blaster can adapt the bench height to optimize the blast, but on a construction project the existing ground and the specified final project grades set limits on any bench-height modification. Table 13.3 gives the relationship between stiffness ratio and the critical blasting factors.

TABLE 13.3 | Stiffness ratio's effect on blasting factors

Stiffness ratio	1	2	3	4 and higher*
Fragmentation	Poor	Fair	Good	Excellent
Air blast	Severe	Fair	Good	Excellent
Flyrock	Severe	Fair	Good	Excellent
Ground vibration	Severe	Fair	Good	Excellent

*Stiffness ratios above 4 yield no increase in benefit.

One of the parameters in both Eqs. [13.1] and [13.2] was the diameter of the explosive, D_e. The diameter of the explosive is limited by the diameter of the blasthole. If it is desirable to drill larger blastholes for economic reasons, the burden distance will be affected. Consider Example 13.1: If the contractor had wanted to use a 5-in. bit, the charge would now be 5 inches in diameter.

Using the 5-in. explosive diameter the burden distance would be

$$B = \left(\frac{2 \times 0.8}{2.8} + 1.5 \right) \times 5 = 9.4 \text{ ft}$$

If the extent of the excavation perpendicular to the face was 90 ft, the original burden distance would require 15 rows ($\frac{90}{6.2} + 1$). Using larger holes, the burden distance is increased to 9.4 ft and only 10 rows would be required ($\frac{90}{9.4} + 1$). By increasing the blasthole diameter, the number of holes that will have to be drilled and loaded has been reduced.

However, the question of bench height has not been considered in either example. If the bench height was limited to 13 ft because that was the specified excavation depth, which size blasthole should be used? The stiffness ratio (SR) for the two blasthole diameters under consideration are

$$3\text{-in. blasthole:} \quad \frac{13 \text{ ft height}}{6.2 \text{ ft burden}} = 2.1 \text{ SR}$$

$$5\text{-in. blasthole:} \quad \frac{13 \text{ ft height}}{9.4 \text{ ft burden}} = 1.4 \text{ SR}$$

Table 13.3 indicates that for the 5-in. blasthole there will be blasting problems. Even the 3-in. blasthole can be expected to yield only fair results, which indicates that the shot should be redesigned.

It would be good to have an SR value of at least 3. Such a value is necessary for the blast to yield good results. Try a 2-in. blasthole and an explosive diameter of 2 in.

$$B = \left(\frac{2 \times 0.8}{2.8} + 1.5 \right) \times 2.0 = 4.1 \text{ ft}$$

$$2\text{-in. blasthole:} \quad \frac{13 \text{ ft height}}{4.1 \text{ ft burden}} = 3.2 \text{ SR}$$

This means more rows will be required but better breakage will result. Drilling cost will be increased, but secondary blasting and handling cost should be minimized.

Three factors affect burden distance: the specific gravity of the rock, the diameter of the explosives, and either the specific gravity of the explosives or the relative bulk strength of the explosive in the case of an emulsion. The underlying geological conditions of a project are a given, and the contractor must work with the site environment. There are many different commercial explosives having various strengths, but across the complete range of explosive strengths, the effect in calculated burden distance would be very small, only 2 to 3 ft. If the burden distance must be altered to achieve a satisfactory stiffness ratio, i.e., an effective blast that does not cause damage, explosive diameter is the parameter to be adjusted.

Stemming

The purpose of stemming is to confine the explosive energy to the blasthole. To function properly, the material used for stemming must lock into the borehole. It is common practice to use drill cuttings as the stemming material. Very fine cuttings or drill cuttings that are essentially dust will not accomplish the desired purpose. It may be necessary in such cases to bring in crushed stone. Likewise, very coarse materials do not make good stemming because they tend to bridge and will be ejected from the hole. To function properly, the stemming material should have an average diameter 0.05 times the diameter of the hole, and it should be angular.

If the stemming distance is too great, there will be poor top breakage from the explosion and backbreak will increase. When the stemming distance is inadequate, the explosion will escape prematurely from the hole (see Fig. 13.3).

FIGURE 13.3 I Explosive energy escaping from the blastholes.

Under normal conditions, properly designed burden and explosive, and good stemming material, a stemming distance, T, of 0.7 times the burden distance, B, will be satisfactory but stemming depth can range from 0.7 to 1.3 B. The stemming equation is

$$T = 0.7 \times B \qquad\qquad [13.4]$$

When drilling dust is used for stemming material, the stemming distance will have to be increased by another 30% of the burden distance because the dust will not lock into the hole.

Sometimes it is necessary to deck load a blasthole. Deck loading is the placing of several charges in the blasthole separated from each other by stemming material. This is necessary when the blasthole passes through a seam or crack in the rock. In such a situation, if the shot were fired with explosive material completely loading the hole, the force of the blast would blow out through the weak zone. Therefore, when such conditions are encountered, a charge is placed below and above the weak zone, and stemming material is tamped between the charges.

Subdrilling

A shot will normally not break to the very bottom of the blasthole. This can be understood by remembering that the second mechanism of breakage is flexural rupture. To achieve a specified bottom grade, it is necessary to drill below the desired floor elevation. This portion of the blasthole below the desired final grade is termed "subdrilling." The subdrilling distance, J, required can be approximated by the formula

$$J = 0.3 \times B \qquad\qquad [13.5]$$

Satisfactory subdrilling depths can range from $0.2B$ to $0.5B$ as results are dependent on the type of rock and its structure.

Subdrilling represents the depth required for explosive placement, not a field drilling depth. During the drilling operation, there will be random drilling depth errors, holes will slough, and material will accidentally fall into some holes. Therefore, for practical reasons, drilling should be to a depth slightly greater than that calculated. An extra foot of drilling depth for holes greater than 12 ft in required depth should be considered for economical reasons. To load explosive from the proper elevation, the blaster can always add cuttings to the hole. It is costly to stop the loading operation and bring a drill back to increase the depth of a hole. It is also costly and hazardous to perform secondary blasting because a hole was not loaded at the proper depth.

Spacing

Initiation timing and the stiffness ratio control the proper spacing of blastholes. When holes are spaced too closely and fired instantaneously, there will be venting of the energy, with the resulting air blast and flyrock. When the spacing is extended, there is a limit beyond which fragmentation will become harsh. Before beginning a spacing analysis, two questions must be answered concerning the shot: (1) Will the charges be fired instantaneously or will delays be used? (2) Is the stiffness ratio greater than 4? An SR of less than 4 is considered a low bench, and a high bench is an SR value of 4 or greater. This means there are four design cases to be considered:

1. Instantaneous initiation, with the SR greater than 1 but less than 4.

$$S = \frac{L + 2B}{3} \qquad\qquad \text{[13.6]}$$

where
   S = Spacing
   L = Bench height

2. Instantaneous initiation, with the SR equal to or greater than 4.

$$S = 2B \qquad\qquad \text{[13.7]}$$

3. Delayed initiation, with the SR greater than 1 but less than 4.

$$S = \frac{L + 7B}{8} \qquad\qquad \text{[13.8]}$$

4. Delayed initiation, with the SR equal to or greater than 4.

$$S = 1.4B \qquad\qquad \text{[13.9]}$$

The actual spacing utilized in the field should be within 15% plus or minus of the calculated value.

EXAMPLE 13.4
_____

It is proposed to load 4-in.-diameter blastholes with bulk ANFO. The contractor would like to use an 8 × 8 drill pattern (8-ft burden and 8-ft spacing). Assuming the burden distance is correct, will the 8-ft spacing be acceptable? The bench height is 35 ft and each hole is to be fired on a separate delay.

Check stiffness ratio, $\dfrac{L}{B}$ for high or low bench:

$$\frac{L}{B} = \frac{35}{8} = 4.4, > 4 \text{ therefore high bench.}$$

Delay timing, therefore use Eq. [13.9].

$$S = 1.4 \times 8 = 11.2 \text{ ft}$$

Range, 11.2 ± 15%: 9.5 ≤ S ≤ 12.9

The proposed spacing of 8 ft does not appear to be sufficient. As a minimum, the pattern should be changed to 8-ft burden × 9.5-ft spacing for the first trial shot in the field.

EXAMPLE 13.5

A project in granite rock will have an average bench height of 20 ft. An explosive having a specific gravity of 1.2 has been proposed. The contractor's equipment can easily drill 3-in.-diameter holes. Assume the packaged diameter of the explosives will be 2.5 in. Delay blasting techniques will be utilized. Develop a blast design for the project.

The specific gravity of granite is between 2.6 and 2.9 (see Table 13.1). Use the average, 2.75.

Using Eq. [13.1], we obtain

$$B = \left(\frac{2 \times 1.2}{2.75} + 1.5 \right) \times 2.5 = 5.9 \text{ ft}$$

Use 6 ft for burden distance. Remember that the numbers calculated here will be used in the field. Make it easy for those actually doing the work out there in the dust, mud, and other wonderful conditions. It is probably best to round the calculated numbers to the nearest foot or half foot before giving the design to the driller or blaster.

The stiffness ratio: $\frac{L}{B} = \frac{20}{6} = 3.3$

From Table 13.3, the stiffness ratio is good.

The stemming depth, from Eq. [13.4], is:

$$T = 0.7 \times 6 = 4.2 \text{ ft}$$

Use 4 ft for stemming.

The subdrilling depth, from Eq. [13.5] is

$$J = 0.3 \times 6 = 1.8 \text{ ft}$$

Use 2 ft for subdrilling.

The spacing, for an SR greater than 1 but less than 4 and using delay initiation, from Eq. [13.8], is

$$S = \frac{20 + (7 \times 6)}{8} = 7.75 \text{ ft}$$

7.75 ± 15%: The range for S is 6.6 to 8.9 ft

As a first trial shot, use a 6-ft burden × 8-ft spacing pattern.

Note that we have attempted to design a blast that will require only integer measurements in the field.

POWDER FACTOR

The amount of explosive required to fracture a cubic yard of rock is a measure of the economy of a blast design. Table 13.4 is a loading density chart that allows the engineer to easily calculate the weight of explosive required for a blasthole. In Example 13.5, the diameter of the explosives was 2.5 in. and the explosive had a density of 1.2. Using Table 13.4, the loading density is found to be 2.55 lb per ft of charge. The powder column length is the total hole length less the stemming, 18 ft in this case [20 ft + 2 ft (subdrilling) − 4 ft (stemming)]. Assuming a primer was not employed, the total weight of explosive used per blast hole would be 18 ft × 2.55 lb per ft = 45.9 lb.

TABLE 13.4 | Explosive loading density chart in pounds per foot of column for given explosive specific gravity

Column diam. (in.)	Explosive specific gravity							
	0.80	**0.90**	**1.00**	**1.10**	**1.20**	**1.30**	**1.40**	**1.50**
1	0.27	0.31	0.34	0.37	0.41	0.44	0.48	0.51
$1\frac{1}{4}$	0.43	0.48	0.53	0.59	0.64	0.69	0.74	0.80
$1\frac{1}{2}$	0.61	0.69	0.77	0.84	0.92	1.00	1.07	1.15
$1\frac{3}{4}$	0.83	0.94	1.04	1.15	1.25	1.36	1.46	1.56
2	1.09	1.23	1.36	1.50	1.63	1.77	1.91	2.04
$2\frac{1}{2}$	1.70	1.92	2.13	2.34	2.55	2.77	2.98	3.19
3	2.45	2.76	3.06	3.37	3.68	3.98	4.29	4.60
$3\frac{1}{2}$	3.34	3.75	4.17	4.59	5.01	5.42	5.84	6.26
4	4.36	4.90	5.45	6.00	6.54	7.08	7.63	8.17
$4\frac{1}{2}$	5.52	6.21	6.89	7.58	8.27	8.96	9.65	10.34
5	6.81	7.66	8.51	9.36	10.22	11.07	11.92	12.77
$5\frac{1}{2}$	8.24	9.27	10.30	11.33	12.36	13.39	14.42	15.45
6	9.81	11.03	12.26	13.48	14.71	15.93	17.16	18.39
$6\frac{1}{2}$	11.51	12.95	14.39	15.82	17.26	18.70	20.14	21.58
7	13.35	15.02	16.68	18.35	20.02	21.69	23.36	25.03
8	17.43	19.61	21.79	23.97	26.15	28.33	30.51	32.69
9	22.06	24.82	27.58	30.34	33.10	35.85	38.61	41.37
10	27.24	30.64	34.05	37.46	40.86	44.26	47.67	51.07

The amount of rock fractured by one blasthole is the pattern area times the depth to grade. For the 6 × 8 ft pattern having a 20-ft depth, each hole would have an affected volume of 35.6 cy $\left(\dfrac{6 \text{ ft} \times 8 \text{ ft} \times 20 \text{ ft}}{27 \text{ cf per cy}} \right)$. The powder factor will be 1.29 lb per cy $\left(\dfrac{45.9 \text{ lb}}{35.6 \text{ cy}} \right)$. With experience, this number provides the engineer a check on the blast design.

In all the examples and discussion up to this point, it has been assumed that only one explosive was used in a blasthole. This is not typically the case; if a hole is loaded with ANFO, it will require a primer to initiate the explosion. When it is expected that the bottom of some holes will be wet, an allowance must be made for a water-resistant explosive, such as a slurry in those locations. In the case of a powder column that is 18 ft, and that will be loaded with ANFO, specific gravity 0.8, a primer will have to be placed at the bottom of the hole. If an 8-in.-long stick of primer, specific gravity 1.3, is used, there will be 216 in. of ANFO and 8 in. of primer. This is about 4% primer.

The determination of wet holes usually has to be made strictly as a percentage guess based on limited project geotechnical information. This percentage is added to the percentage calculated for the primer because the primer is required whether or not the holes are wet. In this case, assuming all dry holes, the weight of explosives based on a 2.5-in. explosive diameter would be

ANFO   1.70 lb per ft $\times$ (208 in. $\div$ 12 in. per ft) = 29.46 lb
Dynamite 2.77 lb per ft $\times$ (8 in. $\div$ 12 in. per ft) = $\underline{\;\;1.85\;lb}$
                                                             31.31 lb

Total per hole is 31.31 lb for 18 ft of powder column.

It is necessary to know the total explosive weight to calculate the powder factor and, as will be discussed later, to check vibration. The breakout as to the amount of agent and the amount of primer is important because the price of the two is considerably different. Primer can cost up to nine times the unit price of agent.

MATERIAL HANDLING CONSIDERATIONS

The economics of handling the fractured rock is a factor that should be considered in blast design. Although it is critical to achieve good breakage, the blast pattern will affect such considerations as type of equipment and the bucket fill factor of the excavators. Appropriate piling of the blasted rock by the shot is dependent on the blast design. To utilize the blast to accomplish both objectives, these principles should be considered:

1. Rock movement will be parallel to the burden dimension.
2. Instantaneous initiation along a row causes more displacement than delayed initiation.
3. Shots delayed row-by-row scatter the rock more than shots fired in a V pattern (see Fig. 13.4).
4. Shots designed in a V-pattern firing sequence (see Fig. 13.4) give maximum piling close to the face.

TRENCH ROCK

In the case of excavating trench rock, the diameter or width of the structural unit, whether a pipe or conduit, is the principal consideration. When considering necessary trench width, the space required for working and backfill placement requirements must be taken into account. Many specifications will specifically address backfill width. Another important consideration is the size of the excavation equipment.

The geology will have a considerable impact on the blast design. Trenches are at the ground surface; usually, they will be extending through soil overburden and weathered unstable rock into solid rock. This nonuniform condition must be recognized. The blaster must check each individual hole to determine the actual depth of rock. Explosives are placed only in rock, not in the overburden.

FIGURE 13.4 I V-pattern (square corner) firing sequence with progressive delays; numbers indicate firing order.

If only a narrow trench in an interbedded rock mass is required, a single row of blastholes located on the trench centerline is usually adequate. Equation [13.1] will provide a first try for hole spacing. The timing of the shot should sequence down the row. The firing of the first hole provides the free face for the progression, which is why Eq. [13.1] is applicable. In the situation where the trench is shallow or there is little overburden, blasting mats laid over the alignment of the trench may be necessary to control flyrock. In the case of a wide trench or when solid rock is encountered, a double row of blastholes is common.

FIGURE 13.5 | A presplit rock face.

PRESPLITTING ROCK

This is a technique of drilling and blasting that breaks rock along a relatively smooth surface, as illustrated in Fig. 13.5. Another technique used to control overbreak and achieve a definite final excavation surface is line drilling. Line drilling is simply the drilling of a single row of unloaded, closely spaced holes along the perimeter of the excavation. This line of holes provides a weak plane into which the blast can break.

Presplit blastholes can be fired either in advance of other operations or with the production shot. The advantages of firing the presplitting shot before other operations are

1. It is easier to coordinate drilling and loading of presplitting with the production shots.
2. Deep cuts can be presplit in one shot with production shots fired later in two or more benches.

If the presplit holes are fired with the production shot, there should be a minimum of a 200 msec delay between the presplit blast and the firing of the nearest production blastholes.

The formula for selecting presplitting hole spacing and determining the amount of explosive with which each hole should be loaded is presented next. Usually, the holes are $2\frac{1}{2}$ to 3 in. in diameter and are drilled along the desired surface at spacings varying from 18 to 36 in., depending on the characteristics of the rock. When the rock is highly jointed the hole spacing must be reduced. There have, however, been some presplit operations using holes as large as $12\frac{1}{4}$ in. in diameter with depths of more than 80 ft. The presplit holes are loaded with one or two sticks of explosive at the bottoms, and then have smaller charges, such as $1\frac{1}{4}$- × 4-in. sticks spaced at 12-in. intervals, to the top of the hole. The sticks may be attached to Primacord with tape, or hollow sticks may be used, which permits the Primacord to pass through the sticks with cardboard tube spacers between the charges. It is important that the charges be less than half the presplit hole diameter and they should not touch the walls of the hole.

When the explosives in these holes are detonated ahead of the production blast, the webs between the holes will fracture, leaving a surface joint that serves as a barrier to the shock waves from the production blast. This will essentially eliminate breakage beyond the fractured surface.

Explosive Load and Spacing

The approximate load of explosive per foot of presplit blasthole is given by

$$d_{ec} = \frac{D_h^2}{28}$$ [13.10]

where

d_{ec} = explosive load in pounds per foot
D_h = diameter of blasthole in inches

When this formula is used to arrive at an explosive loading, the spacing between blastholes can be determined by the equation:

$$Sp = 10D_h$$ [13.11]

where Sp = presplit blasthole spacing in inches

Because of the variations in the characteristics of rocks, the final determination of the spacings of the holes and the quantity of explosive per hole should be determined by tests conducted at the project. Many times field trials will allow the constant in Eq. [13.11] to be increased from 10 to as much as 14.

Presplit blastholes are not extended below grade. In the bottom of the hole, a concentrated charge of two to three times d_{ec} should be placed instead of subdrilling.

Because a presplit blast is meant only to cause fracture, drill cutting can be used as stemming. The purpose is to only momentarily confine the gases and reduce noise. Two to five feet of stemming is normal. This technique has been used successfully with vertical holes and with slanted holes whose slopes are not less than about 1 to 1. It has been used with limited success in tunnel work.

EXAMPLE 13.6

By contract specification, the walls of a highway excavation through rock must be presplit. The contractor will be using drilling equipment capable of drilling a 3-in. hole. What explosive load and hole spacing should be used for the first presplit shot on the project?

Using Eqs. [13.10] and [13.11], one gets

$$d_{ec} = \frac{3^2}{28} = 0.32 \text{ lb/ft}$$

$$Sp = 10 \times 3 = 30 \text{ in.}$$

Bottom load should be $3 \times 0.32 = 0.96$ lb.

SEISMIC EFFECT

Blasting activities can cause varying degrees of vibration that spread through the ground. Though the vibrations diminish in strength with distance from the source, they can achieve audible and feelable ranges in buildings close to the work site. Rarely do these vibrations reach levels that cause damage to structures but the issue of vibration problems is controversial. The case of old, fragile, or historical buildings is a situation where special care must be exercised in controlling vibrations because there is a danger of significant structural damage. The issue of vibration can cause restraints on blasting operations and lead to additional project cost and time. The determination of "acceptable" vibration levels is in many cases very difficult due to its subjective nature with regard to being a nuisance. Humans and animals are very sensitive to vibration, especially in the low frequency range (1 to 100 Hz).

It is the unpredictability and unusual nature of a vibration source, rather than the level itself, that is likely to result in complaints. The effect of intrusion tends to be psychological rather than physiological, and is more of a problem at night, when occupants of buildings expect no unusual disturbance from external sources.

Vibration Sources and Strength Levels

When vibration levels from an "unusual source" exceed the human threshold of perception [peak particle velocity (PPV), 0.008 to 0.012 in. per sec] complaints may occur. In an urban situation, serious complaints are probable when PPV exceeds 0.12 in. per sec [6], even though these levels are much less than what would result from slamming a door in a modern masonry building. People's tolerance will be improved provided that the origin of the vibrations is known in advance and no damage results. It is important to provide people with a motivation to accept some temporary disturbance. Appropriate practice is to avoid vibration-causing activities at night as people are more aware of vibration in their homes during nighttime hours.

Table 13.5 presents a matrix of human response information in relation to both blasting and earthquake motion. It is interesting to note that humans are more sensitive to blasting motion than to that of earthquakes.

There are also published studies comparing the stresses imposed on structures by typical environmental charges and equivalent particle velocities. A 19% change in inside humidity imposes a stress equivalent to about a 2.8 in. per sec

TABLE 13.5 | Human response to motion [7]

Response of humans	Earthquakes (in./sec)	Blasting (in./sec)
Barely perceptible	0.26–0.80	0.01–0.10
Distinctly perceptible	0.46–1.40	0.05–0.50
Strongly perceptible	1.50–5.70	0.50–5.00

particle velocity. A 35% change in outside humidity imposes a stress equivalent to almost a 5.0 in. per sec particle velocity. A 12°F change in inside temperature imposes a stress equivalent to about a 3.3 in. per sec particle velocity. A 27°F change in inside temperature imposes stress equivalent to almost an 8 in. per sec particle velocity and a 23 mph wind differential imposes a stress equivalent to about a 2.2 in. per sec particle velocity. Typical construction blasting creates particle velocities of less than 0.5 in. per sec.

Consequently, it must be remembered that people can perceive very low levels of vibration, and at the same time they are unaware of the silent environmental forces acting on and causing damage to their homes. So even though construction activities actually cause movements significantly less than those created by common natural occurrences, the impact perceived by humans can cause concern.

Therefore, because blasting operations may cause actual or alleged damages to buildings, structures, and other properties located in the vicinity of the blasting operations, it is desirable to examine, and possibly photograph, any structures for which charges of damages may be made following blasting operations. Before beginning blasting operations, seismic recording instruments can be placed in the vicinity of the shots to monitor the magnitudes of vibrations caused by the blasting. The persons responsible for the blasting may conduct the monitoring, or if the company responsible for the blasting carries insurance covering this activity, a representative of the insurance carrier may provide this service.

Vibration Mitigation

Rock exhibits the property of elasticity. When explosives are detonated, elastic waves are produced as the rock is deformed and then regains its shape. The two principal factors that will affect how this motion is perceived at any discrete point are the size of the explosive charge detonated and distance to the charge. It is, therefore, obvious that delays in initiation will reduce the effect because the individual charges are less than the total that would have been fired without the delays.

A series of questions that should be addressed concerning vibration effects are

1. To what extent will vibrations be caused?
2. Are sensitive people or structures in the vicinity?
3. Is damage/intrusion possible?
4. Can site-specific trials be conducted to assess possible damage/intrusion?

Modify blasting design and/or construction method if the answer is yes to number three. Answering these questions requires a clear understanding of blast design and blast location in relation to critical receptors.

The U.S. Bureau of Mines has proposed a formula to evaluate vibration and as a way to control blasting operations:

$$D_s = \frac{d}{\sqrt{W}}$$ [13.12]

where

D_s = scaled distance (nondimensional factor)

d = distance from shot to a structure in feet

W = maximum charge weight per delay in pounds

A scaled distance value of 50 or greater indicates that a shot is safe with respect to vibration according to the Bureau of Mines. Some regulatory agencies require a value of 60 or greater.

The statement can be made that two identical blasts cause different effects. Excessive burden distances will cause high levels of ground vibration per pound of explosive and burden is never constant across a project. Some of the critical factors affecting vibration, which should be carefully considered, are

1. Burden
2. Spacing
3. Subdrilling
4. Stemming depth
5. Type of stemming
6. Bench height
7. Number of decks
8. Charge geometry
9. Powder column length
10. Rock type
11. Rock physical properties
12. Geological features
13. Number of holes in a row
14. Number of rows
15. Row-to-row delays
16. Initiator precision
17. Face angle to structure
18. Explosive energy

Preblast Surveys

Prior to exposing a structure to the vibration caused by blasting operations, it is good practice to make a preblast survey and document its condition. To be of value, the survey must be thorough and accurate. Structures are subjected to many dynamic forces. But because many of these act over long time periods, the effects often go unobserved. Temperature, humidity, wind, soil conditions, and operating equipment can all cause cracking of a structure. In many cases, the effect is seasonal and the cracks will open and close with changing conditions. Besides the condition of the structure, possible contributing soil conditions should be noted during the survey; e.g., drainage problems, pooling of water, and seepage.

SAFETY

An accident involving explosives may easily kill or cause serious injury and property damage. The prevention of such accidents depends on careful planning and faithful observation of proper blasting practices. There are federal and state regulations concerning the transportation and handling of explosives. The manufacturer will provide safety information on specific products. In addition to regulations and product information, there are recommended practices, such as the evacuation of the blast area during the approach of an electrical storm whether electric or nonelectric initiation systems are used. An excellent source for material on recommended blasting safety practices is the Institute of Makers of Explosives in Washington, DC.

The federal Office of Surface Mining requires that states have minimum training and licensing standards for blasters at surface coal mines. Blasters must have at least 40 hr of training and pass a written test. The comprehensive written test covers blasting rules, safety, applications, reporting, and product awareness.

Blast Planning

All blasting operations should be carefully planned. Issues that should be considered in the plans include

1. Information for and education of the public.
2. Blaster-in-charge qualification and designation.
3. Explosive storage and security requirements.
4. Explosive use and handling.
5. Site access control and all-clear procedures.
6. Specific blast design and control procedures.
7. Blast reporting.
8. Vibration and overpressure monitoring.
9. Potential liability and insurance.

Explosive Storage

All explosives, blasting agents, and initiation devices must be stored in magazines that have been constructed, approved, and licensed in accordance with local, state, and federal regulations. Magazines must be kept locked and admittance restricted. Maintaining accurate records of explosive materials is essential to ensure that stocks are being properly accounted for and rotated, and that the oldest stocks are used first.

Misfires

In shooting charges of explosives, one or more charges may fail to explode. This is referred to as a "misfire." It is necessary to dispose of this explosive before excavating the loosened rock. The most satisfactory method is to shoot it if possible.

If electric blasting caps are used, the leading wires should be disconnected from the source of power prior to investigating the cause of the misfire. If the leg wires to the cap are available, test the cap circuit; and if the circuit is satisfactory, try again to set off the charge.

When it is necessary to remove the stemming to gain access to a charge in a hole, the stemming should be removed with a wooden tool instead of a metal tool. If water or compressed air is available, either one can be used with a rubber hose to wash the stemming out of the hole. A new primer, set on top of or near the original charge, can be used to fire the charge.

SUMMARY

The operation referred to as "blasting" is performed to break rock so that it can be quarried for processing in an aggregate production operation, or to excavate a right-of-way. There is no single solution in designing a blast. Rock is not a homogeneous material. There are fracture planes, seams, and changes in burden to be considered. Blast designs use idealized assumptions. Because of these facts, it must always be understood that the theoretical design is only the starting point for blasting operations in the field. Critical learning objectives include:

- An understanding of the difference between an explosive that exhibits a proportional relationship between explosive density and strength, and one that exhibits differing strengths at equal densities.
- An ability to calculate burden distance based on explosive type.
- An ability to design an initial blast including adjusting blasthole size to obtain a satisfactory stiffness ratio.
- An understanding of how initiation affects blasthole spacing.
- An ability to calculate the powder factor.
- An ability to design a presplitting shot.
- An ability to use the U.S. Bureau of Mines formula to check blast vibration.

These objectives are the basis for the problems that follow.

PROBLEMS

13.1   A contractor plans to use bulk ANFO, specific gravity 0.8, to open an excavation in basalt rock. The drilling equipment available will drill a 4-in. blasthole. What is the recommended burden distance for the first trial shot? (Burden 8 ft)

13.2   A contractor plans to use a packaged emulsion having a specific gravity of 1.2 and relative bulk strength of 135 to open an excavation in granite rock, specific gravity 2.7. The drilling equipment available will drill a 4-in. blasthole. The explosives come packaged in $3\frac{1}{2}$-in.-diameter sticks. What is the recommended burden distance for the first trial shot?

13.3   A project in dolomite will have an average bench height of 18 ft. An emulsion relative bulk strength 105, specific gravity 0.8, will be the explosive used on the project. Dynamite sticks 8 in. long having a specific gravity of 1.3 and a diameter exactly equal to the D_e of the emulsion will be used as a detonator. The contractor's equipment can easily drill 3-in.-diameter holes. It is assumed that single rows of no more than 10 holes will be detonated instantaneously. The excavation site has structures within 600 ft. The local regulatory agency specifies a scaled distance factor of no less than 55. Develop a blast design for the project using the maximum permissible integer spacing. (Burden 6 ft, stemming 4 ft, sub-drilling 2 ft, 6 × 11 pattern, explosives 40.2 lb per hole, only three holes can be fired at one time.)

13.4 A project in sandstone, having a specific gravity of 2.0, will have an average bench height of 23 ft. An emulsion relative bulk strength 105, specific gravity 0.8, will be the explosive used on the project. Dynamite sticks 1 ft long having a specific gravity of 1.3 and a diameter exactly equal to the D_e of the emulsion will be used as a detonator. The contractor's equipment can easily drill 4-in.-diameter holes. It is assumed that single rows of no more than 30 holes will be detonated instantaneously. The closest structures to the excavation site are one-half mile away. The local regulatory agency specifies a scaled distance factor of no less than 60. Even though an emulsion is being used because of expected wet conditions, the holes will be lined. The thickness of the lining material is $\frac{1}{4}$ in. Develop the blast design spacing, subdrilling, and stemming dimensions for the project. How many holes can be fired instantaneously? The boss has told you that good fragmentation will be acceptable.

13.5 A project in limestone will have an average bench height of 18 ft. ANFO, specific gravity 0.84, will be the explosive used on the project. Dynamite, specific gravity 1.3, will be the primer explosive used. The dynamite comes in 8-in.-l ong × $1\frac{3}{4}$-in.-diameter sticks. The contractor's equipment can easily drill $3\frac{1}{2}$-in.-diameter holes. It is assumed that single rows of no more than 10 holes will be detonated instantaneously. The excavation site has structures within 900 ft. The local regulatory agency specifies a scaled distance factor of no less than 60. The holes will be lined. The thickness of the lining material is $\frac{1}{4}$ in. Develop a blast design for the project.

13.6 The blasting in Problem 13.5 must be conducted so as to limit overbreakage, develop a presplitting blast plan.

13.7 A materials company is opening a new quarry in a limestone formation. Tests have shown that the specific gravity of this formation is 2.7. The initial mining plan envisions an average bench height of 24 ft. based on the loading and hauling equipment capabilities. Bulk ANFO, specific gravity 0.8, and a primer, specific gravity 1.5, will be the explosives used. The contractor's equipment can drill 6-in.-diameter holes. Delayed initiation will be utilized. Develop a blasting plan for a conservative cost estimate. (Burden 8 ft, stemming 6 ft, subdrilling 2 ft, spacing conservative 9 ft)

13.8 An investigation of a highway project revealed a sandstone formation with steeply dipping bedding into the face. Many weak joints were identified. An analysis of the plan and profile sheets, and the cross sections showed that the average bench height will be about 15 ft. Bulk ANFO, specific gravity 0.8, and primer, specific gravity 1.3, will be the explosives used on the project. The primer comes in 8-in.-long × $1\frac{3}{4}$-in.-diameter sticks. The contractor's equipment can easily drill 6-in.-diameter holes. It is assumed that delayed initiation will be utilized.

a.   Develop a blast design for the project.

b.   If the burden distance is held constant at 5 ft but the hole spacing is varied in 1-ft increments across the range of S developed in part (a), what is the cost per cubic of rock if the ANFO is $0.166 per lb, and dynamite $1.272 per lb?

REFERENCES

1. *Blasters' Handbook,* E. I. du Pont de Nemours & Co., Inc., Wilmington, Del., 1977.

2. *Explosive and Rock Blasting,* Atlas Powder Company, Dallas, Tx., 1987.

3. Harris, Frank, *Ground Engineering Equipment and Methods,* McGraw-Hill, New York, 1983.

4. Hemphill, G. B., *Blasting Operations,* McGraw-Hill, New York, 1981.

5. Konya, C. J., and E. J. Walter, "Blasthole Timing Controls Vibration, Airblast and Flyrock," *Coal Mining,* January 1988.

6. New, Barry M., "Ground Vibration Caused by Construction Work," *Tunneling and Underground Space Technology,* Vol. 5, No. 3, Great Britain, 1990.

7. Oriard, L. L., "The Scale of Effects in Evaluating Vibration Damage Potential," *Proceedings 15th Conference on Explosives and Blasting Technique,* Society of Explosive Engineers, Cleveland, Ohio, 1989.

8. Persson, Per-Anders, Roger Holmberg, and Jaimin Lee, *Rock Blasting and Explosives Engineering,* CRC Press, Inc., Boca Raton, Fla., 1996.

9. Revey, G. F., "To Blast or Not to Blast?" *Practice Periodical on Structural Design and Construction,* American Society of Civil Engineers, Vol. 1, No. 3, August 1996.

10. *Rock Blasting and Overbreak Control,* National Highway Institute, U.S. Department of Transportation, Federal Highway Administration, Pub. No. FHWA-HI-92-001, 1991.

11. *Rock Fragmentation Prediction (Breaker),* Computer Software Package, Precision Blasting Services, Montville, Ohio, 1989.

12. Langefors, U., and B. Kihlström, *The Modern Technique of Rock Blasting,* John Wiley & Sons, Inc., New York, 1983.

BLASTING-RELATED WEBSITES

1. The Institute of Makers of Explosives (IME), www.ime.org. IME is the safety association of the commercial explosives industry in the United States and Canada. It provides technical information and recommendations concerning commercial explosive materials and promotes the safety and protection of employees, users, the public, and the environment throughout all aspects of the manufacture and use of explosive materials. IME is located at 1120 19th Street NW, Suite 310, Washington, DC 20036-3605.

2. Institute of Explosives Engineers (IexpE), //www.iexpe.org/. The Institute exists to represent those companies and individuals who use explosives as part of the everyday tools of their trade. It is located in Great Britain at Centenary Business Centre, Hammond Close, Attleborough Fields, Nuneaton, Warwickshire CV11 6RY.

3. International Society of Explosives Engineers (ISEE), www.isee.org/. ISEE is a professional society dedicated to promoting the safe and controlled use of explosives in mining, quarrying, construction, manufacturing, forestry, and many other commercial pursuits. With more than 4,200 members from more than 85 countries, and with 32 Chapters in the United States, Canada, and Australia, the Society is recognized as a world leader in providing explosives technology and eduction, and promoting public understanding of the benefits of explosives.

14

Aggregate Production

Many types and sizes of crushing and screening plants are used in the construction industry. The capacity of a crusher will vary with the type of stone, size of feed, size of the finished product, and extent to which the stone is fed uniformly into the crusher. The screening process is based on the simple premise that particle sizes smaller than the screen cloth opening size will pass through the screen and that oversized particles will be retained. After stone is crushed and screened, it is necessary to handle it carefully or the large and small particles may separate.

INTRODUCTION

The amount of processing required to produce suitable aggregate materials for construction purposes depends on the nature of the raw materials available and the desired attributes of the end product. Four functions are required to accomplish the desired results:

1. Particle size reduction—crushing.
2. Separation into particle size ranges—sizing/screening.
3. Elimination of undesirable materials—washing.
4. Handling and movement of the crushed materials—storage and transport.

This chapter is devoted primarily to the first three operations: crushing, sizing, and washing. Storage is discussed here but transport has already been examined in Chapters 6, 7, 8, and 10.

In operating a rock excavation and crushing plant, the drilling pattern, the amount of explosives, the size shovel or loader used to load the stone, and the size of the primary crusher should be coordinated to ensure that all stone can be economically used. It is desirable for the loading capacity of the shovel or loader in the pit and the capacity of the crushing plant to be approximately equal. Table 14.1 gives the recommended minimum sizes of jaw and gyratory crushers required to handle the stone being loaded with buckets of the specified capacities.

TABLE 14.1 | Recommended minimum sizes of primary crushers for use with shovel buckets of the indicated capacities

Capacity of bucket [cy (cu m)]		Jaw crusher [in. (mm)]*		Gyratory crusher, size of openings [in. (mm)]†	
$\frac{3}{4}$	(0.575)	28 × 36	(712 × 913)	16	(406)
1	(0.765)	28 × 36	(712 × 913)	16	(406)
$1\frac{1}{2}$	(1.145)	36 × 42	(913 × 1,065)	20	(508)
$1\frac{3}{4}$	(1.340)	42 × 48	(1,065 × 1,200)	26	(660)
2	(1.530)	42 × 48	(1,065 × 1,200)	30	(760)
$2\frac{1}{2}$	(1.910)	48 × 60	(1,260 × 1,525)	36	(915)
3	(2.295)	48 × 60	(1,260 × 1,525)	42	(1,066)
$3\frac{1}{2}$	(2.668)	48 × 60	(1,260 × 1,525)	42	(1,066)
4	(3.060)	56 × 72	(1,420 × 1,830)	48	(1,220)
5	(3.820)	66 × 86	(1,675 × 2,182)	60	(1,520)

*The first number is the width of the opening at the top of the crusher, measured perpendicular to the jaw plates. The second two digits are the width of the opening, measured across the jaw plates.
†The recommended sizes are for gyratory crushers equipped with straight concaves.

Portable Crushing and Screening Plants

Many types and sizes of portable crushing and screening plants are used in the construction industry. When there is a satisfactory deposit of stone near a project that requires aggregate, it frequently will be more economical to set up a portable plant and produce the crushed stone instead of purchasing it from a commercial source. Larger commercial aggregate plants are, however, the major source of crushed stone for the construction industry in metropolitan areas. A typical portable crushing and screening plant is illustrated in Fig. 14.1.

PARTICLE SIZE REDUCTION

GENERAL INFORMATION

Crushers are sometimes classified according to the stage of crushing that they accomplish, such as primary, secondary, tertiary, etc. A primary crusher receives the stone directly from the excavation after blasting, and produces the first reduction in stone size. The output of the primary crusher is fed to a secondary crusher that further reduces the stone size. Some of the stone may pass through four or more crushers before it is reduced to the desired size.

Crushing plants use step reduction because the amount of size reduction accomplished is directly related to the energy applied. When there is a large difference between the size of the feed material and the size of the crushed product, a large amount of energy is required. If there was a concentration of this energy in a single-step process, excessive fines will be generated, and normally there is

FIGURE 14.1 | Portable aggregate plant in operation.

Source: Cedarapids Inc., a Terex Company.

only a limited market for fines. Fines are a non-revenue-producing waste material at many plants. Therefore, the degree of breakage is spread over several
stages as a means of closely controlling product size and limiting waste material.

As stone passes through a crusher, the reduction in size can be expressed as
a reduction ratio: the ratio of crusher feed size to product size. The sizes are usually defined as the 80% passing size of the cumulative size distribution. For a
jaw crusher, the ratio could be estimated as the gape, which is the distance
between the fixed and moving faces at the top, divided by the distance of the
open side setting at the bottom. Thus, if the gape distance between the two faces
at the top is 16 in. and at the bottom the open-side setting is 4 in., the reduction
ratio is 4.

The reduction ratio of a roller crusher could be estimated as the ratio of the
dimension of the largest stone that can be nipped by the rolls, divided by the setting of the rolls that is the smallest distance between the faces of the rolls.

A more accurate measurement of reduction ratio is to use the ratio of size
corresponding to 80% passing for both the feed and the product. Table 14.2 lists
the major types of crushers and presents data on attainable material reduction
ratios.

Crushers are also classified by their method of mechanically transmitted
fracturing energy to the rock. Jaw, gyratory, and roll crushers work by applying
compressive force. As the name implies, impact crushers apply high-speed
impact force to accomplish fracturing. By using units of differing size, crushing
chamber configuration, and speed, the same mechanical type crusher can be
employed at different stages in the crushing operation.

Jaw crushers, however, are typically employed as primary units because
of their large energy-storing flywheels and high mechanical advantage. True

TABLE 14.2 | The major types of crushers

Crusher type	Reduction ratio range
Jaw	
Double toggle	
Blake	4:1–9:1
Overhead pivot	4:1–9:1
Single toggle: Overhead eccentric	4:1–9:1
Gyratory	
True	3:1–10:1
Cone	
Standard	4:1–6:1
Attrition	2:1–5:1
Roll	
Compression	
Single roll	Maximum 7:1
Double roll	Maximum 3:1
Impact	
Single rotor	to 15:1
Double rotor	to 15:1
Hammer mill	to 20:1
Specialty crushers	
Rod mill	
Ball mill	

gyratories are the other crusher type employed as primary unit. These have, in recent years, become the primary unit of choice. A *true* gyratory is a good primary crusher because it provides continuous crushing and can handle slabby material. Jaw crushers do not handle slabby material well. Models of gyratory, roll, and impact crushers can be found in both secondary and tertiary applications.

JAW CRUSHERS

These machines operate by allowing stone to flow into the space between two jaws, one of which is stationary while the other is movable. The distance between the jaws diminishes as the stone travels downward under the effect of gravity and the motion of the movable jaw, until the stone ultimately passes through the lower opening. The movable jaw is capable of exerting a pressure sufficiently high to crush the hardest rock. Jaw crushers are usually designed with the toggle as the weakest part. The toggle will break if the machine encounters an uncrushable object or is subjected to overload. This limits the damage to the crusher.

Double Toggle

The *Blake* type, illustrated in Fig. 14.2, is a double-toggle jaw crusher. The movable jaw is suspended from a shaft mounted on bearings on the crusher frame. Rotating an eccentric shaft raises and lowers the pitman and actuates the two toggles, and these produce the crushing action. As the pitman raises the two toggles, a high pressure is exerted near the bottom of the swing jaw that partially closes

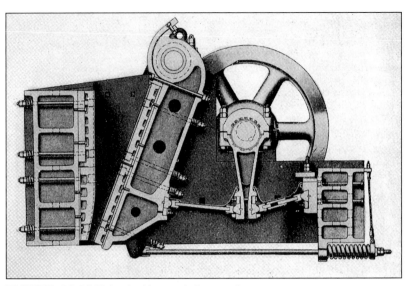

FIGURE 14.2 | Blake double-toggle jaw crusher.

Source: Fiat-Allis Construction Machinery, Inc.

the opening at the bottom of the two jaws. This operation is repeated as the eccentric shaft is rotated.

The jaw plates that are made of manganese steel may be removed, replaced, or, in some cases, reversed. The jaws may be smooth, or, in the event the stone tends to break into slabs, corrugated jaws may be used to reduce the slabbing. The swing jaw may be straight, or it may be curved to reduce the danger of choking.

The movable jaw operates at a fairly slow speed but has a large stroke at its bottom. This large stroke and a conservative nip angle allow Blake-type crushers to handle hard, tough, abrasive rock and still have good capacity.

Table 14.3 gives representative capacities for various sizes of Blake-type jaw crushers. In selecting a jaw crusher, consideration must be given to the size of the feed stone. The top opening of the jaw should be at least 2 in. wider than the largest stones that will be fed to it.

Capacity tables can be based on the open or closed position of the bottom of the swing jaw; therefore, the table should specify which setting applies. The closed position is commonly used for most crushers and is the basis for the values given in Table 14.3. However, Blake-type jaw crushers are often rated based on the open-side setting. The capacity is given in tons per hour based on a standard material unit weight of 100 lb per cf when crushed.

An *overhead pivot* jaw crusher is similar to the Blake type, but by placing the swing jaw's pivot over the centerline of the crushing chamber, there is more stroke at the feed opening. Additionally, this causes the motion to be more in a direction perpendicular to the stationary jaw. By having a higher operating speed this type crusher will have capacities similar to the Blake design.

TABLE 14.3 | Representative capacities of Blake-type jaw crushers, in tons per hour (metric tons per hour) of stone*

Size crusher [in. (mm)]†	Maximum rpm	Maximum hp (kW)	Closed setting of discharge opening [in. (mm)]										
			1 (25.4)	1½ (38.1)	2 (50.8)	2½ (63.5)	3 (76.2)	4 (102)	5 (137)	6 (152)	7 (178)	8 (203)	9 (229)
10 × 6 (254 × 406)	300	15 (11.2)	11 (10)	16 (14)	20 (18)								
10 × 20 (254 × 508)	300	20 (14.9)	14 (13)	20 (18)	25 (23)	34 (31)							
15 × 24 (381× 610)	275	30 (22.4)		27 (24)	34 (31)	42 (38)	50 (45)						
15 × 30 (381 × 762)	275	40 (29.8)		33 (30)	43 (39)	53 (48)	62 (56)						
18 × 36 (458 × 916)	250	60 (44.8)		46 (42)	61 (55)	77 (69)	93 (84)	125 (113)					
24 × 36 (610 × 916)	250	75 (56.0)			77 (69)	95 (86)	114 (103)	150 (136)					
30 × 42 (762 × 1,068)	200	100 (74.6)				125 (113)	150 (136)	200 (181)	250 (226)	300 (272)			
36 × 42 (916 × 1,068)	175	115 (85.5)				140 (127)	160 (145)	200 (181)	250 (226)	300 (272)			
36 × 48 (916 × 1,220)	160	125 (93.2)				150 (136)	175 (158)	225 (202)	275 (249)	325 (294)	375 (339)		
42 × 48 (1,068 × 1,220)	150	150 (111.9)				165 (149)	190 (172)	250 (226)	300 (272)	350 (318)	400 (364)	450 (408)	
48 × 60 (1,220 × 1,542)	120	180 (134.7)					220 (200)	280 (254)	340 (309)	400 (364)	450 (408)	500 (454)	550 (500)
56 × 72 (1,422 × 1,832)	95	250 (186.3)						315 (286)	380 (345)	450 (408)	515 (468)	580 (527)	640 (580)

*Based on the closed position of the bottom swing jaw and stone weighing 100 lb per cf when crushed.
†The first number indicates the width of the feed opening, whereas the second number indicates the width of the jaw plates.

Single Toggle

When the eccentric shaft of the single-toggle crusher (illustrated in Fig. 14.3) is rotated, it gives the movable jaw both a vertical and a horizontal motion. This type of crusher is used quite frequently in portable rock-crushing plants because of its compact size, lighter weight, and reasonably sturdy construction. The capacity of a single toggle crusher is usually rated at the closed side setting and is less than that of a Blake-type unit.

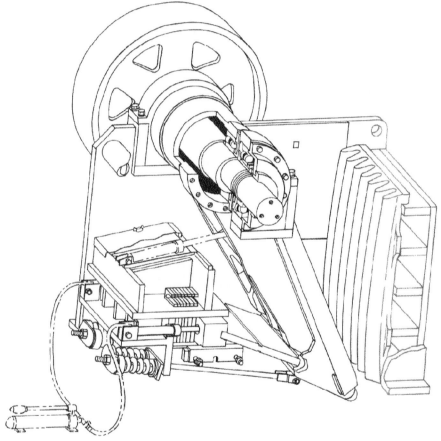

FIGURE 14.3 | Single-toggle-type jaw crusher.

Source: Cedarapids, Inc., a Terex Company.

Sizes of Jaw and Roll Crusher Product

While the setting of the discharge opening of a crusher will determine the maximum-size stone produced, the aggregate sizes will range from slightly greater than the crusher setting to fine dust. Experience gained in the crushing industry indicates that for any given setting for a jaw or roll crusher, approximately 15% of the total amount of stone passing through the crusher will be larger than the setting. If the openings of a screen that receives the output from such a crusher is the same size as the crusher setting, 15% of the output will not pass through the screen. Figure 14.4 presents the percent of material passing or retained on screens having the size openings indicated. The chart can be applied to both jaw- and roll-type crushers. To read the chart, select the vertical line corresponding to the crusher setting. Then move down this line to the number that indicates the size

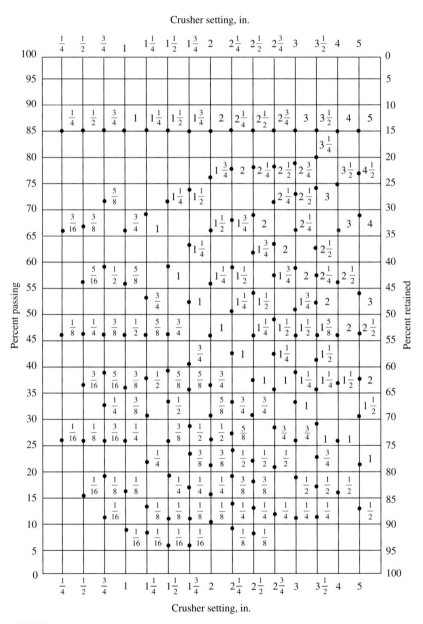

FIGURE 14.4 | Analysis of the size of aggregate produced by jaw and roll crushers.

Source: Universal Engineering Company.

of the screen opening. From the size of the screen opening, proceed horizontally to the left to determine the percent of material passing through the screen or to the right to determine the percent of material retained on the screen.

EXAMPLE 14.1

A jaw crusher with a closed setting of 3 in. is fed shot rock at the rate of 50 tons per hr (tph). Determine the amount of stone produced in tons per hour within the following size ranges: in excess of 2 in.; between 2 and 1 in.; between 1 and $\frac{1}{4}$ in.; and less than $\frac{1}{4}$ in.

From Fig. 14.4, the amount retained on a 2-in. screen is 42% of 50, which is 21 tons per hr. The amount in each of the size ranges is determined as

Size range (in.)	Percent passing screens	Percent in size range	Total output* of crusher (tph)	Amount produced in size range (tph)
Over 2	100–58	42	50	21.0
2–1	58–33	25	50	12.5
$1-\frac{1}{4}$	33–11	22	50	11.0
$\frac{1}{4}$–0	11–0	11	50	5.5
Total		100%		50.0 tph

*This is a closed system; what is fed into the crusher must equal the product. The example is also based on material weighing 100 lb per cf crushed.

Always read the fine print when using a crusher production chart.

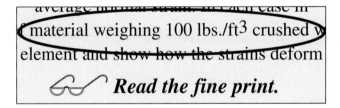

average initial strain. In each case in
material weighing 100 lbs./ft^3 crushed w
element and show how the strains deform
👓 *Read the fine print.*

GYRATORY CRUSHERS

Gyratories are the most efficient of all primaries. A gyrating mantle mounted within a deep bowl characterizes these crushers. They provide continuous crushing action and are used for both primary and secondary crushing of hard, tough, abrasive rock. To protect the crusher from uncrushable objects and overload, the outer crushing surface can be spring-loaded or the mantle height can be hydraulically adjustable.

True Gyratory

A section through a gyratory crusher is illustrated in Fig. 14.5. The crusher unit consists of a heavy cast-iron or steel frame, with an eccentric shaft and driving gears in the lower part of the unit. In the upper part, there is a cone-shaped crushing

FIGURE 14.5 | Section through a gyratory crusher.

chamber, lined with hard-steel or manganese-steel plates called the "concaves." The crushing member includes a hard-steel crushing head mounted on a vertical steel shaft. This shaft and head are suspended from the spider at the top of the frame that is so constructed that some vertical adjustment of the shaft is possible. The eccentric support at the bottom causes the shaft and the crushing head to gyrate as the shaft rotates, thereby varying the width of the space between the concaves and the head. As the rock which is fed in at the top (see Fig. 14.6) of the crushing chamber moves downward, it undergoes a reduction in size until it finally passes through the opening at the bottom of the chamber.

The size of a gyratory crusher is the width of the receiving opening, measured between the concaves and the crusher head. The setting is the width of the bottom opening and may be the open or closed dimension. When a setting is given, it should be specified whether it is the open or closed dimension. Normally, the capacity of a *true* gyratory crusher is based on an open-side setting. The ratio of reduction for true gyratory crushers usually ranges from about 3:1 to 10:1, with an average value around 8:1. A gyratory crusher and the conveyor for delivering the stone are shown in Fig. 14.7.

If a gyratory crusher is used as a primary crusher, the size selected may be dictated by the size of the rock from the blasting operation, or it may be dictated by a desired capacity. When a gyratory crusher is used as a secondary crusher, increasing its speed within reasonable limits may increase the capacity.

FIGURE 14.6 I Feed chamber over the top of a gyratory crusher with a hoe ram for reducing oversize feed material.

FIGURE 14.7 I Gyratory crusher in an aggregate plant.

Table 14.4 gives representative capacities of gyratory crushers, expressed in tons per hour, based on a continuous feed of stone having a unit weight of 100 lb per cf when crushed. Crushers with straight concaves are commonly used as primary crushers, whereas those with nonchoking concaves are commonly used as secondary crushers.

Cone Crushers

Cone crushers are used as secondary or tertiary crushers. They are capable of producing large quantities of uniformly fine crushed stone. A cone crusher differs from a true gyratory crusher in these respects:

1. It has a shorter cone.
2. It has a smaller receiving opening.
3. It rotates at a higher speed, about twice that of a *true* gyratory.
4. It produces a more uniformly sized stone.

Standard models have large feed openings for secondary crushing, and produce stone in the 1 to 4 in. range. The capacity of standard model is usually rated at closed-side setting.

Attrition models are for producing stone having a maximum size of about $\frac{1}{4}$ in. The capacity of an attrition model cone crusher may not be related to closed-side setting.

Figure 14.8 shows the difference between a gyratory and a standard cone crusher. The conical head on the cone crusher is usually made of manganese steel. It is mounted on a vertical shaft and serves as one of the crushing surfaces. The other surface is the concave that is attached to the upper part of the crusher frame. The bottom of the shaft is set in an eccentric bushing to produce the gyratory effect as the shaft rotates.

The maximum diameter of the crusher head can be used to designate the size of a cone crusher. However, it is the size of the feed opening, the width of the opening at the entrance to the crushing chamber, that limits the size of the rocks that can be fed to the crusher. The magnitude of the eccentric throw and the setting of the discharge opening can be varied within reasonable limits. Because of the high speed of rotation, all particles passing through a crusher will be reduced to sizes no larger than the closed-size setting that should be used to designate the size of the discharge opening.

Table 14.5 gives representative capacities for the Symons standard cone crusher, expressed in tons of stone per hour for material having a unit weight 100 lb per cf when crushed.

ROLL CRUSHERS

Roll crushers are used for producing additional reductions in the sizes of stone after the output of the blasting operation has been subjected to one or more stages of prior crushing. A *roll crusher* consists of a heavy cast-iron frame equipped with either one or more hard-steel rolls, each mounted on a separate horizontal shaft.

TABLE 14.4 | Representative capacities of gyratory crushers, in tons per hour (metric tons per hour) of stone*

Size of crusher [in. (cm)]	Approximate power required [hp (kW)]	Open-side setting of crusher [in. (mm)]											
		1½ (38)	1¾ (44)	2 (51)	2¼ (57)	2½ (63)	3 (76)	3½ (89)	4 (102)	4½ (114)	5 (127)	5½ (140)	6 (152)
Straight concaves													
8 (20.0)	15–25 (11–19)	30 (27)	36 (33)	41 (37)	47 (42)								
10 (25.4)	25–40 (19–30)		40 (36)	50 (45)	60 (54)								
13 (33.1)	50–75 (37–56)				85 (77)	100 (90)	133 (120)						
16 (40.7)	60–100 (45–75)						160 (145)	185 (167)	210 (190)				
20 (50.8)	75–125 (56–93)							200 (180)	230 (208)	255 (231)			
30 (76.2)	125–175 (93–130)								310 (281)	350 (317)	390 (353)		
42 (106.7)	200–275 (150–205)										500 (452)	570 (515)	630 (569)
Modified straight concaves													
8 (20.0)	15–25 (11–19)	35 (32)	40 (36)	45 (41)									
10 (25.4)	25–40 (19–30)		54 (49)	60 (54)	65 (59)								
13 (33.1)	50–75 (37–56)				95 (86)	130 (117)							
16 (40.7)	60–100 (45–75)						150 (135)	172 (155)	195 (176)				
20 (50.8)	75–125 (56–93)							182 (165)	200 (180)	220 (199)			
30 (76.2)	125–175 (93–130)								340 (308)	370 (335)	400 (362)		
42 (106.7)	200–275 (150–205)										607 (550)	650 (589)	690 (625)
Nonchoking concaves													
8 (20.0)	15–25 (11–19)	42 (38)	46 (42)										
10 (25.4)	25–40 (19–30)	51 (46)	57 (52)	63 (57)	69 (62)								
13 (33.1)	50–75 (37–56)	79 (71)	87 (79)	95 (86)	103 (93)	111 (100)							
16 (40.7)	60–100 (45–75)			107 (96)	118 (106)	128 (115)	150 (135)						
20 (50.8)	75–125 (56–93)				155 (140)	169 (152)	198 (178)	220 (198)	258 (233)	285 (257)	310 (279)		

*Based on continuous feed and stone weighing 100 lb per cf when crushed.

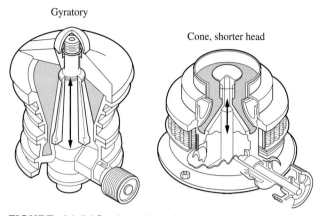

FIGURE 14.8 I Sections through a gyratory and a cone crusher.

Single Roll

With a single roll crusher, the material is forced between a large diameter roller and an adjustable liner. Because the material is dragged against the liner, these crushers are not economical for crushing highly abrasive materials. But they can handle sticky materials.

Double Roll

Roll crushers with two rollers are so constructed that each roll is driven independently by a flat-belt pull or a V-belt sheave. One of the rolls is mounted on a slide frame to permit an adjustment in the width of the discharge opening between the two rolls. The movable roll is spring-loaded to provide safety against damage to the rolls when noncrushable material passes through the machine.

Feed Size

The maximum size of material that may be fed to a roll crusher is directly proportional to the diameter of the rolls. If the feed contains stones that are too large, the rolls will not grip the material and pull it through the crusher. The angle of nip (grip), B in Fig. 14.9, that is constant for smooth rolls, has been found to be 16°45′.

The maximum-size particles that can be crushed are determined as follows. Referring to Fig. 14.9, these terms are defined:

$$R = \text{radius of rolls}$$

$$B = \text{angle of nip } (16°45′)$$

$$D = R \cos B = 0.9575R$$

$$A = \text{maximum-size feed}$$

$$C = \text{roll setting (size of finished product)}$$

TABLE 14.5 | Representative capacities of Symons standard cone crushers, in tons per hour (metric tons per hour) of stone*

Size of crusher [ft (m)]	Size of feed opening [in. (mm)]	Minimum discharge settings [in. (mm)]	1/4 (6.3)	3/8 (9.5)	1/2 (12.7)	5/8 (15.9)	3/4 (19.1)	7/8 (22.3)	1 (25.4)	1 1/4 (31.8)	1 1/2 (38.0)	2 (50.8)	2 1/2 (63.5)
2 (0.61)	2 1/4 (57)	1/4 (5.6)	15 (14)	20 (18)	25 (23)	30 (27)	35 (32)						
2 (0.61)	3 1/4 (82)	3/8 (9.5)		20 (18)	25 (23)	30 (27)	35 (32)	40 (36)	45 (41)	50 (45)	60 (54)		
3 (0.91)	3 7/8 (96)	3/8 (9.5)		35 (32)	40 (36)	55 (50)	70 (63)	75 (68)					
3 (0.91)	5 1/8 (130)	1/2 (12.7)			40 (36)	55 (50)	70 (63)	75 (68)	80 (72)	85 (77)	90 (81)	95 (86)	
4 (1.22)	5 (127)	3/8 (9.5)		60 (54)	80 (72)	100 (90)	120 (109)	135 (122)	150 (136)				
4 (1.22)	7 3/8 (187)	3/4 (19.0)					120 (109)	135 (122)	150 (136)	170 (154)	177 (160)	185 (167)	
4 1/4 (1.29)	4 1/4 (114)	1/2 (12.7)			100 (90)	125 (113)	140 (126)	150 (136)					
4 1/4 (1.29)	7 3/8 (187)	5/8 (15.8)				125 (113)	140 (126)	150 (136)	160 (145)	175 (158)			
4 1/4 (1.29)	9 1/2 (241)	3/4 (19.0)					140 (126)	150 (136)	160 (145)	175 (158)	185 (167)	190 (172)	
5 1/2 (1.67)	7 1/8 (181)	5/8 (15.8)				160 (145)	200 (181)	235 (213)	275 (249)				
5 1/2 (1.67)	8 5/8 (219)	7/8 (22.2)						235 (213)	275 (249)	300 (272)	340 (304)	375 (340)	450 (407)
5 1/2 (1.67)	9 7/8 (248)	1 (25.4)							275 (249)	300 (272)	340 (304)	375 (340)	450 (407)
7 (2.30)	10 (254)	3/4 (19.0)					330 (300)	390 (353)	450 (407)	560 (507)	600 (543)	800 (725)	
7 (2.30)	11 1/2 (292)	1 (25.4)							450 (407)	560 (507)	600 (543)	800 (725)	
7 (2.30)	13 1/2 (343)	1 1/4 (31.7)								560 (507)	600 (543)	800 (725)	900 (815)

*Based on stone weighing 100 lb per cf when crushed.
Source: Courtesy Nordberg Manufacturing Company.

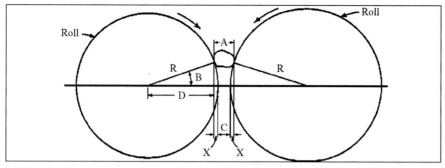

FIGURE 14.9 | Crushing rock between two rolls.

Then

$$X = R - D$$
$$= R - 0.9575R \Rightarrow 0.0425R$$
$$A = 2X + C$$
$$= 0.085R + C \qquad\qquad [14.1]$$

EXAMPLE 14.2

Determine the maximum-size stone that may be fed to a smooth-roll crusher whose rolls are 40 in. in diameter, when the roller setting is 1 in.

$$A = \left(0.085 \times \frac{40 \text{ in.}}{2}\right) + 1 \text{ in.}$$

$$A = 2.7 \text{ in.}$$

Capacity

The capacity of a roll crusher will vary with the type of stone, size of feed, size of the finished product, width of rolls, the speed at which the rolls rotate, and extent to which the stone is fed uniformly into the crusher. Referring to Fig. 14.9, the theoretical volume of a solid ribbon of material passing between the two rolls in 1 min would be the product of the width of the opening times the width of the rolls times the speed of the surface of the rolls. The volume can be expressed in cubic inches per minute or cubic feet per minute (cfm). In actual practice, the ribbon of crushed stone will never be solid. A more realistic volume should approximate one-fourth to *one-third* of the theoretical volume. An equation that can be use as a guide in estimating the capacity is derived using these terms.

$$C = \text{distance between rolls in inches}$$
$$W = \text{width of rolls in inches}$$

S = peripheral speed of rolls in inches per minute

N = speed of rolls in rpm

R = radius of rolls in inches

V_1 = theoretical volume in cubic inches per minute or cfm

V_2 = actual volume, in cubic inches per minute or cfm

Q = probable capacity in tons per hour

Then

$$V_1 = CWS$$

Assume one-third of the theoretical volume then

$$V_2 = \frac{V_1}{3}$$

$$V_2 = \frac{CWS}{3} \text{ cu in. per min}$$

Divide by 1,728 cu in. per cf.

$$V_2 = \frac{CWS}{5,184} \text{ cfm}$$

Assume the crushed stone has a unit weight of 100 lb per cf.

$$Q = \frac{100 \text{ lb/cf} \times (60 \text{ min/hr}) V_2}{2,000 \text{ lb/ton}} \Rightarrow 3 \text{ min/cu ft} \times V_2 \text{ tons per hr}$$

$$= \frac{CWS}{1,728} \text{ tons per hr} \qquad\qquad \textbf{[14.2]}$$

S can be expressed in terms of the diameter of the roll and the speed in rpm:

$$S = 2\pi RN$$

Substituting this value of S in Eq. [14.2] gives

$$Q = \frac{CW\pi RN}{864} \text{ tons per hr} \qquad\qquad \textbf{[14.3]}$$

Table 14.6 gives representative capacities for smooth-roll crushers, expressed in tons of stone per hour for material having a unit weight of 100 lb per cf when crushed. These capacities should be used as a guide only in estimating the probable output of a crusher. The actual capacity may be more or less than the given values.

If a roll crusher is producing a finished aggregate, the reduction ratio should not be greater than 4:1. However, if a roll crusher is used to prepare feed for a fine grinder, the reduction may be as high as 7:1.

TABLE 14.6 | Representative capacities of smooth-roll crushers, in tons per hour (metric tons per hour) of stone*

Size of crusher [in. (mm)][†]	Speed (rpm)	Power required [hp (kW)]	Width of opening between rolls [in. (mm)]						
			$\frac{1}{4}$ (6.3)	$\frac{1}{2}$ (12.7)	$\frac{3}{4}$ (19.1)	1 (25.4)	$1\frac{1}{2}$ (38.1)	2 (50.8)	$2\frac{1}{2}$ (63.5)
16 × 16	120	15–30	15.0	30.0	40.0	55.0	85.0	115.0	140.0
(414 × 416)		(11–22)	(13.6)	(27.2)	(36.2)	(49.7)	(77.0)	(104.0)	(127.0)
24 × 16	80	20–35	15.0	30.0	40.0	55.0	85.0	115.0	140.0
(610 × 416)		(15–26)	(13.6)	(27.2)	(36.2)	(49.7)	(77.0)	(104.0)	(127.0)
30 × 18	60	50–70	15.0	30.0	45.0	65.0	95.0	125.0	155.0
(763 × 456)		(37–52)	(13.6)	(27.2)	(40.7)	(59.0)	(86.0)	(113.1)	(140.0)
30 × 22	60	60–100	20.0	40.0	55.0	75.0	115.0	155.0	190.0
(763 × 558)		(45–75)	(18.1)	(36.2)	(49.7)	(67.9)	(104.0)	(140.0)	(172.0)
40 × 20	50	60–100	20.0	35.0	50.0	70.0	105.0	135.0	175.0
(1,016 × 508)		(45–75)	(18.1)	(31.7)	(45.2)	(63.4)	(95.0)	(122.0)	(158.5)
40 × 24	50	60–100	20.0	40.0	60.0	85.0	125.0	165.0	210.0
(1,016 × 610)		(45–75)	(18.1)	(36.2)	(54.3)	(77.0)	(113.1)	(149.5)	(190.0)
54 × 24	41	125–150	24.0	48.0	71.0	95.0	144.0	192.0	240.0
(1,374 × 610)		(93–112)	(21.7)	(43.5)	(64.3)	(86.0)	(130.0)	(173.8)	(217.5)

Source: Courtesy Iowa Manufacturing Company.

*Based on stone weighing 100 lb per cf when crushed.

[†]The first number indicates the diameter of the rolls, and the second indicates the width of the rolls.

IMPACT CRUSHERS

In impact crushers, the stones are broken by the application of high-speed impact forces. Advantage is taken of the rebound between the individual stones and against the machine surfaces to use fully the initial impact energy. The design of some units additionally seeks to shear and compress the stones between the revolving and stationary parts. Speed of rotation is important to the effective operations of these crushers as the energy available for impact varies as the square of the rotational speed.

Single Rotor

The single-rotor-type impact crusher (see Fig. 14.10) breaks the stone by both the impact action of the impellers striking the feed material and by the impact that results when the impeller-driven material strikes against the aprons within the crusher unit. These units produce a cubical product but are economical only for low-abrasion feeds. The unit's production rate is affected by the rotor speed. The speed also affects the reduction ratio. Therefore, any speed adjustment should be done only after consideration is given to both elements, production and final product.

Double Rotor

These units are similar to the single-rotor models and accomplish aggregate-size reduction by the same mechanical mechanisms. They will produce a somewhat

FIGURE 14.10 I Vertical shaft impact breaker with the top removed.

higher proportion of fines. With both single- and double-rotor crushers, the impacted material flows freely to the bottom of the units without any further size reduction.

Hammer Mills

The hammer mill, which is the most widely used impact crusher, can be used for primary or secondary crushing. The basic parts of a unit include a housing frame; a horizontal shaft extending through the housing; a number of arms and hammers attached to a spool, which is mounted on the shaft; one or more manganese-steel or other hard-steel breaker plates; and a series of grate bars, whose spacing may be adjusted to regulate the width of openings through which the crushed stone flows. These parts are illustrated in the cross section of the crusher shown in Fig. 14.11.

FIGURE 14.11 | Cutaway of hammer mill rock crusher showing breaking action.
Source: Cedarapids Inc., a Terex Company.

As the stone to be crushed is fed into the mill, the hammers, which revolve at a high rpm, strike the particles, breaking them and driving them against the breaker plates, which further reduces their sizes. Final size reduction is accomplished by grinding the material against the bottom grate bars.

The size of a hammer mill may be designated by the size of the feed opening. The capacity will vary with the size of the unit, the kind of stone crushed, the size of the material fed to the mill, and the speed of the shaft. Hammer mills will produce a high proportion of fines and cannot handle wet or sticky feed material.

SPECIAL AGGREGATE PROCESSING UNITS

Rod Mills

To produce fine aggregate, such as sand, from stone that has been crushed to suitable sizes by other crushing equipment, rod or ball mills are frequently used. Sometimes concrete specifications require the use of a homogeneous aggregate regardless of size. If crushed stone is used for coarse aggregate, sand manufactured from the same stone can satisfy the specifications.

A *rod mill* is a circular steel shell lined on the inside with a hard-wearing surface. The mill is equipped with a suitable support or trunnion arrangement at each end and a driving gear at one end. It is operated with its axis in a horizontal position. The rod mill is charged with steel rods, whose lengths are slightly less than the length of the mill. Stone is fed through the trunnion at one end of the mill and flows to the discharge at the other end. As the mill rotates slowly, the stone is constantly subjected to the impact of the tumbling rods that produce the desired grinding. A mill may be operated wet or dry, i.e., with or without water added. The size of a rod mill is specified by the diameter and the length of the shell, such as 8 × 12 ft, respectively.

Ball Mill

A *ball mill* is similar to a rod mill but it uses steel balls, having a distribution of sizes (see Fig. 14.12), instead of rods to supply the impact necessary to grind the stone. Ball mills will produce fine material with smaller grain sizes than those produced by a rod mill. Figure 14.13 shows a ball mill in a plant.

FEEDERS

Compression-type crushers (jaw) are designed to use particle interaction in the crushing process. An underfed compression crusher produces a larger percentage of oversize material, as the necessary material is not present to fully develop interparticle crushing. In an impact crusher, efficient use of interparticle collisions is not possible with an underfed machine. Gyratory-type crushers do not need feeders.

The capacity of compression- and impact-type crushers will be increased if the stone feed is at a uniform rate. Surge feeding tends to overload a crusher, and then the surge is followed by an insufficient supply of stone. Using a feeder ahead of a crusher eliminates most surge feeding problems that reduce crusher

FIGURE 14.12 | Variable size balls for a ball mill.

FIGURE 14.13 | A ball mill in operation.

capacity. The installation of such a feeder may increase the capacity of a jaw crusher as much as 15%.

There are many types of feeders:

1. Apron.
2. Vibrating.
3. Plate.
4. Belt.

Apron feeder. A feeder constructed of overlapping pans that form a continuous belt is referred to as an apron feeder. It will provide a continuous positive discharge of feed material. These feeders have the advantage that they can be obtained in considerable lengths.

Vibrating feeder. There are both simple vibrating feeders activated by a vibrating unit similar to the ones used on horizontal screens (discussed in the next section) and vibrating grizzly feeders. A vibrating grizzly feeder eliminates fines from the material entering the crusher. A vibrating feeder requires less maintenance than an apron feeder.

Plate. By means of rotating eccentrics, a plate can be made to move back and forth on a horizontal plane and used to uniformly feed material to a crusher.

Belt feeder. Operating on the same principal as the apron feeder, the belt feeder is used for smaller sizes of material, usually sand or small-diameter aggregates.

SURGE PILES

A stationary crushing plant can include several types and sizes of crusher, each probably followed with a set of screens and a belt conveyor to transport the stone to the next crushing operation or to storage. A plant can be designed to provide temporary storage for stone between the successive stages of crushing. This plan has the advantage of eliminating or reducing the surge effect that frequently exists when the crushing, screening, and handling operations are conducted as a liner operation.

The stone in temporary storage ahead of a crusher that is referred to as a "surge pile" can be used to keep at least a portion of a plant in operation at all times. Within reasonable limits, the use of a surge pile ahead of a crusher permits the crusher to be fed uniformly at the most satisfactory rate regardless of variations in the output of other equipment ahead of the crusher. The use of surge piles has enabled some plants to increase their final output production by as much as 20%.

Advantages derived from using surge piles are

1. They enhance uniform feed and ensure high crusher efficiency. Even if there are excavation and hauling disruptions, the crushing plant operations will not be interrupted.

2. Should the primary crusher break down, the rest of the plant can continue to operate.
3. Repairs can be made to either the primary or secondary sections of the plant without complete stoppage of production.

Arguments against the use of surge piles are

1. They take up additional storage area.
2. They require the construction of storage bins or reclaiming tunnels.
3. They increase the amount of stone handling.

The decision to use or not use surge piles should be based on an analysis of the advantages and disadvantages for each plant.

CRUSHING EQUIPMENT SELECTION

In selecting crushing equipment, it is essential that certain information be known prior to making the selection. The information needed should include, but will not necessarily be limited to, these items:

1. The kind of stone to be crushed.
2. The required capacity of the plant.
3. The maximum size of the feed stones (information concerning the size ranges of the feed is also helpful).
4. The method of feeding the crushers.
5. The required specified size ranges of the product.

Example 14.3 illustrates a method that can be used to select crushing equipment.

EXAMPLE 14.3

Select a primary and a secondary crusher to produce 100 tph of crushed limestone. The maximum-size stones from the quarry will be 16 in. The quarry stone will be hauled by truck, dumped into a surge bin, and fed to the primary crusher by an apron feeder, which will maintain a reasonably uniform rate of feed. The aggregate will be used on a project whose specifications require the following size distributions:

Size screen opening (in.)		
Passing	**Retained on**	**Percent**
$1\frac{1}{2}$		100
$1\frac{1}{2}$	$\frac{3}{4}$	42–48
$\frac{3}{4}$	$\frac{1}{4}$	30–36
$\frac{1}{4}$	0	20–26

Consider a jaw crusher for the primary and a roll crusher for the secondary crushing. The output of the jaw crusher will be screened and the material meeting the specification sizes removed. The remaining material, greater than $1\frac{1}{2}$ in., will be fed to the roll crusher.

Assume a setting of 3 in. for the jaw crusher. This will give a ratio of reduction of approximately 5:1, which is satisfactory. The jaw crusher must have a minimum top opening of 18 in. (16-in.-maximum-size feed stone plus 2 in.). Table 14.3 indicates a 24- by 36-in. size crusher has a probable capacity of 114 tph based on stone weighing 100 lb per cf when crushed. Figure 14.4 indicates that the product of the crusher will be distributed by these sizes:

Size range (in.)	Percent passing screens	Percent in size range	Total output of crusher (tph)	Amount produced in size range (tph)
Over $1\frac{1}{2}$	100–46	54	100	54.0
$1\frac{1}{2}$–$\frac{3}{4}$	46–26	20	100	20.0
$\frac{3}{4}$–$\frac{1}{4}$	26–11	15	100	15.0
$\frac{1}{4}$–0	11–0	11	100	11.0
Total		100		100.0

As the roll crusher will receive the output from the jaw crusher, the rolls must be large enough to handle 3-in. stone. Assume a setting of $1\frac{1}{2}$ in. From Eq. [14.1] the minimum radius will be 17.7 in. ($3 = 0.085R + 1.5$). Try a 40- by 20-in. (Table 14.6) crusher with a capacity of approximately 105 tph for a $1\frac{1}{2}$-in. setting.

For any given setting, the crusher will produce about 15% stone having at least one dimension larger than the setting. Thus, for a given setting 15% of the stone that passes through the roll crusher will be returned for recrushing. The total amount of stone passing through the crusher, including the returned stone, is determined as follows. Let

$$Q = \text{total amount of stone through the crusher}$$

Then

$$0.15Q = \text{amount of returned stone}$$
$$0.85Q = \text{amount of new stone}$$
$$Q = \frac{\text{amount of new stone}}{0.85}$$
$$= \frac{54 \text{ tph}}{0.85} \Leftrightarrow 63.5 \text{ tph}$$

The 40- by 20-in. roll crusher will handle this amount of stone easily. The distribution of the output of this crusher by size range will be

Size range (in.)	Percent passing screens	Percent in size range	Total output of crusher (tph)	Amount produced in size range (tph)
$1\frac{1}{2}$–$\frac{3}{4}$	85–46	39	63.5	24.8
$\frac{3}{4}$–$\frac{1}{4}$	46–19	27	63.5	17.1
$\frac{1}{4}$–0	19–0	19	63.5	12.1
Total		85		54.0

Now combine the output of each crusher by specified sizes.

Size range (in.)	From jaw crusher (tph)	From roll crusher (tph)	Total amount (tph)	Gradation requirement (percent)	Percent in size range
$1\frac{1}{2}-\frac{3}{4}$	20.0	24.8	44.8	42–48	44.8
$\frac{3}{4}-\frac{1}{4}$	15.0	17.1	32.1	30–36	32.1
$\frac{1}{4}-0$	11.0	12.1	23.1	20–26	23.1
Total	46.0	54.0	100.0		100.0

SEPARATION INTO PARTICLE SIZE RANGES

SCALPING CRUSHED STONE

The term *scalping,* as used in this chapter, refers to a screening operation. Scalping removes, from the main mass of stone to be processed, that stone that is too large for the crusher opening or is small enough to be used without further crushing. Scalping can be performed ahead of a primary crusher, and it represents good crushing practice to scalp all crushed stone following each successive stage of reduction.

Scalping ahead of a primary crusher serves two purposes. It prevents oversize stones from entering the crusher and blocking the opening, and it can be used to remove dirt, mud, or other debris that is not acceptable in the finished product. If the product of the blasting operation will contain oversize stones, it is desirable to remove them ahead of the crusher. Scalping should accomplish this removal. The use of a "grizzly," which consists of a number of widely spaced parallel bars, can be used to scalp material ahead of the primary crusher.

The product of the blasting operation may contain an appreciable amount of stone that meets the specified size requirements. In this event, it may be good economy to remove such stone ahead of the primary crusher, thereby reducing the total load on the crusher and increasing the overall capacity of the plant.

It is usually economical to install a scalper after each stage of reduction to remove specification sizes. This stone may be transported to grading screens, where it can be sized and placed in appropriate storage.

SCREENING AGGREGATE

In all but the most basic and fundamental crushing operations, the crushed rock particles must be separated into two or more particle size ranges. This separation enables us to direct certain selected material to receive additional or special processing, or certain material may be diverted to bypass unnecessary processing. The screening process is based on the simple premise that particle sizes smaller than the screen cloth opening size will pass through the screen and that oversized particles will be retained.

Screens openings can be described by either of two terms: (1) mesh and (2) clear opening. The term "mesh" refers to the number of openings per lineal inch. To count the number of openings to an inch, measure from the center of the screen wire to a point 1-in. distant. Clear opening or "space" is a term that refers to the distance between the inside edges of two parallel wires.

Most specifications covering the use of aggregate stipulate that the different sizes shall be combined to produce a blend having a given size distribution. Persons who are responsible for preparing the specifications for the use of aggregate realize that crushing and screening cannot be done with complete precision, and accordingly they allow some tolerance in the size distribution. The extent of tolerance may be indicated by a statement such as: The quantity of aggregate passing a 1-in. screen and retained on a $\frac{1}{4}$-in. screen shall be not less than 30% or more than 40% of the total quantity of aggregate.

Revolving Screens

Revolving screens have several advantages over other types of screens, especially when they are used to wash and screen sand and gravel. The operating action is slow and simple, and the maintenance and repair costs are low. If the aggregate to be washed contains silt and clay, a scrubber can be installed near the entrance of a screen to agitate the material in water. At the same time, streams of water can be sprayed on the aggregate as it moves through the screen.

Vibrating Screens

The vibrating screens consist of one or more layers or "decks" of open mesh wire cloth mounted one above the other in a rectangular metal box (Fig. 14.14). These are the most widely used aggregate production screens. The vibration is obtained by means of an eccentric shaft, a counterweight shaft, or electromagnets attached to the frame or to the screens.

A unit may be horizontal or inclined with slight slope (20° or less) from the receiving to the discharge end. The vibration, 850 to 1,250 strokes per min., causes the aggregate to flow over the surface of the screen. Normally, large amplitude and slower speed are necessary for large screen openings, and the opposite is necessary for small screen openings. In the case of a horizontal screen, the throw of the vibrations must move the material both forward and upward. For that reason, its line of action is 45° relative to the horizontal.

Most of the particles that are smaller than the openings in a screen will drop through the screen, while the oversize particles will flow off the screen at the discharge end. For a multiple-deck unit, the sizes of the openings will be progressively smaller for each lower deck.

A screen will not pass all material whose sizes are equal to or less than the dimensions of the openings in the screen. Some of this material may be retained on and carried over the discharge end of a screen. The efficiency of a screen may be defined as the ratio of the amount of material passing through a screen divided by the total amount that is small enough to pass through, with the ratio expressed as a percent. The highest efficiency is obtained with a single-deck screen, usually amounting to 90 to 95%. As additional decks are installed, the

FIGURE 14.14 | Triple-deck vibrating screen box.

efficiencies of these decks will decrease, being above 85% for the second deck and 75% for the third deck. Wet screenings will increase screening efficiency but additional equipment is necessary for handling the water.

The capacity of a screen is the number of tons of material that 1 sq ft will separate per hour. The capacity of a screen is *not* the total amount of material that can be fed and passed over its surface, but the rate at which it separates desired material from the feed. The capacity will vary with the size of the openings, kind of material screened, moisture content of the material, and other factors. Because of the factors that affect the capacity of a screen, it will seldom if ever be possible to calculate in advance the exact capacity of a screen. If a given number of tons of material must be passed per hour, it is prudent to select a screen whose total calculated capacity is 10 to 25% greater than the required capacity.

The chart in Fig. 14.15 gives capacities for dry screening that can be used as a guide in selecting the correct size screen for a given flow of material. The capacities given in the chart should be modified by the application of appropriate correction factors. Representative values of these factors are given next.

Efficiency Factors

If a low screening efficiency is permissible, the actual capacity of a screen will be higher than the values given in Fig. 14.15. Table 14.7 gives factors by which the chart values of capacities may be multiplied to obtain corrected capacities for given efficiencies.

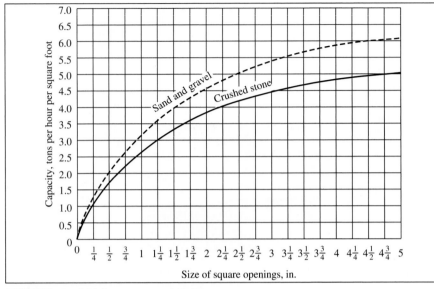

FIGURE 14.15 I Screen-capacity chart.

TABLE 14.7 I Efficiency factors for aggregate screening

Permissible screen efficiency (%)	Efficiency factor
95	1.00
90	1.25
85	1.50
80	1.75
75	2.00

Deck Factors

This is a factor whose value will vary with the particular deck position for multiple-deck screens. The values are given in Table 14.8.

TABLE 14.8 I Deck factors for aggregate screening

For deck number	Deck factor
1	1.00
2	0.90
3	0.75
4	0.60

Aggregate-Size Factors

The capacities of screens given in Fig. 14.15 are based on screening dry material that contains particle sizes such as would be found in the output of a representative crusher. If the material to be screened contains a surplus of small sizes, the capacity of the screen will be increased, whereas if the material contains a surplus of large sizes, the capacity of the screen will be reduced. Table 14.9 gives representative factors, which may be applied to the capacity of a screen to correct for the effect of excess fine or coarse particles.

TABLE 14.9 | Aggregate-size factors for screening

Percent of aggregate less than $\frac{1}{2}$ the size of screen opening	Aggregate-size factor
10	0.55
20	0.70
30	0.80
40	1.00
50	1.20
60	1.40
70	1.80
80	2.20
90	3.00

Determining the Screen Required

Figure 14.15 provides the theoretical capacity of a screen in tons per hour per square foot based on material weighing 100 lb per cf when crushed. The corrected capacity of a screen is given by the equation

$$Q = ACEDG \qquad \textbf{[14.4]}$$

where

Q = capacity of screen, tons per hour

A = area of screen in square feet

C = theoretical capacity of screen in tons per hour per square foot

E = efficiency factor

D = deck factor

G = aggregate-size factor

The minimum area of a screen to provide a given capacity is determined from the equation

$$A = \frac{Q}{CEDG} \qquad \textbf{[14.5]}$$

EXAMPLE 14.4

Determine the minimum-size single-deck screen, having $1\frac{1}{2}$-in.-sq. openings, for screening 120 tph of dry crushed stone, weighing 100 lb per cf when crushed. The screen box is 4 ft wide. A screening efficiency of 90% is satisfactory. An analysis of the aggregate indicates that approximately 30% of it will be less than $\frac{3}{4}$ in. in size. The values of the factors to be used in Eq. [14.5] are

$$Q = 120 \text{ tph}$$

$$C = 3.32 \text{ tph per sf} \quad \text{(Fig. 14.15)}$$

$$E = 1.25 \qquad\qquad \text{(Table 14.7)}$$

$$D = 1.0 \qquad\qquad \text{(Table 14.8)}$$

$$G = 0.8 \qquad\qquad \text{(Table 14.9)}$$

Substituting these values in Eq. [14.5], we get

$$A = \frac{120 \text{ tph}}{3.3 \text{ tph per sf} \times 1.25 \times 1.0 \times 0.8} = 36.4 \text{ sf}$$

In view of the possibility of variations in the factors used, and to provide a margin of safety, it is recommended to select a screen whose total calculated capacity is 10 to 25% greater than the required capacity.

$$A = 36.4 \text{ sf} \times 1.10 \Rightarrow 40.0 \text{ sf}$$

Therefore, as a minimum, use a 4- by 10-ft (40-sf) screen.

Sand Preparation and Classification Machines

When the specifications for sand and other fine aggregates require the materials to meet size gradations, it is frequently necessary to produce the gradations by mechanical equipment. Several types of equipment are available for this purpose. There are mechanical and water flow machines that classify sand into a multiple number of individual sizes. Sand and water are fed to these classifiers at one end of the unit's tank. As the water flows to the outlet end of the tank, the sand particles settle to the bottom of the tank, the coarse ones first and the fine ones last. When the depth of a given size reaches a predetermined level, a sensing paddle will actuate a discharge valve at the bottom of the compartment to permit that material to flow into the splitter box, from which it can be removed and used.

Another machine for handling sand is the screw-type classifier (Fig. 14.16). This unit can be used to produce specification sand. A sand screw is erected so that the material must move up the screw to be discharged. In the case of the screw shown in Fig. 14.16, the motor is at the discharge end. Sand and water are fed into the hopper. As the spiral screws rotate, the sand is moved up the tank to the discharge outlet under the motor. Undesirable material is flushed out of the tank with overflowing water.

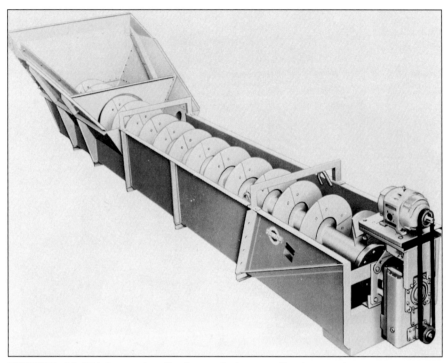

FIGURE 14.16 | Screw classifier for producing specification sand.
Source: Kolberg Manufacturing Corporation.

ELIMINATION OF UNDESIRABLE MATERIALS

LOG WASHERS

When natural deposits of aggregate, such as sand and gravel, or individual pieces of crushed stone contain deleterious material as a part of the matrix or as deposits on the surface of the aggregate, it will be necessary to remove these materials before using the aggregate. One method of removing the material is to pass the aggregate through a machine called a "log washer," which is illustrated in Fig. 14.17. This unit consists of a steel tank with two electric-motor-driven shafts, to which numerous replaceable paddles are attached. When the washer is erected in a plant the discharge end is raised. The aggregate to be processed is fed into the unit at the lower end, while a constant supply of water flows into the elevated end. As the shafts are rotated in opposite directions, the paddles move the aggregate toward the upper end of the tank, while producing a continuous scrubbing action between the particles. The stream of water will remove the undesirable material and discharge it from the tank at the lower end, whereas the processed aggregate will be discharged at the upper end.

FIGURE 14.17 | Log washer for scrubbing coarse aggregate.
Source: Telsmith Division, Barber-Greene Company.

HANDLING CRUSHED-STONE AGGREGATE

SEGREGATION

After stone is crushed and screened to provide the desired size ranges, it is necessary to handle it carefully or the large and small particles may separate, thereby destroying the blend in sizes that is essential to meeting graduation requirements. If aggregate is permitted to flow freely off the end of a belt conveyor, especially at some height above the storage pile, the material will be segregated by sizes. A strong crosswind tends to separate the smaller sizes from the larger sizes.

Specifications covering the production of aggregate frequently stipulate that the aggregate transported by a belt conveyor shall not be permitted to fall freely from the discharge end of a belt. The end of the belt should be kept as low as possible and the aggregate should be discharged through a rock ladder, containing baffles, to prevent segregation. Figure 14.18 shows a rock ladder used to reduce segregation.

FIGURE 14.18 I A rock ladder used to reduce the segregation of aggregate.

SAFETY

GENERAL INFORMATION

Crushing and screening equipment is designed by the manufacturer considering the safety of all operating personnel. It should not be changed or modified in any manner that eliminates the accident prevention devices. All equipment comes from the manufacturer with protective guards, covers, and shields installed around moving parts. These devices protect operators and others working on or near the machines. Even with safety devices, basic safety practices should be followed at all times.

■ Do not remove guards, covers, or shields when the equipment is running.

■ Replace all guards, covers, or shields after maintenance.

■ Never lubricate the equipment when it is in motion.

- Always establish a *positive lockout* of the involved power source before performing maintenance.
- Block parts as necessary to prevent unexpected motion while performing maintenance or repair.
- Do not attempt to remove jammed product or other blockage when the equipment is running.
- Wear appropriate personal protection apparatus: eye protection, hearing protection, breathing mask, and a hard hat.

SUMMARY

In operating a crushing plant, the drilling pattern, the amount of explosives, the size shovel or loader used to load the stone, and the size of the primary crusher should be coordinated to ensure that all stone can be economically used. Jaw crushers have large energy-storing flywheels and provide a high mechanical advantage. Gyratory and cone crushers have a gyrating mantle mounted within a bowl. A roll crusher consists of a heavy cast-iron frame equipped with either one or more hard-steel rolls each mounted on a separate horizontal shaft. In impact crushers, the stones are broken by the application of high-speed impact forces. Manufacturers provide capacity charts for their crushers. These are typically based on a standard stone weight of 100 lb per cf.

Following the crushing operation it is almost always necessary to size the product. This sizing or screening enables us to direct certain selected material to receive additional or special processing, or certain material may be diverted to bypass unnecessary processing. The screening is based on the simple premise that particle sizes smaller than the screen cloth opening size will pass through the screen and that oversized particles will be retained. Critical learning objectives include:

- An understanding and ability to use a manufacturer's crusher capacity chart.
- An ability to calculate roll crusher feed size.
- An ability to design a crushing plant based on required size distribution specifications.
- An ability to calculate required screen sizes.

These objectives are the basis for the problems that follow.

PROBLEMS

14.1 A jaw crusher with a closed setting of 3 in. produces 200 tph of crushed stone. Determine the number of tons per hour produced in each of the following size ranges: in excess of $2\frac{1}{2}$ in.; between $2\frac{1}{2}$ and $1\frac{1}{2}$ in.; between $1\frac{1}{2}$ and $\frac{1}{4}$ in.; less than $\frac{1}{4}$ in. (54 tph in excess of $2\frac{1}{2}$ in.; 54 tph between $2\frac{1}{2}$ and $1\frac{1}{2}$ in.; 70 tph between $1\frac{1}{2}$ and $\frac{1}{2}$ in.; 22 tph less than $\frac{1}{4}$ in.)

14.2 A roll crusher set at 2 in. produces 120 tph of crushed stone. Determine the number of tons per hour produced in each of the following size ranges: in excess of $1\frac{1}{2}$ in.; between $1\frac{1}{2}$ and $\frac{3}{4}$ in.; between $\frac{3}{4}$ in. and $\frac{1}{4}$ in.

14.3 Select a jaw crusher for primary crushing and a roll crusher for secondary crushing to produce 200 tph of limestone rock. The maximum-size stone from the quarry will be 24 in. The stone is to be crushed to the following specifications:

Size screen opening (in.)		
Passing	**Retained on**	**Percent**
2		100
2	$1\frac{1}{4}$	30–40
$1\frac{1}{4}$	$\frac{3}{4}$	20–35
$\frac{3}{4}$	$\frac{1}{4}$	10–30
$\frac{1}{4}$	0	0–25

Specify the size and setting for each crusher selected. (48 × 60 jaw w/ 3″ setting and 24 × 16 roll w/ 2″ setting)

14.4 A jaw crusher and a roll crusher are used in an attempt to crush 140 tph. The maximum-size stone from the quarry will be 18 in. The stone is to be crushed to these specifications:

Size screen opening (in.)		
Passing	**Retained on**	**Percent**
$2\frac{1}{2}$		100
$2\frac{1}{2}$	$1\frac{1}{2}$	30–50
$1\frac{1}{2}$	$\frac{1}{2}$	20–40
$\frac{1}{2}$	0	10–30

Select crushers to produce this aggregate. Can the product of these crushers be processed to provide the desired sizes in the specified percentages? If so, tell how.

14.5 A 30- by 42-in. jaw crusher is set to operate with a $2\frac{1}{2}$-in. opening. The output from the crusher is discharged onto a screen with $1\frac{1}{2}$-in. openings, whose efficiency is 90%. The aggregate that does not pass through the screen goes to a 40- by 20-in. roll crusher, set at $1\frac{1}{4}$ in. Determine the maximum output of the roll crusher in tons per hour for material less than 1 in. There will be no recycle of roll crusher oversize. What amount, in tons per hour, of the roll crusher output is in the range of 1 to $\frac{1}{2}$ in., and what amount is less than $\frac{1}{2}$ in.?

14.6 A portable crushing plant is equipped with these units:

One jaw crusher, size 15 × 30 in.

One roll crusher, size 30 × 22 in.

One set of horizontal vibrating screen, two decks, with $1\frac{1}{2}$- and $\frac{3}{4}$-in. openings.

The specifications require that 100% of the aggregate shall pass a $1\frac{1}{2}$-in. screen and 50% shall pass a $\frac{3}{4}$-in. screen. Assume that 10% of the stone from the quarry will be smaller than $1\frac{1}{2}$ in. and that this aggregate will be removed by passing the quarry product over the screen before sending it to the jaw crusher. The aggregate will weigh 110 lb per cf. Determine the maximum output of the plant, expressed in tons per hour. Include the aggregate removed by the screens prior to sending it to the crushers.

14.7 The crushed stone output from a 36- by 42-in. crusher with a closed opening of $2\frac{1}{2}$ in. is passed over a single horizontal vibrating screen with $1\frac{1}{2}$-in. openings. If the permissible screen efficiency is 90%, use the information in the book to determine the minimum-size screen, expressed in square feet, required to handle the output of the crusher. (45.5 sf minimum)

14.8 The crushed stone output from a 42- by 48-in. jaw crusher with a closed setting of 3 in. is passed over a $2\frac{1}{2}$- and a $1\frac{1}{2}$-in.-openings vibrating screen. The permissible efficiency is 85%. The stone has a unit weight of 110 lb per cf when crushed. Determine the minimum-size screen, in square feet, required to handle the crusher output. The screening unit is 4 ft wide; what will be the nominal screen sizes?

14.9 The crushed stone output from a 36- by 42-in. jaw crusher with a closed setting of 4 in. is to be screened into the following sizes: $2\frac{1}{2}$ to $1\frac{1}{2}$ in.; $1\frac{1}{2}$ to $\frac{3}{4}$ in.; less than $\frac{3}{4}$ in. A three-deck horizontal vibrating screen will be used to separate the three sizes. The stone weighs 115 lb per cf. If the permissible screen efficiency is 90%, determine the minimum-size screen for each deck, expressed in square feet, required to handle the output of the crusher. The screening unit is 4 ft wide; what will be the nominal screen sizes?

REFERENCES

1. Barksdale, Richard D. (ed), *The Aggregate Handbook,* National Stone Association, Washington, D.C., 1991.

2. *Cedarapids Pocket Reference Book,* 13th Pocket Edition, Cedarapids Inc., 916 16th Street NE, Cedar Rapids, IA 52402.

3. Cedarapids, Inc., 909 17th Street NE, Cedar Rapids, IA 52402-5259, www.cedarapids.com/.

4. Kelly, Errol G., and David J. Spottiswood, *Introduction to Mineral Processing,* Wiley, New York, New York, 1982.

5. Midwestern Industries, Inc., P.O. Box 810, Massillon, Ohio 44648-0810, www.midwesternind.com/.

6. Nordberg Group, Lokomonkatu 3, P.O. Box 307, FIN-33101, Tampere, Finland, www.nordberg.com/.

7. *Telsmith Mineral Processing Handbook,* 1976, Telsmith, Inc., P.O. Box 539, Mequon, WI 53092, www.telsmith.com/.

AGGREGATE-RELATED WEBSITES

1. International Center for Aggregates Research (ICAR), www.ce.utexas.edu./org/icar/index.html. The ICAR was established in 1992 by the Aggregates Foundation for Technology, Research and Education (AFTRE). The National Stone Association and the National Aggregates Association (NAA), representing the aggregates industry, established AFTRE to promote research, education, and technology transfer related to aggregates. The Center is operated jointly by the University of Texas at Austin (UT) and Texas A&M University (TAMU).

2. Mining Technology, www.mining-technology.com/. Mining Technology is a British website for the international mining industry. The site includes news on

current projects and developments; an equipment, products, and services guide; exhibition and conference listings; an industry associations directory; and a comprehensive links page to relevant mining resources.

3. National Aggregates Association (NAA), www.nationalaggregates.org/ naa2.htm,1100 Bonifant Street, Suite 400, Silver Spring, MD 20910-3358, Tel.: 301-562-1940 or (800) NAA-1020, Fax: 301-587-9419; 2101 Wilson Blvd., Suite 100, Arlington, VA 22201, Tel.: 703-525-8788, or toll-free at 800-342-1415, Fax: 703-525-7782. The NAA is an international trade association for producers of construction aggregates. The association is involved in product research, government relations, education and training, and regulatory issues. NAA regularly provides members with a wide variety of publications on topics including land reclamation and safety and technical reports on aggregate uses in concrete and bituminous materials.

4. National Slag Association, www.taraonline.com/nationalslagassoc/main.html, 110 W. Lancaster Ave., Suite 2, Wayne, PA 19087-4043, Tel.: 610-971-4840, Fax: 610-971-4841. The National Slag Association is an international trade association of iron and steel slag processors worldwide. Slag is the nonmetallic, commercial product made in furnaces simultaneously with iron and steel and used primarily as a construction aggregate.

5. National Stone Association (NSA), www.aggregates.org/. National Stone Association, 2101 Wilson Blvd., Suite 100, Arlington, VA 22201, toll free, 800-342-1415 or 703-525-8788, Fax: 703-525-7782. Note there was an NAA and NSA merger in 2000. The NSA is the national trade association representing and promoting the wide-ranging interests of crushed stone and other aggregate producers along with manufacturing and services companies that provide equipment and services to the industry. The association provides support in areas including operating productivity improvements, engineering research, safety and health, environmental concerns, and technical issues.

Asphalt Mix Production and Placement

The ability to easily accommodate a pavement section to stage construction and to recycle old pavements, and the fact that mix designs can be adjusted to utilize local materials, are three critical factors favoring the use of asphalt paving materials. An asphalt plant is a high-tech group of machines capable of uniformly blending, heating, and mixing the aggregates and asphalt cement of asphalt concrete, while at the same time meeting strict environmental regulations, particularly in the area of air emissions. Asphalt pavers consist of a tractor, either track or rubber-tired, and a screed. Mat thickness, which is controlled by the screed, can be maintained by using grade sensors or by tracing an external reference with a shoe or ski.

INTRODUCTION

This chapter deals with the equipment and methods used for the production and placement of asphalt pavements. Although the same or similar equipment may in some instances be used for other purposes, such as rollers for compaction of other materials, they will be treated separately in this chapter in relation to asphalt operations.

Asphalt paving materials are produced for the construction of highways, parking lots, and airfield pavements. The ability to easily accommodate a pavement section to stage construction and to recycle old pavements, and the fact that mix designs can be adjusted to utilize local materials, are three critical factors favoring the use of asphalt paving materials. When appraising asphalt production

and paving equipment, consideration must be given to the types of projects anticipated. Some asphalt mixing plants are operated primarily as producers of multiple mixes for FOB (free on board) plant sales or to serve multiple paving spreads
on small jobs. These plants must be able to easily and quickly change their production mix to meet the requirements of multiple customers. Other plants are
high-volume producers serving a single paving spread; this is particularly true of
portable plants that are moved from project to project. Equipment is available that
is specifically designed to meet the needs of both these situations. A constructor
must select the equipment and methods that allow the service flexibility best
suited to the specific project types that are expected to be undertaken.

GLOSSARY OF TERMS

The following glossary defines the important terms that are used in describing
asphalt equipment and procedures.

Aggregate. A mixture of rock materials having specified properties, used
as pavement base and subbase and in the asphalt concrete material.

Alligator cracks. Interconnected cracks in an asphalt pavement that form a
pattern similar to an alligator's skin.

Asphalt paving material. A mixture of asphalt and aggregate to form a
strong, flexible, weather-resistant paving layer in a pavement structure.

Asphaltic concrete. Asphalt paving material prepared in a hot-mix plant
and used for construction of highway or airfield pavements having high
traffic volumes and axle loads.

Automatic feeder control. A device on an asphalt paver that controls the
level of mix in the screw chamber ahead of the screed.

Automatic screed controls. A system that overrides the self-leveling action
of the asphalt paver screed and permits it to pave to a predetermined grade
and slope, using either a rigid or a mobile-type reference.

Baghouse. An asphalt plant component that uses special fabric filter bags
to capture the fine material particles contained in either the dryer or drum
mixer exhaust gases. The captured dust is normally returned to the mix.

Base course. That part of the pavement structure placed directly on the
subbase.

Batch plant. An asphalt mixing plant that proportions and mixes liquid
asphalt and aggregates in individual batches.

Binder. The material in a paving mix used to bind the aggregate particles
together, to prevent the entrance of moisture and to act as a cushioning
agent.

Binder course. A layer of mix placed between the base and the surface
course.

Bleeding. Formation of a film of asphalt on a pavement surface due to
upward movement of liquid asphalt in the mix.

Breakdown. The initial compaction that takes place directly behind the paver and is intended to achieve maximum density in the shortest time frame.

Cold feed system. The equipment used in an asphalt mixing plant to proportion and feed the aggregates prior to drying.

Cold planer (milling machine). A machine that uses a rotating drum with cutting teeth to remove roadway material.

Combination screed. A type of asphalt paver screed that utilizes both a tamper bar and vibration to achieve compaction of the mix.

Cutoff shoe. A device used to reduce, in small increments, the paving width of an asphalt-paving machine.

Drum mix plant. An asphalt mixing plant that combines both the aggregate drying and the mixing functions in a single drum.

Dwell time. The time required for material to pass through a dryer or mixer.

Echelon paving. Using two or more asphalt pavers operating one ahead of another to obtain a hot joint between the mats.

Edge cracks. Longitudinal cracks near the edge of the pavement, sometimes branching toward the shoulder, usually caused by insufficient shoulder support.

Flash point. A test of the volatility of asphalts based on the temperature at which a small flame drawn across the surface of an asphalt cement sample produces a flash at any point on the surface, AASHTO T48 for asphalt cements.

Flights. Metal plates of various shapes placed longitudinally inside the shell of a drum dryer or mixer. As the material moves through the drum, the flights serve to first lift the aggregate and then to drop it through the flame and hot gases.

Friction course. A layer of asphalt paving material usually less than 1 in. thick placed on a structurally sound pavement to improve skid resistance and smoothness.

Grade or string line. A wire or string erected at a specified grade and alignment that is used as a reference for an automatic control system of a paver, cold planer, or fine-grade machine.

Grade reference. An erected string line, curb, gutter, adjacent mat, or mobile averaging device used to provide reference to an automatic control system of a paver, cold planer, or fine-grade machine.

Hot bins. Bins used to store dried aggregates in an asphalt plant prior to proportioning and mixing.

Hot elevator. A bucket elevator used to carry hot, dried aggregate from the dryer to the gradation unit of an asphalt plant.

Hot mix asphalt. An asphalt paving material produced in a central mixing plant. The mix aggregates are dried and heated, and soon afterward combined with hot liquid asphalt. The final combined material is then delivered for placement at high temperatures.

Hot oil heater. A heater for increasing the temperature of heat transfer oil, which in turn is used as the means to provide temperature control for plant process operations and for increasing or maintaining the temperature of liquid asphalt, stored at an asphalt plant.

Hveem. A method of asphalt mix design based on the cohesion and friction of a compacted specimen.

Leveling arms. Two long arms extending forward from each side of the asphalt paver screed and attached to tow points on the tractor. This mechanical connection allows the screed to float on the mix during placement.

Leveling course. A new layer of asphalt paving material placed over a distressed roadway to improve its geometry prior to resurfacing.

Leveling wedges. Patches of asphalt paving material used to level sags and depressions in an old pavement prior to resurfacing.

Lift (mat). A layer of asphalt paving material separately placed and compacted.

Longitudinal crack. A pavement crack that is roughly parallel to the centerline.

Los Angeles abrasion test. A widely used means of determining the resistance of an aggregate to wear and abrasion.

Lute. A type of rake used to smooth out minor surface irregularities in the hot asphalt paving material behind the paver.

Marshall stability. The maximum load resistance that an asphalt paving material test specimen will develop when tested with the Marshall stabilometer at 140°F (60°C).

Porosity. The relative volume of voids in a solid or mixture of solids is termed porosity. Porosity is used to indicate the ability of the solid or the mixture of solids to absorb a liquid such as asphalt.

Prime coat. An application of liquid asphalt material over an untreated base to coat and bond the loose aggregate particles, waterproof the surface, and promote adhesion between the base and the overlying course.

Pugmill. A mechanical device for mixing materials, consisting of paddles attached to rotating shafts.

RAP (reclaimed asphalt pavement). The asphalt paving material removed from an old existing paved surface by cold planing or ripping.

Raveling. The term for the process whereby an asphalt pavement breaks apart and loses aggregate particles from its surface.

Recycling. The reuse of reclaimed asphalt pavement combined with virgin asphalt paving materials to produce asphaltic concrete for new pavement structures.

Reflective cracking. Fissures in an asphalt overlay that exhibit the crack pattern in the underlying pavement structure.

Rejuvenating agent (softening agent). An organic material with chemical and physical attributes that restores desired properties to aged asphalt.

Road widener. A special machine used to place asphalt paving material, aggregate, or stabilized materials adjacent to a roadway structure.

Rubberized asphalt. An asphalt paving material containing powdered or shredded rubber that is introduced into the mix to produce a more resilient pavement.

Ruts. Channels that develop in a pavement as a result of wheel loads being repetitively applied in the same locations.

Screed. That part of an asphalt paver that smooths and compacts the asphalt paving materials.

Self-leveling. The action of a floating asphalt paver screed that permits it to reduce humps and fill in low spots while paving.

Slope. The transverse inclination of a roadway or other surface termed slope.

Surface course. The top or riding surface of a pavement structure.

Superpave. An asphalt concrete mix-design method for determining the aggregate structure and proportions of aggregate and asphalt cement of asphaltic concrete.

Surge bin. A storage bin, usually cylindrical, used to hold hot asphalt paving materials at the plant. A surge bin allows continuous plant operation and faster load-out of trucks.

Tack coat. A light application of liquid asphalt usually emulsified with water. Used to help ensure a bond between the surface being paved and the new mat that is being placed.

Tamping screed. A type of asphalt paver screed that utilizes a tamper bar mounted in front of the screed plate to achieve initial compaction of the mix.

Thickness controls. Manually operated controls usually located at the outer edge of the main asphalt paver screed by which the screed operator can raise or lower the angle of attack of the screed plate to increase or decrease the mat thickness.

Total moisture. The total of both the surface and internal moisture present in an aggregate. Generally expressed as a percentage of the aggregate weight.

Transverse crack. A pavement crack that is roughly perpendicular to the roadway centerline.

Washboarding. Ripples formed transversely across the width of a pavement, or road section.

Weigh hopper. A batch plant component usually located under the hot bins in which the aggregates and asphalt are weighed prior to discharge into the mixing pugmill.

Wet collector. An asphalt plant dust collection system utilizing water and a high-pressure venturi to capture dust particles from the exhaust gas of a dryer or drum mixer. Wet collected dust cannot be returned to the mix.

Windrow. A continuous pile of material placed on a grade or previously placed mat for later pickup or spreading.

Windrow elevator. A device that travels ahead of an asphalt paver that picks up windrowed mix and feeds it into the paver hopper. This allows for more continuous paver operation to be achieved. Windrow elevators are usually attached to the paver.

STRUCTURE OF ASPHALT PAVEMENTS

Asphalt binders offer great flexibility to construct pavements tailored to the requirements of the local situation. Material selection, and production and construction processes vary depending on the type of application.

Traditionally, asphalt pavements were constructed as layered systems, see Fig. 15.1. This structure consists of a prepared subgrade, a granular subbase, a granular base, and an asphalt-bound surface layer. The asphalt surface layer consists of two courses, the surface wearing course and the underlying binder course. The aggregate characteristics and amount of asphalt binder are designed specifically to match the needs of the binder and surfacing courses. Both the binder and surface course materials are produced at a hot mix plant where the aggregates and binder are heated and mixed to produce the hot mix asphalt concrete.

This pavement structure is still used in many applications; however, to meet the needs of increasing traffic demands, there is a tendency to replace the base and subbase layers with asphalt-bound materials. The construction of pavements with full-depth asphalt-bound layers is similar to the construction of pavements with aggregate bases and subbases.

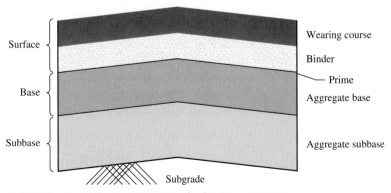

FIGURE 15.1 | Typical structure of a flexible pavement.

Hot mix asphalt concrete provides a quality surface for many applications. However, for low-volume road applications, a lower-cost pavement surface alternative can be used. Two popular surfaces are surface treatments and blade mix surfaces. These surfaces provide a wearing layer for the traffic. They must be placed on a base designed to support the traffic loads.

A surface treatment is constructed over a base by spraying a layer of liquid asphalt on the surface, then covering that binder with a layer of aggregates. The surface is then rolled to orient the aggregates into the liquid binder.

A blade mix surface is constructed by placing a windrow of aggregate materials on the prepared base. A liquid asphalt is metered onto the windrow at the desired application rate. A grader is then used to work the material across the base surface until the binder is thoroughly mixed with the aggregate. The grader then spreads the blended material, and a roller compacts the surface to the required density.

FLEXIBLE PAVEMENTS

Flexible pavements have a ride surface constructed with a combination of asphalt binder and aggregates. These two basic materials can be designed in many ways to construct a pavement surface suitable to the local conditions. Pavements are designed to meet these objectives:

1. Support the axle loads imposed by the traffic.
2. Protect the base and subbase from moisture.
3. Provide a stable-smooth and skid-resistant riding surface.
4. Resist weathering.
5. Provide economy.

The aggregate and asphalt binder that make up the paving material must provide a stable structure capable of supporting the repetitive vertical wheel loads imposed and resisting the kneading mechanism that wheel rotation and movement transmit to the structure. The pavement surface will be subjected to abrasive wear, while the entire section will have to resist structural movement. In addition to providing a structural wearing surface, the asphalt mixture must seal the base, subbase, and subgrade to prevent water intrusion. This is because of the influence of moisture on the strength of soils. Fuel efficiency, comfort, and safety are affected by surface texture. Besides moisture effects, the action of heat and cold can be very destructive to pavements. These objectives fix the performance criteria of a pavement section. The purpose of blending aggregates and asphalt binder is to achieve a final product that satisfies these objectives.

Aggregates

The load applied to an asphalt pavement is primarily carried by the aggregates in the mix. The aggregate portion of a mix accounts for 90 to 95% of the material by weight. Good aggregates and proper gradation of those aggregates are critical to the mix's performance. Ideally, an aggregate gradation should be provided that permits the minimum amount of expensive asphalt binder to be used.

The binder fills most of the voids between the aggregate particles as well as the voids *in* the particle surfaces.

Generally, the project owner will specify a range of allowable gradations for an asphalt concrete mix. Under the Superpave mix design system, the allowable aggregate gradations are designated by the nominal maximum aggregate size. The nominal maximum aggregate size is the sieve size that is one size larger than the first sieve that retains more than 10% of the aggregate mass. The maximum aggregate size is one sieve size larger than the nominal maximum aggregate size. The designated aggregate sizes used in Superpave are 37.5, 25, 19.5, 12.5, and 9.5 mm nominal maximum aggregate size. The gradation control points for a 12.5-mm Superpave aggregate designation are shown in Fig. 15.2. For the production of an asphalt concrete mix, a blend of aggregates must be used such that the combined aggregate gradation will fall within the control points, as shown in Fig. 15.3.

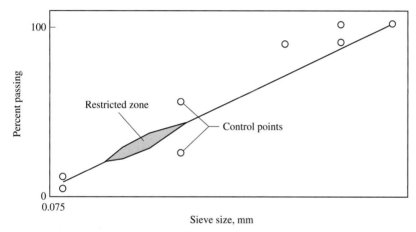

FIGURE 15.2 I Superpave gradation control points for a 12.5-mm aggregate.

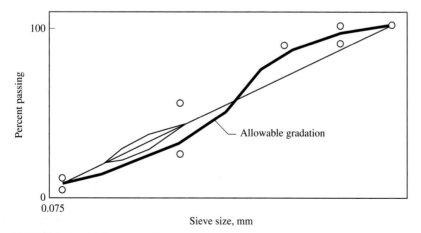

FIGURE 15.3 I Example of an allowable 12.5-mm Superpave-designed aggregate blend.

Figure 15.2 demonstrates that a full spectrum of aggregate particle sizes is required for an asphalt concrete mix. Individual aggregate particle sizes can range from over 25 mm down to particles that pass through a sieve with openings of 0.0075 mm (No. 200 sieve). If a single stockpile was created with the desired gradation, the larger aggregates would tend to separate from the smaller aggregates. Asphalt concrete made from a segregated stockpile would have pockets of small and fine aggregate and the pavement performance would suffer. To reduce the segregation problem, aggregates are stored in stockpiles of similar sizes. A minimum of three stockpiles is required to store coarse, intermediate, and fine aggregates. Frequently a fourth stockpile is used to provide better control of the fine-aggregate characteristics. During the asphalt concrete mix-design process, the needed proportions of each aggregate stockpile are determined. During the mix production, the stockpiles are blended in the predetermined ratios. The method used to combine the aggregates during production depends on the type of production.

Besides gradation, the aggregate properties of cleanliness, resistance to wear and abrasion, texture, porosity, and resistance to stripping (debonding between the asphalt and the aggregate) are important.

The amount of foreign matter, whether soil or organic material, that is present with the aggregate will reduce the load-carrying ability of a pavement. Visual inspection can often identify an aggregate cleanness problem, and washing, wet screening, or other methods, as discussed in Chapter 14, can be employed to correct the situation.

The effects of aggregate surface texture are manifested in the strength of the pavement structure and in the workability of the mix. Strength is influenced by the ability of the individual aggregate particles to "lock" together under load. This ability to lock is enhanced by angular rough-textured particles. Smooth aggregates, such as "river run" gravels and sands, produce a pavement that exhibits a reduced strength compared to one constructed from aggregates having rough surfaces. When necessary, rounded gravels can be crushed to create more angular surfaces. A mix using aggregates having a rough surface will require slightly more asphalt binder.

The porosity of an aggregate affects the amount of asphalt binder required in a mix. More asphalt must be added to a mix containing porous aggregates to make up for that which is absorbed by the aggregates and is not available to serve as binder. Slag and other manufactured aggregates can be highly porous, which will increase the asphalt cost proportion of a mix. However, because of the ability of these materials to resist wear, their use can be justified based on total lifetime project economics.

Some aggregates and asphalts have compatibility problems and the asphalt can separate or strip from the aggregate over the life of the pavement. This problem is evaluated during the mix-design process, and if the tests indicate a potential for stripping, then an antistrip admixture will be specified. Both liquid and powder antistrip materials are available. Liquid antistrip materials are added to the hot asphalt cement during production. The powder antistrip materials, lime or Portland cement, are added to the aggregate. Both liquid and powder antistrip are effective. The selection of either the liquid or powder antistrip may depend on the production capabilities of the asphalt plant.

Asphalts

Asphalt cement is a bituminous material that is produced by distillation of petroleum crude oil. This process can occur naturally and there are several sources of natural asphalt throughout the world, with the Trinidad Lake asphalt being the most famous natural asphalt. However, the vast majority of asphalt is produced by the petroleum industry. Historically, asphalt cement was a waste product that was the "bottom of the barrel" material left over when fuels were extracted from crude oil. As the petroleum industry gained sophistication, the ability to extract higher value materials from the crude oil increased and altered the quality of the asphalt cements. In addition, different crude oil sources have different chemical compositions, which affects the quality of the asphalt cements. Thus, the quality of an asphalt cement is a function of the original crude oil source and the refining process.

Asphalt cement is a viscoelastic material that behaves as a liquid at high temperatures and as an elastic solid at low temperatures. At pavement operating temperatures, asphalt cement has a semisolid consistency. For construction, asphalt cement must be put into a liquid state to mix and coat aggregates. Asphalt cement can either be heated to a liquid condition, approximately 300°F, prior to mixing with the aggregates or converted to a liquid product by dilution or emulsification.

The quality of asphalt cement used for highway construction is controlled through specifications. Specification methods have evolved over time. The earliest specification relied on the ability of the asphalt technologist to detect asphalt quality by chewing on a glob of asphalt. Fortunately, tests have replaced chewing for asphalt cement specifications. These tests evaluate the physical characteristics of the asphalt cement; currently, there are no chemical tests for asphalt cement. There are currently four specification methods for asphalt cement. The most recent, performance grade specifications, are displacing the other methods. However, not all agencies have adopted the performance grade specifications, so construction engineers may be exposed to some of the earlier specifications.

Penetration Grades of Asphalt Cement The earliest codified specification for asphalt cement is the penetration method, AASHTO M20. The primary test for this specification is the penetration test, AASHTO T49. In this test, an asphalt sample, at a temperature of 77°F, is subjected to a needle load, with a 10-g mass and a test time of 5 sec. The distance the needle penetrates into the sample is a measure of the consistency of the asphalt cement. Soft asphalts have high penetration while hard asphalts have little penetration.. The asphalt cement penetration grades are 40–50, 60–70, 85–100, 120–150, and 200–300 pen. The tested asphalt cement is placed into one of these grades based on the average of three penetration tests. In addition to the penetration test, there are a variety of other quality tests to which the asphalt cement must adhere to meet the material specifications.

Viscosity Grades of Asphalt Cement   The next specification method to gain wide acceptance classified the asphalt cement based on viscosity at 140°F. AASHTO M226 specifies two asphalt cement specifications, AC and AR. The difference between the AC and AR specification is that the AC grade designation is based on testing asphalt cement in an unconditioned state, i.e., the asphalt cement is in an "as produced" state. The grade designation for the AR specification is based on testing the asphalt cement after it has been conditioned. The conditioning process simulates the hardening of the asphalt cement during the construction process. The grade designations are AC 2.5, 5, 10, 20, and 40 and AR 500, 1,000, 2,000, 4,000, 8,000, and 16,000. As in the penetration grading method, there are several other tests that the asphalt cement must comply with for classification into one of these grades.

The viscosity and penetration specifications were effective in controlling the quality of the asphalt cement. However, the tests performed for these specifications could not be used to relate the quality of the asphalt cement to the performance of the pavement. In the late 1980s, the federal government sponsored a major research program to enhance highway construction and performance. One of the products of this research was performance grade specification for asphalt cement.

Performance Grades of Asphalt Cement   The performance grade asphalt cement specifications relate engineering measures of asphalt cement properties to specific pavement performance concerns. The penetration and viscosity specifications have required test limits at fixed temperatures. The performance grade specifications use fixed test criteria but the test temperatures for evaluating the asphalt cement are selected to reflect the local construction conditions. The grade specifications for the performance grade method are designated based on the temperature application range of the asphalt cement. The grades in the performance grade system are designated as PGhh-ll, where PG identifies the Performance Grade specification, hh identifies the high temperature application for the asphalt cement in degrees Celsius, and ll identifies the low temperature designation for the asphalt cement in degrees Celsius. The performance grade designations for asphalt cements are shown in Table 15.1.

Table 15.1 demonstrates that there is a wide range of grades available in the performance grade specifications. Not all these grades will be available in a local area. Refineries limit their production to match the market. Generally the state departments of transportation (DOTs) are the largest customers. Therefore the refineries seek to meet the needs of the DOTs. This limits the number of grades of asphalt cements needed. Figure 15.4 shows the four performance grades of asphalt cement specified by the North Carolina Department of Transportation.

TABLE 15.1 | Performance grade asphalt cement classifications

Lower temperature designation	Upper temperature designation						
	46	52	58	64	70	76	82
-10		X		X	X	X	X
-16		X	X	X	X	X	X
-22		X	X	X	X	X	X
-28		X	X	X	X	X	X
-34	X	X	X	X	X	X	X
-40	X	X	X	X	X		
-46	X	X					

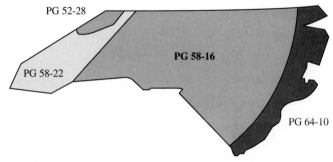

PG 52-28

PG 58-16

PG 58-22

PG 64-10

FIGURE 15.4 | North Carolina Department of Transportation specified performance grade asphalt cements.

Liquid Asphalt Cement To allow construction without having to heat the asphalt cement to 300°F two methods have been developed to reduce asphalt cement to a liquid state: cutbacks and emulsions. Asphalt cutbacks are a blend of asphalt cement and a fuel product. Rapid, medium, and slow cure cutbacks can be formulated by diluting the asphalt with gasoline or naphtha, diesel, and fuel oil, respectively. Emulsions are the other options for preparing a liquid asphalt. Asphalt emulsions are produced by using a colloidal mill to break down asphalt cement into very fine "globules" that are introduced into water that has been treated with an emulsifying agent. The emulsifying agent is a soaplike material that allows the asphalt cement globules to remain suspended in the water. Due to cost, safety, and environmental concerns, asphalt emulsions have largely displaced cutbacks when a liquid asphalt is needed.

Emulsions are manufactured in several grades and types as shown in Table 15.2. In addition to the ASSHTO specified emulsions, many states have specifications for asphalt emulsions suited to their local conditions. State department of transportation specifications control these. Although there is a wide variety of emulsions available, SS-1 and SS-1h are the predominant asphalt emulsions used for prime and tack coats.

The primary applications of emulsion asphalt cements are for tack coats and prime coats during pavement construction, and as the binder for surface

TABLE 15.2 | Types of asphalt emulsions recognized in AASHTO Standards

Particle charge	Set	Emulsion viscosity*		Residual asphalt cement[†]	
		1	**2**	**h**	**s**
Anionic	Rapid	RS-1	RS-2		
			HFRS-2		
	Medium	MS-1	MS-2	MS-2h	
		HFMS-1	HFMS-2	HFMS-2h	HFMS-2s
	Slow	SS-1		SS-1h	
Cationic	Rapid	CRS-1	CRS-2		
	Medium		CMS-2		CMS-2h
	Slow	CSS-1		CSS-1h	

*Refers to the viscosity of the emulsion: 1 has a lower viscosity than 2, as measured by the Saybolt Furol test; HF indicates a high float emulsion, generally used with dusty aggregates or when placing chip seals on grades.

[†]The letter h indicates asphalt cement residue of 40 to 90 penetration, and s indicates asphalt cement residue with more than 200 penetration. If an h or s is not indicated, penetration is 100 to 200.

treatments, slurry seals, and cold patch materials. For tack and prime coats and surface treatments, the binder is applied by spraying through a distributor truck.

ASPHALT CONCRETE

Asphalt cements are used as the binder in paving mixes. The asphalt cement usually represents less than 10% of the mix by weight. However, it serves the very important functions of bonding the aggregate particles together, preventing the entrance of moisture, and acting as a cushioning medium. All asphalt concrete is a blend of aggregate and asphalt cement. However, by varying the aggregate gradation, different types of asphalt concrete can be produced, such as dense-graded, open-graded, and stone matrix asphalt. The most common type of asphalt concrete is dense-graded mix.

Asphalt concrete mix design refers to the process of selecting the asphalt cement, aggregate structure, and proportions of aggregate and asphalt cement that provide the optimum combination of materials for the construction project. Several methods have been developed for mix design, such as Hveem, Marshall, and Superpave. The common feature of all asphalt concrete mix-design procedures is that, through laboratory evaluation and specifications, the pavement construction requirements are defined. These include the grade of the asphalt cement, the specifications governing the quality of the aggregates, and the quality control requirements for the in-place asphalt concrete mixture.

Different mix-design procedures will require different quality control parameters for construction. The most common parameters include

1. Target aggregate gradations.
2. Target asphalt content.
3. Density of the mix.
4. Volume of air voids.

5. Volume of voids in the mineral aggregate.
6. Volume of voids filled with asphalt.
7. Dust to binder ratio.

The mix design process generally consists of estimating a target aggregate gradation. Samples are then prepared at different asphalt contents, and tested with the relevant test methods for the mix-design method. Plots are prepared of the mix-design parameters versus asphalt content. The optimum asphalt content is determined, and construction documents are prepared.

ASPHALT PLANTS

GENERAL INFORMATION

Hot mixed asphalt is produced at a central plant and transported to the paving site in trucks. An *asphalt plant* is a high-tech group of machines capable of uniformly blending, heating, and mixing the aggregates and asphalt cement of asphalt concrete, while at the same time meeting strict environmental regulations, particularly in the area of air emissions. Drum and batch plants are the two most common plant types. There are a few continuous operation plants still in use; however, this technology is no longer manufactured so it is not described here. Drum mix plants are a newer technology than batch plants and generally are more economical to operate. Drum plants were introduced in the 1970s and dominate the new plant market. About 95% of the new plants are the drum type. However, about 70% of operational plants are batch plants. In 2000 it was reported that 5,000 to 6,000 batch plants are still in operation but sales of new batch plants are almost at a standstill.

The technologies of batch and drum plants are described next. While the mixing process between the two plants is distinctly different, there are many similar elements that vary only in detail between the two types of plants. The similar elements are the dust collection, asphalt storage, truck scales, and storage or surge silos. The truck scales are at the loading location at the plant. The empty and loaded weight of the trucks is measured to determine the weight of the load. The truck scales must be calibrated and certified.

BATCH PLANTS

Batch plants date from the beginning of the asphalt industry. Their primary components in the order of material flow are

- Cold feed system.
- Drum dryer.
- Hot elevator.
- Hot screens.

- Hot bins.
- Asphalt-handling system.
- Pugmill mixer.
- Dust collectors.
- Surge silo.

Cold Feed Systems

Cold feed bins provide aggregate surge storage and a uniform flow of properly sized material for mixing. Cold feed systems usually consist of three to six open-top bins mounted together as a single unit (see Fig. 15.5). The size of the bins is balanced with the operating capacity of the plant. The individual bins have steep sidewalls to promote material flow. In the case of sticky aggregates, it may be necessary to have wall vibrators. The individual bins can be fed from sized aggregate stockpiles by front-end loader, clamshell, or conveyor. At the bottom of each bin is a gate for controlling material flow and a feeder unit for metering the flow. The plant operator adjusts the flow of the aggregates from each bin to ensure a sufficient flow of material to keep an adequate charge of aggregates in the hot bins. Belt feeders are the most common equipment for transporting the aggregates from the cold bins to the dryer drum, but vibratory and apron feeds can be found.

Drum Dryer

The purposes of a drum dryer are to heat and dry the aggregates of the mix. Aggregate temperature controls the resulting temperature of the mix. If the aggregate has been heated excessively, the asphalt will harden during mixing. If the aggregates have not been heated adequately, it is difficult to coat them completely with asphalt. Therefore, the aggregate must be heated sufficiently at this step in the process to produce a final mix at the desired temperature (further discussion under "Mixing").

FIGURE 15.5 | Asphalt plant with a six-bin cold feed system.

Aggregates are introduced at the end of the drum dryer opposite the burner and travel through the drum counter to the gas flow. The drum is inclined downward from the aggregate feed end to the burner end. This slope causes the aggregate to move through the drum by gravity. The drum rotates and steel angles, "flights," mounted on the inside lift the aggregate and dump it through the hot gas and burner flame. Finally, the heated aggregate is discharged into the hot (bucket) elevator, which carries it to the screens at the top of the batch plant tower.

Hot Screening

The batch plant vibrating screen unit is usually a $3\frac{1}{2}$-deck arrangement. This enables gradation control of four aggregate sizes into four different hot bins. The screen unit ejects oversized material out of the production cycle. While the screens provide gradation control, they will not function properly unless the proportioning and flow from the cold feed are correct. If the screens are overloaded, material, which should be passing through a screen and into a hot bin, is carried instead into the bin of the next larger aggregate size. Such a situation destroys the mix formulation and must be avoided.

Hot Bins

The aggregates from the hot screens are stored in the hot bins until the plant is ready to make a batch of asphalt concrete. One of the key elements in operating a batch plant is to ensure that the hot bins have sufficient material to feed the pugmill for the production of a batch of asphalt concrete. One of the potential advantages of a batch plant as compared to a drum plant is that the batches are individually blended from the hot bins. This allows the aggregate blend of one batch to be different from the blend of the next batch. However, this is contingent on having the proper aggregates available in the hot bins. Frequently under high-production-rate conditions, the aggregates in the hot bins will not be sized properly for gross changes in the gradations of the mix. Thus, the flexibility of the batch plant is compromised. Large changes in the gradation of the aggregates generally must be accomplished by altering the flow of aggregates from the cold bins so that the proper amount of material is stored in the hot bins.

Weight Hopper

Aggregate from the hot bins is dropped into a weight hopper situated below the bins and above the pugmill. The weight hopper is charged one hot bin at a time to control the gradation of the blended aggregate. The aggregates are weighed cumulatively in the hopper, with the mineral filler added last. After charging, the weight hopper gates are opened to discharge the aggregate into the pugmill.

Asphalt-Handling System

The asphalt cement is stored on site in a heated tank. The asphalt is pumped to the weight tank, ready for discharge into the pugmill. After the aggregates are added to the pugmill, the asphalt cement is pumped through spray bars into the pugmill to coat the aggregates.

Pugmill Mixing

Most plants use a twinshaft pugmill for mixing the batch. To achieve uniform mixing, a pugmill's live zone should be completely filled with mix. The live zone is from the bottom of the box to the top of the paddle arc. The mixing process generally takes about 1 min, 15 sec to load and dry mix, and 45 sec mixing time with the asphalt cement. The actual mixing time needed is evaluated based on inspection of the coating of the coarse aggregate. Plant operation specifications require a sufficient mixing time to fully coat 90 or 95% of the aggregate, depending on the aggregate size.

The capacity of the plant is a function of the pugmill size and the mixing time. Typical batch quantities range from 1.5 to 5 tons. A batch plant with a 5-ton pugmill can produce 300 tons of mix per hour if a continuous operation can be maintained.

The plant is structured so that the discharge gate of the mixer is sufficiently high to allow truck passage directly below for loading (see Fig. 15.6). Alternatively, a hot elevator can be used to transport the mix to surge silos. These silos allow the plant to operate independently of the availability of trucks for loading the asphalt concrete. This is particularly advantageous when the plant is serving jobs with different mix designs. Silos also allow the plant operator to premix and store several batches of asphalt concrete to accommodate an uneven distribution of truck arrivals at the plant.

Because of the influence of temperature on the quality of the mix, the purchaser usually specifies the mixing temperature, measured immediately after discharge from the pugmill. The specification range will vary with the type and grade of the asphalt cement. In the case of dense-graded mixes the range across all asphalt cements is from 225 to 350°F. The range for open-graded mixes is from 180 to 250°F. Mixing should be at the lowest temperature that will achieve complete asphalt coating of the aggregates and still allow for satisfactory workability. In some cases, during cold weather or for long haul distances, the asphalt concrete is heated an extra 10°F to allow for temperature loss during the haul.

Recycling in a Batch Plant

Reclaimed asphalt pavement, RAP, can be added to the virgin aggregate in one of four locations in a batch plant:

1. Weight box.
2. Separate weight hopper.
3. Bucket-elevator.
4. Heat transfer chamber.

Each of these methods requires superheating the virgin aggregates to provide the heat source for heating and drying the RAP. RAP usually has a moisture content

FIGURE 15.6 | Batch plant with truck below pugmill mixer.

Source: Barber-Greene, A Division of Caterpillar Paving Products Inc.

of 3 to 5%, so steam is released as the RAP is heated. The steam released from the RAP contains dust that must be filtered. Adding the RAP to the weight box is the simplest method, but the time for heat transfer is limited, and capturing the dust-laden steam requires additional equipment. Using a separate weight hopper reduces the time to batch slightly and may improve accuracy, but this method has the same disadvantages as adding the RAP directly to the hopper.

DRUM MIX PLANTS

The primary components of a drum mix plant (see Fig. 15.7) are

■ Cold feed system.
■ Asphalt-handling system.
■ Drying and mixing drum.
■ Elevator.
■ Dust collector.
■ Storage silo.

Cold Feed System

In a drum plant, all of the mixing is performed within the drum and the final mix is discharged directly from the drum into storage silos in a continuous manner. There is no opportunity to adjust the aggregate blend, as can be done with a batch plant. Therefore, the aggregate from each of the cold feed bins must be weighed prior to feeding the material into the drum. Since the aggregates are weighed prior to drying, the moisture content of the aggregates in the cold bins must be monitored and the weights adjusted to ensure that the dry mass of the aggregates is correct. Scales mounted on the conveyor belts measure the weights of the aggregates.

FIGURE 15.7 I Drum-mixer of a drum mix asphalt plant.

Mixing Drum

The mixing drum consists of a long tube with flights for tumbling the aggregates and the mix, a burner for heating the aggregates, and a spray bar for applying the asphalt. The basic operation of the drum plant is that the aggregates are metered into one end of the drum. The time the aggregates spend in the drum ranges from 3 to 4 min. During that time, the aggregates must be thoroughly dried and heated to the mixing temperature. Near the discharge end of the drum, the asphalt cement is sprayed onto the aggregates. Automatic controls monitor the aggregate quantity and meter in the proper amount of asphalt cement. The drum rotation, with the flights in the drum, mixes the aggregate, RAP (if used), and asphalt cement.

A drum mix plant drum usually has a slope in the range of $\frac{1}{2}$ to 1 in. per foot of drum length. Rotation speeds are normally 5 to 10 rpm and common diameters are from 3 to 12 ft, with lengths between 15 and 60 ft. The ratio of length to diameter is from 4 to 6. Longer drums are found in recycling applications. The drum slope, length, rotation speed and flights, and the nature of the material control dwell time.

The production rate of the plant is inversely proportional to the moisture content of the aggregate. For example, increasing the moisture content from 3 to 6% in a batch plant with an 8-ft-diameter drum reduces the production rate from 500 to 300 tons per hour.

Originally drum plants were designed as parallel flow operations in which the aggregates and heated air travel in the same direction down the drum. Subsequent designs have increased the production rate of drum plants by using counter airflow arrangements in which the heated exhaust exits from the top of the drum where the aggregates are introduced. Counterflow plants can operate at about 12% higher production rates.

Recycling in a Drum Plant   Drum plants can be used for recycling reclaimed asphalt pavement by placing a feed collar that introduces the RAP between the portion of the drum that heats the aggregate and the area where the asphalt cement is introduced. Due to the asphalt content of the RAP, it cannot be exposed to direct flame. Heating the RAP requires heat transfer from the virgin aggregate. For RAP production, the aggregates are superheated to provide sufficient heat to raise the temperature of the RAP to the production temperature. One manufacturer has developed an innovative double-barrel drum design (Fig. 15.7), where the RAP is introduced in an outer drum to avoid exposing the RAP to the flame.

Storage Silos

Since drum plants produce a continuous flow of asphalt concrete, the output must be stored in silos (Fig. 15.8) for subsequent dispatch into transport trucks. These silos have a bottom dump for directly discharging the asphalt concrete into the haul trucks. The silos are typically insulated to retain heat. Sophisticated silos can

FIGURE 15.8 | Storage silo at a drum mix asphalt plant.

be completely sealed, and even filled with an inert gas to reduce oxidation of the asphalt cement while the asphalt mix is being stored. One of the problems with storage silos is the potential for flow of the asphalt cement from materials at the top of the silo to the bottom. This results in poor quality paving mix.

DUST COLLECTORS

To avoid contributing to air pollution asphalt plants are equipped with dust control systems. The two most commonly utilized systems are the water venturi approach and the cloth filtration "baghouse" system. The wet approach does require the availability of adequate water supplies. This approach introduces water at the point where dust-laden gas moves through the narrow throat of a venturi-shaped chamber. The dust becomes entrapped in the water and is thereby separated from the exhaust gas. A disadvantage of the wet approach is that the collected material cannot be reclaimed for use in the mix.

The dry baghouse systems allow mechanical collection and use of the fines from the aggregates. The system works by forcing the dust-laden gas through fabric filter bags that hang in a baghouse. By using a reverse pulse of air or by mechanically shaking the bags, the collected dust is removed from the filter. The dust falls into hoppers in the bottom of the baghouse and is moved by augers to a discharge vane feeder. The vane feeder is necessary to keep the baghouse airtight.

Filter bags are made of fabrics that can withstand temperatures up to 450°F. But care must be exercised when using a baghouse system, as excessive temperatures can melt the bags and/or cause a fire.

ASPHALT STORAGE AND HEATING

When liquid asphalt is combined with the aggregate for mixing, the temperature of the asphalt should be in the range of 300°F. Therefore, both drum mix and batch plants have heating systems to keep the liquid asphalt at the required temperature. If asphalt is delivered at a cooler temperature, the systems must be capable of raising the temperature. The two methods commonly used for heating liquid asphalt are the direct fire and the hot oil processes.

A direct fire heater consists of a burner that fires into a tube in the asphalt storage tank. With such systems, sufficient asphalt must always be kept in the tank so that the burner tube is always submerged. These systems have a higher thermal efficiency than the hot oil process.

The hot oil system is a two-stage approach. First, transfer oil is heated and then the heated oil is circulated through piping within the asphalt tank.

The storage temperature for asphalt cement varies according to the grade of the asphalt cement. The viscosity of the asphalt must be low enough to permit pumping; since different asphalts have different temperature-viscosity relationships, the storage temperature is different for the different grades of asphalts. Typically the storage temperature is in the range of 320°F for soft asphalts to 350°F for hard asphalts. For performance-grade asphalts, the producer or supplier should provide the storage temperature that will result in the asphalt cement having the proper viscosity for pumping.

RECLAIMING

Existing asphalt pavements represent a large investment in aggregates and asphalt. By reclaiming these materials, using either cold milling (see Fig. 15.9) or ripping methods, much of that investment can be recaptured. Additionally, cold milling allows restoration of the pavement section without the need to

FIGURE 15.9 | Milling machine removing an old asphalt pavement.

change the grade, thereby eliminating the problems associated with raising curbs and drainage structures.

The reclaimed asphalt paving materials are combined with virgin aggregates, additional asphalt, and/or recycling agents in a hot mix plant to produce new paving mixes. In those cases where the pavement sections were ripped up instead of being milled it may be necessary to crush the reclaimed materials to reduce the particle size. The new mix design will have to account for both the graduation of the aggregate in the RAP and the asphalt content of the RAP.

PAVING EQUIPMENT

A paving operation requires several pieces of equipment. These include

- Sweeper/brooms for removing dust from the surface to be paved.
- Trucks for transporting the asphalt from the plant to the construction site.
- Asphalt distributor truck for applying the prime tack, or seal coats.
- Material transfer vehicle (optional).
- Windrow elevator (optional).
- Paver.
- Rollers.

SWEEPER/BROOM

The sweeper/broom is used to remove dust from the surface of existing pavement prior to laying new asphalt. This is done to ensure proper bonding between the new asphalt and the old pavement. When surfacing a prepared base course, the dust layer should be removed either by sweeping with the sweeper or by wetting the base course and recompacting.

HAUL TRUCKS

Three basic types of trucks are used to haul the asphalt concrete from the plant to the job site: dump trucks, live bottom trucks, and bottom dump trucks. Dump trucks and live bottom trucks transfer the mix directly into the paver hopper. The bottom dump trucks are used to place a windrow of material directly on the pavement and can only be used with pavers that have an elevator for lifting the mix from the pavement and transferring it to the spreader box of the paver. Since the windrow material loses temperature rapidly, this paving operation is generally limited to the southwestern United States, where the ambient temperatures are high.

Regardless of the type of truck used, the trucks should be insulated and covered to reduce heat loss during transportation. Before the trucks are loaded, the bed is coated with an approved release agent. In the past, a heavy fuel oil like kerosene was used for this. However, this contaminates the asphalt binder, and this practice is no longer permitted.

ASPHALT DISTRIBUTORS

When applying an asphalt prime, tack, or seal coat, a specially designed distributor truck is utilized (see Fig. 15.10). An asphalt distributor truck requires constant attention to produce a uniform application. It is critical that the asphalt heater and pump be well maintained. All gages and measuring devices such as the pump tachometer, measuring stick, thermometers, and bitumeter must be properly calibrated. Spray bars and nozzles should be clean and set at the proper height above the surface receiving the application. The factors that affect uniform application are

The asphalt spraying temperature.

The liquid pressure across the spray bar length.

The angle of the spray nozzles.

The nozzle height above the surface.

The distributor speed.

Asphalt distributors have insulated tanks for maintaining material temperature and are equipped with burners for heating the asphalt material to the proper application temperature. Either independently powered or PTO-driven discharge pumps are used to maintain continuous and uniform pressure for the full length of the spray bar. The nozzle angle must be set properly, usually 15 to 30° from the horizontal axis of the spray bar, so that the individual spray fans do not interfere or intermix with one another. The height of the nozzle above the surface determines the width of an individual fan. To ensure the proper lap of the fans, the nozzle (spray bar) height must be set and maintained. The relationship between application rate (gallons per square yard) and truck speed is obvious;

FIGURE 15.10 | Using an asphalt distributor hand-sprayer attachment to apply a tack coat.

truck speed must be held constant during the spraying to achieve a uniform application.

The relationship between application rate, truck configuration, and surface area to be covered is given by

$$L = \frac{9 \times T}{W \times R} \qquad\qquad \textbf{[15.1]}$$

where

L = length of covered surface in feet

T = total gallons to be applied

W = spray bar coverage width in feet

R = rate of the application in gallons per square yard (sy)

This equation can be used to estimate the amount of liquid asphalt required for a job. During construction, Eq. [15.1] can be used to check the application rate by solving for R. The actual application rate is compared to the specifications for the job.

Prior to placement of an asphalt mix on a new base, a prime coat is applied to the base. Normal rates of application for prime vary between 0.20 and 0.60 gal per sy. Prime promotes adhesion between the base and the overlying asphalt mix by coating the absorbent base material, whether it is gravel, crushed stone, a stabilized material, or an earthen grade. The prime coat should penetrate about $\frac{1}{4}$ in., filling the voids of the base. The prime coat additionally acts as a waterproof barrier preventing moisture that penetrates the wearing surface from reaching the base.

Tack coats are designed to create a bond between existing pavements, be they concrete, brick, or bituminous material, and new asphalt overlays. They are also applied between successive mats during new construction. A tack coat acts as an adhesive to prevent slippage of the two mats. The tack coat is a very thin uniform coating of asphalt, usually 0.05 to 0.15 gal per sy of diluted emulsion. The tack coat should become sticky within a few hours. An application that is too heavy will defeat the purpose by causing the layers to creep, as the asphalt serves as a lubricant rather than as a tack.

Seal coats consist of an application of asphalt followed by a light covering of fine aggregate, which is rolled in with pneumatic rollers. Application rates are normally from 0.10 to 0.20 gal per sy.

ASPHALT PAVERS

An asphalt paver consists of a tractor, either track or rubber-tired, and a screed (see Fig. 15.11). The tractor power unit has a receiving hopper in the front and a system of slat conveyors to move the mix through a tunnel under the power plant to the rear of the tractor unit. At the rear of the tractor unit, the mix is deposited on the surface to be paved and augers are used to spread the asphalt evenly across the front of the trailing screed. Two tow arms, pin connected to the tractor unit, draw the screed behind the tractor. The screed controls the asphalt placement

FIGURE 15.11 | Track-mounted asphalt paver.

FIGURE 15.12 | Asphalt paver having two twin screws for moving the mix through the tunnel.

width and depth, and imparts the initial finish and compaction to the material. There is one manufacturer that offers a paver having two sets of twin screws to move the mix through the tunnel to the rear of the paver (see Fig. 15.12). The use of the screw conveyors is said to reduce mix segregation.

Pavers can receive mix directly into their front hoppers or can pick up a windrow of material placed in front of the paver. The traditional method for directly loading the hopper has the truck directly dump the asphalt concrete into the paver hopper. Push rollers mounted on the front frame of the paver (see Fig. 15.12), and extending beyond the hopper, push against the wheels of the truck, or a bar on the paver pushes the pusher bar of the truck. The material is then transferred to the hopper by raising the truck bed or by activating the live bottom.

Loading the hopper with individual truckloads often requires that the paver stops intermittently. This can cause problems with constructing a smooth pavement since frequently the paver must wait between truckloads, and mating up to the truck can bump the paver, causing the screed to release material.

Windrow Elevators

A paving operation being served by trucks means that there must be paver/truck load transfers and times when the paver must operate between truckloads or even stop if deliveries are delayed. The windrow elevator (see Fig. 15.13) was developed to address transfer and truck queuing effects on mat quality. Elevators can also improve production by eliminating the paver to truck mating.

Pavers with integral windrow pickup elevators are available, but a separate elevator unit attachable to the front of a regular paver is the most common approach. By using an attachment, there is the flexibility of using the paver in both direct load and windrow situations.

FIGURE 15.13 | Windrow elevator loading an asphalt paver.

Source: Barber-Greene, A Division of Caterpillar Paving Products Inc.

The flight system of the elevator continuously lifts the mix into the hopper from the windrow. For efficient operation the amount of material in the windrow cross section must equal the amount required for the mat cross section. Minor quantity variations are accommodated by the surge capacity of the paver hopper. Conventional windrow machines have limited width and can handle only fairly narrow rows of material. New elevator designs are being introduced having the capacity to handle wider windrow cross sections. These machines will open windrow operations to smaller, limited-area projects and will allow the use of regular dump trucks for hauling.

Material Transfer Devices

Some contractors use material transfer devices to improve paving quality. A material transfer device can receive multiple truckloads of asphalt concrete, remix the asphalt concrete, and deliver it to the paver hopper. Material transfer devices offer several advantages over direct transfer between the haul truck and the paver. By holding several loads of asphalt concrete, the transfer device reduces surge loading of the paver. This allows the paver to continuously operate while the haul trucks transfer the asphalt concrete to the transfer device. The timing of the trucks to the job site is not as critical, since the paver can operate from the stockpile of material in the transfer device. The material transfer device is self-propelled, so the paver does not need to mate up to and push the haul truck. Finally, by remixing the asphalt concrete, thermal segregation is reduced.

Thermal segregation occurs during the haul since the asphalt concrete at the surface of the load cools while the mix in the center of the load retains heat. The self-leveling screed of the paver is sensitive to the stiffness of the mix, which is largely a function of the temperature of the mix. By remixing the asphalt concrete, the stiffness of the asphalt concrete is more uniform, which allows smoother pavement construction.

Screed

The "floating" screed is free to pivot about its pin connections. This pin-connected tow arm arrangement allows the screed to be self-leveling and gives it the ability to compensate for base surface irregularities. The paver's ability to level out irregularities is controlled by the tractor's wheelbase length and by the length of the screed towing arms. Greater lengths of these two components mean smooth transitions across irregularities and therefore a smooth riding surface.

Mat thickness, which is controlled by the screed, can be maintained by using grade sensors tracing an external reference with a shoe or ski. Or the screed can be tied to a specified grade by the use of sensors tracing a stringline. When all the forces acting on the screed are constant, it will ride at a constant elevation above the grade or follow the stringline. However, there are factors that can cause the sensor regulated screed height to vary:

- The screed *angle of attack.*
- The *head of asphalt* in front of the screed.
- The *paver speed.*

The angle created by the plane of the surface upon which the asphalt is being placed and the plane of the screed bottom is known as the "screed angle of attack" (see Fig. 15.14). This angle is the principal mechanical factor affecting variations in mat thickness. It regulates the amount of material passing under the screed in a given distance. When either the screed or the tow points are vertically displaced, the angle of attack is changed. The screed will immediately begin to move, restoring the original angle, but this correction requires about three tow-arm lengths to be accomplished.

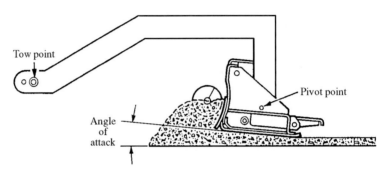

FIGURE 15.14 | Screed angle of attack.

Source: Caterpillar Paving Products Inc.

The asphalt material directly in front of and across the length of the screed is referred to as the head of material. When this material is not held constant by the hopper drag conveyor and the auger feed, the screed angle of attack will be affected and in turn the mat thickness will change. If the head of material becomes too high, the mat thickness will increase as the screed rides up on the excess material with the paver's forward progress. If the volume decreases, the screed moves down, resulting in reduced mat thickness.

Paver speed is linked to the rate at which asphalt mix is delivered from the plant. To produce a smooth mat, forward travel speed should be held constant. Changes in paver speed will affect the screed's angle of attack. Increasing the speed causes the screed to ride down, whereas decreasing the speed has the opposite effect. Additionally, when the paver is stopped, the screed tends to settle into the mat.

Initial mix compaction is achieved by vibration of the screed. Vibrators mounted on the screed are used to impart compaction force to the mat. On many pavers, vibrator speed can be adjusted to match paver speed and mat thickness. Other factors that influence compaction are mix design and placement temperature.

The width of a screed can be changed by stopping the paver and adding fixed width extensions on one or both sides of the basic screed. Some screeds are hydraulically extendible, allowing the paving width to be varied without stopping the paver. Auger additions may be required as screed width is increased to spread material evenly across the entire screed and screed extension width. Most screeds can be adjusted in the horizontal plane to create either a crown or superelevation.

To prevent material from sticking to the screen at the beginning of a paving operation it is necessary to heat the screed. Built-in diesel or propane burner heaters are used to heat the bottom screed plates. Required burner-heating time will vary with air temperature and the type of mix being placed. About 10 min of heating is normal, but care must be exercised as overheating can warp the screed.

Paver Production

Continuous paving operations depend upon balancing paver production with plant production. The critical choke points in the operation, which must be analyzed and managed, are the *plant load-haul unit* and the *haul unit-feed paver* links.

EXAMPLE 15.1

An asphalt plant can produce 324 tons per hour (tph). A project requires paving individual 12-ft lanes with a 2-in. lift averaging 112 lb per sy-in. What average paver speed will match the plant production? How many 20-ton bottom-dump trucks will be required if the total hauling cycle time is 55 min?

$$\frac{324 \text{ tph}}{60 \text{ min/hr}} = 5.4 \text{ ton/min average laydown production}$$

$$\frac{2 \text{ in. (thick)} \times 12 \text{ ft (wide)} \times 1 \text{ ft (length)}}{9 \text{ sq ft (sf)/sy}} = 2.66 \text{ sy-in. per ft paving length}$$

$$\frac{2.66 \text{ sy-in.} \times 112 \text{ lb/sy-in.}}{2,000 \text{ lb/ton}} = 0.149 \text{ ton per ft of paving length}$$

$$\frac{5.4 \text{ ton/min}}{0.149 \text{ ton/ft}} = 36.2 \text{ ft/min, average paver speed}$$

$$20 \text{ tons per truck} \times \frac{60 \text{ min/hr}}{55 \text{ min cycle}} = 21.8 \text{ tph per truck}$$

$$\frac{324 \text{ tph}}{21.8 \text{ tph per truck}} = 14.9 \text{ trucks}$$

therefore, 15 trucks are required.

Another way to analyze the situation would be to consider time. The paver requires a truck every

$$\frac{20 \text{ tons per truck}}{5.4 \text{ ton/min, required for paver}} = 3.7 \text{ min}$$

$$\frac{55 \text{ min (total truck cycle time)}}{3.7 \text{ in (paver requirement)}} = 14.9 \text{ or 15 trucks are required}$$

This is the minimum number of trucks required. However, haul time is rarely consistent, so extra trucks should be assigned to the project to prevent having to stop the paver.

With a windrow operation, delays to individual trucks have a limited effect on paving speed because the windrow storage allows the paver to continue moving. This is an advantage windrowing brings to asphalt laydown operations. However, there is a limit to the windrow length to prevent the stockpile material from cooling below the specified paving temperature. Project management must recognize these temperature restrictions and ensure that windrows are not overextended ahead of the paver laydown operation.

COMPACTION EQUIPMENT

Because of the relationships between pavement air voids and mechanical stability, durability, and water permeability, asphalt pavements are designed based on the mix being compacted to a specified density. Typically for properly designed pavements the air-voids content should be between 3 and 5%. The factors that affect compaction are mix characteristics, lift thickness, mix temperature, and the operational characteristics of the compaction equipment.

Three basic roller types are used to compact asphalt-paving mixes: smooth drum steel wheel (see Fig. 15.15), pneumatic tire, and smooth drum steel wheel vibratory rollers. Pneumatic and smooth drum steel wheel rollers are available with centerpoint articulation. Such articulation allows the front and rear tires or wheels to maintain their overlap throughout the turning range of the machine.

There have been asphalt compaction trials in Australia using a new hot-impact asphalt compactor. This machine uses a rubber belt instead of rollers to transfer the compaction energy to the hot mix asphalt mat. The two most significant advantages that are ascribed to these machines are that they achieve required mat density in fewer passes than conventional rollers and there is uniform compaction

FIGURE 15.15 | Smooth drum steel wheel roller compacting asphalt.

across the full width of the mat without shape loss at the unsupported edge. These machines are still in the development stage but they hold promise as the asphalt compactor of the future [7, 14].

The contact surface area of a steel wheel roller is an arc of the cylindrical wheel surface and the roller width. This contact area will decrease as the rolling progresses and the contact pressure will approach a maximum value. The contact area of a pneumatic tire is an ellipse, influenced by wheel load, tire inflation pressure, and tire sidewall flexure. With a pneumatic roller, the change in contact area as compaction progresses is less dramatic than in the case of a steel wheel roller.

The compaction capability of pneumatic and vibratory steel wheel rollers can be altered to match the construction conditions. With pneumatic rollers, changing the ballast and tire pressure alters the compaction effort. Adjusting the vibrator frequency and amplitude alters the compaction effort of vibratory steel wheel rollers. These capabilities enhance the operational flexibility of both roller types.

Rolling Temperature

Asphalt concrete compaction is achieved by orienting the aggregates into a dense configuration. The hot asphalt binder promotes the reorientation of the aggregates by acting as a lubricant. When the asphalt cools below the softening point, about 180°F, the binder will no longer provide a lubricating action. Below the softening point, the asphalt begins to bind the aggregate into place making further compaction difficult. The temperature at which the mix was placed is the important factor in determining the time available for compaction. The placing temperature is a function of the temperature during production and the heat loss during the haul. To extend the available time, it is good practice to use tarps to cover the mix in the haul trucks (see Fig. 15.16). In extreme cases, it may be necessary to use trucks having insulated bodies.

FIGURE 15.16 | Use of a tarp to cover the asphalt mix in a haul truck.

Once the mix passes through the paver, five factors affect the rate of cooling: air temperature, base temperature, mix laydown temperature, layer thickness, and wind velocity. Higher ambient air temperatures allow more time for compaction. Of greater significance is the temperature of the surface upon which the mix is placed. There is a rapid heat transfer between these two surfaces. The higher the temperature of the mix passing through the paver, the more time will be available for compaction. Probably the most significant factor affecting cooling is layer thickness. As layer thickness increases, the time available for achieving compaction will increase. But thicker layers require more roller passes. High winds can cause the surface of the mat to cool very rapidly.

Rolling should begin at the maximum temperature possible. This maximum temperature is that point at which the mix will support the roller without distorting the asphalt concrete horizontally. Increasing roller drum diameter increases the contact area, which reduces the contact pressure. Hence large-diameter rollers can work on the mat at higher temperatures.

Rolling Steps

Compaction of an asphalt mat is usually viewed in terms of three distinct steps:

- Breakdown rolling.
- Intermediate rolling.
- Finish rolling.

The breakdown step seeks to achieve a required density within a time frame defined by temperature constraints and consistent with paver speed. There is an optimum mix temperature range for achieving proper compaction. If the mix is too hot, the mat will tear and become scarred. If the mix is too cold, the energy requirement for compaction becomes impractical because of the viscous resistance of the asphalt binder. These physical limits define the practical time duration available for the rolling operation.

Sometimes density cannot be achieved by a single roller within the time duration available. In that case, an intermediate rolling step is required to supplement the breakdown step in reaching the required density. Finally, there is a finish-rolling step to remove any surface marks left by the previous rolling or by the paver.

Vibratory steel wheel rollers are usually the roller of choice for breakdown and intermediate rolling because of their adaptability to a range of mixes and differences in mat thickness. With the vibrator turned off, they can also be used for finish rolling. Pneumatic rollers can be used for breakdown and intermediate rolling. However, the high temperature of the mix during the breakdown rolling can cause the mix to stick to the rubber tires, so it is more common to use pneumatic rollers for the intermediate rolling. Steel wheel rollers are used for finish rolling. Since vibratory rollers can be used for each of the compaction steps, many contractors use vibratory rollers exclusively for asphalt concrete compaction.

Roller Capacity

Paver speed and production are set by the asphalt plant capability. The number and type of rollers on the job must be selected to match the placement capacity. Net roller speed, the length of pavement that can be compacted in a unit of time, is influenced by

1. The gross roller speed.
2. The number of passes.
3. The number of laps.
4. The overlap between adjacent laps required to cover the mat width.
5. The extension overedge.
6. The extra passes for joints.
7. The nonproductive travel (overrun for lap change).

Roller capability for a project must match net roller speed with average paver speed. Increasing roller speed reduces compactive effort, so speed alone cannot be used to compensate for production needs. Typical acceptable roller speeds are: 2 to $3\frac{1}{2}$ mph for breakdown, $2\frac{1}{2}$ to 4 mph for intermediate, and 3 to 7 mph for finish rolling.

Vibratory Roller Frequency and Amplitude

Most vibratory rollers allow the operator to select a vibratory frequency and amplitude. The maximum amount of compaction per pass will be achieved by selecting the highest possible frequency. This is because there are more impacts per foot of travel. But for most mixes, the roller should be operated at the lowest possible amplitude. Only in the case of mat thickness greater than 3 in. should a higher amplitude be considered. A high amplitude setting on a thin mat causes the roller to bounce making effective and uniform compaction difficult. In the case of mats having a thickness of 1 in. or less, vibratory rollers should be operated in the static mode.

EXAMPLE 15.2

An asphalt plant will produce 260 tph for a project. The mat will be 12 ft wide, 2 in. thick, and will have a density of 110 lb/sy-in. A vibratory roller with a 66-in.-wide drum will be used for compaction. Assume a 50-min hour efficiency factor for the roller. The overlap between adjacent laps and overedges will be a minimum of 6 in. It is estimated that nonproductive travel will add about 15% to total travel. From a test mat, it was found that three passes with the roller are required to achieve density. To account for acceleration and deceleration when changing directions, add 10% to the average roller speed to calculate a running speed. How many rollers should be used on this project?

$$\frac{12 \text{ ft (wide)} \times 2 \text{ in.} \times 110 \text{ lb/sy-in.}}{9 \text{ sf/sy} \times 2{,}000 \text{ lb/ton}} = 0.147 \text{ ton per ft of paving length}$$

$$\frac{260 \text{ tph}}{60 \text{ min/hr}} = 4.3 \text{ ton/min}$$

Average paving speed,

$$\frac{4.3 \text{ ton/min}}{0.147 \text{ ton/ft}} = 29 \text{ ft/min}$$

Rolling width, 12 ft × 12 in./ft + 2(6 in. each edge) = 156 in.

Effective roller width, 66 in. − 6 in. (overlap) = 60 in.

Number of laps: First lap 66 in.; 156 − 66 = 90 in. remaining
Additional laps,

$$\frac{90 \text{ in.}}{60 \text{ in.}} = 1.5$$

Therefore, three laps will be necessary to cover the 12-ft placement width.

Total number of roller passes, 3 laps × 3 passes per lap = 9 passes

Each pass must cover:

29 ft/min, rolling + 29 ft/min, return + maneuver distance

Total roller distance required to match paver speed

9 × 58 ft/min × 1.15 (nonprod) × 60 min/hr = 36,018 ft/hr

Average roller speed

$$\frac{36,018 \text{ ft/hr}}{50 \text{ min}} = 720 \text{ ft/min or 8.2 mph}$$

Running speed, 8.2 × 1.1 = 9 mph. This speed exceeds the speed limitation for effective compaction of 2 to $3\frac{1}{2}$ mph for breakdown rolling. Therefore, more than one roller is required. For this example, three rollers would allow operation at 3 mph.

Another solution would be to use a roller having a drum width equal to or greater than 84 in. The use of such a roller would reduce the number of laps to 2, and therefore the number of passes to 6.

Effective roller width, 84 in. − 6 in. (overlap) = 78 inches

Number of laps: First lap 84 in.; 156 − 84 = 72 in. remaining

Additional laps

$$\frac{72 \text{ in.}}{78 \text{ in.}} = 0.9$$

The required running speed would then be 6 mph and the project would require only two such rollers.

SAFETY

Personnel working around asphalt plants must be trained to be safety conscious [9]. The potential for a serious fire results from the presence of burner fuel for the dryer, liquid asphalt, and hot oil of the heat transfer system. No open flames or smoking should be permitted in the plant area. All fuel, asphalt, and oil lines should have control valves that can be activated from a safe distance. Both asphalt and hot oil lines should be checked regularly for leaks.

Besides the fire potential many of the mechanical parts are very hot. These will cause severe burns if personnel should come into contact with them. The mechanical system for moving the aggregates, with its belts, sprockets, and chain drives, presents a hazard to personnel if they should become entangled in the system's moving parts. All of these parts should be covered or protected, but they do require maintenance. Maintenance should be performed only when the plant has been completely shut down. The plant operator must be informed that personnel are working on the plant and the lockout switch should be tagged.

Even at the paving spread, the mix is still very hot and direct contact will result in severe burns. Additionally, trucks backing to the paver have limited rear visibility. Personnel must be made aware of these hazards on a regular basis and necessary safety equipment provided.

SUMMARY

Asphalt binders offer great flexibility to construct pavements tailored to the requirements of the local situation. Material selection, production, and construction processes vary depending on the type of application. The load applied to an asphalt pavement is primarily carried by the aggregates in the mix. The aggregate portion of a mix accounts for 90 to 95% of the material by weight. Good aggregates and proper gradation of those aggregates are critical to the mix's performance.

Hot mixed asphalt is produced at a central plant and transported to the paving site in trucks. Drum and batch plants are the two most common plant types. An asphalt paver consists of a tractor power unit and a trailing screed. Two tow arms, pin connected to the tractor unit, draw the screed behind the tractor. The screed controls the asphalt placement width and depth, and imparts the initial finish and compaction to the material.

Because of the relationships between pavement air voids and mechanical stability, durability, and water permeability, asphalt pavements are designed based on the mix being compacted to a specified density. Three basic roller types are used to compact asphalt-paving mixes: smooth drum steel wheel, pneumatic tire, and smooth drum steel wheel vibratory rollers. Critical learning objectives include:

- An understanding of the asphaltic materials.
- An understanding of the mixing process and elements of both batch and drum mix asphalt plants.
- An understanding of equipment required for asphalt paving operations.
- An ability to calculate either coverage or quantity of material for prime, tack, or seal coats.
- An ability based on paver speed to calculate required plant production.
- An ability based on paver speed to calculate required roller capacity.

These objectives are the basis for the problems that follow.

PROBLEMS

15.1 An asphalt plant can produce 300 tph. A project requires paving individual 10-ft lanes with a $1\frac{1}{2}$-in. lift averaging 115 lb per sy-in. What average paver speed will match the plant production? How many 14-ton dump trucks will be required if the total hauling cycle time is 40 min? (52.5 ft/min; 15 trucks)

15.2 An asphalt plant can produce 240 tph. A project requires paving individual 12-ft lanes with a 1-in. lift averaging 112 lb per sy-in. What average paver speed will match the plant production? How many 10-ton dump trucks will be required if the total hauling cycle time is 30 min?

15.3 A plant will produce 440 tph for a project. The mat will be 12 ft wide, 1 in. thick, and will have a density of 112 lb per sy-in. A vibratory roller with a 60-in.-wide drum will be used for compaction. Assume a 50-min hour efficiency factor for the roller. The overlap between adjacent laps and overedges will be 6-in. minimum. It is estimated that nonproductive travel will add about 15% to total travel. From a test mat it was found that three passes with the roller are required to achieve density. To account for acceleration and deceleration when changing directions, add 10% to the average roller speed to calculate a running speed. What is the required roller speed if only one roller is available?

15.4 An asphalt distributor having an 8-ft-long spray bar will be used to apply a prime coat at a rate of 0.3 gal per sy. The road being paved is 16 ft wide and 2,200 ft in length. How many gallons of prime will be required? (1,174 gal)

15.5 An asphalt distributor having a 10-ft-long spray bar will be used to apply a prime coat at a rate of 0.2 gal per sy. The road being paved is 10 ft wide and 5,200 ft in length. How many gallons of prime will be required?

15.6 An asphalt plant will produce 180 tph for a highway project. The mat will be 12 ft wide and $1\frac{1}{2}$ in. thick, and will have a density of 115 lb per sy-in. A vibratory roller with a 66-in.-wide drum will be used for compaction. Assume a 50-min hour efficiency factor for the roller. The overlap between adjacent laps and overedges will be 6-in. minimum. It is estimated that nonproductive travel will add about 15% to total travel. From a test mat, it was found that three passes with the roller are required to achieve density. To account for acceleration and deceleration when changing directions, add 10% to the average roller speed to calculate a running speed. If it is desired to keep the rolling speed under 3 mph, how many rollers are required for the project?

15.7 Cutback asphalt is a mixture of asphalt cement and
   a. Water
   b. Aggregate
   c. A volatile

15.8 Asphalt pavements should contain
   a. No air voids
   b. Some air voids
   c. As many air voids as possible

REFERENCES

1. *Asphalt Construction Handbook,* 6th ed., Barber-Greene, DeKalb, Ill., 1992.

2. *Asphalt Hot-Mix Recycling (MS-20),* Asphalt Institute, Lexington, Ky.

3. *Asphalt Paving Manual for Rubber-Tired Pavers,* Caterpillar Inc., Peoria, Ill., 1987.

4. *Asphalt Plant Manual (MS-3),* Asphalt Institute, Lexington, Ky., 1986.

5. "Asphalt Recycling and Reclaiming '87, Special Report Roads & Bridges," *Roads & Bridges,* October 1987.

6. Astec Inc., P. O. Box 72787, Chattanooga, TN 37407, www.astecinc.com.

7. "Buckets without Borders," *ENR,* Jan. 11, 1999.

8. CMI Corporation, P. O. Box 1985, Oklahoma City, OK 73101. www.cmicorp.com.

9. *Construction of Hot-Mix Asphalt Pavements (MS-22),* 2nd ed., Asphalt Institute, Lexington, Ky.

10. *Design of Hot Mix Asphalt Pavements,* National Asphalt Pavement Association, Lanham, Md., 1991.

11. Foundation for Pavement Preservation, 2025 M Street, NW, Suite 800, Washington, DC 20036, www.pavementpreservation.org.

12. *Hot-Mix Asphalt Paving Handbook,* U.S. Army Corps of Engineers, UN-13(CEMP-ET), July 1991.

13. *Mix Design Methods for Asphalt Concrete and Other Hot-Mix Types (MS-2),* 6th ed., Asphalt Institute, Lexington, Ky.

14. "New Approach Flattens Rollers," *ENR,* February 2, 1998.

15. *OSHA Compliance Manual for Hot Mix Plants,* National Asphalt Pavement Association, Riverdale, Md.

16. Roberts, F. L., P. S. Kandhal, E. R. Brown, D. Lee, and T. W. Kennedy, *Hot Mix Asphalt Materials, Mixtures, Design and Construction,* 2nd ed., NAPA Research and Educational Foundation, Lanham, Md., 1996.

17. Roadtec, 800 Manufacturers Rd., Chattanooga, TN 37405, www.roadtec.com.

18. Rosco Manufacturing Company, P.O. Box B, Madison, SD 57042. www.roscomfg.com.

19. Scherocman, James A., "Compacting Hot-Mix Asphalt Pavements, Part I," *Roads & Bridges,* August 1996.

20. Scherocman, James A., "Compacting Hot-Mix Asphalt Pavements, Part II," *Roads & Bridges,* September 1996.

ASPHALT-RELATED WEBSITES

1. Asphalt Institute: www.asphaltinstitute.org. The Asphalt Institute is a United States–based association of international petroleum asphalt producers, manufacturers, and affiliated businesses. The Asphalt Institute's mission is to promote the use, benefits, and quality performance of petroleum asphalt through engineering, research, and educational activities and through the resolution of issues affecting the industry.

2. U.S. Army Corps of Engineers: www.erdc.usace.army.mil/. Links are provided to the engineering laboratory at the Waterways Experiment Station and the Cold Regions Research and Engineering Laboratory.

3. Asphalt Pavement Alliance: www.asphaltalliance.com. The Asphalt Pavement Alliance is an industry coalition of Asphalt Institute, National Asphalt Pavement Association, and the State Asphalt Pavement Associations. The Alliance is dedicated to enhancing our nation's roads and highways through programs of

education, technology development, and technology transfer on topics relating to hot mix asphalt pavements.

4. National Asphalt Pavement Association (NAPA): www.hotmix.org. The National Asphalt Pavement Association is the national trade association representing the interests of the hot mix asphalt (HMA) industry. NAPA is dedicated to the proposition that high-performance hot mix asphalt pavements are in the best interests of the nation and its mobility as expressed in its world-recognized system of roads, streets, and highways.

5. Asphalt Recycling & Reclaiming Association (ARRA): www.arra.org. Asphalt Recycling & Reclaiming Association members are an integral part of the effort to provide a safe, cost-efficient, and comprehensive network of roads and highways. ARRA presents a wide range of local, regional, and national seminars and conferences to promote the industry and the improvements in technology.

6. International Society for Asphalt Pavements (ISAP): www.asphalt.org. ISAP promotes technical advances in asphalt paving.

16

Concrete Equipment

Concrete consists of Portland cement, water, and aggregate that have been mixed together, placed, consolidated, and allowed to solidify and harden. There are two types of concrete-mixing operations in use: (1) transit mixing and (2) central mixing. Unless the project is in a remote location or is relatively large, the concrete is batched in a central batch plant and transported to the job site in transit-mix trucks. Central-mixed concrete is mixed completely in a stationary mixer and transported to the project in either a truck agitator, a truck mixer operating at agitating speed, or a nonagitating truck.

INTRODUCTION

The Assyrians in 690 B.C. used a mixture of one part lime, two parts sand, and four parts limestone aggregate to create a crude concrete used in building the Bavian canal. In 1824 Joseph Aspdin took out a patent in England on "Portland" cement. That patent marks the beginning of concrete, as we know it today. This manufactured cement consists of limestone and clay burned at temperatures in excess of 2,700°F, and is termed hydraulic in that it will react with water and harden under water. Concrete became widely used in Europe during the late 1800s and was brought to the United States late in that century. Its use continued to spread rapidly as knowledge about it and experience with it grew.

Portland cement concrete is one of the most widely used structural materials in the world for civil works projects (see Fig. 16.1). Its versatility, economy, adaptability, and worldwide availability, and especially its low maintenance requirements, make it an excellent building material. The term "concrete" is applicable for many products but is generally used with Portland cement concrete. It consists of Portland cement, water, and aggregate that have been mixed together, placed, consolidated, and allowed to solidify and harden. The Portland cement and water form a paste, which acts as the glue or binder. When fine aggregate is added (aggregate whose size range lies between the No. 200 mesh

FIGURE 16.1 | Concreting operations during reconstruction of Roosevelt Dam, Arizona.

sieve and the No. 4 sieve), the resulting mixture is termed mortar. Then when coarse aggregate is included (aggregate sizes larger than the No. 4 sieve) concrete is produced. Normal concrete consists of about three-fourths aggregate and one-fourth paste, by volume. The paste usually consists of water-cement ratios between 0.4 and 0.7 by weight. Admixtures are sometimes added for specific purposes, such as to entrain numerous microscopic air bubbles, to improve workability, to impart color, to retard the initial set of the concrete, and to waterproof the concrete.

The operations involved in the production of concrete will vary with the end use of the concrete, but, in general, the operations include

1. Batching the materials.
2. Mixing.
3. Transporting.
4. Placing.
5. Consolidating.
6. Finishing.
7. Curing.

PROPORTIONING CONCRETE MIXTURES

For successful concrete utilization the mixture must be properly proportioned. The American Concrete Institute (ACI) has a number of excellent recommended practices, including one on proportioning concrete mixtures [1]. Detailed treatment of concrete proportioning is beyond the scope of this text, but some practical considerations include

1. Although it takes water to initiate the hydraulic reaction, as a general rule the higher the water-cement ratio, the lower the resulting strength and durability.
2. The more water that is used (which is not to be confused with the water-cement ratio), the higher will be the slump.
3. The more aggregate that is used, the lower the cost of the concrete.
4. The larger the maximum size of coarse aggregate, the less the amount of cement paste that will be needed to coat all the particles and to provide necessary workability.
5. The more the fresh concrete is consolidated the stronger and more durable it becomes.
6. The use of properly entrained air enhances almost all concrete properties with little or no decrease in strength if the mix proportions are adjusted for the air.
7. The surface abrasion resistance of the concrete is almost entirely a function of the properties of the fine aggregate.

FRESH CONCRETE

To the designer, fresh concrete is usually of little importance. To the constructor, fresh concrete is *all-important,* because it is the fresh concrete that must be mixed, transported, placed, consolidated, finished, and cured. To satisfy both the designer and the constructor, the concrete should [2]

1. Be easily mixed and transported.
2. Have minimal variability throughout, both within a given batch and between batches.
3. Be of proper workability so that it can be consolidated, will completely fill the forms, will not segregate, and will finish properly.

To the constructor, workability is an important concrete property. Workability is difficult to define. Like the terms "warm" and "cold," workability depends on the situation. One measure of workability is slump, which is a pseudo measurable value based on an American Society for Testing and Materials standard test (ASTM C143) [3]. The test is very simple to perform. Fresh concrete is placed into a hollow frustrum of a cone, 4 in. in diameter at the top, 8 in. in diameter at the bottom, and 12 in. high. After filling according to a prescribed procedure, the cone is raised from the concrete, allowing the fresh concrete to "slump" down. The amount of slump is measured in inches (or millimeters) from its original height of 12 in., with the stiffest concrete having zero slump and the most fluid concrete having slumps in excess of 8 in. Although the slump measures only one attribute of workability, the flowability of fresh concrete, it is the most widely used measure. Table 16.1 gives the recommended slumps for various types of concrete construction [4].

TABLE 16.1 | Recommended slumps for various types of construction (ACI 211.1)[4]

Types of construction	Slump (in.)	
	Maximum	**Minimum**
Reinforced foundation walls and footing	3	1
Plain footings, caissons, and substructure walls	3	1
Beams and reinforced walls	4	1
Building columns	4	1
Pavements and slabs	3	1
Mass concrete	2	1

BATCHING CONCRETE MATERIALS

Most concrete batches, although designed on the basis of absolute volumes of the ingredients, are ultimately controlled in the batching process on the basis of weight. Therefore, it is necessary to know the weight-volume relationships of all the ingredients. Then each ingredient must be accurately weighed if the resulting mixture is to have the desired properties. It is the function of the batching equipment to perform this weighing measurement.

Cement

For most large projects, the cement is supplied in bulk quantities from cement transport trucks, each holding 25 tons or more, or from railroad cars. Bulk cement usually is unloaded by air pressure from rail cars or special truck trailers and stored in overhead silos or bins. Cement may be supplied in paper bags, each containing 1 cu ft (cf) loose measure and weighing 94 lb net. Bag cement must be stored in a dry place on pallets and should be left in the original bags until used for concrete.

Water

The water that is mixed with the cement to form a paste and to produce hydration must be free from all foreign materials. Organic material and oil may inhibit the bond between the hydrated cement and the aggregate. Many alkalies and acids react chemically with the cement and retard normal hydration. The result is a weakened paste, and the contaminating substance is likely to contribute to deterioration or structural failure of the finished concrete. Required water properties are cleanliness and freedom from organic material, alkalies, acids, and oils. In general, water that is acceptable for drinking may be used for concrete.

Aggregates

To produce concrete of high quality, aggregates should be clean, hard, strong, durable, and round or cubical in shape. The aggregates should be resistant to abrasion from weathering and wear. Weak, friable, or laminated particles of aggregate, or aggregate that is too absorptive, are likely to cause deterioration of the concrete.

Moisture Condition   For determination of mix proportions, the aggregate should be in a saturated surface dry condition, or an adjustment must be made in the water-cement ratio to compensate for the different amount of water contained in the aggregate.

Mix Proportions

The mix specifications will define specific requirements for the materials that constitute the desired concrete product. Typical requirements include:

1. Maximum size aggregate (i.e., $1\frac{1}{2}$ in.).
2. Minimum cement content (sacks per cubic yard or pound per cubic yard).
3. Maximum water–cement ratio (by weight or in gallons per sack of cement).

When computing the material quantities for a concrete mix, these are useful factors and information:

1. The average specific gravity of cement is 3.15.
2. The average specific gravity of coarse or fine aggregate is 2.65.
3. Water weighs 62.4 lb per cf.
4. One cubic foot of water equals 7.48 U.S. gallons.
5. One gallon of water weighs 8.33 lb.
6. Usually the proportion of fine aggregate varies between 25 and 45% of total aggregate volume.

The absolute volume of any ingredient in cu ft is expressed by

$$\text{Volume (cf)} = \frac{\text{Weight of the ingredient (lb)}}{\text{Specific gravity of the ingredient} \times 62.4 \text{ lb/cf}} \quad \textbf{[16.1]}$$

Example 16.1 illustrates a method of calculating quantities of cement, aggregate, and water.

EXAMPLE 16.1

Determine the quantities of materials required per cubic yard to create a concrete mix. The specifications require a maximum size aggregate of $1\frac{1}{2}$ in., a minimum cement content of 6 sacks per cy, and a maximum water–cement ratio of 0.65.

6 sacks $\times$ 94 lb/sack = 564 lb cement per cy

$$\text{Volume cement} = \frac{564 \text{ lb/cy}}{3.15 \times 62.4 \text{ lb/cf}} \Rightarrow 2.87 \text{ cf per cy}$$

Water per cubic yard: 564 lb $\times$ 0.65 = 366.6 lb

$$\text{Volume water} = \frac{366.6 \text{ lb/cy}}{1.0 \times 62.4 \text{ lb/cf}} \Rightarrow 5.88 \text{ cf per cy}$$

Assume 6% air voids.

Volume of air: 27.0 cf/cy $\times$ 0.06 = 1.62 cf per cy

Volume of aggregate:

27.0 − 2.87 (cement) − 5.88 (water) − 1.62 (air) = 16.63 cf per cy

Volume of sand: assume fine aggregate 35%

0.35 × 16.63 cf per cy = 5.82 cf per cy

Weight of fine aggregate:

5.82 cf per cy × 2.65 × 62.4 lb/cf = 962.4 lb per cy

Volume of coarse aggregate:

16.63 − 5.82 = 10.81 cf per cy

Weight of coarse aggregate:

10.81 cf per cy × 2.65 × 62.4 lb/cf = 1,787.5 lb per cy

Batching Control

Usually, concrete specifications require the concrete to be batched with aggregate having at least two size ranges (coarse and fine), but up to six ranges can be required. Aggregate from each size range must be accurately measured. The aggregate, water, cement, and admixtures (if used) are introduced into a concrete mixer and mixed for a suitable period of time until all the ingredients are adequately blended together. Most modern plants have performance data on their mixers to show that they can adequately mix an 8-cy batch of concrete in 1 min.

To control the batching, close tolerances are maintained. The Concrete Plant Manufacturers Bureau publishes the Concrete Plant Standards (CPMB 100-96), which outlines plant tolerances [6]. Batching controls are that part of the batching equipment that provides the means of controlling the batching device for an individual material. They may be mechanical, hydraulic, pneumatic, electrical, or a combination of these means. Table 16.2 presents the permissible tolerances in accordance with CPMB 100-96. Batch plants are available in three categories: (1) manual, (2) semiautomatic, and (3) fully automatic.

Manual Controls   Manual batching is when batching devices are activated manually with accuracy of the batching operation being dependent on the operator's visual observation of a scale or volumetric indicator. The batching devices may be actuated by hand or by pneumatic, hydraulic, or electrical power assists.

Semiautomatic Controls   When activated by one or more starting mechanisms, a semiautomatic batcher control starts the weighing operation of each material and stops automatically when the designated weight has been reached.

Automatic Controls   When activated by a single starting signal, an automatic batcher control starts the weighing operation of each material and stops automatically when the designated weight of each material has been reached and interlocked in such a manner that the

1. Charging device cannot be actuated until the scale has returned to zero balance.

TABLE 16.2 | Typical batching tolerances (CPMB 100-96)

Ingredient	Individual batchers and cumulative batchers with a tare compensated control	Cumulative batchers without a tare compensated control
Cement and other cementitious materials*	±1% of the required weight of materials being weighed or ±0.3% of scale capacity, whichever is greater	±1% of the required cumulative weight of materials being weighed or ±0.3% of scale capacity, whichever is greater
Aggregates	±2% of the required weight of material being weighed or ±0.3% of scale capacity, whichever is greater	±1% of the required cumulative weight of materials being weighed or ±0.3% of scale capacity, whichever is greater
Water	±1% of the required weight of material being weighed or ±0.3% of scale capacity, whichever is greater	
Admixtures	±3% of the required weight of material being weighed or ±0.3% of scale capacity, or ± the minimum dosage rate per 100 lb (45.4 kg) of cement, whichever is greater	±3% of the required cumulative weight of material being weighed or ±0.3% of scale capacity, or ± the minimum dosage rate per 100 lb (45.4 kg) of cement as it applies to each admixture, whichever is greater

*Other cementitious materials are considered to include fly ash, ground granulated blast furnace slag, and other natural or manufactured pozzolans.

2. Charging device cannot be actuated if the discharge device is open.

3. Discharge device cannot be actuated if the charging device is open.

4. Discharge device cannot be actuated until the indicated material is within the applicable tolerances.

MIXING CONCRETE

There are two types of concrete-mixing operations in use: (1) transit mixing and (2) central mixing (see Fig. 16.2). Today, unless the project is in a remote location or is relatively large, the concrete is batched in a central batch plant and transported to the job site in transit-mix trucks, often referred to as *ready-mixed* concrete trucks (see Fig. 16.3). This type of concrete is controlled by ASTM specification C94 [5], and there is a national organization promoting its use (National Ready Mixed Concrete Association [Website 9]).

Ready-Mixed Concrete

Increasingly, concrete is proportioned at a central location and transported to the purchaser in a fresh state, mixed en route. This type of concrete is termed "ready-mixed concrete" or "truck-mixed concrete." It is concrete that is completely mixed in a truck mixer (see Fig. 16.3), with 70 to 100 revolutions at a speed sufficient to mix the concrete completely. Obviously, to be useful, ready-mixed concrete must be available within a reasonable distance from the project. At remote locations and locations requiring large quantities of concrete, generally concrete plants are set up on site.

FIGURE 16.2 | Central-mixed concrete plant, mixer drum is on the far right end of the plant.

FIGURE 16.3 | Transit-mix truck transporting ready-mixed concrete.

The specifications for the batch plant and the transit-mixer transport trucks are covered in detail in ASTM C94 [5]. Of particular importance is the elapsed time from the introduction of water to the placement of the concrete. ASTM C94 allows a maximum of $1\frac{1}{2}$ hours, or before the drum has revolved 300 revolutions, whichever comes first.

FIGURE 16.4 | A large transit-mix truck for transporting ready-mixed concrete.

Transit mixers are available in several sizes up to about 14 cy (see Fig. 16.4), but the most popular size is 8 cy. They are capable of thoroughly mixing the concrete with about 100 revolutions of the mixing drum (see Fig. 16.5). Mixing speed is generally 8 to 12 rpm. This mixing during transit usually results in stiffening the mixture, and ASTM C94 allows the addition of water at the job site to restore the slump, followed by remixing. This has caused problems and raised questions concerning the uniformity of ready-mixed concrete. ACI 304 [1] recommends that some of the water be withheld until the mixer arrives at the project site (especially in hot weather), then the remaining water be added and an additional 30 revolutions of mixing be required. To offset any stiffening, small amounts of additional water are permitted, *provided the design water-cement ratio is not exceeded.* The uniformity requirements of ready-mixed concrete are given in Table 16.3.

Ready-mixed concrete can be ordered in several ways, including

1. *Recipe batch.* The purchaser assumes responsibility for proportioning the concrete mixture, including specifying the cement content, the maximum allowable water content, percentage air, and the admixtures required. The purchaser may also specify the amounts and type of coarse and fine aggregate. Under this approach, the purchaser assumes full responsibility for the resulting strength and durability of the mixture, provided that the stipulated amounts are furnished as specified.

2. *Performance batch.* The purchaser specifies the requirements for the strength of the concrete, and the manufacturer assumes full responsibility for the proportions of the various ingredients that go into the batch.

3. *Part performance and part recipe.* The purchaser generally specifies a minimum cement content, the required admixtures, and the strength requirements, allowing the producer to proportion the concrete mixture within the constraints imposed.

FIGURE 16.5 | Sectional view through the drum of a transit-mix truck.

TABLE 16.3 | Uniformity requirements for ready-mixed concrete (ASTM C94) [5]

Test	Requirements, expressed as maximum permissible difference in results of tests of samples taken from two locations in the concrete batch
Weight per cubic foot calculated to an air-free basis	1.0 lb/cf is greater
Air content, volume percent of concrete	1.0%
Slump	
If average slump is 4 in. or less	1.0 in.
If average slump is 4 to 6 in.	1.5 in.
Coarse aggregate content, portion by weight retained on No. 4 sieve	6.0%
Unit weight of air-free mortar based on average for all comparative samples tested	1.6%
Average compressive strength at 7 days for each sample, based on average strength of all comparative test specimens	7.5%

Today most purchasers of concrete use the third approach, part performance and part recipe, as it ensures a minimum durability while still allowing the ready-mixed concrete supplier some flexibility to supply the most economical mixture.

Central-Mixed Concrete

This is concrete mixed completely in a stationary mixer and transported to the project in either a truck agitator, a truck mixer operating at agitating speed, or a nonagitating truck. Plants usually have mixers capable of mixing up 8 cy of concrete in each batch (although plants have been built with mixers capable of mixing 15 cy of concrete in each batch) and can produce more than 200 cy of concrete per hour. The mixer either tilts to discharge the concrete into a truck or a chute is inserted into the mixer to catch and discharge the concrete. To increase efficiency, many large plants have two mixer drums.

In determining the quantities needed and the output for a given plant, one should include any delays in productivity resulting from reduced operating factors.

PLACING CONCRETE

Once the concrete arrives at the project site, it must be moved to its final position without segregation and before it has achieved an initial set. This movement may be accomplished in several ways, depending on the horizontal and vertical distance of the movement and other constraints that might be imposed. Methods include buckets or hoppers, chutes and drop pipes, belt conveyors, and concrete pumps.

Buckets

Normally, properly designed bottom-dump buckets (see Fig. 16.6) permit concrete placement at the lowest practical slump. Care should be exercised to prevent the concrete from segregating as a result of discharging from too high above the surface or allowing the fresh concrete to fall past obstructions. Gates should be designed so that they can be opened and closed at any time during the discharge of the concrete.

Manual or Motor-Propelled Buggies

Hand buggies and wheelbarrows are usually capable of carrying from 4 to 9 cf of concrete, and thus are suitable on many projects, provided there are smooth and rigid runways upon which to operate. Hand buggies are safer than wheelbarrows because they have two wheels rather than one. Hand buggies and wheelbarrows are recommended for distances less than 200 ft, whereas power-driven or motor-driven buggies—with capacities up to around 14 cf—can traverse up to 1,000 ft economically.

Chutes and Drop Pipes

Chutes are often used to transfer concrete from a higher elevation to a lower elevation. They should have a round bottom, and the slope should be steep enough

FIGURE 16.6 | Concrete bucket being used for the placement of concrete in formwork.

for the concrete to flow continuously without segregation. Drop pipes are used to transfer the concrete vertically down. The top 6 to 8 ft of the pipe should have a diameter at least 8 times the maximum aggregate size and may be tapered so that the lower end is approximately 6 times the maximum aggregate size [1]. Drop pipes are used when concrete is placed in a wall or column to avoid segregation caused by allowing the concrete to free-fall through the reinforcement.

Belt Conveyors

Belt conveyors can be classified into three types: (1) portable or self-contained conveyors, (2) feeders or series conveyors, and (3) side-discharge or spreader conveyors. All types provide for the rapid movement of fresh concrete but must have proper belt size and speed to achieve the desired rate of transportation. Particular attention must be given to points where the concrete leaves one conveyor and either continues on another conveyor or is discharged, as segregation can occur. The optimum concrete slump for conveyed concrete is from $2\frac{1}{2}$ to 3 in. [4]. Figure 16.7 shows a belt conveyor mounted on a track carrier being used to place low-slump roller-compacted concrete. The carrier was fed concrete from the plant by a belt conveyor system.

Concrete Pumps

The placement of concrete through rigid or flexible lines is not new. However, pumping was not used extensively until the 1930s when German pumping

FIGURE 16.7 | Low-slump concrete being placed by a belt conveyor mounted on a track carrier.

equipment was introduced in this country. The pump is an extremely simple machine. By applying pressure to a column of fresh concrete in a pipe, the concrete can be moved through the pipe if a lubricating outer layer is provided and if the mixture is properly proportioned for pumping. To work properly, the pump must be fed concrete of uniform workability and consistency. Today concrete pumping is one of the fastest growing specialty contracting fields in the United States, as perhaps one-fourth of all concrete is placed by pumping. Pumps are available in a variety of sizes, capable of delivering concrete at sustained rates of 10 to 150 cy per hr. Effective pumping range varies from 300 to 1,000 ft horizontally, or 100 to 300 ft vertically [8], although occasionally pumps have moved concrete more than 5,000 ft horizontally and 1,000 ft vertically.

Pumps require a steady supply of *pumpable* concrete to be effective. Today there are three types of pumps being manufactured: (1) piston pumps, (2) pneumatic pumps, and (3) squeeze pressure pumps. They are shown diagrammatically in Figs. 16.8a, 16.8b, and 16.8c, respectively. Most piston pumps currently contain two pistons, with one retracting during the forward stroke of the other to give a more continuous flow of concrete. The pneumatic pumps normally use a reblending discharge box at the discharge end to bleed off the air and to prevent segregation and spraying. In the case of squeeze pressure pumps, hydraulically powered rollers rotate on the flexible hose within the drum and squeeze the concrete out at the top. The vacuum keeps a steady supply of concrete in the tube from the receiving hopper.

Pumps can be mounted on trucks, trailers, or skids. The truck-mounted pump and boom combination is particularly efficient and cost-effective in saving labor and eliminating the need for pipelines to transport the concrete.

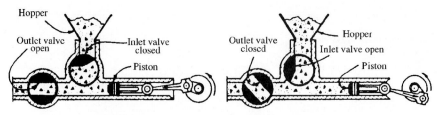

Inlet valve opens while outlet valve is closed and concrete is drawn into cylinder by gravity and piston suction. As piston moves forward inlet valve closes, outlet valve opens, and concrete is pushed into pump line.

(*a*)

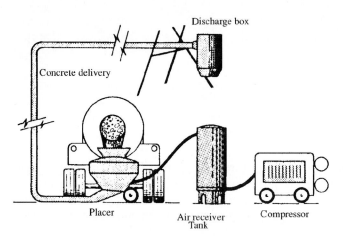

Compressor builds up air pressure in tank, which forces concrete in placer through the line.

(*b*)

Vacuum maintained in pumping chamber returns tube to normal shape, aiding in continous flow of concrete.

Rotating rollers

Material hose

Planetary drive

Collecting hopper

Rollers press concrete through tube into material hose.

Pumping tube

Rotating blades push concrete toward pumping tube.

(*c*)

FIGURE 16.8 | (a) Schematic drawing of a piston-type concrete pump (from ACI 304.2R) [8]. (b) Schematic drawing of a pneumatic-type concrete pump (from ACI 304.2R) [8]. (c) Schematic drawing of a squeeze pressure-type concrete pump (from ACI 304.2R) [8].

Hydraulically operated and articulated, booms come in lengths up to 100 ft but there are a few larger units (see Fig. 16.9).

Successful pumping of concrete is no accident. A common fallacy is to assume that any placeable concrete will pump successfully. The basic principle of pumping is that the concrete moves as a cylinder of concrete. To pump concrete successfully, a number of rules should be carefully followed. These are

1. Use a minimum cement factor of 517 lb of cement per cubic yard of concrete ($5\frac{1}{2}$ sacks per cy).

2. Use a combined gradation of coarse and fine aggregate that ensures *no* gaps in sizes that will allow paste to be squeezed through the coarser particles under the pressures induced in the line. This is the most often overlooked aspect of good pumping! In particular, it is important for the fine aggregate to have at least 5% passing the No. 100 sieve and about 3% passing the No. 200 sieve (see gradations given in reference [8]). Line pressures of 300 psi are common, and they can reach as high as 1,000 psi.

3. Use a minimum pipe diameter of 5 in.

4. Always lubricate the line with cement paste or mortar before beginning the pumping operating.

5. Ensure a steady, uniform supply of concrete, with a slump of between 2 and 5 in. as it enters the pump.

6. Always presoak the aggregates before mixing them in the concrete to prevent their soaking up mix water under the imposed pressure. This is especially important when aggregates are used that have a high absorption (such as structural lightweight aggregate).

FIGURE 16.9 | Pumping concrete in front of the screed during bridge deck construction.

7. Avoid the use of reducers in the conduit line. One common problem is the use of a 4- to 5-in. reducer at the discharge end so that workers will have only a 4-in. flexible hose to move around. This creates a constriction and significantly raises the pressure necessary to pump the concrete.

8. Never use aluminum lines. Aluminum particles will be scraped from the inside of the pipe as the concrete moves through and will become part of the concrete. Aluminum and Portland cement react, liberating hydrogen gas that can rupture the concrete—with disastrous results.

CONSOLIDATING CONCRETE

Concrete, a heterogeneous mixture of water and solid particles in a stiff condition, will normally contain a large quantity of voids when placed. The purpose of consolidation is to remove these entrapped air voids. The importance of proper consolidation cannot be overemphasized, as entrapped air can render the concrete totally unsatisfactory. Entrapped air can be reduced in two ways—use more water or consolidate the concrete. Figure 16.10 shows qualitatively the benefits of consolidation, especially on low-water-content concrete.

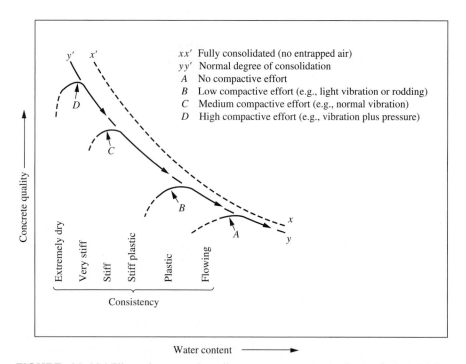

FIGURE 16.10 | Effect of compaction effort on concrete quality (from ACI 309R) [9].

Consolidation is normally achieved through the use of mechanical vibrators. There are three general types [9]: (1) internal, (2) surface, and (3) form vibrators. Internal, or spud vibrators as they are often called, have a vibrating casing or head

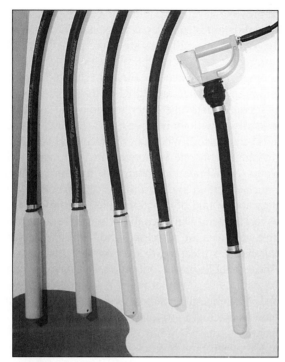

FIGURE 16.11 | Different size handheld concrete vibrators.

that is immersed into the concrete and vibrates at a high frequency (often as high as 10,000 to 15,000 vibrations per min) against the concrete. Currently these vibrations are the rotary type and come in sizes from $\frac{3}{4}$ in. to 7 in. (see Fig. 16.11), each with an effective radius of action [9]. They are powered by electric motors or compressed air.

Surface vibrators exert their effects at the top surface of the concrete and consolidate the concrete from the top down. They are used mainly in slab construction, and there are four general types: (1) the vibrating screed, (2) the pan-type vibrator, (3) the plate or grid vibratory tamper, and (4) the vibratory rolling screed. These surface vibrators operate in the range from 3,000 to 6,000 vibrations per min.

Form vibrators are external vibrators attached to the outside of the form or mold. They vibrate the form that in turn vibrates the concrete. These types of vibrators are generally used in large precast concrete plants.

Recommended Vibration Practices

Internal vibration is generally best suited for ordinary construction provided the section is large enough for the vibrator to be manipulated. As each vibrator has an effective radius of action, vibrator insertions should be vertical at about $1\frac{1}{2}$ times the radius of action. The vibrator should never be used to move concrete laterally, as segregation can easily occur. The vibrator should be rapidly inserted to the bottom of the layer (usually 12- to 18-in.-maximum-lift thickness) and at least 6 in. into the previous layer. It should then be held stationary for about 5 to 15 sec until the consolidation is considered adequate. The vibrator should then be withdrawn slowly. Where several layers are being placed, each layer should be placed while the preceding layer is still plastic.

Vibration accomplishes two actions. First, it "slumps" the concrete, removing a large portion of air that is entrapped when the concrete is deposited. Then continued vibration consolidates the concrete, removing most of the remaining entrapped air. Generally, it will not remove entrained air. The question concerning overvibration is often raised: When does it occur and how harmful is it? The fact is that on low-slump concrete (concrete with less than a 3-in. slump) it is almost impossible to overvibrate the concrete with internal vibrators. When in doubt as to how much vibration to impart to low-slump concrete, vibrate it some more. The same cannot be said of concrete whose slump is 3 in. or more. This concrete can be overvibrated, which results in segregation as a result of coarse aggregate moving away from the vibrating head. Here the operator should note the presence of air bubbles escaping to the concrete surface as the vibrator is inserted. When these bubbles cease, vibration is generally complete and the vibrator should be withdrawn. Another point of caution concerns surface vibrators. They too can overvibrate the concrete at the surface, significantly weakening it if they remain in one place too long.

Another concern is the vibration of reinforcing steel. Such vibration *improves* the bond between the reinforcing steel and the concrete, and thus is desirable. The undesirable side effects include damage to the vibrator and possible movement of the steel from its intended position.

Finally, *revibration* is the process whereby the concrete is vibrated again after it has been allowed to remain undisturbed for some time. Such revibration can be accomplished at any time the running vibrator will sink of its own weight into the concrete and liquefy it momentarily [9]. Such revibration will improve the concrete through increased consolidation.

FINISHING AND CURING CONCRETE

It cannot be stated too strongly that *any* work done to a concrete surface after it has been consolidated will weaken the surface. All too often, concrete finishers overlook this fact and manipulate the surface of the concrete, sometimes even adding water, to produce a smooth, attractive surface. On walls and columns, an attractive surface may be desirable and the surface strength may not be too important, but on a floor slab, sidewalk, or pavement the surface strength is very important. On the latter types of surfaces, only the absolute minimum finishing necessary to impart the desired texture should be permitted, and the use of "jitterbugging" (the forcing of coarse aggregate down into the concrete with a steel grate tool) should not be permitted, as the surface can be weakened significantly. Furthermore, each step in the finishing operation, from first floating to the final floating or troweling, should be delayed as long as possible. This duration is limited by the necessity to finish the concrete to the desired grade and surface smoothness while it can still be worked (still in a plastic state). In no case should finishing commence if any free bleed water has not been blotted up, nor should neat cement or mixtures of sand and cement be worked into such surfaces to dry them up.

Along with placement and consolidation, proper curing of the concrete is extremely important. Curing encompasses all methods whereby the concrete is assured of adequate time, temperature, and supply of water for the cement to continue to hydrate. The time normally required is 3 days, and optimum temperatures are between 40 and 80°F. As most concrete is batched with sufficient water for hydration, the only problem is to ensure that the concrete does not dry out. This can be accomplished by ponding with water (for slabs), covering with burlap or polyethylene sheets, or spraying with an approved curing compound. Curing is one of the least costly operations in the production of quality concrete, and one that is all too frequently overlooked. Concrete, if allowed to dry out during the curing stage, will attempt to shrink. The developing bonds from the cementitious reaction will attempt to restrain the shrinkage from taking place. But the end result is *always* the same: The shrinkage wins out and a crack forms, as the shrinkage stresses are always higher than the tensile strength of the concrete. Proper curing does reduce the detrimental effects of cracking and develops the intended strength of the concrete.

SLIPFORM PAVING

Slipform pavers (see Fig. 16.12) perform the functions of spreading, vibrating, striking off, consolidating, and finishing the concrete pavement to the prescribed cross section and profile with a minimum of handwork required. The name "slipform" is derived from the fact that the side forms of the machine slide forward with the paver and leave the slab edges unsupported. Slipform paving has definitely been proven as a speedy and economical method for producing smooth and durable concrete pavement.

FIGURE 16.12 I Slipform paver spreading, vibrating, striking off, consolidating, and finishing concrete pavement.

The basic operating process consists of molding the plastic concrete to the desired cross section and profile under a single relatively large screed. This is accomplished with a full-width screed that is maintained at a predetermined elevation and cross slope with hydraulic jacks that are actuated through an automatic control system. The control is referenced to offset grade lines pre-erected parallel to the planned profile for each pavement edge. Concrete is delivered into the front of the paver. It is then internally vibrated as it flows under the machine.

Some pavers operate with the concrete deposited directly on the base (see Fig. 16.13a) and spread it transversely with an auger before striking it off with a moldboard. A considerable amount of continuously reinforced concrete pavement is constructed throughout the country. The general practice, in both formed and slipformed paving of continuously reinforced pavements, is to place the steel on the base supported at the proper height ahead of the paving operation (see Fig. 16.14). In that case slipform pavers use an attached belt conveyor (see Fig. 16.13b) to place the concrete in front of the paver as the trucks hauling the concrete cannot maneuver on the base. Most slipform pavers also have automatic dowel inserters that either push or vibrate the dowels into the fresh concrete in the exact position with minimum disturbance to the concrete. All pavers are equipped with a final float-finisher.

Pavement Joints

Typically joints are provided in the concrete pavement at intervals and having dimensions (depth and width) as specified in the project plans. Contraction joints in the transverse direction across the paving lanes are normally saw cut into the pavement. Manufacturers of saws provide (see Fig. 16.15) machines adaptable to any joint situation. Sawing of contraction joints is done as soon as it is possible to get on the new pavement without damaging the surface.

Longitudinal center joints are commonly formed by the use of a continuous polyethylene strip the paver disperses as it moves along. Expansion joints, when required, are generally installed by hand methods.

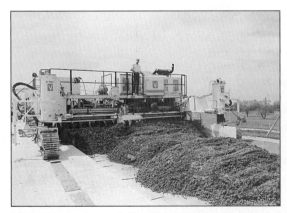

(a) Concrete dumped directly on the base ahead of the paver.

(b) Side conveyor used to place the concrete on the base ahead of the paver.

FIGURE 16.13 I Placement of concrete ahead of the slipform paver.

FIGURE 16.14 I Reinforcing steel placed on the subbase at the proper height ahead of the paving operation.

FIGURE 16.15 I Diamond-bladed saw for cutting pavement joints and freshly cut contraction joint. Note the cut is not made completely through the slab.

Curing Pavement Concrete

Curing is accomplished by spraying a waterproof membrane-curing compound on the pavement. Usually a separate machine follows the paver and applies the compound after all finishing is complete.

Smoothness

The ultimate inspector for a paving project is the driving public. Drivers are acutely aware of pavement smoothness or roughness. Agencies that purchase paving projects specify roadway smoothness and measure performance with "profileometer" (see Fig. 16.16). To construct smooth pavements requires attention to several controlling factors.

> *Concrete.* Many of the problems occurring in pavements concern the mix. The paver cannot cope with variations in slump and delivery of concrete. The concrete exerts forces on the paver as it spreads and forms the plastic mix; when those forces vary, smoothness is affected. A consistent head of material ahead of the paver is necessary for a quality project (a smooth pavement).
>
> *Workability.* Workability usually means slump, but other components affect workability. Entrained air, while adding to the durability of the concrete, also acts as a lubricating agent within the mix to reduce friction between particles, aiding workability.
>
> *Density.* Vibration is a vital part of the paving process, but vibration is not meant to be a transport medium to move the plastic concrete. Concrete

FIGURE 16.16 | A profileometer being used to measure pavement smoothness.

should be moved by auger, conveyor, or other mechanical means prior to vibration. The plastic mix should be placed uniformly in front of the paver.

Sensors. The automatic control system and the stringline that provides intelligence for the operation of the controls cannot be neglected. The stringline must be set accurately, maintained, and checked constantly. Sensor response must be adjusted to prevent too quick or too violent a response. A dampened response results in a smooth surface.

It is recommended that dual stringline operation of the paving process be used whenever possible. The smoothest pavement results are obtained by use of fixed elevation control on both sides of the paver—dual stringline.

Attitude. The machine attitude is the attack angle of the screed in relation to the concrete. If the angle of attack is forced to change due to "bulldozing" piles of concrete mix, improper sensor adjustment, or varying head of concrete, bumps will result. Such conditions cause uncontrolled lift and drop because of the hydraulic action of the plastic concrete.

Weight and traction. The principle that makes a slipform paver work is the consolidation of concrete in a confined space. Weight, therefore, is paramount to controlled consolidation of slipformed concrete. Paver weight needs to be spread over the width of the concrete. Traction is related to weight and power. Power must be sufficient to move the loaded machine and provide the necessary energy to the working tools.

ROLLER-COMPACTED CONCRETE

Roller-compacted concrete or RCC is a lean-mix concrete of zero-slump consistence. RCC is a design-flexible material that allows the use of local, usually less expensive, aggregates. The material was first used in the United States for dam construction, and then for paving large railroad, truck, and logging yards.

Typically, RCC is delivered to the site in dump trucks or scrapers, though belt conveyor systems have also been used [10]. It is spread with dozers or motor graders, and compacted with vibratory steel-drum rollers. Asphalt pavers have also been used on a few projects to place RCC.

PLACING CONCRETE IN COLD WEATHER

When concrete is placed in cold weather, some provision must be made to keep the concrete above freezing during the first few days after it has been placed. Specifications generally require that the concrete be kept at not less than 70°F for 3 days or not less than 50°F for 5 days after placement. ACI 306R contains guidelines for concreting in cold weather [16]. Preheating the water is generally the most effective method of providing the necessary temperature for placement.

PLACING CONCRETE IN HOT WEATHER

When the temperature of fresh concrete exceeds around 85 to 90°F, the resulting strength and durability of the concrete can be reduced. Therefore, most specifications require the concrete to be place at a temperature less than 90°F. When concrete is placed in hot weather, the ingredients should be cooled before mixing. ACI 305R contains guidelines for mixing and placing concrete in hot weather [17]. Methods of cooling include using ice instead of water in the mix and cooling the aggregate with liquid nitrogen.

SHOTCRETING

Shotcreting is mortar or concrete conveyed through a hose and pneumatically projected at high velocity onto a surface [11]. The force of the concrete impacting on the surface compacts the mixture. Usually, a relatively dry mixture is used, and thus it is able to support itself without sagging or sloughing, even for vertical and overhead applications. Shotcrete, or gunite, as it is more commonly known, is used most often in special applications involving repair work, thin layers, or fiber-reinforced layers. Figure 16.17 illustrates the use of shotcrete to line a pond.

There are two methods of producing shotcrete: (1) the dry-mix process and (2) the wet-mix process. The dry-mix process, in which the cement and damp sand are thoroughly mixed and carried to the nozzle, where water is introduced

FIGURE 16.17 | Using the shotcrete method to place concrete lining for a pond.
Source: Challenge-Cook Bros., Inc.

under pressure and intimately mixed with the cement and sand before being jetted onto the surface, has been used successfully for more than 50 years. The wet-mix process involves mixing all the ingredients including water before being delivered through a hose under pressure to the desired surface. Using the wet-mix process, aggregates up to $\frac{3}{4}$ in. in size have been shotcreted. With either process there is some rebound of the mortar or concrete, resulting in a loss of usually 5 to 10%.

FLY ASH

Fly ash is produced as a by-product from the burning of coal. It is the fine residue that would "fly" out the stack if it were not captured by environmental control devices. Modern coal burning plants pulverize the coal before burning it so that 100% will pass the No. 200 sieve. In the furnace, the coal is heated to in excess of 2,700°F, melting all the incombustibles. Because of the strong induced air currents in the furnace, the molten residue from the coal becomes spherical in shape and the majority is carried out at the top of the furnace. Captured in massive electrostatic precipitators or bag houses, the fly ash is extremely fine (often finer than Portland cement). Fly ash has been found to be an excellent mineral admixture in Portland cement concrete, improving almost all properties of the concrete. The Environmental Protection Agency has ruled that fly ash must be allowed in all concrete construction involving federal funds [12].

There are basically two types of fly ash: (1) class F from bituminous coal and (2) class C from subbituminous and lignitic coal. Their quality is governed by ASTM C618. Portland cement, when it combines with water, releases calcium hydroxide (the white streaks noticeable adjacent to exposed concrete around cracks). This calcium hydroxide does not contribute to strength or durability. By introducing fly ash, the calcium hydroxide and fly ash chemically combine in a process called pozzolanic action. It takes a relatively long time (compared with the cementitious action involved between cement and water), but the resulting concrete is stronger, less permeable, and more durable than before. And if the designer can wait for the strength development, fly ash may be used to replace part of the Portland cement, resulting in a lower-cost product.

In addition to strength and durability improvements, fly ash also improves the workability of concrete, primarily because of its spherical shape. It has been used to improve the pumpability of concrete mixes, and finishers report that fly ash concrete is easier to finish.

The previous discussion applies to both class F and class C fly ashes. However, *some* class C fly ashes possess significant cementitious properties themselves, probably because of their relatively high calcium contents. Thus, when used as an admixture, extremely high strengths can be obtained, as illustrated by the use of class C fly ash to achieve 7,500 psi concrete in the Texas Commerce Tower in Houston [14]. These fly ashes can replace significant portions of Portland cement with no detrimental effects, as demonstrated with 25% replacement percentages on several concrete pavements in Texas [15].

Fly ash poses two problems to the concrete designer and constructor. One is that while it has been shown to be a valuable addition to concrete and can result in significant cost saving, it is a by-product from the burning of coal. This means that it can vary in its properties from day to day, and thus a good quality management program is needed to ensure that only high-quality fly ash is used. Furthermore, because fly ash produced by different plants will vary in quality, it is mandatory that each ash be tested prior to its use to ensure that it has the desired quality. Second, adding fly ash means that the concrete batch has five major ingredients rather than four. The chances for a mistake being made in the batching are increased, making it desirable to increase the quality control efforts.

SUMMARY

To successfully produce concrete, the mixture must be properly proportioned. The mix specifications will define specific requirements for the materials that constitute the desired concrete product. Ready-mixed concrete or truck-mixed concrete is proportioned at a central location and transported to the purchaser in a fresh state. It is completely mixed in the truck mixer. Central-mixed concrete is completely mixed in a stationary mixer and transported to the project in either a truck agitator, a truck mixer operating at agitating speed, or a nonagitating truck.

Slipform pavers perform the functions of spreading, vibrating, striking off, consolidating, and finishing the concrete pavement to the prescribed cross section and profile with a minimum of handwork required. The basic operating process consists of molding the plastic concrete to the desired cross section and profile under screed. Critical learning objectives include:

■ An understanding of the fresh concrete properties that are important to the constructor.
■ An ability to calculate the quantities required to produce a concrete mix.
■ An understanding of the methods used to transport fresh concrete.
■ An understanding of the processes used to produce a concrete pavement.

These objectives are the basis for the problems that follow.

PROBLEMS

16.1 A concrete batch calls for the following quantities per cubic yard of concrete, based on saturated surface-dry conditions of the aggregate. Determine the required weights of each solid ingredient and the number of gallons of water required for a 6.5-cy batch. Also determine the wet unit weight of the concrete in pounds per cubic foot. (cement 3,666 lb; fine aggregate 9,230 lb; coarse aggregate 11,960 lb; water 1,841 lb; wet unit weight 152 pcf)

Cement	6.0	bags
Fine aggregate	1,420	lb
Coarse aggregate	1,840	lb
Water	34	gal

16.2 In Problem 16.1, assume that the fine aggregate contains 7% free moisture, by weight, and the coarse aggregate contains 3% free moisture by weight. Determine the required weights of cement, fine aggregate, coarse aggregate, and volume of added water for an 8-cy batch.

16.3 Determine the quantities of materials required per cubic yard to create a concrete mix. The specifications require a maximum size aggregate of 1 in., a minimum cement content of 5 sacks per cy, and a maximum water–cement ratio of 0.60. Assume 6% air voids. (Cement 470 lb per cy; water 4.52 cf per cy; fine aggregate 1,068 lb per cy; coarse aggregate 1,986 lb per cy)

16.4 Determine the quantities of materials required per cubic yard to create a concrete mix. The specifications require a maximum size aggregate of $1\frac{1}{2}$ in., a minimum cement content of $5\frac{1}{2}$ sacks per cy, and a maximum water–cement ratio of 0.63. Assume 5% air voids.

16.5 A concrete retaining wall whose total volume will be 735 cy is to be constructed by using job-mixed concrete containing the following quantities per cubic yard, based on surface-dry sand and gravel:

Cement	6 bags
Sand	1,340 lb
Gravel	1,864 lb
Water	33 gal

The sand and gravel will be purchased by the ton weight, including any moisture present at the time they are weighed. The gross weights, including the moisture present at the time of weighing, are

Item	Gross weight (lb/cy)	Percent moisture by gross weight
Sand	2,918	5
Gravel	2,968	3

It is estimated that 8% of the sand and 6% of the gravel will be lost or not recovered in the stockpile at the job. Determine the total number of tons each of sand and gravel required for the project.

REFERENCES

1. ACI Committee 304, *304R-89: Guide for Measuring, Mixing, Transporting and Placing Concrete,* American Concrete Institute, ACI International, PO Box 9094, Farmington Hills, MI 48333 (1989). http://aci-int.org/*Manual of Concrete.*

2. Mindess, Sidney, and J. F. Young, *Concrete,* Prentice–Hall, Englewood Cliffs, N.J., 1981.

3. ASTM Committee C9, "Test for Slump of Portland Cement Concrete," *Annual Book of ASTM Standards,* Vol. 04.02, published annually.

4. ACI Committee 211, "Standard Practice for Selecting Proportions for Normal, Heavy weight, and Mass Concrete," ACI 211.2, *ACI Manual of Concrete Practice, Part 1,* American Concrete Institute, Detroit, Mich., published annually.

5. ATSM Committee C94, "Standard Specification for Ready-Mixed Concrete," *Annual Book of ASTM Standards,* Vol. 04.02, published annually.

6. *Concrete Plant Standards,* CPMB 100-96, of the Concrete Plant Manufacturers Bureau, 900 Spring Street, Silver Spring, Md., 20910, 1996.

7. Ledbetter, Bonnie S., W. B. Ledbetter, and Eugene H. Boeke: "Mixing, Moving, and Mashing Concrete—75 Years of Progress," *Concrete International,* pp. 69–76, November 1980.

8. ACI Committee 304, "Placing Concrete by Pumping Methods," ACI 304.2R, *ACI Manual of Concrete Practice,* Part 2, American Concrete Institute, Detroit, Mich., published annually.

9. ACI Committee 309, "Standard Practice for Consolidation of Concrete," ACI 309R, *ACI Manual of Concrete Practice,* Part 2, American Concrete Institute, Detroit, Mich., published annually.

10. Stewart, Rita F., and Cliff J. Schexnayder, "Construction Techniques for Roller-Compacted Concrete," Roller-Compacted Concrete Pavements and Concrete Construction, Transportation Research Record 1062, Transportation Research Board, National Research Council, Washington, D.C., 1986, pp. 32–37.

11. ACI Committee 506, "Recommended Practice for Shotcreting," ACI 506R, *ACI Manual of Concrete Practice,* Part 5, American Concrete Institute, Detroit, Mich., published annually.

12. EPA, "Guidelines for Federal Procurement of Cement and Concrete Containing Fly Ash," *Federal Register,* Vol. 48, No. 20, pp. 4230–4253, January 28, 1983.

13. McKerall, W. C., and W. B. Ledbetter, "Variability and Control of Class C Fly Ash," *Cement, Concrete, and Aggregates,* CCAGOP, Vol. 14, No. 2, pp. 87–93, Winter 1982, American Society for Testing and Materials, Philadelphia.

14. Cook, James E., " Research and Application of High Strength Concrete Using Class C Fly Ash," *Concrete International,* July 1982.

15. Ledbetter, W. B., D. J. Teague, R. L. Long, and B. N. Banister: "Construction of Fly Ash Test Sites and Guidelines for Construction," Research Report 240-2, Texas Transportation Institute, Texas A&M University, College Station, Tx., October 1981, 111p.

16. ACI Committee 306, "Cold Weather Concreting," ACI 306R, *ACI Manual of Concrete Practice,* Part 2, American Concrete Institute, Detroit, Mich., published annually.

17. ACI Committee 305, "Hot Weather Concreting," ACI 305R, *ACI Manual of Concrete Practice,* Part 2, American Concrete Institute, Detroit, Mich., published annually.

18. Morgen Manufacturing Company, P.O. Box 160, Yankton, SD 57078-0160.

19. Allentown Pneumatic Gun, Master Builders, Inc., 421 Schantz Road, Allentown, PA 18104.

20. American Coal Ash Association (ACAA), 2760 Eisenhower Avenue, Suite 304, Alexandria, VA 22314, Tel: (703) 317-2400, Fax: (703) 317-2409. The ACAA works with a wide variety of end-user industries and with federal and state government agencies to develop materials specifications and guidelines for use of coal combustion by-products and for clean coal technology by-products as well.

CONCRETE-RELATED WEBSITES

1. American Concrete Institute (ACI), www.aci-int.org. P.O. Box 9094, Farmington Hills, MI 48333-9094, Tel: (248) 848-3700, Fax: (248) 848-3701. The American Concrete Institute is a technical and educational consensus organization whose purpose is to further engineering and technical education, scientific investigation and research, and development of standards for the design and construction of

concrete structures. The institute gathers, correlates, and disseminates information for the improvement of the design, construction, manufacture, use, and maintenance of concrete products and structures.

2. American Concrete Pavement Association (ACPA), www.pavement.com. 5420 Old Orchard Road, Suite A100, Skokie, IL 60077-1059, Tel: (847) 966-2272, Fax: (847) 966-9970. The American Concrete Pavement Association is the national professional organization of the concrete paving industry. Its purposes are to improve the quality and performance of its product.

3. American Concrete Pumping Association, www.concretepumping.com/. 7695 Kinneytuck Court, Lewis Center, OH 43035, Tel: (740) 548-2351, Fax: (740) 548-2352. The objective of the American Concrete Pumping Association is to promote concrete pumping as the method of choice of placing concrete in a safe and expedient manner. The association promotes this concrete placement method through development of videotapes on concrete pumping and its magazine, *Concrete Pumping,* plus a newsletter and seminars.

4. The American Shotcrete Association (ASA), www.shotcrete.org/. 38800 Country Club Drive, Farmington Hills, MI 48331, Tel: (248) 848-3780, Fax: (248) 848-3740, E-mail: inform@shotcrete.org. The American Shotcrete Association supports and promotes the training of those in the shotcrete industry in the proper methods, materials, and technique to obtain high-quality shotcrete.

5. American Society of Concrete Contractors (ASCC), www.ascconc.org/. 38800 Country Club Drive, Farmington Hills, MI 48331-3411, Tel: (248) 848-3710, Fax: (248) 848-3711. The American Society of Concrete Contractors is dedicated to improving concrete construction quality, productivity, and safety. Members of ASCC are concrete contractors, material suppliers, equipment manufacturers, and others involved in concrete construction. Activities include an extensive *Safety Manual,* concrete and safety hotlines, safety videos, safety bulletins, troubleshooting newsletters, and the *Contractor's Guide to Quality Concrete Construction.*

6. Concrete Reinforcing Steel Institute (CRSI), www.crsi.org/. 933 N. Plum Grove Road, Schaumburg, IL 60173-4758, Tel: (847) 517-1200, Fax: (847) 517-1206. CRSI is a national trade association representing producers and fabricators of reinforcing steel, epoxy coaters, bar support and splice manufacturers, and other related associates and interested professional architects and engineers. All are supported through research and engineering activities and work on specifications, building codes, and engineering services.

7. Concrete Sawing and Drilling Association (CSDA), www.csda.org. 6089 Frantz Road, Suite 101, Dublin, OH 43017, Tel: (614) 798-2252, Fax: (614) 798-2255. The Concrete Sawing and Drilling Association is a nonprofit trade association of contractors, manufacturers, and affiliated members from the concrete construction and renovation industry. The CSDA mission is to promote professional sawing and drilling methods. Concrete cutting with diamond tools offers many benefits including reduced downtime; precision cutting; maintenance of structural integrity; reduced noise, dust, and debris; limited-access cutting; and the ability to cut heavily reinforced concrete.

8. National Precast Concrete Association (NPCA), www.precast.net/. 10333 N. Meridian Street, Suite 272, Indianapolis, IN 46290, Tel: (317) 571-9500 or 800/366-7731, Fax: (317) 571-0041. The National Precast Concrete Association pursues quality control programs through plant certification and emphasizes

promotional activities structured to increase the market share for those products produced by member companies.

9. National Ready Mixed Concrete Association (NRMCA), www.nrmca.org/. 900 Spring Street, Silver Spring, MD 20910. Tel: (301) 587-1400 or (888) 84-NRMCA, Fax: (301) 585-4219. The National Ready Mixed Concrete Association is an international trade association for producers of ready-mixed concrete. The association is involved in engineering advances, technical research, educational seminars, workshops and conferences, and regulatory issues. NRMCA provides a large publications list from which members may choose training manuals and articles on such topics as safety, financial management, maintenance, and driver education.

10. Portland Cement Association (PCA), www.portcement.org. 5420 Old Orchard Road, Skokie, IL 60077, Tel: (847) 966-6200, Fax: (847) 966-8389. The Portland Cement Association represents cement companies in Canada and the United States. The PCA serves as the nucleus of the cement industry's work in research, promotion, education, and public affairs. PCA's wholly owned subsidiary, Construction Technology Laboratories, Inc., provides a wide range of research, testing, and consulting services.

11. Precast/Prestressed Concrete Institute (PCI), www.pci.org. 175 West Jackson Boulevard, Suite 1859, Chicago, IL, 60604-2801, Tel: (312) 786-0300, Fax: (312) 786-0353. The Precast/Prestressed Concrete Institute is dedicated to fostering greater understanding of precast/prestressed concrete and to serve as the focal point for the advancement of that industry through technical research and marketing support.

12. U.S. Army Corps of Engineers, www.erdc.usace.army.mil/. Links are provided to the engineering laboratory at the Waterways Experiment Station and the Cold Regions Research and Engineering Laboratory.

17

Cranes

Construction cranes are generally classified into two major families: (1) mobile cranes and (2) tower cranes. Because cranes are used to hoist and move loads from one location to another, it is necessary to know the lifting capacity and working range of a crane selected to perform a given service. The rated load for a crane as published by the manufacturer is based on ideal conditions. Load charts can be complex documents listing numerous booms, jibs, and other components that may be employed to configure the crane for various tasks. It is critical that the chart being consulted be for the actual crane configuration that will be used.

MAJOR CRANE TYPES

Cranes are a broad class of construction equipment used to hoist and place loads. Each type of crane is designed and manufactured to work economically in a specific site situation.

Construction cranes are generally classified into two major families: (1) mobile cranes and (2) tower cranes. Mobile cranes are the machines of choice in North America, as contractors have traditionally favored them over tower cranes. Tower cranes are usually used in North America only when job-site conditions make mobile crane movement impossible, or for high-rise construction. They are, however, the machines that dominate the construction scene in Europe, whether in the big cities or in rural areas.

The most common mobile crane types are

1. Crawler.
2. Telescoping-boom truck mounted.
3. Lattice-boom truck mounted.
4. Rough-terrain.
5. All-terrain.

6. Heavy lift.
7. Modified cranes for heavy lift.

The most common tower crane types are

1. Top slewing.
2. Bottom slewing.

Some of the mobile-type machines in their basic configuration can have different front-end operating attachments that allow the unit to be used as an excavator or a pile driver, or in other specialized tasks. Such diverse usages are discussed in Chapters 4, 8, 12, 18, and 19.

The Power Crane and Shovel Association (PCSA) has conducted and supervised both studies and tests which provide considerable information related to the performance, operating conditions, production rates, economic life, and cost of owning and operating these units. The association has participated in establishing and adopting standards that are applicable to this equipment. This information has been published in technical bulletins and booklets.

Some of the information from the PCSA is reproduced in this book, with permission of the association. Items of particular interest are cited in the references at the end of the chapter.

MOBILE CRANES

CRAWLER CRANES

The *full revolving superstructure* of this type unit is mounted on a pair of continuous, parallel crawler tracks. Many manufacturers have different option packages available that permit configuration of the crane to a particular application, standard lift, tower unit, or duty cycle. Units in the low to middle range of lift capacity have good lifting characteristics and are capable of duty-cycle work such as handling a concrete bucket. Machines of 100-ton capacity and above are built for lift capability and do not have the heavier components required for duty-cycle work. The universal machines incorporate heavier frames, have heavy-duty or multiple clutches and brakes, and have more powerful swing systems. These designs allow for quick changing of drum laggings that vary the torque/speed ratio of cables to the application. Figure 17.1 illustrates a crawler crane on a bridge project.

The crawlers provide the crane with travel capability around the job site. The crawler tracks provide such a large ground contact area that soil failure under these machines is only a problem when operating on soils having a low bearing capacity. Before hoisting a load the machine must be leveled and ground settlement considered. If soil failure or ground settlement is possible, the machine can be positioned and leveled on mats. The distance between crawler tracks affects stability and lift capacity. Some machines have a feature whereby

FIGURE 17.1 | Crawler crane on a bridge project.

the crawlers can be extended. For many machines, this extension of the crawlers can be accomplished without external assistance.

To relocate a crawler crane between projects requires that it be transported by truck, rail, or barge. As the size of the crane increases, the time and cost to dismantle, load, investigate haul routes, and reassemble the crane also increase. The durations and costs can become significant for large machines. Relocating the largest machines can require 15 or more truck trailer units. These machines usually have lower initial cost per rated lift capability, but movement between jobs is more expensive. Therefore, crawler-type machines should be considered for projects requiring long duration usage at a single site.

Many new models utilize modular components to make dismantling, transporting, and assembling easier. Quick-disconnect locking devices and pin connectors have replaced multiple-bolt connections.

Most crawler crane models have a fixed-length lattice boom (Fig. 17.1), which is also the crane type discussed in this section. A lattice boom is cable-suspended, and therefore acts as a compression member, *not* a bending member like a telescoping hydraulic boom. However, new small-size crawler models exist that are equipped with a telescoping boom. Some of these models are equipped with rubber tracks to make them suitable for urban works and movement on asphalt pavements.

TELESCOPING-BOOM TRUCK-MOUNTED CRANES

There are truck cranes (see Fig. 17.2) that have a self-contained telescoping boom. Most of these units can travel on the public highways between projects under their own power with a minimum of dismantling. Once the crane is leveled at the new work site, it is ready to work without setup delays. These machines, however, have higher initial cost per rated lift capability. If a job requires crane utilization for a few hours to a couple of days, a telescoping truck crane should be given first consideration because of its ease of movement and setup.

The multisection telescoping boom is a permanent part of the full revolving superstructure. In this case, the superstructure is mounted on a multiaxle truck/carrier. There are three common power and control arrangements for telescoping-boom truck cranes:

1. A single engine as both the truck and crane power source, with a single, dual-position cab used both for driving the truck and operating the crane.
2. A single engine in the carrier but with both truck and crane operating cabs.
3. Separate power units for the truck and the superstructure. This arrangement is standard for the larger capacity units.

FIGURE 17.2 | Telescoping truck crane on a building site.

FIGURE 17.3 | A large lattice-boom truck crane with extended outriggers on large mats.

Telescoping-boom truck cranes have extendable outriggers for stability. In fact, many units cannot be operated safely with a full reach of boom unless the outriggers are fully extended and the machine is raised so that the tires are clear of the ground. In the case of the larger machines, the width of the outriggered vehicle may reach 40 ft, which necessitates careful planning of the operation area. Additionally, these heavy machines transfer, through the outriggers, extremely high loads to the ground. This high ground loading must be considered vis-à-vis the soil-bearing capacity. Large-size timber or steel mats that are used to spread the load over a larger ground area further increase the overall vehicle width. These outrigger space considerations are also a concern when using large lattice-boom truck cranes (see Fig. 17.3).

LATTICE-BOOM TRUCK-MOUNTED CRANES

As with the telescoping-boom truck crane, the lattice-boom truck crane has a full revolving superstructure mounted on a multiaxle truck/carrier. The advantage of this machine is the *lattice boom.* The lattice-boom structure is lightweight. This reduction in boom weight means additional lift capacity, as the machine predominantly handles hoist load and less weight of boom. The lattice boom does take longer to assemble. The lightweight boom will give a less expensive lattice-boom machine the same hoisting capacity as a larger telescoping-boom unit. Figure 17.4 shows a lattice-boom truck crane handling a precast panel on a building project.

The disadvantage of these units is the time and effort required to disassemble them for transport. In the case of the larger units, it may be necessary to

FIGURE 17.4 | A large lattice-boom truck crane.

remove the entire superstructure. Additionally, a second crane is often required for this task. Some newer models are designed so that the machine can separate itself without the aid of another crane.

ROUGH-TERRAIN CRANES

These cranes are mounted on two-axle carriers (Fig. 17.5). The operator's cab may be mounted in the upper works, allowing the operator to swing with the load. However, on many models, the cab is located on the carrier. This is a simpler design because controls do not have to be routed across the turntable. In turn, these units have a lower cost.

These units are equipped with unusually large wheels and closely spaced axles to improve maneuverability at the job site. They further earn the right to their name by their high ground clearance, as well as their ability to move on slopes of up to 50 or 70%, depending on the particular make and manufacturer. Most units can travel on the highway but have maximum speeds of only about 30 mph. In the case of long moves between projects, they should be transported on low-bed trailers.

They are sometimes referred to as "cherry pickers." This comes from their use during World War II in handling bombs, as the slang name for a bomb was a cherry.

Many units now have joystick controls. A *joystick* allows the operator to manipulate four functions simultaneously. The most common models are in the

FIGURE 17.5 | Rough-terrain crane.

18- to 50-ton capacity range and typically are employed as utility machines. They are primarily lift machines but are capable of light, intermittent duty-cycle work.

ALL-TERRAIN CRANES

The *all-terrain crane* is designed with an undercarriage capable of long-distance highway travel. Yet the carrier has all-axle drive and all-wheel steer, crab steering, large tires, and high ground clearance. It has dual cabs, a lower cab for fast highway travel, and a superstructure cab that has both drive and crane controls. The machine can, therefore, be used for limited pick-and-carry work. By combining job-site mobility and transit capability, this machine is appropriate when multiple lifts are required at scattered project sites or at multiple work locations on a single project. Because this machine is a combination of two features, it has a higher cost than an equivalent capacity telescoping truck crane or a rough-terrain crane. But an all-terrain machine can be positioned on the project without the necessity of having other construction equipment prepare a smooth travel way as truck cranes would require. Additionally the all-terrain crane does not need a lowboy to haul it between distant project sites, as would a rough-terrain machine.

FIGURE 17.6 | Lampson LTL-1200, with 370 ft of boom and 120 ft of jib setting a 215-ton bridge at 180-ft radius.

Source: Neil F. Lampson, Inc.

HEAVY-LIFT CRANES

These are machines that provide lift capacities in the 600- through 2000-short-ton* range. These cranes consist of a boom and counterweight, each mounted on independent crawlers that are coupled by a stinger (see Fig. 17.6). This configuration utilizes a vertical strut and inclined mast to decrease compressive forces in the boom.

MODIFIED CRANES FOR HEAVY LIFTING

These are basically systems that significantly increase the lift capacity of a crawler crane. A crane's capacity is limited by one of two factors: (1) structural strength or (2) tipping moment. If you add a counterweight to prevent tipping when hoisting a heavy load, there is a point when the machine is so overbalanced that without a load it would tip backward. At some point, even with sufficient counterweight, the boom is put into such high compression that it will give way at the butt. Manufacturers, understanding both the need of users to make occasional heavy lifts and the users' reluctance to buy a larger machine for a onetime

---

*A short ton equals 2,000 lb as opposed to a metric or long ton, which equals 2,240 lb.

use, have developed systems which provide the capability while maintaining machine integrity. The four principal systems available are

1. Trailing counterweight.
2. Extendable counterweight.
3. Ring system.
4. Guy derrick.

Trailing Counterweight

The base crane does not carry the trailing counterweight. Instead, the additional counterweight is mounted on a wheeled platform behind the crane, with the platform pin connected to the crane (see Fig. 17.7). The system utilizes a mast positioned behind the *boom,* with the boom suspension lines mounted at the top of

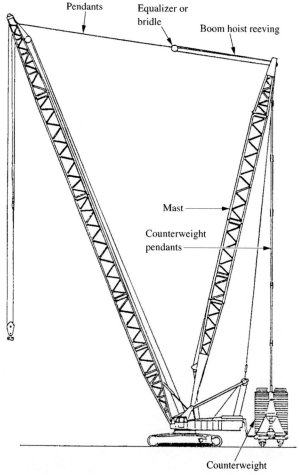

FIGURE 17.7 I Trailing counterweight crane modification for heavy lifting.
Source: Construction Safety Association of Ontario.

the mast. This increases the angle between the boom and the suspension lines, thereby decreasing the compressive forces on the boom.

Extendable Counterweight

One manufacturer offers a machine with a counterweight system that can be extended away from the rear of the machine to match the leverage with the requirements of the lift.

Ring System

With the ring system, a large circular turntable ring is created outside the base machine (see Fig. 17.8). The heavy counterweight system is supported on this ring. There are auxiliary pin-connected frames at the front and rear of the base machine. These allow both the boom/mast foot and the counterweight to be moved away from the machine. Using rollers or wheels, the auxiliary frames ride the ring. The base crane is really only a power and control source.

FIGURE 17.8 | Crawler crane ring system modification for heavy lifting.

The new hydraulic self-erecting systems allow a crawler crane to be modified to a ringer configuration in only three days. This ability makes the system very competitive for jobs lasting more than just a few days.

Because of their size and weight, ringer systems render cranes mounted on them practically no longer "mobile." To partly overcome this limitation, one manufacturer offers a feature by which the crane can lift its own ring, crawl to a new location, and lower the ring to the ground again.

Guy Derrick

The guy-derrick configuration will immobilize the crane. But a sevenfold increase in lifting capacity can be achieved when a high vertical mast tied off with guy cables is mounted on the base crane.

CRANE BOOMS

Most cranes come equipped with standardized booms that are designed to optimize their performance over a range of applications. However, cranes may also utilize optional boom configurations that allow them to adapt to specific lifting conditions. Optional booms and differing boom tops allow a machine to accommodate load clearance, longer reach, or increased lift capacity requirements (see Fig. 17.9).

LIFTING CAPACITIES OF CRANES

Because cranes are used to hoist and move loads from one location to another, it is necessary to know the lifting capacity and working range of a crane selected to perform a given service. Figure 17.10 shows typical crane lifting capacities for four specific crawler cranes of varying sizes. The lifting capacities of units made by different manufacturers will vary from the information in the figure. Individual manufacturers and suppliers will furnish machine-specific information in literature describing their machines.

When a crane lifts a load attached to the hoist line that passes over a sheave located at the boom point of the machine, there is a tendency to tip the machine over. This introduces what is defined as the *tipping condition*. With the crane on a firm, level supporting surface in calm air, it is considered to be at the point of tipping when a balance is reached between the overturning moment of the load and the stabilizing moment of the machine.

During tests to determine the tipping load for wheel-mounted cranes, the outriggers should be lowered to relieve the wheels of all weight on the supporting surface or ground. The radius of the load is the horizontal distance from the axis of rotation of the crane to the center of the vertical hoist line or tackle with the load applied. The *tipping load* is the load that produces a tipping condition at a specified radius. The load includes the weight of the item being lifted plus the weights of the hooks, hook blocks, slings, and any other items used in hoisting the load, including the weight of the hoist rope located between the boom-point sheave and the item being lifted.

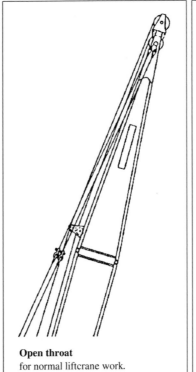

Open throat
for normal liftcrane work.

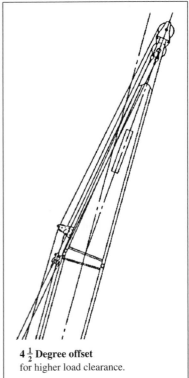

4 $\frac{1}{2}$ Degree offset
for higher load clearance.

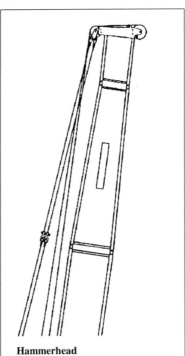

Hammerhead
for heavy lifts and superior
load clearance.

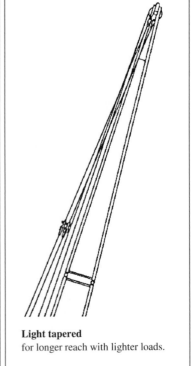

Light tapered
for longer reach with lighter loads.

FIGURE 17.9 | Optional crane booms and boom tops.

Source: Manitowoc Engineering Co.

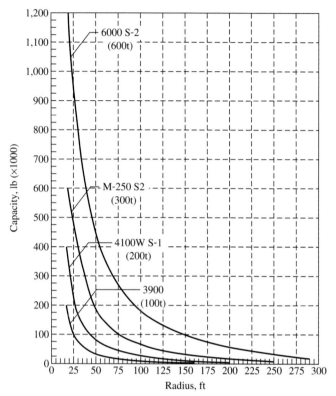

FIGURE 17.10 | Safe lifting capacities for four crawler cranes.

Source: Manitowoc Engineering Co.

RATED LOADS FOR LATTICE- AND TELESCOPIC BOOM CRANES

The rated load for a crane as published by the manufacturer is based on ideal conditions. Load charts can be complex documents listing numerous booms, jibs, and other components that may be employed to configure the crane for various tasks. It is critical that the chart being consulted be for the actual crane configuration that will be utilized. *Interpolation between the published values IS NOT permitted;* use the next lower value. Rated loads are based on ideal conditions, a level machine, calm air (no wind), and no dynamic effects.

A partial safety factor with respect to tipping is introduced by the PCSA rating standards that state that the rated load of a lifting crane shall not exceed the following percentages of tipping loads at specified radii [9]:

1. Crawler-mounted machines, 75%.
2. Rubber-tire-mounted machines, 85%.
3. Machines on outriggers, 85%.

It should be noted that there are other groups that recommend rating criteria. The Construction Safety Association of Ontario recommends that for rubber-tire-mounted machines a factor of 75% be utilized.

One manufacturer is producing rubber-tire-mounted cranes having intermediate outrigger positions. For intermediate positions greater than one-half the fully extended length, the manufacturer is using a rating based on 80% of the tipping load. For intermediate positions less than one-half the fully extended length, a rating based on 75% is used. At this time, there is no standard for this type machine.

Load capacity will vary depending on the quadrant position of the boom with respect to the machine's undercarriage. In the case of crawler cranes, the three quadrants that should be considered are

1. Over the side.
2. Over the drive end of the tracks.
3. Over the idler end of the tracks.

Crawler crane quadrants are usually defined by the longitudinal centerline of the machine's crawlers. The area between the centerlines of the two crawlers is considered over the end and the area outside the crawler's centerline is considered over the side.

In the case of wheel-mounted cranes, the quadrants of consideration will vary with the configuration of the outrigger locations. If a machine has only four outriggers, two on each side, one located forward and one to the rear, the quadrants are usually defined by imaginary lines running from the superstructure center of rotation through the position of the outrigger support. In such a case, the three quadrants to consider are

1. Over the side.
2. Over the rear of the carrier.
3. Over the front of the carrier.

Some wheel-mounted cranes have an outrigger directly in the front, or there can be other machine-specific outrigger configurations. Therefore, the best practice is to consult the manufacturer's specifications.

The important point is that the rated load should be based on the direction of minimum stability for the mounting, unless otherwise specified. The minimum stability condition restricts the rated load because the crane must both raise and swing loads. The swinging motion will cause the boom to move through various quadrants, changing the load's effect on the machine. Further, it should be remembered that the rating is based on the fact that the outriggers are fully extended.

Rated loads are based on the assumption that the crane is in a level position (for the full 360° of swing). When a crane is not level even small variations significantly affect lifting capacity. In the case of a short-boom machine operating at minimum radius, 3° out of level can result in a 30% loss in capacity. For long-boom machines, the loss in capacity can be as great as 50% [8].

Another important consideration with modern cranes is that tipping is not always the critical capacity factor. At short radii, capacity may be dependent on boom or outrigger strength and structural capacity, and at long radii pendant tension can be the controlling element. Manufacturers' load charts will limit the rated capacity to values below the minimum critical condition taking into account all possible factors.

Table 17.1 illustrates the kind of information issued by the manufacturers of cranes. The crane in this example is described as a 200-ton, nominal rating, crawler-mounted cable-controlled crane with 180 ft of boom. It is important to realize that the lifting capacity by which mobile cranes are identified (although not necessarily named) does not represent a standard classification method, which is why the wording *nominal* is added to the description. In most cases, if a ton rating is used it refers to lifting capacity with a basic boom and minimum radius. Some manufacturers use a load moment rating classification system. The designation would be "tm." As an example, a Demag AC 650 (reflecting 650-ton lifting capacity, in this case referring to a metric ton) is named by a big crane rental company AC 2000 (reflecting 2000 tm max. load moment).

TABLE 17.1 | Lifting capacities in pounds for a 200-ton, nominal rating, crawler crane with 180 ft of boom*

Radius (ft)	Capacity (lb)	Radius (ft)	Capacity (lb)	Radius (ft)	Capacity (lb)
32	146,300	80	39,200	130	17,900
36	122,900	85	35,800	135	16,700
40	105,500	90	32,800	140	15,500
45	89,200	95	30,200	145	14,500
50	76,900	100	27,900	150	13,600
55	67,200	105	25,800	155	12,700
60	59,400	110	23,900	160	11,800
65	53,000	115	22,200	165	11,100
70	47,600	120	20,600	170	10,300
75	43,100	125	19,200	175	9,600

*Specified capacities based on 75% of tipping loads.
Source: Manitowoc Engineering Co.

The capacities located in the upper portion of a load chart (Table 17.2), usually defined by either a bold line or by shading, represent structural failure conditions. Operators can feel the loss of stability prior to a tipping condition. But in the case of a structural failure, there is no sense of feel to warn the operator; therefore, load charts must be understood and all lifts must be in strict conformance with the ratings.

While the manufacturer will consider crane structural factors when developing a capacity chart for a particular machine, operational factors affect absolute capacity in the field. The manufacturer's ratings can be thought of as valid for a *static* set of conditions. A crane on a project operates in a *dynamic* environment, lifting, swinging, and being subjected to air currents and temperature variations. The load chart provided by the manufacturer does not take into

TABLE 17.2 | Lifting capacities in pounds for a 25-ton truck-mounted hydraulic crane*

Load radius (ft)	Lifting capacity (lb)† Boom length (ft)						
	31.5	40	48	56	64	72	80
12	50,000	45,000	38,700				
15	41,500	39,000	34,400	30,000			
20	29,500	29,500	27,000	24,800	22,700	21,100	
25	19,600	19,900	20,100	20,100	19,100	17,700	17,100
30		14,500	14,700	14,700	14,800	14,800	14,200
35			11,200	11,300	11,400	11,400	11,400
40			8,800	8,900	9,000	9,000	9,000
45				7,200	7,300	7,300	7,300
50				5,800	5,900	6,000	6,000
55					4,800	4,900	4,900
60					4,000	4,000	4,000
65						3,100	3,300
70							2,700
75							2,200

*Specified crane capacities based on 85% of tipping loads.
†The loads appearing below the solid line are limited by the machine stability. The values appearing above the solid line are limited by factors other than machine stability.

account these dynamic conditions. Factors that will greatly affect actual crane capacity on the job are:

1. Wind forces on the boom or load.
2. Swinging the load.
3. Hoisting speed.
4. Stopping the hoist.

These dynamic factors should be carefully considered when planning a lift.

Rated Loads for Hydraulically Operated Cranes

The rated tipping loads for hydraulic cranes are determined and indicated as for cable-controlled cranes. However, in the case of hydraulic cranes, the critical load rating is sometimes dictated by hydraulic pressure limits instead of tipping. Therefore, load charts for hydraulic cranes represent the controlling-condition lifting capacity of the machine and the governing factor may not necessarily be tipping. The importance of this is that the operator cannot use his physical sense of balance (machine feel) as a gauge for safe lifting capability.

WORKING RANGES OF CRANES

Figure 17.11 shows graphically the height of the boom point above the surface supporting the crane and the distance from the center of rotation for the crane for various boom angles for the crane whose lifting capacities are given in Table 17.1.

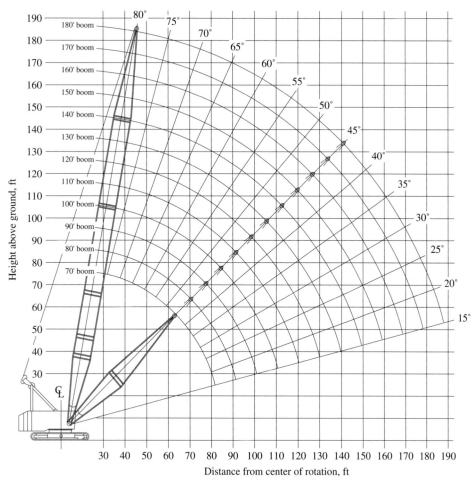

FIGURE 17.11 | Working ranges for a 200-ton crawler crane, nominal rating.

Source: Manitowoc Engineering Co.

The maximum length of the boom may be increased to 180 ft. The length of the boom is increased by adding sections at or near midlength of the boom, usually in 10-ft, 20-ft, or 40-ft increments.

Using the information in Fig. 17.11, determine the minimum boom length that will permit the crane to lift a load 34 ft high to a position 114 ft above the surface on which the crane is operating. The length of the block, hook, and slings that is required to attach the hoist rope to the load is 26 ft. The location of the project will require the crane to pick the load from a truck at a distance of 70 ft from the center of rotation of the crane. Thus the operating radius will be 70 ft.

To lift the load to the specified location, the minimum height of the boom point of the crane must be at least 114 + 34 + 26 = 174 ft above the ground supporting the crane. An examination of the diagram in Fig. 17.11 reveals that for a radius of 70 ft, the height of the boom point for a 180-ft-long boom is high enough.

If the block, hook, and slings weigh 5,000 lb, determine the maximum net weight of the load that can be hoisted. Using Table 17.1, we find that for a boom length of 180 ft and a radius of 70 ft the maximum total load is 47,600 lb. If the weight of the block, hook, and slings is deducted from the total load, the net weight of the lifted object will be 42,600 lb, which is the maximum safe weight of the lifted object.

TOWER CRANES

CLASSIFICATION

Tower cranes provide high lifting height and good working radius, while taking up a very limited area. These advantages are achieved at the expense of low lifting capacity and limited mobility, as compared to mobile cranes. The three common tower crane configurations are (1) a special vertical boom arrangement on a mobile crane, (2) a mobile crane superstructure mounted atop a tower (see Fig. 17.12a), or (3) a vertical tower with a jib. The latter description is often referred to in the United States as the European type (see Fig. 17.12b), but it is the type perceived elsewhere as a "tower crane" when this term is used with no further details. Tower cranes of this latter type usually fall within one of two categories:

1. *Top-slewing (fixed tower)* tower cranes (Fig. 17.12b) have fixed towers and a swing circle mounted at the top, allowing only the jib, tower top, and operator cabin to rotate.

2. *Bottom slewing (slewing tower)* tower cranes (Fig. 17.13) have the swing circle located at the base, and both the tower and jib assembly rotate relative to the base.

The main differences between these two categories are reflected in setup and dismantling procedures and in lifting height. Bottom-slewing tower cranes essentially erect themselves using their own motors, in a relatively short and simple procedure. They are often referred to as "self-erecting" or "fast-erecting" cranes. This is achieved, however, at the expense of service height, as dictated by the telescoping tower (mast) that, because of its revolving base, cannot be braced to a permanent structure. On the other hand, setting up and dismantling top-slewing tower cranes requires more time, is more complicated, and can be a costlier procedure. Erection of a top-slewing tower crane requires the assistance

(a) Mobile crane superstructure mounted atop a tower.

(b) Tower crane; static base, fixed tower, European type.

FIGURE 17.12 | Tower crane configurations.

FIGURE 17.13 | Bottom-slewing tower crane.

of other equipment, but the crane can reach extreme heights. Consequently, the bottom-slewing models are suitable mainly for shorter-term service of low-rise buildings, while top-slewing cranes commonly serve high-rise buildings on jobs requiring a crane for a long duration.

In the mobile crane culture of the United States, tower cranes are usually the machines of choice when

1. Site conditions are restrictive.
2. Lift height and reach are great.
3. There is no need for mobility.

In Europe, though, where these machines were extensively introduced after World War II, they are seen on all types of projects: next to low-rise buildings on spacious sites, on road construction jobs, and even on utility projects. Additionally, long rail tracks are assembled to provide crane mobility on the job site. On all these project types, where tower cranes are the choice although not the necessity, economy is accomplished by low operating cost compensating for possibly higher setup cost.

OPERATION

Vertical tower cranes can be mounted on a mobile crane substructure, a fixed base, or a traveling base, or can be configured to climb within the structure being constructed.

Mobile Cranes Rigged with Vertical Towers

Crawler- and truck-mounted tower cranes use pinned jibs extending from special booms that are set vertically. A crawler-mounted tower crane can travel over firm level ground after the tower is erected, but it has only limited ability to handle loads while moving. A truck-mounted tower crane must have its outriggers extended and down before the tower is raised. Therefore, it cannot travel with a load and the tower must be dismantled before the crane can be relocated.

Fixed-Base Tower Cranes

The fixed-base-type crane, commonly of the top-slewing configuration, has its tower mounted on an engineered foundation block (see Fig. 17.14). Using another crane, the tower is erected to its full height at the beginning of the project or the crane can have the mechanical capability, usually hydraulic, to raise itself, allowing for the addition of structural sections to the tower. There is a vertical limit known as the *maximum free-standing* height to which these cranes can safely rise above a base (typically 150 ft for average-size top-slewing cranes, and up to 250 ft for the larger-size machines). If it is necessary to raise the tower above this limiting height, lateral bracing must be provided. Even when bracing is provided, there is a *maximum braced height* tower limit (although 1,000-ft-high top-slewing cranes are not particularly exceptional). These limits are dictated by the structural capacity of the tower frame and are machine-specific.

FIGURE 17.14 | Engineered foundation block for a fixed-base-type tower crane.

Common dimensions and capacities for top-slewing tower cranes are

1. Jib length: 100 to 200 ft
2. Section length: 10 to 20 ft
3. Base dimensions: 13 by 13 to 20 by 20 ft
4. Tower cross section: 4 by 4 to 8 by 8 ft
5. Maximum lifting capacity at end of jib: 2,000 to 10,000 lb
6. Hoisting and trolleying speed: 80 ft/min

A tower that is assembled using other equipment at the beginning of a project cannot have a height greater than its maximum free-standing height as there is no structure in place to which the lateral bracing can be connected. If, after the building of the structure progresses, it is necessary to raise the crane's tower, the procedure would require removal of the crane's superstructure from the tower frame before a tower section could be added. This would involve other equipment such as a mobile crane and would be a very costly proposition.

The raising of self-erecting towers to greater vertical heights is fairly easy and economical. A self-erecting tower crane has a short section termed the "guide section" or "climbing cage" of the hydraulically operated erecting tower that is situated below the slewing ring for this purpose. The erecting operation is a three-step procedure:

1. The crane hoists a new section of tower and moves it next to its tower.
2. The guide section hydraulically jacks up the slewing ring and jibs, and the new section of tower is inserted and positioned on the previously erected tower.
3. The hydraulic jacks are released and the slewing ring and jibs are repositioned and fastened to the extended tower.

Traveling-Tower Cranes

The ballasted base of this type of tower crane is usually set on a pair of fixed rails (see Fig. 17.15). This allows the crane to move along the rails with a load. The advantage is the increased coverage of the work area that can be achieved. Maximum grade for crane rails is model-specific, but usually not greater than 1%. Traveling-tower cranes can be either of the top-slewing or bottom-slewing type, with heights commonly not exceeding 150 ft for the former and 100 ft for the latter.

Climbing-Frame Tower Crane

The climbing-frame-type crane is supported by the floors of the building that it is being used to construct. The reactions of both the weight of the crane and the

FIGURE 17.15 | Traveling-tower cranes set on a pair of fixed rails.

loads lifted are transmitted to the host structure. The crane will have only a short tower because it moves vertically as construction progresses. This climbing movement, which causes work interruptions, is done incrementally, every few floors, depending on tower height. The taller the tower, the less frequent the climbing procedure.

Vertical movement of the climbing-frame-type crane is by a system of hydraulically activated rams and latchings. Normally, the crane is initially mounted on a fixed base, and as the work progresses, it transfers to the climbing frame mounted on the structure. A typical floor section cannot safely support the load imposed by the operating crane; therefore, it is imperative for the structural designer to consider the loads imposed by the crane in the area of this opening. Even when consideration for the crane loads is taken into account in the building design, it will usually be necessary to use shoring for several floors below the tower crane frame.

At the end of construction, there will be a tower crane at the top of the structure with no means of lowering itself. Removal must be by external methods, such as a mobile crane or by using a derrick. Because of the heights involved and the possible physical interference of the completed structure, the dismantling operation must be carefully planned. If two or more tower cranes were used on the project and there is overlapping hook coverage, it may be possible to use the crane with the higher hook height to help dismantle the lower crane. Under circumstances where none of these techniques can be employed, an expensive solution is to use a helicopter.

Jib Configurations

The horizontal jib arrangement (see Fig. 17.16) for a top-slewing crane is commonly known as a "hammer head," though the proper terminology is *saddle jib.* This type of jib is fixed in a horizontal position by pendants. Actually, there are two jibs: (1) the forward, main jib with a load block hung from a trolley that moves along the jib to change operating radius and (2) an opposite rear counterweight jib—the counter jib. The operator's cab is usually directly below the main jib, either at the top of the tower above the slewing ring, or attached to the side of the tower.

Bottom-slewing cranes do not have a counterjib, thus all the counterweight is placed on the rotating tower base (see Figs. 17.13 and 17.15). The operator's cab is directly below the jib, at the top of the tower. Often these bottom-slewing cranes are operated from a control stand at the tower's base or by remote control and not from a top cab.

If the jib is pinned at its base and supported by cables that are used to control its angle of inclination, it is known as a *luffing jib.* The crane's hook radius is varied by controlling the jib inclination. There are also fixed luffing-jib arrangements. These have the jib supported by jib pendants at a fixed angle of inclination, and hook radius is varied by use of a trolley traveling along the angled jib. A saddle-jib machine can handle loads closer to the tower than a luffing-jib machine, a factor that can be important in crane selection. The nomenclature for a tower crane is shown in Fig. 17.16.

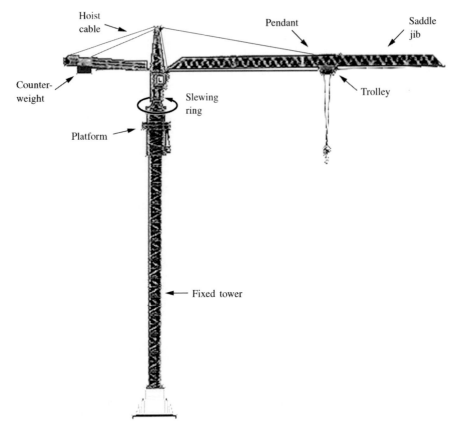

FIGURE 17.16 | Nomenclature for a tower crane.

TOWER CRANE SELECTION

The use of a tower crane requires considerable planning because the crane is a fixed installation on the site for the duration of the major construction activities. From its fixed position, it must be able to cover all points from which loads are to be lifted and to reach the locations where the loads must be placed. Therefore, when selecting a crane for a particular project, the engineer must ensure that the weight of the loads can be handled at their corresponding required radius. Individual tower cranes are selected for use based on

1. Weight, dimension, and lift radii of the heaviest loads.
2. Maximum free-standing height of the machine.
3. Maximum braced height of the machine.
4. Machine-climbing arrangement.
5. Weight of machine supported by the structure.
6. Available head room, which can be developed.
7. Area that must be reached.

8. Hoist speeds of the machine.
9. Length of cable the hoist drum can carry.

The vertical movement of material during the construction process creates an *available head room* clear-distance requirement. This distance is defined as the vertical distance between the maximum achievable crane-hook position and the uppermost work area of the structure. The requirement is set by the dimensions of those loads that must be raised over the uppermost work area during the building process. For practical purposes and safety, hook height above the serviced building top should never be smaller than 20 ft. When selecting a tower crane for a very tall structure, a climbing-type crane may be the only choice capable of meeting the available headroom height requirement.

RATED LOADS FOR TOWER CRANES

Table 17.3 is a capacity chart for a climbing tower crane having a maximum reach of 218 ft. This particular crane can have a stationary free-standing height such that there is 212 ft of clear hook height. While hook or lift height does not affect capacities directly, there is a relation when hoist speed is considered. Information concerning this relationship between hoist-line speed and load capacity is shown in Table 17.4.

Tower cranes are usually powered by alternating current (ac) electric motors, producing only low-level noises for city-friendly operation. A crane having a

TABLE 17.3 | Lifting capacities in pounds for a tower crane

Jib model	L1	L2	L3	L4	L5	L6	L7	
Maximum hook reach	**104′ 0″**	**123′ 0″**	**142′ 0″**	**161′ 0″**	**180′ 0″**	**199′ 0″**	**218′ 0″**	**Hook reach**
	27,600	27,600	27,600	27,600	27,600	27,600	27,600	10′ 3″
	27,600	27,600	27,600	27,600	27,600	27,600	27,600	88′ 2″
	27,600	27,600	27,600	27,600	27,600	27,600	25,800	94′ 6″
	27,600	27,600	27,600	27,600	27,600	25,800	24,200	101′ 0″
	27,600	27,600	27,600	27,600	26,800	24,900	23,400	104′ 0″
		27,600	27,600	27,600	25,200	23,600	22,200	109′ 8″
		27,600	27,600	25,600	23,300	21,800	20,500	117′ 8″
		27,000	27,000	25,100	22,800	21,300	20,100	120′ 0″
Lifting capacities		26,300	26,300	24,300	22,200	20,700	19,500	123′ 0″
in pounds,			24,800	22,800	20,800	19,300	18,300	130′ 0″
two-part line			22,400	20,700	18,700	17,400	16,400	142′ 0″
				19,500	17,600	16,300	15,400	150′ 0″
				18,800	16,800	15,700	14,800	155′ 0″
				17,900	16,200	15,100	14,200	161′ 0″
					15,200	14,200	13,300	170′ 0″
					14,200	13,200	12,400	180′ 0″
						12,300	11,600	190′ 0″
						11,700	10,800	199′ 0″
							10,200	210′ 0″
							9,700	218′ 0″

(*Continued on next page*)

TABLE 17.3 I Lifting capacities in pounds for a tower crane (*continued*)

Jib model	L1	L2	L3	L4	L5	L6	L7	
Maximum hook reach	100' 9"	119' 9"	138' 9"	157' 9"	176' 9"	195' 9"	214' 9"	Hook reach
	55,200	55,200	55,200	55,200	55,200	55,200	55,200	13' 6"
	55,200	55,200	55,200	55,200	55,200	55,200	55,200	48' 9"
	55,200	55,200	55,200	55,200	55,200	55,200	51,400	51' 0"
	55,200	55,200	55,200	55,200	55,200	51,500	48,500	53' 6"
	55,200	55,200	55,200	55,200	51,300	48,300	45,600	56' 6"
	55,200	55,200	55,200	50,700	47,100	44,600	42,100	60' 6"
	46,200	46,200	46,200	42,800	39,700	37,400	35,200	70' 0"
	39,400	39,400	39,400	36,500	34,100	31,900	29,900	80' 0"
	34,600	34,600	34,600	31,900	29,700	27,700	26,100	90' 0"
	30,700	30,700	30,700	28,200	26,100	24,100	22,600	100' 9"
Lifting capacities in pounds, four-part line		27,800	27,800	25,600	23,600	21,700	20,300	110' 0"
		25,400	25,400	23,200	21,300	19,600	18,300	119' 9"
			23,100	21,100	19,300	17,700	16,400	130' 0"
			21,300	19,400	17,800	16,300	15,100	138' 9"
				17,600	16,200	14,700	13,600	150' 0"
				16,400	15,100	13,800	12,700	157' 9"
					13,600	12,400	11,400	170' 0"
					12,900	11,800	10,800	176' 9"
						11,500	10,600	180' 0"
						10,700	9,800	190' 0"
						10,200	9,300	195' 9"
							9,100	200' 0"
							8,300	210' 0"
							8,100	214' 9"

Counterweights

Jib	L1	L2	L3	L4	L5	L6	L7
105-HP hoist unit AC	37,200 lb	47,600 lb	50,800 lb	37,200 lb	40,800 lb	44,000 lb	54,400 lb
165-HP hoist unit AC	34,000 lb	44,000 lb	47,600 lb	34,000 lb	40,800 lb	40,800 lb	50,800 lb

Source: Morrow Equipment Company, L.L.C.

higher motor horsepower can achieve higher operating speeds. When considering the production capability of a crane for duty-cycle work, hoist line speed and the effect motor size has on speed, as shown by Table 17.4, can be very important. This is especially true for high-rise construction, where travel time of the hook between loading and unloading areas is the most significant part of the crane's cycle time, as opposed to low-rise construction, where travel time is insignificant compared to load rigging and unrigging times. If a project requires operating speeds that are higher than provided by an existing crane, replacing the crane's motors with more powerful ones is an alternative option to bringing in another crane.

Hoist-cable configuration is another factor affecting lifting speed. Cranes can usually be rigged with one of two hoist-line configurations, a two-part line or a four-part line. The four-part-line configuration provides a greater lifting

TABLE 17.4 | Effect of hoist line speed on lifting capacities of a tower crane

105 HP-AC, Eddy Current Brake, with four-speed remote controlled gear box, recommended service = 224 amp					
(1) Trolley, two-part line			**(2) Trolleys, four-part line**		
Gear	**Maximum load (lb)**	**Maximum speed (fpm)**	**Gear**	**Maximum load (lb)**	**Maximum speed (fpm)**
1	27,600	100	1	55,200	50
2	15,700	200	2	31,400	100
3	9,300	300	3	18,600	150
4	5,500	500	4	11,000	250

165 HP-AC, Eddy Current Brake, with four-speed remote controlled gear box, recommended service = 250 amp					
(1) Trolley, two-part line			**(2) Trolleys, four-part line**		
Gear	**Maximum load (lb)**	**Maximum speed (fpm)**	**Gear**	**Maximum load (lb)**	**Maximum speed (fpm)**
1	27,600	160	1	55,200	80
2	17,600	250	2	32,200	125
3	10,600	400	3	21,200	200
4	6,200	630	4	12,400	315

Source: Morrow Equipment Company, L.L.C.

capacity than a two-part line within the structural capacity constraints of the tower and jib configuration. The maximum lifting capacity of the crane will be increased by 100% with the four-part-line configuration. However, the increased lifting capacity is acquired with a resulting loss of 50% in vertical hoist speed.

Examination of the Table 17.3 load chart illustrates these points. The first portion of the table is for a crane rigged with a two-part line. Considering an L7 jib model, a crane so rigged could lift 27,600 lb at a radius of 10 ft 3 in., 25,800 lb at a radius of 94 ft 6 in., and 10,200 lb at a radius of 210 ft 0 in. This same crane rigged with a four-part-line arrangement and an L7 jib can lift 55,200 lb at a radius of 13 ft 6 in., 26,100 lb at a radius of 90 ft 0 in., and 8,300 lb at a radius of 210 ft 0 in.

When the operating radius is less than about 90 ft, the crane has a greater lifting capacity with a four-part-line than the two-part-line arrangement. However, when the operating radius exceeds 90 ft, the crane has a slightly greater lifting capacity with the two-part line. This is because of the increased weight of the four-part rigging system and the fact that structural capacity is the critical factor affecting load-lifting capability. However, when operating at a radius of less than 90 ft, the hoisting system controls the load-lifting capability.

Tower crane load charts are usually structured assuming that the weight of the hook block is part of the crane's dead weight. But the rigging system is taken as part of the lifted load. When calculating loads, the Construction Safety Association of Ontario recommends that a 5% working margin be applied to computed weight.

EXAMPLE 17.2

Can the tower crane, whose load chart is shown in Table 17.3, lift a 15,000-lb load at a radius of 142 ft? The crane has an L7 jib and a two-part line hoist. The slings that will be used for the pick weigh 400 lb.

Weight of load	15,000 lb	
Weight of rigging	400 lb	(slings)
	15,400 lb	
	× 1.05	working margin
Required capacity	16,170 lb	

From Table 17.3 the maximum lifting capacity at a 142-ft hook reach is 16,400 lb.

$$16,400 \text{ lb} > 16,170 \text{ lb}$$

Therefore, the crane can safely make the lift.

RIGGING

RIGGING BASICS

A crane is designed to pick (or lift) a load through the use of a hoisting mechanism using ropes. The load must be properly attached to the crane by a rigging system. To properly attach the load, it is necessary to determine the forces that will affect the job, and then to select and arrange the equipment that will move the load safely. The forces involved in rigging will vary with the method of connection, and the effects of motion. After correctly determining the weight and center of gravity of the load, the rigger must analyze the loading situation by proper application of mechanical laws and by resolving load movement-induced stresses.

Weight

The most important step in any rigging operation is to correctly determine the weight of the load. If this information cannot be obtained from the shipping papers, design plans, catalog data, or other dependable sources, it may be necessary to calculate the weight. It is good practice to verify the load weight as stated in the documents. Weights and properties of structural members can be obtained from:

1. Manual of Steel Construction, American Institute of Steel Construction (www.aisc.org).
2. Cold Formed Steel Design Manual, American Iron and Steel Institute (www.steel.org).
3. Aluminum Design Manual, American Aluminum Association (www.aluminum.org).

Center of Gravity

The center of gravity of an object is that location where the object will balance when lifted. When the object is suspended freely from a hook, this point will always be directly below the hook. Thus, a load that is slung above and through the center of gravity will be in equilibrium. It will not tend to slide out of the hitch or become unstable.

One way to determine the center of gravity of an odd-shaped object is to break up the shape into simple masses, and determine the resultant balancing load and its location at a point where the weights multiplied by their respective lever arms are in balance. The location of such a point can be computed for any quantity by using 1 for the lever arm of the larger weight (W_1). To be in equilibrium, $W_1 \times 1$ must equal the small weight (W_2) times the unknown arm length (see Fig. 17.17), and the unknown arm is then obtained from the equation:

$$\text{Unknown lever arm} = \frac{W_1 \times 1}{W_2} \qquad \text{[17.1]}$$

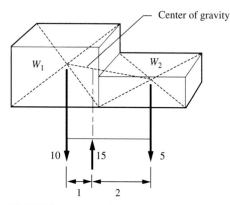

FIGURE 17.17 | Calculating center of gravity.

EXAMPLE 17.3

The larger part of an odd-shaped load (see Fig. 17.17) weighs 10 tons. The other part of the shape weighs 5 tons. The larger part has a 4 × 4 × 4 ft square shape. The smaller part is 4 ft wide by 4 ft long by 2 ft high. Determine the center of gravity in the long dimension of the object.

$$\text{Unknown lever arm} = \frac{10 \times 1}{5} \Rightarrow 2 \text{ ft}$$

Therefore, the center of gravity of the object is located 1 ft from the center of the larger part in the direction of the smaller part or if speaking of the total object the center of gravity in the long dimension is 3 ft from the edge of the larger part. The center of gravity in the short dimension is along the centerline of that dimension.

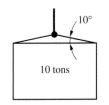

FIGURE 17.18 |
Stresses induced in a
set of slings.

Stresses

To calculate the stress developed by the load on a rigging arrangement, it must be remembered that all forces must be in equilibrium. If a 10-ton load is supported by a set of slings in such a manner that the individual sling legs make a 10° angle with the load (see Fig. 17.18), the sling is stressed 24 tons and there is a 23.3-ton horizontal reaction. Changing the sling angle to 45° will reduce the stress in the sling to 7 tons and the horizontal reaction to 4.8 tons.

Thus, it is apparent that when the rigging arrangement creates small sling angles, the resulting sling stress produced will be considerably greater than the load. The solution in such situations is to use a supplemental compression member—a spreader bar (see Fig. 17.4). The use of a spreader bar will allow for greater sling angles and reduce the induced sling stresses.

Mechanical Laws

The hook-block and tackle lifting mechanism acts as a lever with the fulcrum at the side of the block and the rope fixed at the boom point. These blocks provide a mechanical advantage for lifting the load.

Motion

Lifting operations involve loads in motion. The inertia of an object at rest or in motion that takes place when there is a decrease or increase in speed increases the stresses sustained by the rigging. As a hoist line starts to move, the load must accelerate from zero to the normal speed of ascent. This requires an additional force over and above the weight of the load. This force will vary with the rate of change in velocity.

A load lifted very slowly induces little additional stress in the load line. A load accelerated quickly by mechanical power may put twice as much stress on the line as the weight being lifted.

Factor of Safety

It is virtually impossible for the rigger to evaluate all the variables that can affect the rigging and lifting of a load. To compensate for unforeseen influences, a factor of safety is usually applied to the materials being utilized. This factor is defined as the usual breaking strength of a material divided by the allowable load weight. In the case of plow-steel cable, the factor is 10; for manila rope, it is 5. But always use the manufacturer's suggested safety factor. If the provided rating is in terms of breaking strength, divide that value by the factor of safety to arrive at the safe working load.

SLINGS

The safety and efficiency of a lift depend on the working attachments that secure the load to the crane hook. Generally called slings (see Fig. 17.4), these working attachments may be wire rope, chain, or nylon-web straps. When using a sling to lift, remember that the capacity of the sling depends on its material and size, the configuration in which it's used, its type of end terminals, and the angle that the legs make with the load.

Wire Rope

The capacity of a wire rope sling is based on the nominal strength of the wire rope. Factors affecting the overall strength of the sling include attachment or splicing efficiency, the wire rope's construction, and the diameter of the hook over which the eye of the sling is rigged. Consider the wire rope sling's ability to bend without distortion, and to withstand abrasive wear and abuse. Kinking, for example, causes serious structural damage and loss of strength.

Chains

Only alloy steel chain is suitable for slings used in overhead lifting—predominantly grade 80 or grade 100 alloy chain. Chain slings are ideal for rugged loads that would destroy other types of slings. Chain slings can have one to four legs connected to a master link.

Synthetic Web

Synthetic web slings are good for use on expensive loads, highly finished parts, fragile parts, and delicate equipment. They have less tendency than wire rope or chain slings to crush fragile objects. Because they are flexible, they tend to mold themselves to the shape of the load, thus gripping a load securely. Synthetic slings are elastic and stretch more than wire rope or chain, better-absorbing heavy shocks and cushioning loads.

Sling Inspection

Frequent sling inspection is essential to safe lifting operations.

Wire Rope Slings   If damage from any of these is visible, consider removing the sling from service:

1. Kinking, crushing, or any other damage resulting in distortion of the rope structure.
2. Severe corrosion of the rope or end attachments.
3. Severe localized abrasion or scraping.

Chain Slings   Make a link-by-link inspection for

1. Excessive wear.
2. Twisted, bent, or cut links.
3. Cracks in the weld area or any portion of the link.
4. Stretched links.

Synthetic Slings   If damage from any of these is visible, remove the sling from service:

1. Acid or caustic burns.
2. Melting or charring of any part of the sling.
3. Holes, tears, cuts, or snags.

SAFETY

CRANE ACCIDENTS

Crane accident data is limited because typically only deaths and injuries are reported. Property damage incidents are usually not reported, except to insurance carriers. The seriousness of a crane accident, however, is self-evident. Table 17.5 presents OSHA crane fatality data.

TABLE 17.5 | OSHA crane fatality accident data 1984–1994

Cause of accident	Crane operator	Other worker	Unknown	Total	Percent (%)
Boom or crane contact with power lines	17	179	2	198	39
Assembly/disassembly	2	51	5	58	12
Boom buckling	6	34	1	41	8
Crane overturned	23	12	2	37	7
Rigging failure	3	33	0	36	7
Overloading	8	14	0	22	4
Struck by moving load	1	21	0	22	4
Two-blocking*	1	10	0	11	2
Other	4	72	1	77	17
Total	65	426	11	502	100

*Two-blocking occurs when the lower load block or hook makes contact with the upper load block, boom point, or boom point machinery of the crane.

SAFETY PLANS AND PROGRAMS

There should be a corporate-level generalized crane safety program and a project-specific crane safety plan.

Crane Safety Program

The company crane safety plan should address

1. Equipment inspection.
2. Hazard analysis—concern for the public, power lines, etc.
3. Crane location.

4. Crane movements.

5. Definition of lifts—critical, production, general.

Crane Safety Plan

On a site- or project-specific basis, the crane safety plan addresses the same topics as the crane safety program. Equipment inspection methods and standards are set forth in detail. Operator aids, such as load indicators and anti-two-block devices, must be inspected daily prior to crane operation. Hazard analysis considers dangers identified in the contract documents and those associated with location of and access to the work—electric lines; ground conditions; and weather, wind, and cold. Crane location discusses crane setup locations with attention to unusual support or interference problems. Crane movement defines procedures for controlling crane movement. For each individual critical lift, a separate written lift plan should be prepared identifying equipment, load properties, personnel, location, and load path. The goal of lift planning is to eliminate as many variables as possible from the lifting equation. Every critical lift should be listed in the crane safety plan. Production lifts do not require a plan for each individual pick but a plan considering appropriate site-specific parameters for each type of production lift should be prepared. General lifts need not be listed.

ZONES OF RESPONSIBILITY

Because of various project situations, assignment of responsibilities can vary but there are three general categories of responsibility.

Rigging

Rigging personnel attach the load to the hook and perform other ground- or structure-based operations. Rigging personnel are responsible for the stability of the load. That responsibility involves

1. Verifying the actual weight of the load and communicating that information to the crane operator.

2. Attaching (rigging) the load using suitable lifting gear.

3. Signaling or directing the movement of the load by communication with the operator.

Operator

The crane operator controls the lift. The operator may abort the lift at any time from initial pickup to final placement. The operator is responsible for all crane movements from the hook upward as well as swing and travel motions. The operator's responsibility involves

1. Confirming from which individual directions will be given. Other than the case of a stop signal, an operator should respond only to the signals from the designated signal person.

2. Confirming that the configuration of the crane is appropriate for the load that is to be lifted, and in conformance with the load chart.
3. Being aware of the site conditions above, at, and below the ground.
4. Confirming the weight of the load.
5. Knowing the location and destination of the load.

Lift Director

The lift director is responsible for the entire lift and must ensure that there is full compliance with the Crane Safety Plan and the appropriate lift plan. The lift director is specifically responsible for

1. Ensuring that each of the other parties, riggers, operator, and signal persons understands their functions.
2. Ensuring that a signal person is assigned. If multiple signal persons are required, a thorough briefing on the transition between signalers with the crane operator is required.
3. Making a definite and clear assignment of the outrigger duties and responsibilities to avoid misunderstandings concerning the status of the outrigger operation.
4. Ensuring that everyone knows to whom they should communicate concerns about anything they observe that affects safety.

SUMMARY

Cranes are a broad class of construction equipment used to hoist and place material and machinery. The most common mobile crane types are (1) crawler, (2) telescoping-boom truck mounted, (3) lattice-boom truck mounted, (4) rough-terrain, (5) all-terrain, (6) heavy-lift, and (7) modified cranes for heavy lifting. The most common tower crane types are (1) top-slewing and (2) bottom-slewing. Tower cranes provide high lifting height and good working radius, while taking up a very limited area. These advantages are achieved at the expense of low lifting capacity and limited mobility, as compared to mobile cranes.

The rated load for a crane, as published by the manufacturer, is based on ideal conditions. Load charts can be complex documents listing numerous booms, jibs, and other components that may be employed to configure the crane for various tasks. It is critical that the chart being consulted be for the actual crane configuration that will be used. *Interpolation between the published values IS NOT permitted;* use the next lower value. Rated loads are based on ideal conditions, a level machine, calm air (no wind), and no dynamic effects. Critical learning objectives include:

■ An understanding of the basic mobile and tower crane types.
■ An understanding of load charts and their limitations.
■ An ability to read a mobile crane load chart.
■ An ability to read a tower crane load chart.
■ An understanding of responsibilities assigned by lift safety plans and programs.

These objectives are the basis for the problems that follow.

PROBLEMS

17.1 Using Fig. 17.10, select the minimum size crane required to unload pipe weighing 166,000 lb per joint and lower it into a trench when the distance from the centerline of the crane to the trench is 50 ft. (300-ton crane)

17.2 Using Figure 17.11, select the minimum length boom required to hoist a load of 80,000 lb from a truck at ground level and place it on a platform 76 ft above the ground. The minimum allowable vertical distance from the bottom of the load to the boom point of the crane is 42 ft. The maximum horizontal distance from the center of rotation of the crane to the hoist line of the crane when lifting the load is 40 ft.

17.3 High Lift Construction Co. has determined that the heaviest load to be lifted on a project they anticipate bidding weighs 14,000 lb. From the proposed tower crane location on the building site, the required reach for this lift will be 150 ft. The crane will be equipped with an L5 jib and a two-part hoist line (see the Table 17.3 load chart). This critical lift is of a piece of limestone facing and will require a 2,000-lb spreader bar attached to a 300-lb set of slings. If assembled in the proposed configuration, can the crane safely make the pick? (17,600 lb capacity > 17,115 lb load. Therefore, the crane can safely make the pick.)

17.4 Low Ball Construction Co. has determined that the heaviest load to be lifted on one of their projects weighs 22,000 lb. From the tower crane location on the building site, the required reach for this lift will be 100 ft. The crane is equipped with an L7 jib and a four-part hoist line (see the Table 17.3 load chart). This critical lift is of a piece of mechanical equipment and will require a 1,000-lb spreader bar attached to a 200-lb set of slings. If assembled in the proposed configuration, can the crane safely make the pick?

REFERENCES

1. *Articulating Boom Cranes,* ASME B30.22-1993, an American National Standard, The American Society of Mechanical Engineers, 345 East 47th Street, New York, 1994.

2. Bates, Glen E., and Robert M. Hontz, *Exxon Crane Guide, Lifting Safety Management System,* Specialized Carriers & Riggers Association, Fairfax, Va., 1998.

3. *Below-the-Hook Lifting Devices,* ASME B30.20-1993, an American National Standard, The American Society of Mechanical Engineers, 345 East 47th Street, New York, 1994.

4. *Crane Handbook,* Construction Safety Association of Ontario, 74 Victoria St., Toronto, Ontario, Canada M5C 2A5, 1990.

5. "Crane Load Stability Test Code—SAE J765," in *SAE Recommended Practice Handbook,* Society of Automotive Engineers, Inc., Warrendale, Pa., 1980.

6. *Crane Safety on Construction Sites,* ASCE Manuals and Reports on Engineering Practice No. 93, American Society of Civil Engineers, Reston, Va., 1998.

7. *Hammerhead Tower Cranes,* ASME B30.3-1990, an American National Standard, The American Society of Mechanical Engineers, 345 East 47th Street, New York, 1992.

8. *Mobile Crane Manual,* Construction Safety Association of Ontario, 74 Victoria St., Toronto, Ontario, Canada M5C 2A5, 1993.

9. *Mobile Power Crane and Excavator and Hydraulic Crane Standards, PCSA Standard 1,* Power Crane and Shovel Association, a Bureau of Construction Industry Manufacturers Association, 111 East Wisconsin Avenue, Milwaukee, WI 53202, 1968, www.cimanet.com/.

10. *Rigging Manual,* Construction Safety Association of Ontario, 74 Victoria St., Toronto, Ontario, Canada M5C 2A5, 1992.

11. Shapira, Aviad, and Clifford J. Schexnayder, "Selection of Mobile Cranes for Building Construction Projects," *Construction Management & Economics* (United Kingdom), Vol. 17, No. 4, 1999, pp. 519–527.

12. Shapiro, Howard I., Jay P. Shapiro, and Lawrence K. Shapiro, *Cranes and Derricks,* 2d ed., McGraw-Hill Book Company, New York, 1991.

13. Shapira, Aviad, and Jay D. Glascock, "Culture of Using Mobile Cranes for Building Construction," *Journal of Construction Engineering and Management,* ASCE, Vol. 122, No. 4, 1996, pp. 298–307.

14. Rossnagel, W. E., Lindley R. Higgins, and Joseph A. MacDonald, *Handbook of Rigging for Construction and Industrial Operations,* 4th ed., McGraw-Hill, 1988.

15. Link-Belt Construction Equipment Company, P.O. Box 13600, 2651 Palumbo Drive, Lexington KY 40583-3600, www.linkbelt.com/.

16. Manitowoc Cranes, Inc., P.O. Box 70, Manitowoc, WI 54221, www.manitowoc.com/.

17. Grove Worldwide, 1565 Buchanan Trail East, P.O. Box 21, Shady Grove, PA 17256, www.groveworldwide.com/.

18. Liebherr-America, Inc., 4100 Chestnut Ave., P.O. Box Drawer O., Newport News, VA 23605, www.liebherr.com.

19. POTAIN, 18 rue de Charbonnières B.P. 173 69132 ECULLY Cedex FRANCE, www.potain.com/.

20. American Crane Corporation, A Subsidiary of Terex Lifting, 202 Raleigh Street, Wilmington, NC 28412, www.american-crane.com.

CRANE AND LIFTING-RELATED WEBSITES

1. Construction Safety Association of Ontario, www.csao.org/. 74 Victoria St., Toronto, Ontario, Canada M5C 2A5. Provides health and safety education, consultation, and information to workers and management in the construction industry. Keeps the industry well informed about health and safety through educational programs and material such as publications and videotapes. Researches the health and safety impact of construction procedures, equipment, and materials. Assists the industry in developing and implementing standards, procedures, and regulations for health and safety.

2. Cranes Today, www.cranestodaymagazine.com/. Cranes Today on the web is a gateway to the lifting industry on-line and industry news.

3. Power Crane and Shovel Association (PCSA), www.cimanet.com/. PCSA is a bureau of Construction Industry Manufacturers Association, 111 East Wisconsin Avenue, Milwaukee, WI 53202. The Power Crane and Shovel Association provides services tailored to meet the needs of the lattice-boom and truck crane industry. Bureau member companies are considered the global leaders in the

promotion of worldwide harmonization of crane standards. PCSA explores technological questions and legislative and regulatory concerns (domestic and worldwide) that affect manufacturers of cranes. It also promotes the standardization and simplification of terminology and classification of cranes for worldwide harmonization.

4. Society of Automotive Engineers (SAE), www.sae.org/jsp/index.jsp. SAE World Headquarters, 400 Commonwealth Drive, Warrendale, PA 15096-0001, Tel: (877) 606-7323 (United States and Canada only) or (724) 776-4970 (outside the United States and Canada), Fax: Customer Service: (724) 776-0790, Headquarters: (724) 776-5760. The Society of Automotive Engineers is a resource for technical information and expertise used in designing, building, maintaining, and operating self-propelled vehicles. SAE publishes thousands of technical papers and books each year, and leading-edge periodicals and Internet and CD-ROM products, too. The SAE Cooperative Research Program helps facilitate projects that benefit the mobility industry as a whole. Numerous meetings and expositions provide worldwide opportunities to network and share information. The SAE also offers a full complement of professional development activities such as seminars, workshops, and continuing education programs.

5. Specialized Carriers and Rigging Association (SC&RA), www.scranet.org/. 2750 Prosperity Avenue, Suite 620, Fairfax, Virginia 22031-4312, Tel: (703) 698-0291, Fax: (703) 698-0297, E-mail info@scranet.org. Specialized Carriers and Rigging Association provides information about safe transporting, lifting, and erecting oversize and overweight items.

Draglines and Clamshells

Draglines and clamshells are both bucket attachments hung from a lattice-boom crane. A dragline works, as the name implies, by dragging a dragline-type bucket toward the machine. The greatest advantage of a dragline over other machines is its long reach for digging and dumping. The dragline is designed to excavate below the level of the machine. The clamshell bucket is designed to excavate material in a vertical direction. It works like an inverted jaw with a biting motion. The clamshell is capable of working at, above, and below ground level.

INTRODUCTION

Draglines and clamshells are both bucket attachments hung from a lattice-boom crane. The terms dragline and clamshell refer to the particular type of bucket used and to the digging motion of the bucket. A dragline works, as the name implies, by dragging a dragline-type bucket toward the machine. The clamshell bucket is designed to excavate material in a vertical direction. It works like an inverted jaw with a biting motion. With both types of excavators, the buckets are attached to the crane only by cables. Therefore, the operator has no positive control of the bucket, as with hydraulic excavators.

Draglines and clamshell machines belong to a group that is frequently identified as the Power Crane and Shovel Association (PCSA) family [1]. This association has conducted and supervised studies and tests that have provided considerable information related to the performance, operating conditions, production rates, economic life, and cost of owning and operating these machines. The association has participated in establishing and adopting certain standards that are applicable to the equipment. The results of the studies, conclusions and actions, and the standards have been published in technical bulletins and booklets. Some of that information published by the Power Crane and Shovel Association is reproduced in this book, with the permission of the association.

DRAGLINES

GENERAL INFORMATION

The dragline is a versatile machine capable of a wide range of operations. It can handle materials that range from soft to medium hard. The greatest advantage of a dragline over other machines is its long reach for digging and dumping. A dragline does not have the positive digging force of a hydraulic shovel or hoe. The bucket breakout force is derived strictly from its own weight. Therefore, it can bounce, tip over, or drift sideward when it encounters hard material. These weaknesses are particularly noticeable with smaller machines and lightweight buckets.

Draglines are used to excavate material and load it into hauling units, such as trucks or tractor-pulled wagons, or to deposit it in levees, dams, and spoil banks near the pits from which it is excavated. The dragline is designed to excavate below the level of the machine. A dragline usually does not have to go into a pit or hole to excavate. It operates adjacent to the pit while excavating material from the pit by casting its bucket. This is very advantageous when earth is removed from a ditch, canal, or pit containing water. If the material is hauled with trucks, they do not have to go into the pit and maneuver through mud. However, when possible, it is better to position the haul units in the pit. Positioning the haul units in the excavation below the dragline will reduce hoist time and increase production.

Frequently it is possible to use a dragline with a long boom to dispose of the earth in one operation if the material can be deposited along the canal or near the pit. This eliminates the need for hauling units, reducing the cost of handling the material.

Crawler-mounted draglines (Fig. 18.1) can operate over soft ground conditions that would not support wheel- or truck-mounted equipment. The travel speed of a crawler machine is very slow, frequently less than 1 mph, and it is necessary to use auxiliary hauling equipment to transport the unit from one job to another. Wheel- and truck-mounted lattice-boom cranes may also be rigged with dragline attachments, but that is not a common occurrence.

There are also very large draglines, 5 cubic yards (cy) and up. These are usually walking machines (Fig. 18.2). They have a base that rests on the ground and carries the turntable on its upper side. Cams operate the walking shoes on each side of the machine. As the cam rotates, the shoes contact the ground, raise the machine, and support its weight as a movement or step is made. These large machines are used in strip mining and levee work.

DESCRIPTION OF A DRAGLINE

Dragline components (Fig. 18.3) consist of a drag bucket and a fairlead assembly. Wire ropes are used for the boom suspension, drag, bucket hoist, and dump

FIGURE 18.1 I Crawler-mounted dragline excavating a ditch.

FIGURE 18.2 I Walking dragline.

lines. The fairlead guides the drag cable onto the drum when the bucket is being loaded. The hoist line that operates over the boom-point sheave is used to raise and lower the bucket. In the digging operation, the drag cable is used to pull the

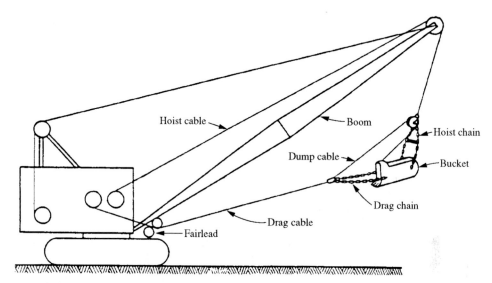

FIGURE 18.3 | Basic parts of a dragline.

bucket through the material. When the bucket is raised and moved to the dump point, releasing the tension on the drag cable empties it.

Size of a Dragline

The size of a dragline is indicated by the size of the bucket, expressed in cubic yards. However, most draglines may handle more than one size bucket, depending on the length of the boom utilized and the unit weight of the material excavated. The relationship between bucket size and boom length and angle is presented in Fig. 18.4. The dragline boom may be angled relatively low when operating. However, boom angles of less than 35° from the horizontal are seldom advisable because of the possibility of tipping the machine. Because the maximum lifting capacity of a dragline is limited by the force that will tilt the machine over, it is necessary to reduce the size of the bucket when a long boom is used or when the excavated material has a high unit weight. When excavating wet-sticky material and casting onto a spoil bank, the chance of tipping the machine increases because of material sticking in the bucket. In practice, the combined weight of the bucket and its load should produce a tilting force not greater than 75% of the force required to tilt the machine. A longer boom, with a smaller bucket, will be used to increase the digging reach or the dumping radius when it is not desirable to bring in a larger machine.

If the material is difficult to excavate, the use of a smaller bucket that will reduce the digging resistance may permit an increase in the production.

Typical working ranges for a dragline that will handle buckets varying in size from $1\frac{1}{4}$ to $2\frac{1}{2}$ cy are given in Table 18.1 (see Fig. 18.5 for the dimensions given in the table).

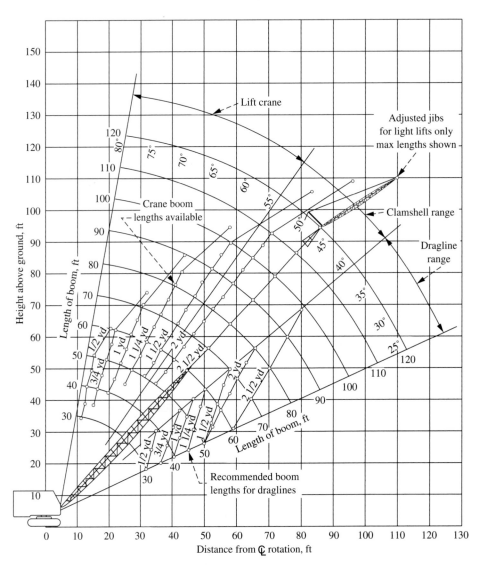

FIGURE 18.4 I Working ranges of dragline and clamshell machines.

Operation of a Dragline

Swinging the empty bucket to the digging position while at the same time slacking off the draglines and the hoist lines starts the excavating cycle. There are separate drums on the crane unit for each of these cables so that they can be coordinated into a smooth operation. Digging is accomplished by pulling the bucket toward the machine while regulating the digging depth by means of the tension maintained in the hoist line. When the bucket is filled, the operator takes in on the hoist line while playing out the dragline. The bucket is so constructed

TABLE 18.1 | Typical dragline working ranges with maximum counterweights

J, boom length 50 ft						
Capacity (lb)*	12,000	12,000	12,000	12,000	12,000	12,000
K, boom angle (degrees)	20	25	30	35	40	45
A, dumping radius (ft)	55	50	50	45	45	40
B, dumping height (ft)	10	14	18	22	24	27
C, max. digging depth (ft)	40	36	32	28	24	20
J, boom length 60 ft						
Capacity (lb)*	10,500	11,000	11,800	12,000	12,000	12,000
K, boom angle (degrees)	20	25	30	35	40	45
A, dumping radius (ft)	65	60	55	55	52	50
B, dumping height (ft)	13	18	22	26	31	35
C, max. digging depth (ft)	40	36	32	28	24	20
J, boom length 70 ft						
Capacity (lb)*	8,000	8,500	9,200	10,000	11,000	11,800
K, boom angle (degrees)	20	25	30	35	40	45
A, dumping radius (ft)	75	73	70	65	60	55
B, dumping height (ft)	18	23	28	32	37	42
C, max. digging depth (ft)	40	36	32	28	24	20
J, boom length 80 ft						
Capacity (lb)*	6,000	6,700	7,200	7,900	8,600	9,800
K, boom angle (degrees)	20	25	30	35	40	45
A, dumping radius (ft)	86	81	79	75	70	65
B, dumping height (ft)	22	27	33	39	42	47
C, max. digging depth (ft)	40	36	32	28	24	20
D, digging reach	Depends on working conditions and operator's skills with bucket					

*Combined weight of bucket and material must not exceed capacity.

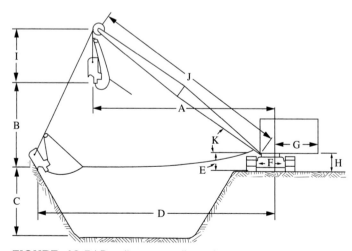

FIGURE 18.5 | Dragline range dimensions.

that it will not dump its contents until the dragline tension is released. Hoisting, swinging, and dumping of the loaded bucket follow in that order; then the cycle is repeated. An experienced operator can cast the excavated material beyond the end of the boom.

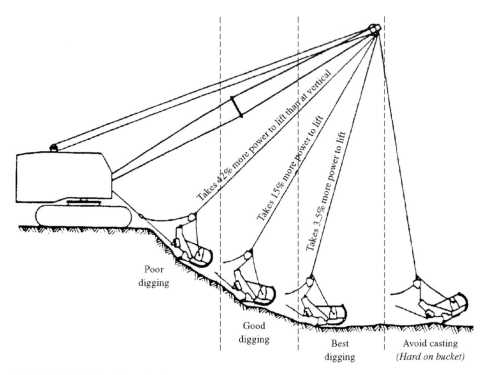

Poor
digging

Good
digging

Best
digging

Avoid casting
(Hard on bucket)

Takes 42% more power to lift than at vertical

Takes 15% more power to lift

Takes 3.5% more power to lift

FIGURE 18.6 I Dragline digging zones.

When compared to a hydraulic excavator, the dragline bucket is more difficult to accurately control when dumping. Therefore, when loading haul units with a dragline, larger units should be used to reduce the spillage. A size ratio equal to at least five to six times the capacity of the dragline bucket is recommended.

Figure 18.6 shows the dragline digging zones. The work should be planned to permit most of the digging to be done in the zones that permit the best digging, with the poor digging zone used as little as possible.

DRAGLINE PRODUCTION

Dragline production will vary with these factors:

1. Type of material being excavated.
2. Depth of cut.
3. Angle of swing.
4. Size and type of bucket.
5. Length of boom.
6. Method of disposal, casting, or loading haul units.
7. Size of the hauling units, when used.
8. Skill of the operator.

TABLE 18.2 | Approximate dragline digging and loading cycles for various angles of swing*

Size of dragline bucket (cy)	Easy digging light moist clay or loam angle of swing (degrees)				Sand or gravel angle of swing (degrees)				Good common earth angle of swing (degrees)			
	45	90	135	180	45	90	135	180	45	90	135	180
$\frac{3}{8}$	16	19	22	25	17	20	24	27	20	24	28	31
$\frac{1}{2}$	16	19	22	25	17	20	24	27	20	24	28	31
$\frac{3}{4}$	17	20	24	27	18	22	26	29	21	26	30	33
1	19	22	26	29	20	24	28	31	23	28	33	36
$1\frac{1}{4}$	19	23	27	30	20	25	29	32	23	28	33	36
$1\frac{1}{2}$	21	25	29	32	22	27	31	34	25	30	35	38
$1\frac{3}{4}$	22	26	30	33	23	28	32	35	26	31	36	39
2	23	27	31	35	24	29	33	37	27	32	37	41
$2\frac{1}{2}$	25	29	34	38	26	31	36	40	29	34	40	44

*Time is in seconds with no delays when digging at optimum depths of cut and loading trucks on the same grade as the excavator.
Source: Power Crane and Shovel Association.

9. Physical condition of the machine.

10. Job conditions.

Table 18.2 gives approximate dragline digging and loading cycles for various angles of swing.

The output of a dragline should be expressed in bank measure cubic yards (bcy) per hour. This quantity is best obtained from field measurements. It can be estimated by multiplying the average loose volume per bucket by the number of cycles per hour and dividing by 1 plus the swell factor for the material, expressed as a fraction. For example, if a 2-cy bucket, excavating material whose swell is 25%, will handle an average loose volume of 2.4 cy, the bank-measure volume will be 1.92 cy, $\frac{2.4}{1.25}$. If the dragline can make 2 cycles per min, the output will be 3.84 bcy per min (2 × 1.92) or 230 bcy per hour. This is an ideal 60-min hour peak output that will not be sustainable over the duration of a project.

Optimum Depth of Cut

A dragline will produce its greatest output if the job is planned to permit excavation at the optimum depth of cut where possible. Table 18.3 gives the optimum depth of cut for various sizes of buckets and classes of materials based on using short-boom draglines. Ideal outputs of short-boom draglines, expressed in bcy, for various classes of materials, when digging at the optimum depth, with a 90° swing, and no delays are presented in the table. The upper figure is the optimum depth in feet and the lower number is the ideal output in cubic yards.

Effect of Depth of Cut and Swing Angle on Production

The outputs of draglines given in Table 18.3 are based on digging at optimum depths with a swing angle of 90°. For any other depth or swing angle the ideal output of the machine must be adjusted by an appropriate depth-swing factor.

TABLE 18.3 | Optimum depth of cut and ideal outputs of short-boom draglines*

	Size of bucket [cy (cu m)]†								
Class of material	$\frac{3}{8}$ **(0.29)†**	$\frac{1}{2}$ **(0.38)†**	$\frac{3}{4}$ **(0.57)†**	**1 (0.76)†**	**1$\frac{1}{4}$ (0.95)†**	**1$\frac{1}{2}$ (1.14)†**	**1$\frac{3}{4}$ (1.33)†**	**2 (1.53)†**	**2$\frac{1}{2}$ (1.91)†**
Moist loam or light sandy clay	5.0	5.5	6.0	6.6	7.0	7.4	7.7	8.0	8.5
	(1.5)‡	(1.7)‡	(1.8)‡	(2.0)‡	(2.1)‡	(2.2)‡	(2.4)‡	(2.5)‡	(2.6)‡
	70	95	130	160	195	220	245	265	305
	(53)§	(72)§	(99)§	(122)§	(149)§	(168)§	(187)§	(202)§	(233)§
Sand and gravel	5.0	5.5	6.0	6.6	7.0	7.4	7.7	8.0	8.5
	(1.5)	(1.7)	(1.8)	(2.0)	(2.1)	(2.2)	(2.4)	(2.5)	(2.6)
	65	90	125	155	185	210	235	255	295
	(49)	(69)	(95)	(118)	(141)	(160)	(180)	(195)	(225)
Good common earth	6.0	6.7	7.4	8.0	8.5	9.0	9.5	9.9	10.5
	(1.8)	(2.0)	(2.4)	(2.5)	(2.6)	(2.7)	(2.8)	(3.0)	(3.2)
	55	75	105	135	165	190	210	230	265
	(42)	(57)	(81)	(104)	(127)	(147)	(162)	(177)	(204)
Hard, tough clay	7.3	8.0	8.7	9.3	10.0	10.7	11.3	11.8	12.3
	(2.2)	(2.5)	(2.7)	(2.8)	(3.1)	(3.3)	(3.5)	(3.6)	(3.8)
	35	55	90	110	135	160	180	195	230
	(27)	(42)	(69)	(85)	(104)	(123)	(139)	(150)	(177)
Wet, sticky clay	7.3	8.0	8.7	9.3	10.0	10.7	11.3	11.8	12.3
	(2.2)	(2.5)	(2.7)	(2.8)	(3.1)	(3.3)	(3.5)	(3.6)	(3.8)
	20	30	55	75	95	110	130	145	175
	(15)	(23)	(42)	(58)	(73)	(85)	(100)	(112)	(135)

*In cubic yards (cubic meters) bank measure (bcy) per 60-min hour.
†These values are the sizes of the buckets in cubic meters (cu m).
‡These values are the optimum depths of cut in meters (m).
§These values are the optimum ideal outputs in cubic meters (cu m).

The effect of the depth of cut and swing angle on dragline production is given in Table 18.4. In the table, the percentage of optimum depth of cut is obtained by dividing the actual depth of cut by the optimum depth for the given material and bucket, then multiplying the result by 100.

TABLE 18.4 | Effect of the depth of cut and swing angle on dragline production

Percentage of optimum depth	Angle of swing (degrees)							
	30	**45**	**60**	**75**	**90**	**120**	**150**	**180**
20	1.06	0.99	0.94	0.90	0.87	0.81	0.75	0.70
40	1.17	1.08	1.02	0.97	0.93	0.85	0.78	0.72
60	1.24	1.13	1.06	1.01	0.97	0.88	0.80	0.74
80	1.29	1.17	1.09	1.04	0.99	0.90	0.82	0.76
100	1.32	1.19	1.11	1.05	1.00	0.91	0.83	0.77
120	1.29	1.17	1.09	1.03	0.98	0.90	0.82	0.76
140	1.25	1.14	1.06	1.00	0.96	0.88	0.81	0.75
160	1.20	1.10	1.02	0.97	0.93	0.85	0.79	0.73
180	1.15	1.05	0.98	0.94	0.90	0.82	0.76	0.71
200	1.10	1.00	0.94	0.90	0.87	0.79	0.73	0.69

Production Estimating

Hourly production rates for cranes with dragline attachments are given in Table 18.3. These rates are based on optimum cutting depth, 90° swing angle, type of soil, and maximum efficiency. Table 18.4 gives correction factors for different depths of excavation and swing angles. Refer to Table 4.3 for soil conversion factors. Overall efficiency should be based on past experiences.

Step 1. Determine the estimated production from Table 18.3, based on proposed bucket size and the type of material.

Step 2. Determine the percent of optimum depth of cut, using the following formula and data from Table 18.3:

Percentage of optimum depth of cut

$$= \frac{\text{Actual depth of cut}}{\text{Optimum depth of cut (Table 18.3)}} \times 100 \quad \textbf{[18.1]}$$

Step 3. Determine the depth of cut/swing angle correction factor from Table 18.4, using the calculated percent of optimum depth of cut and the planned angle of swing. In some cases, it may be necessary to interpolate between Table 18.4 values.

Step 4. Determine an overall efficiency factor based on the job conditions. Draglines are seldom productively working for more than 45 min in 1 hr.

Step 5. Determine production rate. Multiply the estimated production by the depth/swing correction factor and the efficiency factor:

Step 6. Determine soil conversion, if needed (Table 4.3).

Step 7. Determine total hours:

$$\text{Total hours} = \frac{\text{cubic yards moved}}{\text{Production rate/hr}} \quad \textbf{[18.2]}$$

EXAMPLE 18.1

A 2-cy short-boom dragline is to be used to excavate hard, tough clay. The depth of cut will be 15.4 ft, and the swing angle will be 120°. Determine the probable production of the dragline. There are 35,000 bcy of material to be excavated. How long will the project require?

Step 1. Determine the estimated production from Table 18.3, based upon a 2-cy bucket size and hard, tough clay material: 195 bcy.

Step 2. Determine the percent of optimum depth of cut, Eq. [18.1]. Optimum depth of cut Table 18.3: 11.8 ft.

$$\text{Percentage of optimum depth of cut} = \frac{15.4 \text{ ft}}{11.8 \text{ ft}} \times 100 \Rightarrow 130\%$$

Step 3. Determine the depth of cut/swing angle correction factor from Table 18.4:

Percentage of optimum depth of cut = 130%
Swing angle = 120°
Depth of cut/swing angle correction factor = 0.89

Step 4. Determine an overall efficiency factor based on the job conditions. Draglines seldom work at better than a 45-min hour:

$$\text{Efficiency factor} = \frac{45 \text{ min}}{60 \text{ min}} \Rightarrow 0.75$$

Step 5. Determine production rate. Multiply the estimated production by the depth/swing correction factor and the efficiency factor:

$$\text{Production} = 195 \times 0.89 \times 0.75 \Rightarrow 130 \text{ bcy per hr}$$

Step 6. Determine soil conversion, if needed (Table 4.3). Not necessary in this example.

Step 7. Determine total hours, Eq. [18.2]:

$$\text{Total hours} = \frac{35,000 \text{ bcy}}{130 \text{ bcy per hr}} \Rightarrow 269 \text{ hr}$$

EFFECT OF BUCKET SIZE AND BOOM LENGTH ON DRAGLINE PRODUCTION

In selecting the size and type bucket to use on a project, one should match the size of the dragline and bucket properly to obtain the best action and the greatest operating efficiency that will produce the greatest output of material. Buckets are generally available in three types: (1) light duty, (2) medium duty, and (3) heavy duty. Light-duty buckets are used for excavating materials that are easily dug, such as sandy loam, sandy clay, or sand. Medium-duty buckets are used for general excavating service in excavating clay, soft shale, or loose gravel. Heavy-duty buckets are used for mine stripping, handling blasted rock, and excavating hardpan and highly abrasive materials. Buckets are sometimes perforated to permit excess water to drain from the loads. Figure 18.7 shows a medium-duty dragline bucket.

Table 18.5 gives representative capacities, weight, and dimensions for dragline buckets.

The normal size of a dragline bucket is based on its struck capacity that is expressed in cubic feet (cf). In selecting the most suitable size bucket for use with a given dragline, it is desirable to know the weight of the loose material to be handled, expressed in pounds per cubic foot (lb per cf). While it is desirable to use the largest bucket possible in the interest of increasing production, a careful analysis should be performed to make sure the combined weight of the load and the bucket does not exceed the recommended safe load for the dragline. The importance of this analysis is illustrated by referring to the information given in Table 18.1.

FIGURE 18.7 I Medium-duty dragline bucket.

TABLE 18.5 I Representative capacities, weights, and dimensions of dragline buckets

Size (cy)	Struck capacity (cf)	Weight of bucket (lb)			Dimension (in.)		
		Light duty	Medium duty	Heavy duty	Length	Width	Height
$\frac{3}{8}$	11	760	880		35	28	20
$\frac{1}{2}$	17	1,275	1,460	2,100	40	36	23
$\frac{3}{4}$	24	1,640	1,850	2,875	45	41	25
1	32	2,220	2,945	3,700	48	45	27
$1\frac{1}{4}$	39	2,410	3,300	4,260	49	45	31
$1\frac{1}{2}$	47	3,010	3,750	4,525	53	48	32
$1\frac{3}{4}$	53	3,375	4,030	4,800	54	48	36
2	60	3,925	4,825	5,400	54	51	38
$2\frac{1}{4}$	67	4,100	5,350	6,250	56	53	39
$2\frac{1}{2}$	74	4,310	5,675	6,540	61	53	40
$2\frac{3}{4}$	82	4,950	6,225	7,390	63	55	41
3	90	5,560	6,660	7,920	65	55	43

EXAMPLE 18.2

Assume that the material to be handled has a loose weight of 90 lb per cf. The use of a 2-cy medium-duty bucket will be considered. If the dragline is to be operated with an 80-ft boom at a 40° angle, the maximum safe load will be 8,600 lb. The approximate weight of the bucket and its load will be

Bucket, from Table 18.5	=	4,825 lb
Earth, 60 cf at 90 lb per cf	=	5,400 lb
Combined weight	=	10,225 lb
Maximum safe load	=	8,600 lb

As this weight will exceed the safe load of the dragline, it will be necessary to use a smaller bucket. Try a $1\frac{1}{2}$-cy bucket, whose combined weight will be

Bucket, from Table 18.5	= 3,750 lb
Earth, 47 cf at 90 lb per cf	= 4,230 lb
Combined weight	= 7,980 lb
Maximum safe load	= 8,600 lb

If a $1\frac{1}{2}$-cy bucket is used, it may be filled to heaping capacity, without exceeding the safe load of the dragline.

If a 70-ft boom at a 40° angle, whose maximum safe load is 11,000 lb, will provide sufficient working range for excavating and disposing of the earth, a 2-cy bucket may be used and filled to heaping capacity. The ratio of the output resulting from the use of a 70-ft boom and a 2-cy bucket, compared with a $1\frac{1}{2}$-cy bucket, should be approximately

Output ratio (60 cf/47 cf) × 100	= 127%
Increase in production	= 27%

This does not consider the cycle time effect of the different boom lengths.

Example 18.2 illustrates the importance of analyzing a job prior to selecting the size dragline and bucket to be used. Random selection of equipment can result in a substantial increase in the cost of material handling. Efficient handling of materials is a key issue in heavy construction.

EXAMPLE 18.3

This example illustrates a method of analyzing a project to determine the size dragline required for digging a canal. Select a crawler-mounted dragline to excavate 210,000 bcy of common earth having a loose weight of 80 lb per cf. The dimensions of the canal will be

Bottom width	20 ft
Top width	44 ft
Depth	12 ft
Side slopes	1:1

The excavated earth will be cast into a levee along one side of the canal, with a berm of at least 20 ft between the toe of the levee and the nearest edge of the canal. The cross-sectional area of the canal will be

$$\frac{(20 + 44)}{2} \times 12 = 384 \text{ sq ft (sf)}$$

If the earth swells 25% when it is excavated, the cross-sectional area of the levee will be

$$384 \text{ sf} \times 1.25 = 480 \text{ sf}$$

The levee dimensions will be

Height	12 ft
Base width	64 ft
Crest width	16 ft
Side slope	2:1

The total width from the outside of the levee to the farside of the canal will be

Width of levee = 64 ft
Width of berm = 20 ft
Width of canal = 44 ft
     Total = 128 ft

or the machine will have to reach 64 ft, assuming it will turn 180° during the digging and dumping operation.

With a boom angle of 30°, a dragline having 70 ft of boom will be required (Fig. 18.3) to provide the necessary digging and dumping reaches, and to permit adequate dumping height and digging depth.

The project must be completed in 1 yr. Assume that weather conditions, holidays, and other major losses in time will reduce the operating time to 44 weeks of 40 hr each, or a total of 1,760 working hours. The required production per working hour will be 119 bcy. It should be possible to operate with a 150° average swing angle. The efficiency factor should be a 45-min hour.

The required production divided by the efficiency factor is

$$\frac{133}{45/60} = 177 \text{ bcy/hr}$$

Assume the depth-swing factor will be 0.81. The required ideal production is

$$\frac{177}{0.81} = 219 \text{ bcy/hr}$$

Table 18.3 indicates a $1\frac{3}{4}$-cy medium-duty bucket will meet the need. The combined weight of the bucket and load will be

Weight of load, 53 cf at 80 lb per cf	= 4,240 lb
Weight of bucket	= 4,030 lb
Total weight	= 8,270 lb
Maximum safe load, Table 18.1	= 9,200 lb

The equipment selected should be checked to verify whether it will produce the required production:

Ideal output, 210 bcy per hr

Percent of optimum depth, $\dfrac{12.0}{9.5} \times 100 = 126$

Depth-swing factor, 0.82

Efficiency factor, 45-min hour, or 0.75

Probable output, 210 bcy per hour $\times$ 0.82 $\times$ 0.83 = 129 bcy per hr

Thus the equipment should produce the required output of 113 cy, with a slight surplus capacity.

EFFECTIVE DRAGLINE OPERATIONS

A dragline should produce from 60 to 90% of a hydraulic shovel of the same size when used to load haul units positioned on the same level. This production should increase to between 70 and 100% if the haul units are positioned in the excavation. Important points to consider when planning a dragline operation include

- Although the dragline bucket can be easily cast beyond the length of the boom, position the machine to eliminate unnecessary casting and hoisting.
- Use heavy timber mats for work on soft ground. The mats should be kept level and as clean as possible.
- When positioning a dragline on a job, ensure access for maintenance personnel, operating personnel, and haul equipment.
- Excavate the working area in layers, not in trenches, and keep it sloped upward toward the machine.
- Never guide the dragline bucket by swinging the superstructure while digging. This puts side stress on the boom that can cause failure. Do not swing the boom until the bucket has been raised clear of the ground.

CLAMSHELLS

GENERAL INFORMATION

The clamshell machine consists of a clamshell bucket hung from a lattice-boom crane (Fig. 18.8). The clamshell bucket consists of two scoops hinged together. A clamshell cannot be operated off a jib boom. Clamshell drum laggings may be the same as those used for the crane, or they may be changed to meet speed and pull requirements of the clamshell. The same wire ropes used for the crane (hook) operation can, usually, be used for clamshell operations. However, two additional lines must be added: a secondary hoist line and the tag line. The tag line is a small diameter cable with spring tension winder used to prevent the clamshell bucket from twisting during operation. The tag line and winder, like the clamshell bucket, are interchangeable with any make or model in the same size range. Being spring-loaded, the tag line does not require operator control and does not attach to the crane's operating drums. The winder is usually mounted on the lower part of the boom.

The clamshell is a vertically operated attachment capable of working at, above, and below ground level. The clamshell bucket is attached to the hoist line instead of a hook block. The clamshell can dig in loose to medium stiff soils. The length of the boom determines the height a clamshell can reach. The length

Figure 18.8 | A crane with a clamshell bucket unloading aggregate from a barge.

of wire rope the cable drums can accommodate limits the depth a clamshell can reach. A clamshell's lifting capacity varies greatly. Factors such as boom length, operating radius, size of clam bucket, and unit weight of material excavated determine a clamshell's safe lifting capacity.

The holding, closing, and tag lines control the bucket. At the start of the digging cycle, the bucket rests on the material to be dug, with the shells open. As the closing line is reeved in, the two shells of the bucket are drawn together, causing them to dig into the material. The weight of the bucket, which is the only crowding action available, helps the bucket penetrate the material. The holding and closing lines then raise the bucket and swing it to the dumping point. The bucket is opened to dump by releasing the tension on the closing line.

Clamshells are used primarily for handling loose materials such as sand, gravel, crushed stone, coal, and shells, and for removing materials from vertical excavations such as cofferdams, pier foundations, sewer manholes, and sheet-lined trenches.

CLAMSHELL BUCKETS

Clamshell buckets are available in various sizes, and in heavy-duty types for digging, medium-weight types for general-purpose work, and lightweight types

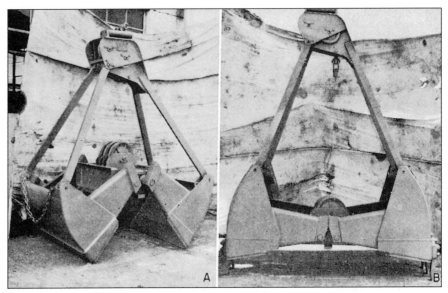

FIGURE 18.9 | (a) Wide rehandling clamshell bucket. (b) Heavy-duty clamshell bucket.

for rehandling light materials. Manufacturers supply buckets either with removable teeth or without teeth. Teeth are used in digging the harder types of materials but are not required when a bucket is used for rehandling purposes. Figure 18.9 illustrates a rehandling and a heavy-duty digging bucket.

The capacity of a clamshell bucket is usually given in cubic yards. A more accurate capacity is given as water-level, plate-line, or heaped-measure, generally expressed in cubic feet. The water-level capacity is the capacity of the bucket if it were hung level and filled with water. The plate-line capacity indicates the capacity of the bucket following a line along the tops of the clams. The heaped capacity is the capacity of the bucket when it is filled to the maximum angle of repose for the given material. In specifying the heaped capacity the angle of repose usually is assumed to be 45°. The term "deck area" indicates the number of square feet covered by the bucket when it is fully open. Table 18.6 gives representative specifications for medium-weight general-purpose-type buckets furnished by one manufacturer.

PRODUCTION RATES FOR CLAMSHELLS

Because of the variable factors that affect the operations of a clamshell, it is difficult to give production rates that are dependable. These factors include the difficulty of loading the bucket, the size load obtainable, the height of lift, the angle of swing, the method of disposing of the load, and the experience of the operator. For example, if the material must be discharged into a hopper, the time required to spot the bucket over the hopper and discharge the load will be greater

TABLE 18.6 | Representative specifications for medium-weight general-purpose-type clamshell buckets

	Size (cy)								
	$\frac{3}{8}$	$\frac{1}{2}$	$\frac{3}{4}$	**1**	$1\frac{1}{4}$	$1\frac{1}{2}$	$1\frac{3}{4}$	**2**	$2\frac{1}{2}$
Capacity (cf)									
Water-level	8.0	11.8	15.6	23.2	27.6	33.0	38.0	47.0	52.0
Plate-line	11.0	15.6	21.9	32.2	37.6	43.7	51.5	60.0	75.4
Heaped	13.0	18.8	27.7	37.4	45.8	55.0	64.8	74.0	90.2
Weights (lb)									
Bucket only	1,662	2,120	2,920	3,870	4,400	5,310	5,440	6,000	7,775
Counterweights	230	300	400	400	400	500	500	600	600
Teeth	180	180	180	180	180	190	266	300	390
Complete	2,072	2,600	3,500	4,450	4,980	6,000	6,206	6,900	8,765
Dimensions									
Deck area (sf)	13.7	16.0	21.8	24.0	29.0	33.4	36.6	40.0	44.6
Width	2'6"	2'6"	3'0"	3'0"	3'5"	3'9"	4'0"	4'3"	4'6"
Length, open	5'5"	6'5"	7'3"	7'10"	8'5"	9'0"	9'2"	9'4"	9'11"
Length, closed	4'9"	5'7"	6'3"	6'9"	7'1"	7'6"	7'11"	8'0"	9'3"
Height, open	7'1"	7'10"	9'1"	9'9"	10'3"	10'9"	10'3"	11'6"	13'0"
Height, closed	5'9"	6'4"	7'4"	7'10"	8'3"	8'9"	8'9"	9'3"	10'4"

than when the material is discharged onto a large spoil bank. Example 18.4 illustrates one method of estimating the probable output of a clamshell.

EXAMPLE 18.4

A $1\frac{1}{2}$-cy rehandling-type bucket, whose empty weight is 4,300 lb, will be used to transfer sand from a stockpile into a hopper 25 ft above the ground. The angle of swing will average 90°. The average loose capacity of the bucket is 48 cf. The specifications for the crane unit are given in Table 18.7.

TABLE 18.7 | Representative performance specifications for a clamshell rigged crane

Speeds	
Travel	0.9 mph maximum
Swing	3 rpm maximum
Rated single-line speed	
Lift clamshell	166 fpm
Dragline	157 fpm
Magnet	200 fpm
Third drum (standard travel)	185 fpm
Third drum (independent travel)	127 fpm
Rated line pulls (with standard engine)	
Lift clamshell	29,600 lb
Dragline	31,400 lb
Magnet	24,800 lb
Third drum	25,500 lb

Performance figures are based on machine equipped with standard engine.

Time per cycle (approx.)

Loading bucket $\qquad$ = 6 sec

Lifting and swing load $\dfrac{25 \text{ ft} \times 60 \text{ sec per min}}{166 \text{ ft per min}}$ = 9 sec*

Dump load $\qquad$ = 6 sec

Swing back to stock pile $\qquad$ = 4 sec

Lost time, accelerating, etc. $\qquad$ = $\underline{4 \text{ sec}}$

$\qquad$ Total cycle time $\qquad$ = 29 sec or 0.48 min

Maximum number of cycles per hour, $\dfrac{60 \text{ min}}{0.48 \text{ min cycle}}$ = 125

Maximum volume per hour (125 cycle $\times$ 48 cf)/ 27 $\quad$ = 222 lcy

If the unit operates at a 45-min hour efficiency, the probable production will be

$$222 \times \frac{45}{60} = 167 \text{ lcy/hr}$$

If the same equipment is used with a general-purpose bucket to dredge muck and sand from a sheet-piling cofferdam partly filled with water, requiring a total vertical lift of 40 ft, and the muck must be discharged into a barge, the production rate previously determined will not apply. It will be necessary to lift the bucket above the top of the dam prior to starting the swing, which will increase the time cycle. Because of the nature of the material, the load will probably be limited to the water-filled capacity of the bucket, which is 33 cf. The time per cycle should be about

Loading bucket $\qquad$ = 8.0 sec

Lifting, $\dfrac{40 \text{ ft} \times 60 \text{ sec per min}}{166 \text{ ft per min}}$ = 14.5 sec

Swinging, 90° at 3 rpm $\dfrac{.25 \text{ rev.} \times 60 \text{ sec per min}}{3 \text{ rev. per min}}$ = 5.0 sec

Dump load $\qquad$ = 4.0 sec

Swing back $\qquad$ = 4.0 sec

Lowering bucket $\dfrac{40 \text{ ft} \times 60 \text{ sec per min}}{350 \text{ ft per min}}$ = 7.0 sec

Lost time, accelerating, etc. $\qquad$ = $\underline{10.0 \text{ sec}}$

Total cycle time $\qquad$ = 52.5 sec or 0.875 min

Maximum number of cycles per hr, 60 min ÷ 0.9-min cycle = 67

Maximum volume per hr $\dfrac{67 \text{ cycle} \times 33 \text{ cf}}{27}$ $\qquad$ = 82 lcy

If the unit operates at a 45-min hour efficiency, the probable production will be

$$82 \text{ lcy} \times \frac{45}{60} = 62 \text{ lcy/hr}$$

---

* A skilled operator should lift and swing simultaneously. If this is not possible, as when coming out of a cofferdam, additional time should be allowed for swinging the load.

EFFECTIVE CLAMSHELL OPERATIONS

Efficient use of the clamshell means an efficient digging, hoisting, swinging, and dumping cycle. Large buckets may increase cycle time. Select the correct bucket size for the machine.

- Position the unit on level ground.
- Position the unit so digging operations are at the same radius as the dumping operation. This will avoid wasting production time by having to raise and lower the boom.
- Remove bucket teeth when working in soft materials.

SUMMARY

The dragline is a versatile machine capable of a wide range of operations. It can handle materials that range from soft to medium hard. The greatest advantage of a dragline over other machines is its long reach for digging and dumping. A dragline does not have the positive digging force of a hydraulic shovel or hoe. The bucket breakout force is derived strictly from its own weight.

The clamshell machine consists of a clamshell bucket hung from a lattice-boom crane. The clamshell bucket consists of two scoops hinged together. The clamshell is a vertically operated attachment capable of working at, above, and below ground level. The clamshell can dig in loose to medium stiff soils. The length of the boom determines the height a clamshell can reach. Critical learning objectives would include:

- An understanding of how a dragline operates and when it should be employed on a project.
- An ability to calculate a production estimate for a dragline.
- An understanding of how a clamshell operates and when it should be employed on a project.
- An ability to calculate a production estimate for a clamshell.

These objectives are the basis for the problems that follow.

PROBLEMS

18.1 Determine the probable production in cubic yard bank measure (bcy) for a 2-cy dragline when excavating and casting good common earth. The average depth of dig will be 12 ft, and the average angle of swing will be 120°. The efficiency factor will be a 45-min hour. (155 bcy per hr)

18.2 Determine the probable production in cubic yard bank measure (bcy) for a 2-cy dragline when excavating and casting tough clay. The average depth of dig will be 9 ft, and the average angle of swing will be 140°. The efficiency factor will be a 45-min hour.

18.3 Determine the largest capacity medium-duty dragline bucket that can be used with a dragline equipped with an 80-ft boom when the boom is operating at an angle of 45°. The earth will weigh 98 lb per cu ft loose measure. (Maximum size bucket is $1\frac{3}{4}$ cy)

18.4 Determine the largest capacity heavy-duty dragline bucket that can be used with a dragline equipped with a 70-ft boom when the boom is operating at an angle of 40°. The earth will weigh 92 lb per cf loose measure.

18.5 A $2\frac{1}{2}$-cy dragline with a standard short boom is operated at 60% optimum depth of cut and a 90° angle of swing. It is used to excavate light sandy clay. The average bucket load, expressed in bcy, can be taken to be 0.80 of the "struck" capacity in cubic feet, as listed in Table 18.5. On the assumption that the data from Tables 18.3 and 18.4 are applicable, calculate the peakrate cycle time for the dragline in minutes.

REFERENCES

1. Power Crane and Shovel Association, A Bureau of Construction Industry Manufacturers Association, 111 East Wisconsin Avenue, Milwaukee, WI 53202.

2. "Crane Load Stability Test Code—SAE J765," in *SAE Recommended Practice Handbook,* Society of Automotive Engineers, Inc., Warrendale, Pa., 1967.

3. Stewart, Rita F., and Cliff J. Schexnayder: "Production Estimating for Draglines," *Journal of Construction Engineering and Management,* American Society of Civil Engineers, Vol. 111, No. 1, March 1985, pp. 101–104.

4. Nichols, Herbert L., Jr., and David A. Day, *Moving the Earth: The Workbook of Excavation,* 4th Ed., McGraw-Hill, New York, 1998.

DRAGLINE- AND CLAMSHELL-RELATED WEBSITES

1. Power Crane and Shovel Association (PCSA), www.cimanet.com/. PCSA is a bureau of Construction Industry Manufacturers Association, 111 East Wisconsin Avenue, Milwaukee, WI 53202. The Power Crane and Shovel Association provides services tailored to meet the needs of the lattice-boom and truck-crane industry. Bureau member companies are considered the global leaders in the promotion of worldwide harmonization of crane standards. PCSA explores technological questions and legislative and regulatory concerns (domestic and worldwide) that affect manufacturers of cranes. It also promotes the standardization and simplification of terminology and classification of cranes for worldwide harmonization.

2. Society of Automotive Engineers (SAE), www.sae.org/jsp/index.jsp. SAE World Headquarters, 400 Commonwealth Drive, Warrendale, PA 15096-0001, Tel: (877) 606-7323 (United States and Canada only) or (724) 776-0790 (outside the United States and Canada), Fax: Customer Service: (724) 776-0790, Headquarters: (724) 776-5760. The Society of Automotive Engineers is a resource for technical information and expertise used in designing, building, maintaining, and operating self-propelled vehicles. SAE publishes thousands of technical papers and books each year, and leading-edge periodicals and Internet and CD-ROM products, too. The SAE Cooperative Research Program helps facilitate projects that benefit the mobility industry as a whole. Numerous meetings and expositions provide worldwide opportunities to network and share information. The SAE also offers a full complement of professional development activities such as seminars, workshops, and continuing education programs.

19

Piles and Pile-Driving Equipment

Piles can be classified on the basis of either their use or the materials from which they are made. On the basis of use, there are two major classifications: (1) sheet and (2) load bearing. Sheet piles are used primarily to retain or support earth. Load-bearing piles, as the name implies, are used primarily to transmit structural loads. The function of a pile hammer is to furnish the energy required to drive a pile. Pile-driving hammers are designated by type and size.

INTRODUCTION

This chapter deals with the selection of piles and the equipment used to drive piles (see Fig. 19.1). Load-bearing piles, as the name implies, are used primarily to transmit structural loads through soil formations with inadequate supporting properties and into or onto soil stratum capable of supporting the loads. If the load is transmitted to the soil through skin friction between the surface of the pile and the soil, the pile is called a *friction pile*. If the load is transmitted to the soil through the lower tip, the pile is called an *end-bearing pile*. Many piles depend on a combination of friction and end bearing for their supporting strengths.

GLOSSARY OF TERMS

The following glossary defines the terminology used in describing piles, pile-driving equipment, and pile-driving methods.

> *Allowable pile load.* The load permitted on any vertical pile applied concentrically in the direction of its axis. It is the least value determined from the capacity of the pile structural member, the allowable bearing pressure on the soil strata underlying the pile tip, the resistance to penetration, the capacity demonstrated by load test, or the basic maximum load prescribed by the building code.

FIGURE 19.1 | Crawler crane driving 12-in. concrete piles using a single-acting air hammer and hydraulic leads.

Source: Tidewater Construction Corporation.

Anchor pile. A pile connected to a structure by one or more ties to furnish lateral support or to resist uplift.

Anvil block. A driving head.

Augered or drilled pile. A concrete pile, which may or may not be belled at the bottom, cast-in-place in an augered hole.

Batter pile. A pile driven at an inclination from the vertical.

Bearing capacity. Allowable pile load as limited by the provision that the pressures developed in the materials along the pile and below the pile tip shall not exceed the allowable soil-bearing values.

Brace pile. A battered pile connected to a structure in a way that resists lateral forces.

Brooming. The splintering of a wood pile at the butt or tip caused by excessive or improper driving. It can be controlled by the use of a pile ring; a driving cap with cushion block; or in the case of the pile tip, a metal shoe.

Butt of a pile. The larger end of a tapered pile; usually the upper end of a driven pile. Also a general term for the upper portion of a pile.

Cast-in-place pile. A concrete pile (with or without internal reinforcing) that is constructed in its permanent location in the ground.

Cushion block. Material inserted between the pile hammer and the pile-driving cap or pile to minimize damage to the pile. Typical materials used for cushioning include micarta, steel, aluminum, coiled cable, and wood.

Cutoff. The prescribed elevation at which the top of a driven pile is cut. Also the portion of pile removed from the upper end of the pile after driving.

Displacement pile. A solid pile or a hollow pile driven with its lower end closed. The volume of the pile displaces the soil laterally and vertically.

Downdrag. Negative friction of soil gripping a pile in the case of settling soils. This condition adds load to the installed pile.

Driving cap or helmet. A steel cap placed over the pile butt to prevent damage to the pile during driving.

Embedment. The length of pile from the ground surface, or from the cutoff below the ground to the tip of the pile.

End-bearing pile. A pile that supports the imposed load by tip bearing on a firm stratum.

Engineering News formula. A simple but frequently inadequate formula for estimating the load-carrying ability of a driven pile.

Follower. A member interposed between the pile hammer and the pile to transmit blows when the top of the pile is below the reach of the hammer.

Friction pile. A pile that supports an imposed load by friction developed between the surface of the pile and the soil through which it is embedded.

Guides. The parallel rails of the pile leads that form a pathway for the hammer.

Heave. The uplifting of the soil surface between or near driven piles. This is caused by the displacement of soil by the pile volume.

Helmet. A driving cap to protect the butt of a pile.

Jetting. The use of high-pressure water jets to place a pile or facilitate the driving of a pile.

Overdriving. Driving a pile in a manner that damages the pile material. This is often the result of continued hammering after the pile meets refusal.

Penetration. Gross penetration is the downward axial movement of the pile per hammer blow. Net penetration is the gross penetration less the rebound, i.e., the net downward movement of the pile per hammer blow.

Pile bent. Two or more piles driven in a group and fastened together by a cap or bracing.

Pile-driving cap. A forged or cast steel piece designed to fit over and around the butt of the pile during driving. A cap is used to prevent damage to the butt of the pile during driving. The terms hood and bonnet are sometimes used in reference to a pile-driving cap.

Pile-driving shoe. A metal shoe placed on the pile tip to prevent damage to the pile and to improve driving penetration.

Pile tip. The lower—and in the case of timber piles usually the smaller— end of a pile.

Set. Net penetration of the pile per hammer blow.

Sheet pile. A pile used to form a continuous wall. Sheet piles have interlocking grooves along their edges so that they can be driven to form a solid wall. The purpose of sheet piling is to resist lateral earth pressure.

Soldier pile. An H or wide-flange (WF) member driven at intervals of a few feet into which horizontal lagging is placed as an excavation proceeds.

Spud. A short member driven and then removed so as to create a hole for inserting a pile. This may be done when the pile is too long for placing directly in the pile leads or to break through a crust of hard material. The term also refers to a vertical member, usually operated hydraulically, that passes through a floating frame and is lowered into the lake or river bottom to hold the frame in position during a work operation, as in the case of a floating pile-driving rig. The spud is then raised when it is necessary to reposition the frame.

Strapping. Banding a timber pile with steel to prevent splitting while driving.

Template. A rigid frame used to guide and hold a pile in the proper position during driving.

Tension pile. A pile designed to resist uplift.

Test pile. A pile driven to ascertain driving conditions on a project and to aid in the selection of appropriate pile lengths for the project. The test also provides information concerning the load-carrying capacity of project piles.

TYPES OF PILES

Piles can be classified on the basis of either their use or the materials from which they are made. On the basis of use, there are two major classifications: (1) *sheet* and (2) *load bearing*.

Sheet piling is used primarily to resist the flow of water and loose soil. Typical uses include cutoff walls under dams, cofferdams, bulkheads, and trench sheeting. On the basis of the materials from which they are made, sheet pilings can be classified as *wood, steel, concrete,* or *composite*.

Considering both the type of the material from which they are made and the method of constructing and driving them, load-bearing piles may be classified as

1. Timber
   a. Untreated.
   b. Treated with a preservative.
2. Concrete
   a. Precast-prestressed.
   b. Cast-in-place with shells.
   c. Augered cast-in-place.
3. Steel
   a. H section.
   b. Steel pipe.
4. Composite
   a. Concrete and steel.
   b. Plastic with steel pipe core.

   Sheet piles can be classified as

1. Steel.
2. Prestressed concrete.
3. Timber.

Each type of load-bearing pile has a place in engineering practice, and for some projects more than one type may seem satisfactory. It is the responsibility of the engineer to select the pile type that is best suited for a given project, taking into account all the factors that affect both installation and performance. Factors that will influence the decision are

1. Type, size, and weight of the structure to be supported.
2. Physical properties of the soil stratum at the site.
3. Depth to a stratum capable of supporting the piles.
4. Variations across the site in the depth to a supporting stratum.
5. Availability of materials for piles.
6. Number of piles required.
7. Driving equipment.
8. Comparative in-place costs.
9. Durability required.

10. Types of structures adjacent to the project.

11. Depth and kind of water, if any, above the ground into which the piles will be driven.

12. Construction nuisances, particularly noise and vibration caused by driving.

SITE INVESTIGATION AND A TEST PILE PROGRAM

For projects of intermediate to large scale, a thorough site investigation can be very cost-effective. The geotechnical information gathered from borings can be used to determine the soil characteristics and the depths to strata capable of supporting the design loads. The number of blows per foot, from geotechnical tests such as the standard penetration test, is normally recorded during the soil sampling operations and can be extremely valuable for use in predicting pile lengths. From this information, pile types, sizes, and capacities may be chosen. Once a pile type has been selected, or if several types are deemed practical for use on a particular project, a test pile program should be conducted.

Pile lengths to be used for the test pile program are generally somewhat longer than the anticipated lengths as determined from the borings. This permits driving the test piles to greater depths if necessary and allows additional pile length for installing the load test apparatus. Several test piles should be driven and carefully monitored at selected locations within the project area. Dynamic testing equipment and testing techniques, discussed in further detail in this chapter, can be utilized to gather significant information pertaining to the combined characteristics of the pile, the soil, and the pile-driving equipment. This information will greatly enhance the engineer's ability to predict the required pile lengths for the project and the number of blows per foot required to obtain the desired bearing capacity.

Depending on the size of the project, one or more of the test piles would be selected for load testing. To apply a load to the selected piles, static test weights, water tanks, or reaction piles and jacks may be employed. Static weights or water tanks allow the loading weight to be incrementally applied directly to the test pile by either adding additional weights or water. In the case of the reaction pile method, steel H piles are used as reaction piles. They are driven in relatively close proximity to the pile to be tested. The magnitude of the test load to be applied and the friction characteristics of the soil determine the number of reaction piles necessary.

A reaction frame or beam is attached to the reaction piles and spans over the top of the test pile in order that the test load can be applied using a hydraulic jack (see Fig. 19.2). The jack is located between the reaction beam and the top of the test pile. As the load is applied by the jack, the reaction beam transfers the load to the test pile, putting it in compression and putting the reaction piles in tension.

Calibrated hydraulic gauges are used to measure the amount of load applied. The test load is applied in increments over a period of time and the pile is continuously monitored for movement. In the case of either type test, direct load or

FIGURE 19.2 | Load test of a 54-in. concrete cylinder pile utilizing reaction piles, beams, and a jack.

Source: Tidewater Construction Corporation.

reaction, the magnitude of the applied test load is normally two to three times the design-bearing capacity of the pile. Any sudden or rapid movement of the pile indicates a failure of the pile.

Once a test pile program has been completed, pile lengths and supporting bearing capacities can be predicted with reasonable accuracy. Where precast concrete piles are determined to be an economical alternative at a particular site, a test pile program can verify with suitable accuracy the required pile lengths. Therefore, when reliable geotechnical and load test data are available, piles can be cast to calculated lengths with little risk of their being either too short or too long.

TIMBER PILES

Timber piles are made from the trunks of trees. Such piles are available in most sections of the country and the world. Pine piles are reasonably available in lengths up to 80 ft, while Douglas fir piles are available in lengths in excess of 100 ft from the Pacific Northwest. Southern pine and Douglas fir timber piles are the most common materials for use as marine fendering systems in the United States.

Preservative treatments, such as salt or creosote, are used to reduce the rate of decay and to fight the attacks of marine borers. Standard C3 of the American Wood Preservers' Association (AWPA) requires 20 pounds per cubic foot (pcf) creosote in timbers used in marine waters. These preservatives and the treatment technique should be carefully selected as they may have environmentally detrimental effects. There are new creosote treating processes, AWPA Standard P1/P13, that minimize the amount of residual creosote that occurs on the surface of the treated product. The Environmental Protection Agency has approved creosote-treated timber piling and timbers for marine water applications.

Untreated timber piles last only a couple of years. Although rarely specified, untreated piles can be used as an economically sacrificial member for temporary construction purposes. Treated piling will last for 50 yr in northern marine waters, and in southern waters they have a useful life of 20 yr. Creosote-treated timber piles at the W. P. Franklin Lock and Dam in Florida were in excellent condition after 28 yr of service. Southern waters are defined as south of Cape Hatteras on the East Coast and below San Francisco on the West Coast.

The advantages of timber piles include

1. The more popular lengths and sizes are available on short notice.
2. They are economical in cost.
3. They are handled easily, with little danger of breakage.
4. After driving they can easily be cut off to any desired length.
5. Usually they can be extracted easily in the event removal is necessary.

The disadvantages of timber piles include

1. It may be difficult to economically obtain piles sufficiently long and straight.

2. They may be difficult or impossible to drive into hard formations.
3. They are difficult to splice when increased lengths are necessary.
4. While they are satisfactory when used as friction piles, they are not usually suitable for use as end-bearings piles under heavy loads.
5. The duration over which they maintain their structural capacity may be short unless the piles are treated with a preservative and preservatives may have undesirable environmental effects.

PRECAST-PRESTRESSED CONCRETE PILES

Precast-prestressed concrete piles are normally manufactured at established plants using approved methods in accordance with the PCI MNL-116-85 *Manual for Quality Control* [13]. Specifications for many projects, such as those used by state highway departments, require the piles to be manufactured at PCI certified plants. However, for projects with a large quantity of piles, transportation costs from existing manufacturing plants may be significant. Therefore, it may be cost-effective to set up a casting facility on the job or in the general vicinity of the project. The establishment of such a facility requires a substantial investment in special equipment and casting forms, as well as a sufficient amount of space for the casting beds, curing area, and storage yard.

Square and octagonal piles are cast in horizontal forms on casting beds, whereas cylinder piles are cast in cylindrical forms and then centrifugally spun. After the piles are cast, they are normally steam cured until they have reached sufficient strength to allow them to be removed from the forms. Under controlled curing conditions and utilizing concrete with compressive strengths of 5,000 psi or greater, the piles may be removed from the forms in as little as 24 hr or as soon as the concrete has developed a compressive strength of 3,500 psi. The piles are then stored and allowed to cure for 21 days or more before reaching driving strengths.

Prestressed concrete piles are reinforced with either $\frac{1}{2}$-in. or $\frac{7}{16}$-in. 270-ksi* high-strength stress relieved or low relaxation tendons or strands. The number and type of strand are determined by the design properties of the pile. In addition, the piles are reinforced with spiral reinforcing. The amount of spiral reinforcing is increased at the ends of the pile to resist cracking and spalling during driving. Piling for marine applications may require that the spirals be epoxy coated. Pile dowels may be cast in the tops of the piles for uplift reinforcing, or dowel sleeves may be cast in the top of the pile and the dowels can then be grouted into the pile after driving. Figure 19.3 shows the details for a typical 12-in. sq prestressed concrete pile.

The square and octagonal piles are traditionally cast on beds 200 to 600 ft long in order that multiple piles may be cast and prestressed simultaneously (see Fig. 19.4). The prestressing strand will be as long as the casting bed. Bulkheads will be placed in the forms as determined by the desired pile lengths. Utilizing stressing jacks, each strand will be pretensioned to between 20 and 35 kips prior

---

* Note that ksi is kips per square inch, where 1 kip = 1,000 lb.

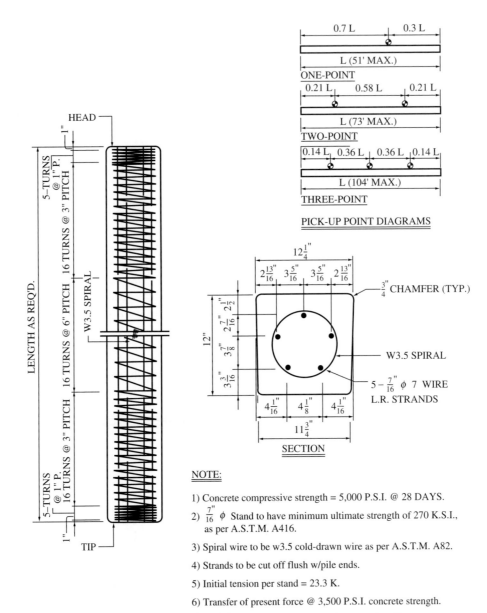

PICK-UP POINT DIAGRAMS

SECTION

NOTE:

1) Concrete compressive strength = 5,000 P.S.I. @ 28 DAYS.

2) $\frac{7}{16}$ ϕ Stand to have minimum ultimate strength of 270 K.S.I., as per A.S.T.M. A416.

3) Spiral wire to be w3.5 cold-drawn wire as per A.S.T.M. A82.

4) Strands to be cut off flush w/pile ends.

5) Initial tension per stand = 23.3 K.

6) Transfer of present force @ 3,500 P.S.I. concrete strength.

FIGURE 19.3 I Typical details of a 12-in. square prestressed concrete pile.

Source: Bayshore Concrete Products, Chesapeake, Inc.

to the concrete placement. Immediately following the concrete placement, the piles are covered with curing blankets and steam is introduced. The steam raises the air temperature at a rate of 30 to 60°F per hour until a maximum temperature of 140 to 160°F is reached. The curing continues and the prestressing forces are not released until the concrete has attained a minimum compressive strength of 3,500 psi.

FIGURE   19.4 | Casting bed for 12-in. prestressed concrete piles showing bulkheads and strands.

Source: Tidewater Construction Corporation.

Cylinder piles are cast in short sections of up to 16 ft in length with prestressing sleeves or ducts cast into the wall of the pile. The wall thickness can vary from 5 in. up to 6 in., depending on the design properties of the pile. Once the concrete is placed in the form, the concrete and the form are centrifugally spun, causing the concrete to consolidate. After the sections have been steam cured and the concrete has obtained sufficient strength, the sections are assembled to the proper pile lengths. The prestressing strand is pulled through the sleeves or ducts and then tensioned with jacks. Finally, the tensioned strands are pressure grouted in the sleeves.

The precast-prestressed piles may be transported by truck to land-based projects or they can be moved by barge in the case of marine projects. For handling concrete piles, care must be exercised to prevent breakage or damage due to flexural stresses. Long piles should be picked up at several points to reduce the unsupported lengths (see Fig. 19.5). When piles are stored or transported, they must be supported continuously or at the pickup points.

Precast concrete piles can be cast in any desired size and length. To support the trestle bents of the Chesapeake Bay Bridge Tunnel project across the mouth of the Chesapeake Bay in Virginia, 2,500 cylinder piles 54 in. in diameter totaling approximately 320,000 lin ft were used. These cylinder piles were cast at a nearby casting yard constructed specifically for the project. The piles were loaded onto barges that were towed to the bridge site. Driving was performed primarily with a 150-ton barge-mounted Whirley crane. The driving barge was outfitted with four jack-up spuds to allow it to hold position and to be jacked-up out of the water when the crane was driving piles. For projects as large as this, the large investment in special equipment is justified, but for a small project such an investment would probably be cost-prohibitive. Thus the maximum-size concrete piles that can be used on a project may be determined by economy of construction equipment—a constructability issue.

One of the disadvantages of using precast-prestressed concrete piles, especially for a project where different lengths of piles are required, is the difficulty of reducing or increasing the lengths of piles. It is for this reason that on major piling projects, it is cost-effective to perform a site investigation program and to conduct a test pile program. Such a program aids in predicting the correct lengths for the precast concrete piles.

If a pile proves to be too long, it is necessary to cut off the excess length. This is done after a pile is driven to its maximum penetration by chipping the concrete away from the reinforcing steel (typically this is a jack hammer operation), cutting the reinforcing with a gas torch, and then removing the surplus length of concrete core. Hydraulic concrete-crushing machines can also be used to cut off the excess length of pile. Such a method requires substantially less time and is less expensive than the chipping method. But either operation represents a waste of material and time, and adds to project cost.

When a precast concrete pile does not develop sufficient driving resistance to support the design load, it may be necessary to increase the length and drive the pile to a greater depth. Unless the reinforcing dowels extend above the top of the pile, it will be necessary to drill holes in the top of the pile to allow the insertion and grouting of reinforcing dowels. Then the added length of pile can be mated to the original.

There are a number of suggested rules for driving prestressed concrete piles.

1. Adequate cushioning material must be used between the pile driver's steel helmet or cap and the head or top of the concrete pile. Usually, a wood cushion is used and may vary in thickness from 4 to 8 in., depending on

FIGURE 19.5 | Barge-mounted Whirley crane using a three-point pickup to lift a 36-in. cylinder pile off a barge and into the leads.

Source: Tidewater Construction Corporation.

the length of the pile and the characteristics of the soil. This is a very economical way of reducing driving stresses in the pile.

2. Driving stresses may be reduced by using a hammer with a heavy ram and a low impact velocity or large stroke. Driving stresses are proportional to the ram impact velocity.

3. Care must be taken when driving piles through soils or soil layers with low resistance. If such soils are anticipated or encountered, it is important to reduce the ram velocity or stroke of the hammer to avoid critical tensile stresses in the pile.

4. In the case of cylinder piles, it is important to prevent the soil plug inside the pile from rising to an elevation above the level of the existing soil on the outside of the pile, thus creating unequal stresses in the pile. This should be monitored, and, in the event the plug does rise inside the pile, it should be excavated down to the level of the existing soil on the outside of the pile.

5. The pile-driving helmet should fit loosely around the top of the pile so that the pile may rotate slightly without binding within the driving head. This will prevent the development of torsional stress.

6. The top of the pile must be square or perpendicular to the longitudinal axis of the pile to eliminate eccentricity that causes stress.

7. The ends of the prestressing strand or reinforcing must either be cut off flush with the top of the pile or the driving helmet must be designed so that the reinforcing threads through the pile cap in order that the hammer's ram does not directly contact the reinforcing during driving. The driving energy must be delivered to the top of the concrete.

The advantages of concrete precast piles include

1. They have high resistance to chemical and biological attacks.

2. They have high load-carrying capacity.

3. In the case of hollow cylinder piles, a pipe can be installed along the center of the pile to facilitate jetting.

The disadvantages of precast concrete piles include

1. It is difficult to reduce or increase the length.

2. Large sizes require heavy and expensive handling and driving equipment.

3. The inability to quickly obtain piles by purchase may delay the starting of a project.

4. Possible breakage of piles during handling or driving produces a delay hazard.

CAST-IN-PLACE CONCRETE PILES

As the name implies, cast-in-place concrete piles are constructed by depositing the freshly mixed concrete in the ground and letting it cure there. The two principal methods of constructing such piles are

1. Driving a metallic shell and leaving it in the ground, and then filling the shell with concrete.
2. Driving a metallic shell and filling the resulting void with concrete as the shell is pulled from the ground.

There are several modifications for each of these two methods. The more commonly used piles constructed by these two methods are described in the next sections.

Raymond Step-Taper Concrete Piles

The step-taper pile is installed by driving a spirally corrugated steel shell, made up of sections 4, 8, 12, and 16 ft long, with successive increases in diameter for each section. A corrugated sleeve at the bottom of each section is screwed into the top of the section immediately below it. Piles of the necessary length, up to a maximum of 80 ft, are obtained by joining the proper number of sections at the job site. The shells are available in various gauges of metal to fit different job conditions. The bottom of the shell, whose diameter can be varied from $8\frac{5}{8}$ to $13\frac{3}{8}$ in., is closed prior to driving by a flat steel plate or a hemispherical steel boot.

After a shell is assembled to the desired length, a step-tapered rigid-steel core or mandrel is inserted and the shell is driven to the desired penetration. The core is removed, and the shell is filled with concrete.

Monotube Piles

The monotube pile is obtained by driving a fluted, tapered steel shell, closed at the tip with an 8-in.-diameter driving point, to the desired penetration. The shell is driven without a mandrel, inspected, and filled with concrete. The desired length of shell is obtained by welding extensions to a standard-length shell. Figure 19.6 shows a crawler crane driving one of these piles. Pile dimensions and additional information can be found at www.monotube.com.

Augered Cast-in-Place Piles

Another form of cast-in-place concrete pile is the augered cast-in-place pile. These piles differ from the previously discussed cast-in-place piles since they do not require a shell or pipe. A continuous flight, hollow shaft auger with a minimum inside diameter of $2\frac{1}{2}$ in. is rotated into the soil to a predetermined tip elevation or refusal, whichever occurs first. As the auger is slowly rotated and withdrawn from the hole, grout is injected under pressure through the hollow shaft. The rate of injection is carefully coordinated with the auger withdrawal rate to ensure that the hole is completely filled with grout. The displaced soil is removed as the auger is rotated and withdrawn.

The concrete can be placed to the ground elevation and then cut off to the required elevation. Sometimes it is necessary to build up the pile to an elevation above ground level. A single reinforcing bar may be installed through the hollow core of the auger or reinforcing bars may be placed after the auger is completely withdrawn. Augered cast-in-place piles having diameters up to 24 in. and lengths greater than 125 ft have been successfully installed.

FIGURE 19.6 I Driving No. 5 gauge, 12-in. monotube piles up to 200 ft long with a 37,000 ft-lb steam hammer.

Source: The Union Metal Manufacturing Company.

The advantages of augered cast-in-place piles include

1. Noise levels during construction are minimized, as there is no requirement to drive a casing.
2. No casing is required.
3. There is little or no detrimental vibration to adjacent structures during construction.

4. They can be installed in areas that have low overhead restrictions.
5. They can be installed in areas of minimum clearance and alongside existing structures.
6. There is little or no detrimental soil upheaval or displacement.
7. Pile splicing is eliminated.

The disadvantages of augered cast-in-place piles include

1. They require very careful placement of the concrete to ensure a structurally sound shaft.
2. Soil and groundwater conditions can greatly affect installation times and cost.
3. Vertical or horizontal ground movement from adjacent pile construction can damage completed piles.
4. There are insufficient data concerning pile behavior under seismic loading conditions.
5. Because of the construction technique, no penetration resistance correlation can be made about pile capacity.
6. Their use in instances where uplift forces can be encountered requires installation of reinforcing steel and the process of installing the steel is somewhat difficult.

Franki Driven Cast-in-Situ Pile

The Franki pressure-injected footing technique is illustrated in Fig. 19.7. Also see www.franki.co.za/products/driven.html or Frankipile Geofranki Soil Tech reference [8]. As illustrated in Fig. 19.7a, a steel drive tube of the desired diameter and length is fitted with an expendable steel boot to close the bottom end. Using a pile-driving hammer of adequate size, usually with 50,000 to 100,000 ft-lb per blow, the tube is driven to the desired depth. As illustrated in Fig. 19.7b, after the tube is driven to the desired depth, the hammer is removed and a charge of dry concrete is dropped into the tube and compacted with a drop hammer to form a compact watertight plug. Following the step in Fig. 19.7b, the tube is raised slightly and held in that position while repeated blows of the drop hammer expel the plug of concrete from the lower end of the tube, leaving in the tube sufficient concrete to prevent the intrusion of water or soil. Additional concrete is dropped into the tube and forced by the drop hammer to flow out of the bottom of the tube to form an enlarged bulb or pedestal, as illustrated by the step in Fig. 19.7c.

The final step in this operation consists of raising the tube in increments. After each incremental upward movement, additional dry concrete is dropped into the tube and is hammered with sufficient energy to force it to flow out of the tube and fill the exposed hole (Fig. 19.7d). This operation cycle is repeated until the hole is filled to the desired elevation. The shaft may be reinforced by inserting a spirally wound steel cage, embedded in the bulb and extended upward for the full height of the shaft.

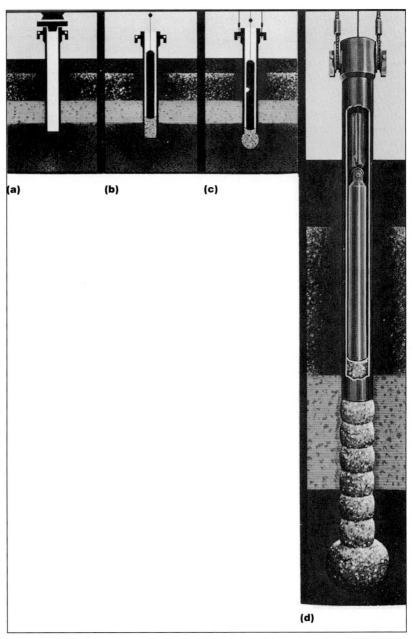

FIGURE 19.7 | Steps in providing Franki pressure-injected footing. (a) The drive tube is driven to the desired depth. (b) Concrete is dropped into the drive tube and compacted. (c) The drop hammer expels concrete from the tube.
(d) Additional steps complete the footing.

Advantages and Disadvantages of Cast-in-Place Concrete Piles

The advantages of cast-in-place concrete piles include

1. The lightweight shells can be handled and driven easily.
2. Variations in length do not present a serious problem. The length of a shell can be increased or decreased easily.
3. The shells can be shipped in short lengths and assembled at the job site.
4. Excess reinforcement to resist stresses caused only by handling the pile is eliminated.
5. The danger of breaking a pile while driving is eliminated.
6. Additional piles can be provided quickly if they are needed.

The disadvantages of cast-in-place concrete piles include

1. A slight movement of the earth around an unreinforced pile may break it.
2. An uplifting force acting on the shaft of an uncased and unreinforced pile may cause it to fail in tension.
3. The bottom of a pedestal pile may not be symmetrical.

STEEL PILES

In constructing foundations that require piles driven to great depths, steel piles probably are more suitable than any other type (see Fig. 19.8).

Steel H Section Piles

Steel piles may be driven to great depths through poor soils to bear on a solid rock stratum. They can be driven through hard materials to a specified depth. The great strength of steel combined with small displacement of soil permits a large portion of the energy from a pile hammer to be transmitted to the bottom of a pile. As a result, one can drive steel piles into soils that cannot be penetrated by any other type of pile. However, in spite of the great strength of these piles sometimes it is necessary to drill pilot holes into compacted sand ahead of steel H piles to obtain the specified penetration. By weld-splicing sections together, lengths in excess of 200 ft can be driven.

Because the steel H piles can be driven in short lengths and additional lengths then welded on top of the previously driven section, they can be utilized more readily in situations where there are height restrictions that limit the length of piles that can be driven in one piece.

Steel-Pipe Piles

These piles are installed by driving pipes to the desired depth and, if desired, filling them with concrete. A pipe can be driven with the lower end closed with a plate or steel driving point, or the pipe can be driven with lower end open. Pipes varying in diameter from 6 to 42 in. have been driven in lengths varying from a few feet to several hundred feet.

FIGURE 19.8 | Diesel hammer driving steel H pile on 1:1 batter.
Source: Pilco, Inc., www.pileco.com/.

A closed-end pipe pile is driven in any conventional manner, usually with a pile hammer. If it is necessary to increase the length of a pile, two or more sections can be welded together or sections can be connected by using an inside sleeve for each joint. This type of pile is particularly advantageous for use on jobs when headroom for driving is limited and short sections must be added to obtain the desired total length.

An open-end pipe pile is installed by driving the pipe to the required depth and then removing the soil from the inside. Removal methods include bursts of compressed air, a mixture of water and compressed air, or the use of an earth auger or a small orange-peel bucket. Finally, the hollow pipe is filled with concrete.

Because open-end pipe piles offer less driving resistance than closed-end piles, a smaller pile hammer can be used. The use of a light hammer is desirable when piles are driven near a structure whose foundation might be damaged by vibrations from the blows of a large hammer. Open-end piles can be driven to depths that could never be reached with closed-end piles.

COMPOSITE PILES

Several types of composite piles are available. These are usually developed and offered to meet the demands of special situations. Two of the most common situations that cause problems are hard ground at the project site and warm marine environments. Hard ground causes problems with applying the energy necessary to drive the pile and at the same time being careful not to destroy the pile. Marine environments subject the pile to marine borer attacks and salt attacks on metal, therefore special piling protective measures are usually specified.

Concrete and Steel Composite Piles

When extremely hard soils or soil layers are encountered, it may be cost-effective to consider the use of a composite concrete and steel pile. The top portion of the pile would be a prestressed concrete pile and the tip would be a steel H pile embedded into the end of the concrete pile. This composite design is suggested for marine applications, where the concrete pile section offers resistance to deterioration and the steel pile tip enables penetration of hard underlying soils.

Plastic with Steel-Pipe Core Piles

Since the mid-eighties composite plastic pilings have been available for special applications [6]. These piles are immune to marine borer attacks, eliminating the need for creosote treatments or special sheathings in marine environments. Their abrasive resistance makes them excellent for fender system use.

Varying the amount of steel in the pile core can alter the piling's physical properties of toughness, resilience, and specific gravity. These pilings can be manufactured in a variety of shapes and colors. Plastic pilings 24 in. in diameter and up to 50 ft long have been developed having steel-pipe cores between 12 and 16 in. in diameter. Using a larger pipe section increases stiffness. On a fender system at the Port of Los Angeles, 13-in.-diameter 70-ft-long piles with a 6-in. steel-pipe core were used (Fig. 19.9). A steel-driving tip or shoe is welded to the end of the pile before driving. The tip aids driving and servers to seal the end of the pile. A diagram of a composite plastic and steel pile and a driving shoe is shown in Figure 19.10.

FIGURE 19.9 | Composite plastic and steel core pile fender system at the Port of Los Angeles.

Source: Plastic Pilings Inc.

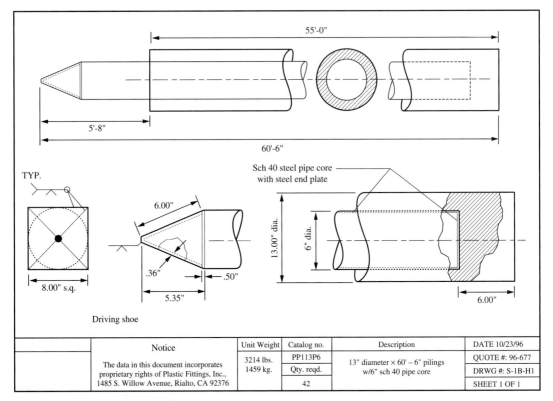

	Notice	Unit Weight	Catalog no.	Description	DATE 10/23/96
	The data in this document incorporates proprietary rights of Plastic Fittings, Inc., 1485 S. Willow Avenue, Rialto, CA 92376	3214 lbs. 1459 kg.	PP113P6	13" diameter × 60′ – 6" pilings w/6" sch 40 pipe core	QUOTE #: 96-677
			Qty. reqd.		DRWG #: S-1B-H1
			42		SHEET 1 OF 1

FIGURE 19.10 | Diagram of a 13–in.-diameter × 60 ft 6 in. sch 40 pipe core composite plastic and steel pile.

Source: Plastic Pilings Inc.

SHEET PILES

Sheet piles are used primarily to retain or support earth. They are commonly used for bulkheads and cofferdams (see Fig. 19.11), and when excavation depths or soil conditions require temporary or permanent bracing to support the lateral loads imposed by the soil or by the soil and adjacent structures. Supported loads would include any live loads imposed by construction operations. Sheet piles can be made of steel, concrete, or timber. Each of these types can support limited loads without additional bracing or tieback systems. When the depth of support is large or when the loads are great, it is necessary to incorporate a tieback or bracing system with the sheet piles.

Steel Sheet Piles

In the United States, steel sheet piling is manufactured in both flat and Z section sheets, which have interlocking longitudinal edges. A few years ago there were two established producers offering eight shapes. Currently, there are more than

FIGURE 19.11 | Circular sheet piling cells used to create a cofferdam.

10 domestic and foreign manufacturers listing almost 200 sheet piling sections in their catalogs. Most steel sheet piles are supplied in standard ASTM 328 grade steel, but high-strength low-alloy grades ASTM A572 and ASTM A690 are available for use where large loads must be supported or where corrosion is a concern. In marine applications, epoxy-coated piles can be used to resist corrosion.

The flat sections are designed for interlocking strength that makes them suitable for the construction of cellular structures. The Z sections are designed for bending, which makes them more suitable for use in the construction of retaining walls, bulkheads, and cofferdams, and for use as excavation support. Figure 19.12 shows various configurations of domestically produced steel sheet piles and their corresponding properties. Several overseas manufacturers produce steel sheet piles with properties similar to the domestic sheets. Those piles sometimes utilize a different interlocking design than the domestic sheets. Careful attention should be given to this design feature to ensure that sheets will perform in the intended usage and that the interlocking system allows the sheets to be threaded and driven easily.

Sheet piles are driven individually or in pairs, frequently with the use of a vibratory hammer and a guide frame or template. It is usually best to drive in pairs whenever possible. The template or guide should be at least half as high as the sheets since it is critical to keep the sheets plumb in both planes. It is very difficult to work with damaged sheets, and they should be avoided if possible. Before threading and driving a sheet into the system it is good practice to strike it with a sledge hammer to remove dirt trapped in the interlocks.

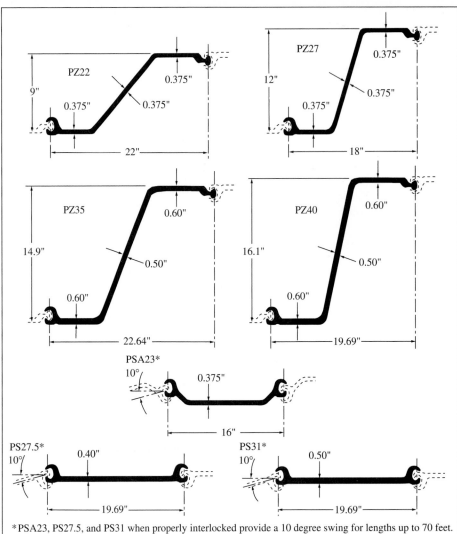

*PSA23, PS27.5, and PS31 when properly interlocked provide a 10 degree swing for lengths up to 70 feet. The dimensions given on this page are nominal.

Properties and Weights

| Section designation | Area, sq in. | Nominal width, in. | Weight, lb | | Moment of inertia, in.[4] | Section modulus, in.[3] | | Surface area sq ft per lin. ft of bar | |
			Per lin. ft of bar	Per sq ft of wall		Single section	Per lin. ft of wall	Total area	Nominal coating area*
PZ22	11.86	22	40.3	22.0	154.7	33.1	18.1	4.94	4.48
PZ27	11.91	18	40.5	27.0	276.3	45.3	30.2	4.94	4.48
PZ35	19.41	22.64	66.0	35.0	681.5	91.4	48.5	5.83	5.37
PZ40	19.30	19.69	65.6	40.0	805.4	99.6	60.7	5.83	5.37
PSA23	8.99	16	30.7	23.0	5.5	3.2	2.4	3.76	3.08
PS27.5	13.27	19.69	45.1	27.5	5.3	3.3	2.0	4.48	3.65
PS31	14.96	19.69	50.9	31.0	5.3	3.3	2.0	4.48	3.65

*Excludes socket interior and ball of interlock.

FIGURE 19.12 | Domestically produced steel sheet piles and their corresponding properties and weights.

Source: Bethlehem Steel Corporation.

Automatic sheet-pile-threading devices can be used to thread the new sheet into the previously threaded or driven sheet. These devices enable the sheets to be threaded without the assistance of a person working at the top of the previously placed sheet. If the steel sheet piling is used for temporary support, the individual sheets can be extracted ("pulled") using a vibratory hammer after construction and backfilling has been completed.

When using a vibratory hammer, the vibrator should be brought to speed before beginning the driving. Similarly, when extracting a pile do not begin pulling until the vibrator has started. In the case of extraction, it is often best to drive the pile a little first before beginning the pulling procedure.

Driving Steel Sheet Piling

Always carefully study the soil borings to anticipate "possible" driving conditions. The presence of boulders or stumps can make driving very difficult. Successful steel sheet pile installation depends on following these commonsense guidelines.

- Work only with undamaged sheets; piles must be straight or driving will be difficult.
- Interlocks must always be free of any dirt, sand, mud, or other debris.
- Always set up a template system. In addition to driving a straight wall, a template or guide system will also aid in keeping sheet pile plumb when excessive driving conditions exist or when an obstruction is encountered.
- Whenever possible it is recommended to drive sheets in pairs.
- Crane boom length must always be long enough to thread additional sheets—normally at least twice the length of the sheets being driven plus a few feet extra.
- Whenever possible, attempt to drive sheets with the male interlock, ball or thumb, etc. leading. This will aid in eliminating the possibility of the sheet developing a "soil plug." (The interlock filling with soil, sand, mud, etc.)
- Never overdrive. When sheets are bending, bouncing, or vibrating with no penetration, it could be an indication of overdriving, or that the sheet has hit an obstruction. However, it may indicate that a larger vibratory driver/extractor or hydraulic impact hammer is needed.

Many of these guidelines are also applicable to prestressed concrete sheet piles.

Prestressed Concrete Sheet Piles

Concrete sheet piles are best suited for applications where corrosion is a concern such as marine bulkheads. The prestressed concrete sheets are precast in thicknesses ranging from 6 to 24 in. and usually have widths of 3 to 4 ft. Figure 19.13 shows the dimensions and properties of various size concrete sheet piles. Conventional steam, air, or diesel hammers can be used to drive concrete sheet piles. However, jetting is often required to attain the proper tip elevation. Once the piles have been driven, the slots between sections are filled with grout.

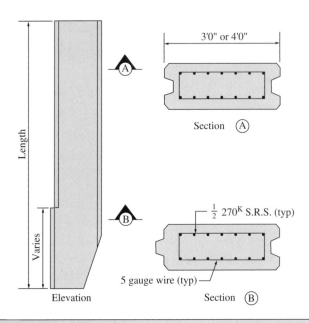

Sheet pile properties (per L.F. of width)						
T Thickness	Area in.^{2}	I, in.^{4}	S, in.^{3}	Max allowable Bending moment (ft kips) $f'c = 5,000$ psi (2)	$f'c = 6,000$ PSI	Approx. weight per (1) l.f.
6"	72"	216	72	6.0	7.2	75#
8"	96"	512	128	10.6	12.8	100#
10"	120"	1,000	200	16.6	20.0	125#
12"	144"	1,728	288	24.0	28.8	150#
16"	192"	4,096	512	42.7	51.2	200#
18"	216"	5,832	648	54.0	64.8	225#
20"	240"	8,000	800	66.7	80.0	250#
24"	288"	13,824	1,152	96.0	115.2	300#

I - Moment of inertia.
S - Section modulus.

(1) Weights based on 150 pcf of regular concrete.
(2) Based on 'o' psi allowable stress on tensile face and 0.4f'c allowable stress on compression face.

FIGURE 19.13 | Prestressed concrete sheet piles and their corresponding properties and weights.

Source: Bayshore Concrete Products, Chesapeake, Inc.

Timber Sheet Piles

Where supported loads are minimal, timber sheet piles can be used. Commonly, timbers of 3 to 4 in. thickness are used in an overlapping pattern three timbers thick. Such a pattern is commonly called "Wakefield sheeting." The timbers are driven with a light hammer or jetted into place. Timber wales and piles can be used to add support to the system. Traditionally timber sheets have been used for bulkheads and to construct groins. When used in permanent marine applications, they should be pressure treated to resist deterioration and borers.

THE RESISTANCE OF PILES TO PENETRATION

In general, the forces that enable a pile to support a load also cause the pile to resist the efforts made to drive it. The total resistance of a pile to penetration will equal the sum of the forces produced by skin friction and end bearing. The relative portions of the resistance contributed by either skin friction or end bearing may vary from almost 0 to 100%, depending more on soil type than on the type of pile. A steel H pile driven to refusal in stiff clay should be classified as a skin-friction pile, whereas the same pile driven through a mud deposit to rest on solid rock should be classified as an end-bearing pile.

Numerous tests have been conducted to determine values for skin friction for various types of piles and soils. A representative value for skin friction can be obtained by determining the total force required to cause a small incremental upward movement of the pile using hydraulic jacks with calibrated pressure gauges.

The value of the skin friction is a function of the coefficient of friction between the pile and the soil and the pressure of the soil normal to the surface of the pile. However, for a soil, such as some clays, the value of the skin friction may be limited to the shearing strength of the soil immediately adjacent to the pile. Consider a concrete pile driven into a soil that produces a normal pressure of 100 psi on the vertical surface of the pile. This is not an unusually high pressure for certain soils such as compacted sand. If the coefficient of friction is

$$0.25, \text{ the value of the skin friction will be } 0.25 \times 100 \text{ psi} \times \frac{144 \text{ sq in.}}{\text{sf}} =$$

3,600 psf (pounds per square foot). Table 19.1 gives representative values of skin friction on piles. The data presented in Table 19.1 is intended as a qualitative guide, not as correct information to be used in any and all cases.

The magnitude of end-bearing pressure can be determined by driving a button-bottom-type pile and leaving the driving casing in place. A second steel pipe, slightly smaller than the driving casing, is lowered onto the concrete button. The force applied through the second pipe to drive the button into the soil is a direct measure of the supporting strength of the soil. This is true because there is no skin friction on the inside pipe.

TABLE 19.1 | Approximate allowable value of skin friction on piles*

Material	Skin friction [psf (kg/sq m)]		
	Approximate depth		
	20 ft (6.1m)	60 ft (18.3 m)	100 ft (30.5 m)
Soft silt and dense muck	50–100	50–120	60–150
	(244–488)	(244–586)	(273–738)
Silt (wet but confined)	100–200	125–250	150–300
	(488–976)	(610–1,220)	(738–1,476)
Soft clay	200–300	250–350	300–400
	(976–1,464)	(1,220–1,710)	(1,476–1,952)
Stiff clay	300–500	350–550	400–600
	(1,464–2,440)	(1,710–2,685)	(1,952–2,928)
Clay and sand mixed	300–500	400–600	500–700
	(1,464–2,440)	(1,952–2,928)	(2,440–3,416)
Fine sand (wet but confined)	300–400	350–500	400–600
	(1,464–1,952)	(1,710–2,440)	(1,952–2,928)
Medium sand and small gravel	500–700	600–800	600–800
	(2,440–3,416)	(2,928–3,904)	(2,928–3,904)

*Some allowance is made for the effect of using piles in small groups.

PILE HAMMERS

The function of a pile hammer is to furnish the energy required to drive a pile. Pile-driving hammers are designated by type and size. The hammer types commonly used include

1. Drop.
2. Single-acting steam or compressed air.
3. Double-acting steam or compressed air.
4. Differential-acting steam or compressed air.
5. Diesel.
6. Hydraulic.
7. Vibratory drivers.

For each of the first six pile hammer types listed the driving energy is supplied by a falling mass that strikes the top of the pile. The size of a drop hammer is designated by its weight, whereas the size of each of the other hammers is designated by theoretical energy per blow, expressed in foot-pounds (ft-lb).

Drop Hammers

A *drop hammer* is a heavy metal weight that is lifted by a hoist line, then released and allowed to fall onto the top of the pile. Because of the high dynamic forces, a pile cap is positioned between the hammer and the pile head. The pile cap serves to uniformly distribute the blow to the pile head and to serve as a "shock absorber." The cap contains a cushion block that is commonly fabricated from wood.

The hammer may be released by a trip and fall freely, or it may be released by loosening the friction band on the hoisting drum and permitting the weight of the hammer to unwind the rope from the drum. The latter type of release reduces the effective energy of a hammer because of the friction loss in the drum and rope. Leads are used to hold the pile in position and to guide the movement of the hammer so that it will strike the pile axially.

Standard drop hammers are made in sizes that vary from about 500 to 3,000 lb. The height of drop or fall most frequently used varies from about 5 to 20 ft. The maximum recommended drop height varies with the pile type, 15 ft for timber piles and 8 ft for concrete piles. When a large energy per blow is required to drive a pile, it is better to use a heavy hammer with a small drop than a light hammer with a large drop. The analogy would be trying to drive a large nail with a tack hammer.

Drop hammers are suitable for driving piles on remote projects that require only a few piles and for which the time of completion is not an important factor. A drop hammer normally can deliver four to eight blows per minute.

The advantages of drop hammers include

1. Small investment in equipment.
2. Simplicity of operation.
3. Ability to vary energy per blow by varying the height of fall.

The disadvantages of drop hammers include

1. Slow rate of driving piles.
2. Danger of damaging piles by lifting a hammer too high.
3. Danger of damaging adjacent buildings as a result of the heavy vibration caused by a hammer.
4. Unable to use directly for underwater driving.

Single-Acting Steam/Air Hammers

A single-acting steam/air hammer (see Fig. 19.14) has a freely falling weight, called a "ram," that is lifted by steam or compressed air, whose pressure is applied to the underside of a piston that is connected to the ram through a piston rod. When the piston reaches the top of the stroke, the steam or air pressure is released and the ram falls freely to strike the top of a pile. The energy supplied by this type of hammer is delivered by the heavy weight striking with a low velocity, due to the relatively short fall distance, usually 3 ft; but fall can vary from 1 to 5 ft for specific hammers. Whereas a drop hammer may strike 4 to 8 blows per minute, a single-acting steam/air hammer will strike 40 to 60 blows per minute when delivering the same energy per blow.

A pile cap is used with a single-acting steam/air hammer. The cap is mated to the case of the hammer. This cap is commonly called an "anvil" or a "helmet."

Single-acting steam/air hammers may be open or enclosed. These hammers are usually used in a pile lead, though they can be fitted to operate free hung. The hammers are available in sizes varying from a few thousand to 1,000,000 ft-lb of

FIGURE 19.14 I Raymond 60X steam hammer driving 54-in. cylinder pile.

Source: Tidewater Construction Corporation.

energy per blow. Tables 19.2a through 19.2c give data on several of the more popular sizes of single-acting steam/air hammers. The length of the stroke and the energy per blow for this type of hammer may be decreased slightly by reducing the steam or air pressure below that recommended by the manufacturer. The reduced pressure has the effect of decreasing the height to which the piston will rise before it begins to fall.

The advantages of single-acting steam/air compared with drop hammers include

1. Greater number of blows per minute permits faster driving.
2. Greater frequency of blows reduces the increase in skin friction between blows.
3. The heavier ram falling at lower velocity transmits a greater portion of the energy to driving piles.
4. Reduction in the velocity of the ram decreases the danger of damage to piles during driving.
5. The enclosed types may be used for underwater driving.

The disadvantages of single-acting steam/air hammers compared with drop hammers include the following:

1. They require more investment in equipment such as a steam boiler or air compressor.
2. They are more complicated, with higher maintenance cost.
3. They require more time to set up and take down.
4. They require a larger crew to operate the equipment.
5. They require a crane having a greater lifting capacity.

Double-Acting Steam/Air Hammers

In the double-acting steam/air hammer, steam or air pressure is applied to the underside of the piston to raise the ram; then during the downward stroke steam is applied to the topside of the piston to increase the energy per blow. Thus, with a given weight of ram, it is possible to attain a desired amount of energy per blow with a shorter stroke than with a longer single-acting hammer. The number of blows per minute will be approximately twice as great as for a single-acting hammer with the same energy rating.

Double-acting hammers commonly deliver 95 to 300 blows per minute. These hammers do not require cushion blocks. The ram strikes upon an alloy steel anvil that fits the pile head.

The lighter ram and higher striking velocity of the double-acting hammer may be advantageous when driving light- to medium-weight piles into soils having normal frictional resistance. It is claimed that the high frequency of blows will keep a pile moving downward continuously, thus preventing static skin friction from developing between blows. However, when heavy piles are driven, especially into soils having high frictional resistance, the heavier weight and

TABLE 19.2a | Specifications for air or steam pile-driving hammers

Rated energy	Model	Manufacturer	Type	Style	Blows per min	Wt. of striking parts	Total weight (lb)	Hammer length (ft-in.)	Jaw dimensions	Boiler HP required (ASME)	Steam consump. (lb/hr)	Air consump. (cfm)	Inlet pressure (psi)	Inlet size (in.)
1,800,000	6300	Vulcan	SGL.-ACT.	Open	42	300,000	575,000	30'0"	22" × 144"(M)	2,804	43,873	19,485(A)	235	2@6"
1,582,220	MRBS 12500	Menck	SGL.-ACT.	Open	36	275,580	540,130	35'9"	Cage	2,400	52,910	26,500	171	2@6"
867,960	MRBS 8000	Menck	SGL.-ACT.	Open	38	176,370	330,690	30'10"	Cage	1,380	30,860	15,900	171	8"
750,000	5150	Vulcan	SGL.-ACT.	Open	46	150,000	275,000	26'3$\frac{1}{2}$"	22" × 120"(M)	1,317	45,426	9,535(A)	175	2@6"
500,000	5100	Vulcan	SGL.-ACT.	Open	48	100,000	197,000	27'4"	22" × 120"(M)	1,043	35,977	7,620(A)	150	2@5"
499,070	MRBS 4600	Menck	SGL.-ACT.	Open	42	101,410	176,370	27'5"	Cage	850	19,840	9,900	142	6"
325,480	MRBS 3000	Menck	SGL.-ACT.	Open	42	66,135	108,025	25'0"	Cage	520	12,130	6,000	142	5"
325,000	5650	Conmaco	SGL.-ACT.	Open	45	65,000	139,300	23'0"	18$\frac{3}{4}$" × 100"	606	20,907	—	160	3@4"
300,000	3100	Vulcan	SGL.-ACT.	Open	60	100,000	195,500	23'3"	18$\frac{3}{4}$" × 88"(M)	900	30,153	6,644(A)	130	3@4"
300,000	560	Vulcan	SGL.-ACT.	Open	47	62,500	134,060	23'0"	18$\frac{3}{4}$" × 88"(M)	606	20,897	4,427(A)	150	2@5"
200,000	540	Vulcan	SGL.-ACT.	Open	48	40,900	102,980	22'7"	14" × 80"(M)	409	14,126	3,022(A)	130	2@5"
189,850	MRBS 1800	Menck	SGL.-ACT.	Open	44	38,580	64,590	22'5"	Cage	295	7,060	3,700	142	4"
180,000	360	Vulcan	SGL.-ACT.	Open	62	60,000	124,830	19'0"	18$\frac{3}{4}$" × 88"(M)	506	17,460	3,736(A)	130	2@4"
180,000	060	Vulcan	SGL.-ACT.	Open	62	60,000	128,840	19'0"	18$\frac{3}{4}$" × 88'(M)	506	17,460	3,736(A)	130	2@4"
150,000	5300	Conmaco	SGL.-ACT.	Open	46	30,000	62,000	20'9$\frac{1}{2}$"	14" × 80"(M)	234	12,296	2,148(A)	160	4"
150,000	530	Vulcan	SGL.-ACT.	Open	42	30,000	57,680	20'5"	10$\frac{1}{2}$" × 54"(M)	234	8,064	1,711	150	3"
120,000	340	Vulcan	SGL.-ACT.	Open	60	40,000	98,180	18'7"	14" × 80"(M)	354	12,230	2,628(A)	120	2@3"
120,000	040	Vulcan	SGL.-ACT.	Open	60	40,000	87,673	17'11"	14" × 80"(M)	354	12,230	2,628(A)	120	2@3"
93,340	MRBS 850	Menck	SGL.-ACT.	Open	45	18,960	27,890	19'8"	Cage	150	3,530	1,950	142	3"
90,000	030	Vulcan	SGL.-ACT.	Open	54	30,000	55,410	16'5"	10$\frac{1}{4}$" × 54"(M)	201	6,944	1,471(A)	150	3"
90,000	300	Conmaco	SGL.-ACT.	Open	55	30,000	55,390	16'10"	11$\frac{1}{4}$" × 56"(F)	201	6,944	1,833(A)	150	3"
81,250	8/D	Raymond	SGL.-ACT.	Open	40	25,000	34,000	19'4"	10$\frac{1}{4}$" × 25"	172	5,950	—	135	3"
75,000	30X	Raymond	SGL.-ACT.	Open	70	30,000	52,000	19'1"	—	246	8,500	—	150	3"
60,000	S-20	MKT	SGL.-ACT.	Closed	60	20,000	38,650	15'5"	— × 36"	190	—	—	150	3"
60,000	020	Vulcan	SGL.-ACT.	Open	59	20,000	43,785	14'8"	10$\frac{1}{4}$" × 54"(M)	161	5,563	1,195(A)	120	3"
60,000	200	Conmaco	SGL.-ACT.	Open	60	20,000	44,560	15'0"	11$\frac{1}{4}$" × 56"(F)	161	7,500	1,634(A)	120	3"
56,875	5/0	Raymond	SGL.-ACT.	Open	44	17,500	26,450	16'9"	10$\frac{1}{4}$" × 25"	100	4,250	—	150	3"
50,200	200-C	Vulcan	DIFFER.	Open	98	20,000	39,000	13'11"	11$\frac{1}{4}$" × 37"	260	8,970	1,746(A)	142	4"
48,750	016	Vulcan	SGL.-ACT.	Open	58	16,250	33,340	13'8"	10$\frac{1}{4}$" × 54"(M)	121	4,182	899(A)	120	3"
48,750	4/0	Raymond	SGL.-ACT.	Open	46	15,000	23,800	16'1"	—	85	—	—	120	2$\frac{1}{2}$"
48,750	150-C	Raymond	DIFFER.	Open	95–105	15,000	32,500	15'9"	—	—	—	—	120	3"

(continued)

Rated energy	Model	Manufacturer	Type	Style	Blows per min	Wt. of striking parts	Total weight (lb)	Hammer length (ft-in.)	Jaw dimensions	Broiler HP required (ASME)	Steam consump. (lb/hr)	Air consump. (cfm)	Inlet pressure (psi)	Inlet size (in.)
48,750	160	Conmaco	SGL.-ACT.	Open	60	16,250	33,200	13'8"	$11\frac{1}{4}$" × 42"(F)	121	6,950	1,290(A)	120	3"
46,350	MS-500	MKT	SGL.-ACT.	Open	40	11,300	15,550	16'8"	— × 26"	64	2,200	1,060	115	3"
45,200	MRBS 500	Menck	SGL.-ACT.	Open	48	11,020	15,210	16'8"	— × 26"	90	2,200	1,100	142	$2\frac{1}{2}$"
42,000	014	Vulcan	SGL.-ACT.	Open	59	14,000	29,590	13'8"	$10\frac{1}{4}$" × 54"(M)	111	3,844	829(A)	110	3"
42,000	140	Conmaco	SGL.-ACT.	Open	60	14,000	30,850	13'10$\frac{1}{4}$"	$11\frac{1}{4}$" × 42"(F)	111	6,920	1,282(A)	110	3"
41,280	160D	Conmaco	DIFFER.	Open	103	16,000	35,400	13'7$\frac{1}{2}$"	$11\frac{1}{4}$" × 42"(F)	237	8,175	1,550(A)	160	3"
40,600	3/0	Raymond	SGL.-ACT.	Open	50	12,500	21,000	15'7"	$10\frac{1}{4}$" × 25"	—	3,000	—	120	$2\frac{1}{2}$"
37,500	S-14	MKT	SGL.-ACT.	Closed	60	14,000	31,700	13'7"	— × 36"	155	—	—	100	3"
37,375	115	Conmaco	SGL.-ACT.	Open	50	11,500	20,830	14'2"	$9\frac{1}{4}$" × 32"(C)	99	3,425	910(A)	120	$2\frac{1}{2}$"
37,375	115	Conmaco	SGL.-ACT.	Open	50	11,500	20,250	15'0"	$9\frac{1}{4}$" × 26"(K)	116	3,980	1,060(A)	120	$2\frac{1}{2}$"
36,000	140-C	Vulcan	DIFFER.	Open	103	14,000	27,984	12'3"	$11\frac{1}{4}$" × 32"	211	7,279	1,425(A)	140	3"
36,000	140D	Conmaco	DIFFER.	Open	103	14,000	31,200	12'3"	$11\frac{1}{4}$" × 42"(F)	211	7,279	1,425(A)	140	3"
32,885	100C	Vulcan	DIFFER.	Open	103	10,000	22,200	14'0"	$9\frac{1}{4}$" × 26"	180	6,210	1,245(A)	140	$2\frac{1}{2}$"
32,500	100	Conmaco	SGL.-ACT.	Open	50	10,000	19,280	14'2"	$9\frac{1}{4}$" × 32"(C)	85	2,945	820(A)	100	$2\frac{1}{2}$"
32,500	100	Conmaco	SGL.-ACT.	Open	50	10,000	18,700	15'0"	$9\frac{1}{4}$" × 26"(K)	101	3,425	950(A)	100	$2\frac{1}{2}$"
32,500	2/0	Raymond	SGL.-ACT.	Open	50	10,000	18,550	15'0"	$10\frac{1}{4}$" × 25"	—	2,400	—	110	2"
32,500	010	Vulcan	SGL.-ACT.	Open	57	10,000	19,500	14'11"	$10\frac{1}{4}$" × 40"(M)	101	3,498	753(A)	105	$2\frac{1}{2}$"
32,500	S-10	MKT	SGL.-ACT.	Closed	55	10,000	22,380	14'1"	— × 30"	130	—	1,000	80	$2\frac{1}{2}$"
27,121	BB3000	Brons	SGL.-ACT.	Closed	42-65	6,615	12,790	18'6"	9" × 26"	—	—	706	90	2"
26,000	80	Conmaco	SGL.-ACT.	Open	50	8,000	17,280	14'2"	$9\frac{1}{4}$" × 32"(C)	75	2,580	730(A)	85	$2\frac{1}{2}$"
26,000	80	Conmaco	SGL.-ACT.	Open	50	8,000	16,700	15'0"	$9\frac{1}{4}$" × 26"(K)	127	3,000	850	85	$2\frac{1}{2}$"
26,000	85C	Vulcan	DIFFER.	Open	111	8,525	19,020	12'7"	$9\frac{1}{4}$" × 26"	180	6,210	1,245(A)	128	$2\frac{1}{2}$"
26,000	08	Vulcan	SGL.-ACT.	Open	50	8,000	16,750	14'10"	$9\frac{1}{4}$" × 26"	127	4,380	880(A)	83	$2\frac{1}{2}$"
26,000	5-B	MKT	SGL.-ACT.	Closed	55	8,000	18,300	14'4"	— × 26"	120	4,140	850	80	$2\frac{1}{2}$"
24,450	80-C	Vulcan	DIFFER.	Open	111	8,000	17,885	12'1"	$9\frac{1}{4}$" × 26"	180	6,210	1,245(A)	120	$2\frac{1}{2}$"
24,450	80-CHYD	Raymond	DIFFER.	Open	110-120	8,000	17,780	11'10"	—	N/A	N/A	—	5,100	—
24,450	80-C	Raymond	DIFFER.	Open	95-105	8,000	17,885	12'2"	$10\frac{1}{4}$" × 25"	80	—	—	120	$2\frac{1}{2}$"
24,375	0	Raymond	SGL.-ACT.	Open	50	7,500	16,000	15'0"	— × 26"	—	—	750	110	2"
24,375	0	Vulcan	SGL.-ACT.	Open	50	7,500	16,250	15'0"	$9\frac{1}{4}$" × 26"	125	9,320	841(A)	80	$2\frac{1}{2}$"
24,000	C-826	MKT	COMPOUND	Closed	85-95	8,000	17,750	12'2"	— × 26"	120	—	875	125	$2\frac{1}{2}$"
19,500	65-C	Raymond	DIFFER.	Open	110	6,500	14,675	11'8"	$9\frac{1}{4}$" × 19"	—	3,100	—	120	2"
19,500	I-S	Raymond	SGL.-ACT.	Open	58	6,500	12,500	12'9"	$7\frac{1}{2}$" × 28$\frac{1}{4}$"	—	1,500	—	100	$1\frac{1}{2}$"
19,500	06(106)	Vulcan	SGL.-ACT.	Open	60	6,500	11,200	13'0"	$8\frac{1}{4}$" × 20"	94	3,230	625(A)	100	2"

(continued)

TABLE 19.2a | (*Continued*)

Rated energy	Model	Manufac-turer	Type	Style	Blows per min	Wt. of striking parts	Total weight (lb)	Hammer length (ft-in.)	Jaw dimensions	Broiler HP required (ASME)	Steam consump. (lb/hr)	Air consump. (cfm)	Inlet pressure (psi)	Inlet size (in.)
19,500	65	Conmaco	SGL.-ACT.	Open	60	6,500	12,100	13′0″	$9\frac{1}{4}$″ × 26″(C)	94	3,230	625(A)	100	2″
19,500	65	Conmaco	SGL.-ACT.	Open	60	6,500	11,200	13′0″	$8\frac{1}{4}$″ × 20″(K)	94	2,300	625(A)	100	2″
19,500	65-CHYD	Raymond	DIFFER.	Open	130	6,500	14,615	12′1″	—	N/A	N/A	—	5,000	—
19,200	65-C	Vulcan	DIFFER.	Open	117	6,500	14,886	12′1″	$8\frac{1}{4}$″ × 20″	152	5,244	991(A)	150	2″
19,150	11B3	MKT	DBL.-ACT.	Closed	95	5,000	14,000	11′2″	— × 26″	126	—	900	100	$2\frac{1}{2}$″
19,150	1100	BSP	DBL.-ACT.	Closed	95	5,000	14,000	11′2″	— × 26″	126	—	900	90	$2\frac{1}{2}$″
16,250	S-5	MKT	SGL.-ACT.	Closed	60	5,000	12,460	13′3″	— × 24″	85	—	600	80	2″
16,000	C-5(STM)	MKT	DBL.-ACT.	Closed	100–110	5,000	11,880		— × 26″	80	—	—	100	$2\frac{1}{2}$″
15,100	50-C	Vulcan	DIFFER.	Open	120	5,000	11,782		$8\frac{1}{4}$″ × 20″	125	4,312	880(A)	120	2″
15,000	1(106)	Vulcan	SGL.-ACT.	Open	60	5,000	9,700		$8\frac{1}{4}$″ × 20″	81	2,794	565(A)	80	2″
15,000	1	Raymond	SGL.-ACT.	Open	60	5,000	11,000	12′9″	$7\frac{1}{2}$″ × $28\frac{1}{4}$″	—	1,400	500	80	$1\frac{1}{2}$″
15,000	50	Conmaco	SGL.-ACT.	Open	60	5,000	10,600	13′0″	$9\frac{1}{4}$″ × 26″(C)	81	2,794	565(A)	80	2″
15,000	50	Conmaco	SGL.-ACT.	Open	60	5,000	9,700	13′0″	$8\frac{1}{4}$″ × 20″(K)	81	1,925	565(A)	80	2″
14,200	C-5(AIR)	MKT	COMPOUND	Closed	100–110	5,000	11,880	8′9″	— × 26″	—	—	585	100	$2\frac{1}{2}$″
13,560	BB-1500	Brons	SGL.-ACT.	Closed	42–65	3,307	6,285	17′7″	8″ × 20″	—	—	353	90	$1\frac{1}{2}$″
13,100	10B3	MKT	DBL.-ACT.	Closed	105	3,000	10,850	9′2″	— × 24″	104	—	750	100	$2\frac{1}{2}$″
13,100	1000	BSP	DBL.-ACT.	Closed	105	3,000	10,850	9′2″	— × 24″	104	—	750	90	$2\frac{1}{2}$″
8,750	900	BSP	DBL.-ACT.	Closed	145	1,600	7,100	8′2″	— × 20″	85	—	600	90	2″
8,750	9B3	MKT	DBL-ACT.	Closed	145	1,600	7,000	8′4″	$8\frac{1}{2}$″ × 20″	85	—	600	100	2″
7,260	30-C	Vulcan	DIFFER.	Open	133	3,000	7,036	8′11″	$7\frac{1}{4}$″ × 19″	70	2,412	488	120	$1\frac{1}{2}$″
7,260	2	Vulcan	SGL.-ACT.	Open	70	3,000	6,700	11′7″	$7\frac{1}{4}$″ × 19″	49	1,690	336(A)	80	$1\frac{1}{2}$″
4,700	700N	BSP	DBL.-ACT.	Closed	225	850	6,500	5′5″	— × 15″	—	—	600	90	2″
4,150	7	MKT	DBL.-ACT.	Closed	225	800	5,000	6′1″	— × 21″	65	—	450	100	$1\frac{1}{2}$″
4,000	DGH-900	Vulcan	DIFFER.	Closed	328	900	5,000	6′9″	VARIES	75	2,620	580(A)	78	$1\frac{1}{2}$″
3,000	600N	BSP	DBL.-ACT.	Closed	250	500	3,800	5′0″	— × 15″	—	—	365	90	$1\frac{1}{2}$″
2,500	6	MKT	DBL.-ACT.	Closed	275	400	2,900	5′3″	— × 15″	45	—	400	100	$1\frac{1}{4}$″
1,200	500N	BSP	DBL.-ACT.	Closed	330	200	2,000	3′11″	— × 12″	—	—	250	90	$1\frac{1}{4}$″
1,000	5	MKT	DBL.-ACT	Closed	300	200	1,500	4′7″	6″ × 11″	35	—	250	100	$1\frac{1}{4}$″
386	DGH-100D	Vulcan	DIFFER.	Closed	303	100	786	4′2″	$4\frac{1}{4}$″ × $8\frac{3}{4}$″	5	—	74	60	1″

(A) adiabatic compression, (C) Conmaco cable hammer, (K) Conmaco key hammer, (F) female jaws special short cylinder, (M) male jaws.

TABLE 19.2b | Specifications for diesel pile-driving hammers

Energy range (ft-lb)	Model	Manufac-turer	Single/double acting	Blows per min.	Piston weight (lb)	Total weight (lb)	Maximum Stroke (ft-in.)	Total length (ft-in.)	Width between jaws (in.)	Fuel used (gph)
280,000–	K150	Kobe	Single	45–60	33,100	80,500	8'6"	29'8"	Cage	16–20
161,300–80,600	D62–02	Delmag	Single	36–53	13,670	28,000	12'8"	17'9"	32	5.3
141,000–63,360	MB70	Mitsubishi	Single	38–60	15,840	46,000	8'6"	19'6"	—	7–10
135,200	MH72D	Mitsubishi	Single	38–60	15,900	44,000	8'6"	19'6"	32	7–10
117,175–62,566	D55	Delmag	Single	36–47	11,860	26,300	9'10"	17'9"	32	5.54
105,600–	K60	Kobe	Single	42–60	13,200	37,500	8'0"	24'3"	42	6.5–8.0
105,000–48,400	D46–02	Delmag	Single	37–53	10,100	19,900	10'8"	17'3"	32	3.3
92,752	KC45	Kobe	Single	39–60	9,920	24,700	9'4"	17'11"	—	4.5–5.5
91,100–	K45	Kobe	Single	39–60	9,900	25,600	9'2"	18'6"	36	4.5–5.5
87,000–43,500	D44	Delmag	Single	37–56	9,460	22,440	9'2"	15'10"	32	4.5
84,300	MH45	Mitsubishi	Single	42–60	9,920	24,500	8'6"	17'11"	37	4.0–5.8
84,000–37,840	M43	Mitsubishi	Single	40–60	9,460	22,660	8'10"	16'3"	37	4.0–5.8
83,100–38,000	D36–02	Delmag	Single	37–53	7,900	17,700	10'8"	17'3"	32	3.0
79,500–	J44	IHI	Single	42–70	9,720	21,500	8'2"	14'10"	37	6.86
79,000–	K42	Kobe	Single	40–60	9,260	24,000	8'6"	17'8"	36	4.5–5.5
78,800–	B45	BSP	Double	80–100	10,000	27,500	—	19'3"	36	5.5
73,780–30,380	D36	Delmag	Single	37–53	7,940	17,780	9'3"	14'11"	32	3.7
72,182	KC35	Kobe	Single	39–60	7,720	17,400	9'4"	16'10"	—	3.2–4.3
70,800–	K35	Kobe	Single	39–60	7,700	18,700	9'2"	17'8"	30	3.0–4.0
65,600–	MH35	Mitsubishi	Single	42–60	7,720	18,500	8'6"	17'3"	32	3.4–5.3
64,000–29,040	M33	Mitsubishi	Single	40–60	7,260	16,940	8'0"	13'2"	32	3.4–5.3
63,900–	B35	BSP	Double	80–100	7,700	21,200	—	18'5"	36	4.5
63,500–	J35	IHI	Single	72–70	7,730	16,900	8'3"	14'6"	32	4.76
63,000–42,000	DE70/50B	MKT	Single	40–50	7,000	14,600	10'6"	15'10"	26	3.3
62,900–31,800	D30–02	Delmag	Single	38–52	6,600	13,150	10'7"	17'2"	26	1.7
60,100–	K32	Kobe	Single	40–60	7,050	17,750	8'6"	17'8"	30	2.75–3.5
54,200–23,870	D30	Delmag	Single	39–60	6,600	12,346	8'3"	14'2"	26	2.9
51,518–	KC25	Kobe	Single	39–60	5,510	12,130	9'4"	16'10"	—	2.4–3.2
50,700–	K25	Kobe	Single	39–60	5,510	13,100	9'3"	17'6"	26	2.5–3.0
48,400–24,600	D22–02	Delmag	Single	38–52	4,850	11,400	10'7"	17'2"	26	1.6
46,900	MH25	Mitsubishi	Single	42–60	5,510	13,200	8'6"	16'8"	28	2.4–3.7
45,700–	B25	BSP	Double	80–100	5,510	15,200	—	17'9"	30	3.5

(continued)

TABLE 19.2b | (Continued)

Energy range (ft-lb)	Model	Manufacturer	Single/double acting	Blows per min.	Piston weight (lb)	Total weight (lb)	Maximum Stroke (ft-in.)	Total length (ft-in.)	Width between jaws (in.)	Fuel used (gph)
45,000–30,000	DE70/50B	MKT	Single	40–50	5,000	12,600	10'6"	15'10"	26	3.3
45,000–30,000	DE50B	MKT	Single	40–50	5,000	12,000	10'6"	14'9"	26	3.0
45,000–30,000	DA55B	MKT	Single	40–50	5,000	18,300	10'9"	17'4"	26	2.7
45,000–20,240	M23	Mitsubishi	Single	42–60	5,060	11,220	8'10"	14'1"	26	2.4–3.7
45,000–	660	ICE	Double	84	7,564	23,423	—	15'1"	30	3.25
41,300–	K22	Kobe	Single	40–60	4,850	12,350	9'2"	17'6"	26	2.0–2.75
39,780–	D22	Delmag	Single	42–60	4,850	11,150	8'2"	14'2"	26	3.44
39,100	J22	IHI	Single	42–70	4,850	10,800	10'0"	14'0"	26	3.2
38,200–31,200	DA55B	MKT	Double	78–82	5,000	18,300	—	17'4"	26	3.0
36,000–24,000	DE40	MKT	Single	40–50	4,000	11,275	10'6"	15'0"	26	3.0
31,000–17,700	520	ICE	Double	80–84	5,070	12,545	—	16'6"	26	1.35
28,100–	MH15	Mitsubishi	Single	42–60	3,310	8,400	8'6"	16'1"	26	1.3–2.1
27,100–	D15	Delmag	Single	40–60	3,300	6,615	8'3"	13'11"	20	1.75
26,200–	B15	BSP	Double	80–100	3,300	9,000	—	17'0"	26	2.5
26,000–11,800	M14S	Mitsubishi	Single	42–60	2,970	7,260	8'9"	13'7"	26	1.3–2.1
25,200–16,800	DE30/20B	MKT	Single	40–50	2,800	7,250	10'0"	15'4"	20	2.0
25,200–16,800	DA35B	MKT	Single	40–50	2,800	10,000	10'9"	17'0"	20	1.7
25,200–	DE30	MKT	Single	40–50	2,800	8,125	10'9"	15'0"	20	2.0
24,400–	K13	Kobe	Single	40–60	2,860	7,300	8'6"	16'8"	26	.57–2.0
22,500–	D12	Delmag	Single	40–60	2,750	6,050	8'2"	13'11"	20	2.11
21,000–16,000	DA35B	MKT	Double	78–82	2,800	10,000	—	17'0"	20	2.0
19,840–7,700	440	ICE	Double	86–90	4,000	9,840	—	13'9"	20	1.6
18,000–12,000	DE20	MKT	Single	40–50	2,000	6,325	9'5"	13'3"	20	1.6
18,000–	312	ICE	Double	100–105	3,857	10,375	—	10'9"	26	1.1
16,000–12,000	DE30/20B	MKT	Single	40–50	2,000	6,450	10'0"	15'4"	20	2.0
9,100	D5	Delmag	Single	40–60	1,100	2,730	8'3"	12'2"	19	1.32
8,800–	DE10	MKT	Single	40–50	1,100	3,100	8'0"	12'2"	10"BP	0.9
8,100–	180	ICE	Double	90–95	1,725	4,550	—	11'7"	18	0.65
3,630–	D4	Delmag	Single	50–60	836	1,360	4'4"	7'9"	Beam	0.21
1,815–	D2	Delmag	Single	60–70	484	792	4'1"	6'9"	Beam	0.075

TABLE 19.2c I Specifications for hydraulic vibratory pile-drivers/extractors

Dynamic forces (tons)	Model	Manu-facturer	Fre-quency (vpm)	Ampli-tude (in.)	HP	Max. pull extraction (tons)	Pile clamp force (tons)	Suspended weight (lb)	Shipping weight (lb)	Height (ft-in.)	Depth (ft-in.)	Width	Throat width (in.)
182	V-36	MKT	1600	0.75	550	80	80	18,800	36,300	13-1	1-0	12-0	14
145.4	812	ICE	750–1500	$\frac{1}{2}$–1	330	40	100	14,700	30,200	9-0	2-0	8-0	12
139	50H1	PTC	1500	1.25	370	44							
100.5	V-20	MKT	1650	0.66	295	40	75	12,500	23,900	5-3	1-2		14
111.4	4000	Foster	1400	0.72	299	40	100/200	18,800	32,300	9-10	1-10	9-10	12
78.3	V-16	MKT	1750	0.47	161	50	75	11,700	20,600	5-3	1-2		14
71.0	V-14	MKT	1500	0.32	140	50	75	10,000	29,500	5-3	1-2		14
65.2	416	ICE	800–1600	$\frac{1}{4}$–1	175	40	100	12,200	26,200	8-9	1-10	8-0	12
55	20H6	PTC	1500	0.88	185	22							
48.5	1700	Foster	1400	0.39	147	30	80/100	12,900	26,900	7-0	1-10		12
38.8	14H2	PTC	1500	0.85	120	16.5							
36.4	216	ICE	800–1600	$\frac{1}{4}$–$\frac{3}{4}$	115	20	50	4,500	12,500	6-6	5-0	3-11	12
35.2	1200	Foster	1425	0.34	85	20	60	6,700	11,670	5-0	1-11		12
34.4	7H4	PTC	2000	0.50	115	16.5							
30.0	V-5	MKT	1450	0.50	59	20	31	6,800	10,800	5-4	1-2		14

slower velocity of a *single-acting* hammer will transmit a greater portion of the rated energy into driving the piles. Table 19.2 includes dimensions and other data for several of the more popular sizes of double-acting hammers.

The advantages of double-acting compared with single-acting hammers include

1. The greater number of blows per minute reduces the time required to drive piles.
2. The greater number of blows per minute reduces the development of static skin friction between blows.
3. Piles can be driven more easily without leads.

The disadvantages of double-acting compared with single-acting hammers include

1. The relative light weight and high velocity of the ram make this type of hammer less suitable for use in driving heavy piles into soils having high frictional resistance.
2. The hammer is more complicated.

Differential-Acting Steam/Air Hammers

A differential-acting steam/air hammer is a modified single-acting hammer in that steam/air pressure used to lift the ram is not exhausted at the end of the upward stroke but is valved over the piston to accelerate the ram on the downstroke. The number of blows per minute is comparable with that for a double-acting hammer, whereas the weight and the equivalent free fall of the ram are comparable with those of a single-acting hammer. Thus, it is claimed that this type of hammer has the advantages of the single- and double-acting hammers. These hammers require the use of a pile cap with cushioning material and a set of leads.

It is reported that this hammer will drive a pile in one-half the time required by the same-size single-acting hammer and in doing so will use 25 to 35% less steam or air. These hammers are available in open or closed types. Table 19.2a gives dimensions and data for these hammers. The values given in the table for rated energy per blow are correct provided the steam or air pressure is sufficient to produce the indicated normal blows per minute.

Diesel Hammers

A diesel pile-driving hammer (see Fig. 19.15) is a self-contained driving unit that does not require an external source of energy such as a steam boiler or an air compressor. In this respect it is simpler and more easily moved from one location to another than a steam hammer. A complete unit consists of a vertical cylinder, a piston or ram, an anvil, fuel- and lubricating-oil tanks, a fuel pump, injectors, and a mechanical lubricator.

After a hammer is placed on top of a pile, the combined piston and ram are lifted to the upper end of the stroke and released to start the unit operating.

FIGURE 19.15 | Diesel hammer driving sheet piling.
Source: Pileco, Inc.

Figure 19.16 illustrates the operation of a diesel hammer. As the ram nears the end of the downstroke, it activates a fuel pump that injects the fuel into the combustion chamber between the ram and the anvil. The continued downstroke of the ram compresses the air and the fuel to ignition heat. The resulting explosion drives the pile downward and the ram upward to repeat its stroke. The energy per blow, which can be controlled by the operator, may be varied over a wide range. Table 19.2b lists the specifications for several makes and models of diesel hammers.

Open-end diesel hammers deliver 40 to 55 blows per minute. The closed-end models operate at 75 to 85 blows per minute. In the United States, diesel hammers are almost always used in a set of leads, yet, in other parts of the world, they are commonly used free-hanging. These hammers require a pile cap with "live" cushioning material.

A time plot of the force applied to the pile for a diesel hammer is quite different from those for the previously discussed hammers. In the case of a diesel

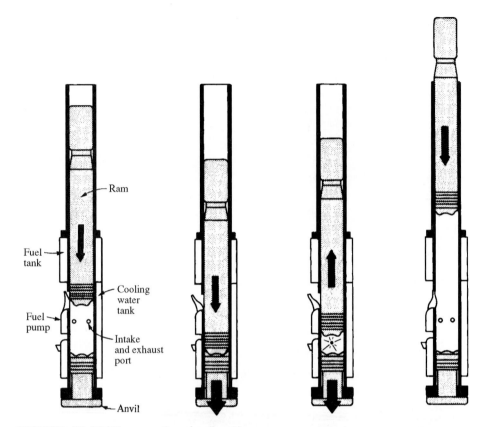

Ram

Fuel tank

Cooling water tank

Fuel pump

Intake and exhaust port

Anvil

FIGURE 19.16 | The operation of a diesel hammer.

Source: L. B. Foster Company.

hammer, the force begins to build as soon as the falling ram closes the exhaust ports of the cylinder and compresses the trapped air. With the explosion of the fuel at the bottom of the stroke, there is a force spike that decays as the ram travels upward. The important point is that the loading to the pile spans time and changes magnitude across the time span.

The advantages of diesel compared with steam/air hammers include

1. The hammer needs no external source of energy. Thus, it is more mobile and it requires less time to set up and start operation.
2. The hammer is economical to operate. The rated fuel consumption for a 24,000-ft-lb hammer is 3 gal per hour (gph) when it is operating. Because a hammer does not operate continuously, the actual consumption is less.
3. The hammer is convenient to operate in remote areas. Because the hammer uses diesel oil as a source of energy, it is not necessary to provide a boiler, water for steam, and fuel.

4. The hammer operates well in cold weather. Diesel hammers have been used at temperatures well below 0°F, where it would be difficult or impossible to provide steam.

5. The hammer is light in weight when compared with the weight of a steam hammer of equal rating.

6. Maintenance and service are simple and fast.

7. The energy per blow increases as the driving resistance of a pile increases.

8. Because the resistance of a pile to driving is necessary for continuous operation of a diesel hammer, this hammer will not operate if a pile breaks or falls out from under the hammer.

9. Because of the low velocity in easy driving, and because the piston reacts to the impact needed for each blow by rebounding up its cylinder, a diesel hammer is less likely to batter the piles when driving them.

10. The energy per blow and the number of blows per minute can be varied easily to permit a diesel hammer to operate most effectively for an existing condition [16].

The disadvantages of diesel compared to other hammers include

1. It is difficult to determine the energy per blow for this hammer because the height to which the piston ram will rise following the explosion of the fuel in the combustion chamber is a function of the driving resistance. For this reason, there is uncertainty about the accuracy of applying dynamic pile-driving formulas to diesel hammers.

2. The hammer may not operate well when driving piles into soft ground. Unless a pile offers sufficient driving resistance to activate the ram, the hammer will not operate.

3. The number of strokes per minute is less than for a steam hammer. This is especially true for a diesel hammer with either an open end or top.

4. The length of a diesel hammer is slightly greater than the length of a steam hammer of comparable energy rating.

Hydraulic Hammers

These hammers operate on the differential pressure of hydraulic fluid instead of steam or compressed air used by conventional hammers. Dynamic pile-driving equations in current use are applicable to these hammers.

Another type of hydraulic pile hammer that can be used for driving and extracting steel H piles and steel sheet piles incorporates a gripping and pushing or pulling technique. This pile driver grips the pile and then pushes the pile down approximately 3 ft. At the end of the downstroke, the pile is released and the gripper slides up the pile 3 ft to begin the process of another push. The equipment can be used in reverse for extracting piles. These drivers develop up to 140 tons of pressing or extracting force, are compact, make minimal noise, and cause very little vibration. They are well suited for driving piles in areas where there is restricted overhead space since piles may be driven in short lengths and spliced.

Vibratory Pile Drivers

Vibratory pile hammers have demonstrated their effectiveness in speed and economy in driving piles into certain types of soil. These drivers are especially effective when the piles are driven into water-saturated noncohesive soils. The drivers may experience difficulty in driving piles into dry sand, or similar materials, or into cohesive soils that do not respond to the vibrations.

The drivers are equipped with horizontal shafts, to which eccentric weights are attached. As the shafts rotate in pairs, in opposing directions, at speeds that can be varied in excess of 1,000 rpm, the forces produced by the rotating weights produce vibrations. The vibrations are transmitted to the pile because it is rigidly connected to the driver by clamps. From the pile the vibrations are transmitted into the adjacent soil. The agitation of the soil materially reduces the skin friction between the soil and the pile. This is especially true when the soil is saturated with water. The combined dead weight of the pile and the driver resting on the pile will drive the pile rapidly.

Figure 19.17 illustrates the basic principle of the rotating weights, using six shafts. As noted in the figure, the two inner shafts, with lighter weights, rotate at twice the speeds of the top and bottom shafts. During each revolution of two top and bottom shafts, the forces contributed by all weights will act downward at 0 and

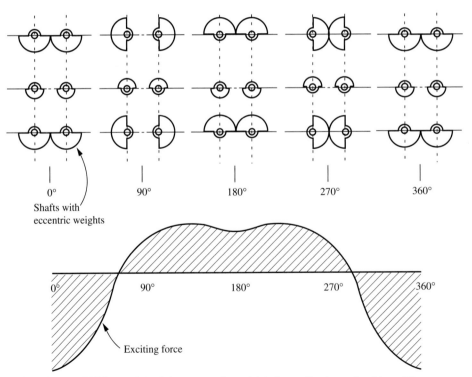

FIGURE 19.17 | Operation of the eccentric weights for a vibratory pile driver. Diagram shows how exciting force of a six-shaft vibrator varies with the position of the eccentric weights attached to the shafts.

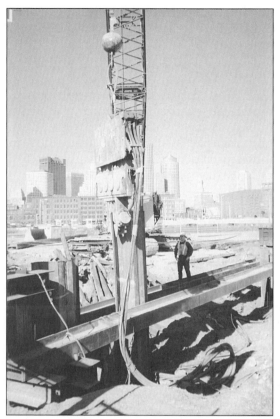

FIGURE 19.18 | A vibratory driver driving steel H pile.

360°, whereas at 180° the forces tend to counteract each other, as indicated in the exciting force curve. Figure 19.18 illustrates a vibratory driver driving steel piling.

Leads are rarely employed with vibratory drivers. The driver is powered either electrically or hydraulically; therefore, a generator or hydraulic power pack is needed as an energy source. Because leads are not required, a smaller crane can usually be employed to handle vibratory driver work.

Performance Factors for Vibratory Drivers

There are five performance factors that determine effectiveness of a vibratory driver.

1. *Amplitude.* This is the magnitude of the vertical movement of the pile produced by the vibratory unit. It may be expressed in inches or millimeters.

2. *Eccentric moment.* The eccentric moment of a vibratory unit is a basic measure or indication of the size of a driver. It is the product of the weight of the eccentricities multiplied by the distance from the center of rotation

of the shafts to the center of gravity of the eccentrics. The heavier the eccentric weights and the farther they are from the center of rotation of the shaft, the greater the eccentric moment of the unit.

3. *Frequency.* This is expressed as the number of vertical movements of the vibrator per minute, which is also the number of revolutions of the rotating shafts per minute. Tests conducted on piles driven by vibratory drivers have indicated that the frictional forces between the piles being driven and the soil into which they are driven are at minimum values when frequencies are maintained in the range of 700 to 1,200 vibrations per min. In general, the frequencies for piles driven into clay soils should be lower than for piles driven into sandy soils.

4. *Vibrating weight.* The vibrating weight includes the vibrating case and the vibrating head of the vibrator unit, plus the pile being driven.

5. *Nonvibrating weight.* This is the weight of that part of the system that does not vibrate, including the suspension mechanism and the motors. Nonvibrating weights push down on and aid in driving the piles.

METHODS OF SUPPORTING AND POSITIONING PILES DURING DRIVING

When driving piles, it is necessary to have a method that will position the pile in the proper location with the required alignment or batter and which will support the pile during driving (see Fig. 19.19). The following methods are utilized to accomplish such alignment and support.

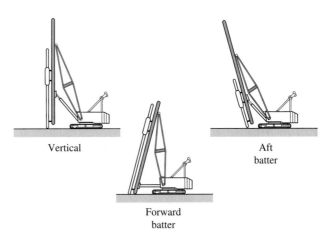

Vertical

Forward batter

Aft batter

FIGURE 19.19 | Pile alignment nomenclature.

Fixed Leads

Normally, a set of leads consists of a three-sided steel lattice frame similar in construction to a crane boom with one side open. These are referred to as U-type leads (see Figs. 19.1 and 19.6). Typical technical data for this type of leads can be found at www.pileco.com/u-type_lead_tech_info.html. The open side allows positioning of the pile in the leads and under the hammer. The leads have a set of rails or guides for the hammer. Running on the rails, the hammer is lifted above the pile height when a new pile is threaded into the leads. During driving, the hammer descends along the lead rails as the pile moves downward into the ground. When driving batter piles, the leads are positioned at an angle.

The term "fixed" is used to indicate that the leads are held in a fixed position by the pile-driving rig or by some other means. The bottom of the leads is commonly attached to the crane or to the driving platform to assure proper positioning of the pile during driving.

Swing Leads

Leads that are not attached at their bottom to the crane or driving platform are known as "swing leads." Such an arrangement allows the driving rig to position a pile at a location farther away than would be possible with fixed leads. This is not generally the preferred method of driving a pile since it is more difficult to position the pile accurately and to maintain vertical alignment during driving. If for any reason the pile tends to twist or run off the intended alignment, it is difficult to control the pile with swing leads.

Hydraulic Leads

To control pile position, hydraulic leads use a system of hydraulic cylinders connected between the bottom of the leads and the driving rig (see Fig. 19.20). This system allows the operator to position the pile very quickly and accurately. Hydraulic leads are extremely useful in driving batter piles since the system can rapidly and easily adjust the angle of the leads for the required batter. The system is more costly than standard fixed leads but any contractor who is regularly involved in pile driving operations quickly recovers that dollar difference because of the increased driving productivity.

Templates

Many times a template is used to support and hold the pile in position during driving. Templates are usually constructed from steel pipe or beams and may have several levels of framing to support long piles or piles on a batter (see Fig. 19.21). In marine work where access to the pile groups is restricted, templates are used regularly. Frequently, a set of leads can be fixed or attached to a template beam and the combined systems are used to support and guide the pile during driving. Templates or guide frames are commonly used when driving sheet piling. When driving sheet pile cells, circular templates are used to maintain proper alignment.

JETTING PILES

The use of a water jet to assist in driving piles into sand or fine gravel frequently will speed the driving operation. The water, which is discharged through a nozzle

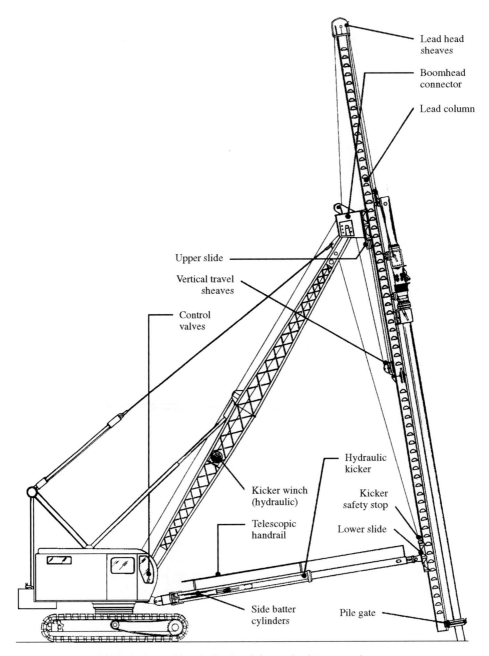

FIGURE 19.20 | Vertical travel leads (hydraulic) attached to a crawler crane.

Source: Berminghammer Corporation Limited.

at the lower end of a jet pipe, keeps the soil around the pile in agitation, thereby reducing the resistance due to skin friction. Successful jetting requires a plentiful supply of water at a pressure high enough to loosen the soil and to remove it from

FIGURE 19.21 I Barge-mounted Whirley crane driving concrete piles on a batter utilizing a template.

Source: Tidewater Construction Corporation.

the hole ahead of the penetration of the pile. Commonly used jet pipes vary in size from 2 to 4 in. in diameter, with nozzle diameters varying from $\frac{1}{2}$ to $1\frac{1}{2}$ in. The water pressure at the nozzle may vary from approximately 100 to more than 300 psi, with the water quantity usually varying from 300 to 500 gallons per minute (gpm), but as high as 1,000 gpm in some instances.

Although some piles have been jetted to final penetration, this is not considered good practice primarily because it is impossible to determine the safe supporting capacity of a pile so driven. Most specifications require that piles shall be driven the last few feet without the benefit of jetting. The Foundation Code of the city of New York requires a contractor to obtain special permission prior to jetting piles and specifies that piles shall be driven the last 3 ft with a pile hammer.

SPUDDING AND PREAUGERING

At proposed pile locations where there is evidence of previous construction or when it is suspected the preexisting foundations may interfere with driving operations, it may be prudent to preauger or predrive a short H pile section to below the depth of the suspected interference. If there is evidence of very hard overlying soil strata, it may be necessary to preauger or predrill through the strata prior to beginning the pile-driving operation.

Preaugering should also be considered where piles are to be driven through an earth fill and into natural soil. This is often the case at bridge abutments where the embankment fill is placed prior to construction of the bridge. The depth of the preaugering should coincide with the depth of the placed fill in order that the piles develop their full bearing capacity in the natural soil.

DRIVING PILES BELOW WATER

If it is necessary to drive piles below water, either of two methods can be used. When the driving unit is a drop hammer, an open-type steam hammer, or a diesel hammer, the pile is driven until the top is just above the surface of the water. Then a follower is placed on top of the pile, and the driving is continued through the follower. The follower can be made of wood or steel and must be strong enough to transmit the energy from the hammer to the pile.

When the driving unit is an enclosed steam hammer, the driving can be continued below the surface of the water, without a follower. It will, however, be necessary to install an exhaust hose to the surface of the water for the steam. Additionally, about 60 cubic feet per minute (cfm) of compressed air must be supplied to the lower part of the hammer housing to prevent water from flowing into the casing and around the ram. The compressed air should be at a pressure to match the depth that the hammer is below the water. This would be about $\frac{1}{2}$ psi for each foot of depth below the water surface.

SELECTING A PILE-DRIVING HAMMER

Selecting the most suitable pile-driving hammer for a given project involves a study of several factors. Factors to be considered include the size and type of piles, the number of piles, the character of the soil, the location of the project, the topography of the site, the type of rig available, and whether driving will be done on land or in water. A pile-driving contractor is usually concerned with selecting a hammer that will drive the piles for a project at the lowest practical cost. As most contractors must limit their ownership to a few representative sizes and types of hammers, a selection should be made from hammers already owned, unless conditions are such that it is economical or necessary to purchase or rent an additional size or type. Naturally, more consideration should be given to the selection of a hammer for a project that requires several hundred piles than for a project that has only a few piles.

As stated, the function of a pile hammer is to furnish the energy required to drive a pile. This energy is supplied by a weight that is raised and permitted to drop onto the top of a pile, under the effect of gravity alone or with steam/air acting during the downward stroke. The theoretical energy per blow will equal the product of the weight times the equivalent free fall. Since some of this energy is lost in friction as the weight travels downward, the net energy per blow will be less than the theoretical energy. The actual amount of net energy depends on the efficiency of the particular hammer. The efficiencies of the pile hammers vary from 50 to 100%.

Table 19.3 gives recommended sizes of hammers for different types and sizes of piles and driving resistances. The sizes are indicated by the theoretical foot-pounds of the energy delivered per blow. The theoretical energy per blow given in Table 19.2 is correct provided the hammer is operated at the designated number of strokes per minute.

TABLE 19.3 | Recommended sizes of hammers for driving various types of piles*

Length of piles (ft)	Depth of penetration	Weight of various types of piles (lb/lin. ft)						
		Steel sheet†			Timber		Concrete	
		20	30	40	30	60	150	400
Driving through ordinary earth, moist clay, and loose gravel, normal frictional resistance								
25	$\frac{1}{2}$	2,000	2,000	3,600	3,600	7,000	7,500	15,000
	Full	3,600	3,600	6,000	3,600	7,000	7,500	15,000
50	$\frac{1}{2}$	6,000	6,000	7,000	7,000	7,500	15,000	20,000
	Full	7,000	7,000	7,500	7,500	12,000	15,000	20,000
75	$\frac{1}{2}$	—	7,000	7,500	—	15,000	—	30,000
	Full	—	—	12,000	—	15,000	—	30,000
Driving through stiff clay, compacted sand, and gravel, high frictional resistance								
25	$\frac{1}{2}$	3,600	3,600	3,600	7,500	7,500	7,500	15,000
	Full	3,600	7,000	7,000	7,500	7,500	12,000	15,000
50	$\frac{1}{2}$	7,000	7,500	7,500	12,000	12,000	15,000	25,000
	Full	—	7,500	7,500	—	15,000	—	30,000
75	$\frac{1}{2}$	—	7,500	12,000	—	15,000	—	36,000
	Full	—	—	15,000	—	20,000	—	50,000

*Size expressed in foot-pounds of energy per blow.
†The indicated energy is based on driving two steel sheet piles simultaneously. In driving single piles, use approximately two-thirds the indicated energy.

In general, it is sound practice to select the largest hammer that can be used without overstressing or damaging a pile. As already shown, when a large hammer is used, a greater portion of the energy is effective in driving the pile that produces a higher operating efficiency. Therefore, the hammer sizes given in Table 19.3 should be considered as the minimum sizes. In some instances hammers as much as 50% larger may be used advantageously.

Consideration must also be given to the capabilities of the crane that will be employed to handle the hammer, leads, and pile. The crane must be able to handle this total load at the maximum reach that will be required by the project site conditions. Therefore, while it is important to know the hammer's rated energy and speed (blows per minute) when examining the question of capability to drive the required piles, the total operating weight specification is important when selecting the crane that will support the hammer (see Fig. 19.22).

The final question to be considered is how much boom will the crane require? This is dependent on the length of the piles and the operating length of the hammer. The hammer specifications will provide this information (see Fig. 19.23). To the sum of these two dimensions should also be added an allowance for positioning the pile under the hammer.

Working specifications

Rated energy..42,000 ft-lb (5,807 kg-m)
Minimum energy..16,000 ft-lb (2,212 kg-m)
Stroke at rated energy ....10' 3" (312 cm)
Maximum obtainable stroke...................................10' 5" (318 cm)
Speed (blows per minute) 37-55
Bearing based on EN formula................................210 tons (190 tons)

Weights

Bare hammer..7,610 lbs (3,452 kg)
Ram...4,088 lbs (1,854 kg)
Anvil ....545 lbs (247 kg)
Typical operating weight with cap ....8,710 lb (3,950 kg)

FIGURE 19.22 | Typical pile hammer working specifications.

Dimensions of hammer

Width (side to side)...20" (508 cm)
Depth..29" (737 cm)
Centerline to front ....13 3/4" (349 mm)
Centerline to rear..15 1/4" (387 mm)
Length (hammer only)..16' 1" (490 cm)
Operating length (top of ram to top of pile).............27' 9" (846 cm)

FIGURE 19.23 | Typical pile hammer dimensions specification.

CALCULATING PILE-SUPPORTING STRENGTH

There are many pile-driving equations, each of which is intended to give the supporting strength of a pile. The equations are empirical, with coefficients that have been determined for certain existing or assumed conditions under which it was developed. None of the equations give dependable values for the supporting strength of the piles for all the varying conditions that exist on foundation jobs.

It is not within the scope of this book to analyze the various pile-driving equations or the theory related to them. For a more comprehensive study of this subject it is suggested that the reader consult the books listed at the end of this chapter. Perhaps the most popular equation in the United States is the *Engineering News* equation. Its popularity seems due primarily to its simplicity rather than its accuracy. For the three types of pile-driving hammers in common use, the *Engineering News* equation has these forms. For a drop hammer,

$$R = \frac{2WH}{S + 1.0} \qquad \textbf{[19.1]}$$

For a single-acting steam hammer,

$$R = \frac{2WH}{S + 0.1} \qquad \textbf{[19.2]}$$

For a double- and differential-acting steam hammer,

$$R = \frac{2E}{S + 0.1} \qquad \textbf{[19.3]}$$

where

R = safe load on a pile in pounds
W = weight of a falling mass in pounds
H = height of free fall for mass W in feet
E = total energy of ram at bottom of its downward stroke in foot-pounds
S = average penetration per blow for last 5 or 10 blows in inches

DYNAMIC FORMULAS

Considerable progress has been made in the conduct of tests and studies analyzing the properties of piles. In 1964 the Ohio Department of Transportation sponsored a research program at the Case Western Reserve University, Cleveland, Ohio, whose objective was to develop reliable techniques to predict static capacities for load-bearing piles. One of the systems developed by this research was a Pile-Driving Analyzer, a programmed field computer used to analyze a pile during the driving operation.

The Pile-Driving Analyzer processes the strain and acceleration signals from two strain transducers and two accelerometers, which are attached to the pile during driving. For each hammer blow, the analyzer converts the analog strain and acceleration signals into digital force and velocity data plotted against time. These traces are displayed on an oscilloscope during driving so that the quality of the data can be instantly evaluated. This allows for immediate correction of the driving operations if necessary. Specific data from each hammer blow can be selected from a menu of 36 possible dynamic quantities for print out on a paper tape. Additionally, the signals can be recorded on a magnetic tape or digitally stored so that further analyses of the data can be performed, if necessary.

Wave Equation Analysis of Piles

As a result of the ability to test a pile dynamically during the driving process, a new generation of formula has been developed to predict pile load-bearing capacities. One analysis method makes use of a differential equation describing the wave propagation process. This equation, commonly known as the "wave equation," can be written in the form:

$$\frac{\partial^2 u}{\partial t^2} = \frac{E \partial^2 u}{p \partial x^2}$$

[19.4]

where

x = a position on the rod

u = the displacement of the rod at point x

t = time

E = the elastic modulus of the rod material

p = the mass density of the rod

Solutions of the wave equation have been employed for the analysis of pile driving, but this is a difficult process to apply because of the boundary conditions encountered during driving. Working under contract for the Federal Highway Administration, Rausche and Goble developed a pile wave equation program named WEAP. This program is in the public domain and operates on a wide variety of computer platforms.

This wave equation program models the entire pile-driving system. The pile hammer, hammer cushion, helmet, and pile are each represented by a series of masses and springs. The size of each of the masses is determined from the weight of the piece of the system represented. For example, if the pile is divided into discrete lengths of 5 ft, then the equivalent mass element will have the same mass as 5 ft of pile. Similarly, the discrete spring would have the same stiffness as 5 ft of pile. Some of the elements of the system are naturally discrete. An example of a discrete element is the hammer cushion. It has the function of behaving like a spring to soften the effect of the impact on the ram and to protect the ram from damage. It can be easily represented by a spring in the computer model. The helmet is a very compact element that contributes little or no flexibility to the system, so it can be treated as only a mass element.

Wave Equation Limitations

A wave equation analysis requires input assumptions that can significantly affect the program results. Potential sources of error include assumptions about hammer performance, the hammer and pile cushion properties, the soil resistance distribution, as well as soil quake and damping characteristics. However, insight into these assumptions can be obtained through dynamic measurement and analysis.

Dynamic measurements of force, velocity, and energy at the point on the pile where the transducers and accelerometers are attached can readily be compared to the wave equation values computed for a corresponding model pile segment. Adjustments to the wave equation input parameters can then be made depending on the agreement between the measured and computed values. This approach is the simplest use of the data available from the dynamic measurements and is an easy way to "calibrate" the wave equation, thereby reducing potential errors.

CAPWAP Analysis

In pile-driving analysis, three unknowns exist: (1) pile forces, (2) pile motions, and (3) pile boundary conditions. If any two of the three are known, the remaining unknown can be determined. CAPWAP is a computer program first developed by Rausche at Case Western Reserve University. The CAPWAP program uses the continuous pile model and applies the records of force and velocity obtained with the Pile-Driving Analyzer to an assumed soil model consisting of the soil resistance distribution, quakes, and damping characteristics at each soil segment and at the pile toe. With this input data of pile motion and assumed boundary conditions, the program computes a force wave trace at the pile head. This computed force wave trace is compared to the force wave trace measured in the field by the Pile-Driving Analyzer.

Therefore, with pile force and pile motion quantified, the CAPWAP program and an experienced engineer can determine the boundary conditions through a trial-and-error process of signal matching. The boundary conditions include the pile capacity, the soil resistance distribution, and soil quake and damping characteristics. The CAPWAP model uses this iteration process of computer runs matched to field-measured force wave traces to predict pile conditions and capacity.

SUMMARY

Load-bearing piles are used primarily to transmit structural loads through soil formations with inadequate supporting properties into or onto soil stratum capable of supporting the loads. If the load is transmitted to the soil through skin friction between the surface of the pile and the soil, the pile is called a *friction pile*. If the load is transmitted to the soil through the lower tip, the pile is called an *end-bearing pile*.

The function of a pile hammer is to furnish the energy required to drive a pile. Pile-driving hammers are designated by type and size. When driving piles, it is necessary to have a method that will position the pile in the proper location with the required alignment or batter and which will support the pile during driving.

There are many pile-driving equations, each of which is intended to give the supporting strength of a pile. None of the equations give dependable values for the supporting strength of the piles for all the varying conditions that exist on foundation jobs. Critical learning objectives would include:

■ An understanding of the different pile types and the advantages of each.
■ An understanding of the different types of pile-driving hammers and the advantages of each.
■ An understanding of the equations used to calculate pile support strength and the limitations of those equations.

These objectives are the basis for the problems that follow.

PROBLEMS

19.1 The falling ram of a drop hammer used to drive a timber pile is 5,000 lb. The free-fall height during driving was 19 in., and the average penetration for the last eight blows was $\frac{1}{2}$ in. per blow. What is the safe rated load using the *Engineering News* equation?

19.2 If the hammer in Problem 19.1 had been a single-acting steam type, what will be the safe rated load using the *Engineering News* equation?

19.3 If the hammer in Problem 19.1 had been a double-acting steam type having a rated total energy of ram at the bottom of a stroke of 19,150 ft-lb, what will be the safe rated load using the *Engineering News* equation?

19.4 If a BSP 1000 double-acting steam hammer is used to drive a pile and the average penetration per blow for the last 10 blows is $\frac{1}{4}$ in., what will be the safe rated load using the *Engineering News* equation?

REFERENCES

1. Adams, James, James Dees, and James Graham, "'Clean' Creosote Timber," *The Military Engineer,* Vol. 88, No. 578, pp. 25, 26, June–July 1996.

2. Bayshore Concrete Products, P.O. Box 230, Cape Charles, VA 23310, www.solartown.com/bayshore/.

3. Berminghammer Corporation Limited, Wellington St. Marine Terminal, Hamilton, Ontario L8L 4Z9, Canada, www.berminghammer.com/.

4. Bethlehem Steel Corporation, Piling Products, Bethlehem, PA 18016, www.bethsteel.com/.

5. Bruce, R. N., Jr., and D. C. Hebert, "Splicing of Precast Prestressed Concrete Piles: Part 1—Review and Performance of Splices," *Journal of the Prestressed Concrete Institute,* Vol. 19, No. 5, September–October, 1974.

6. *Construction Productivity Advancement Research (CPAR) Program,* USACERL Technical Report 98/123, U.S. Army Corps of Engineers, Construction Engineering Research Laboratories, September, 1998.

7. *Design of Pile Foundations, Technical Engineering and Design Guides as Adapted from the US Army Corps of Engineers, No. 1,* American Society of Civil Engineers, New York, 1993.

8. Frankipile Geofranki Soil Tech, Johannesburg, P.O. Box 39075, Bramley 2018, 688 Main Pretoria Road, Wynberg, Sandton, SA, Tel: (011) 887-2700, Fax: (011) 887-0958.

9. Gendron, G. J., "Pile Driving: Hammers and Driving Methods," *Highway Research Record No. 333,* Transportation Research Board, Washington, D.C., 1970.

10. Goble, G. G., K. Fricke, and G. E. Likens, Jr., "Driving Stresses in Concrete Piles," *Journal of Prestressed Concrete Institute* Vol. 21, pp. 70–88, January–February 1976.

11. Goble, G. G., G. E. Likens, and F. Rausche, "Bearing Capacity of Piles from Dynamic Measurements," Final Report, Department of Civil Engineering, Case Western University, Cleveland, Ohio, March 1975.

12. "Large-Diameter Composite Unit Stands in for 40 Timbers," *ENR,* February 21, 2000, pp. 32–35.

13. "Manual for Quality Control," PCI MNL-116-85, Prestressed Concrete Institute, 201 N. Wells St., Chicago, IL 60606.

14. McClelland, B., J. A. Focht, Jr., and W. J. Emrich, "Problems in Design and Installation of Offshore Piles," *Journal of the Soil Mechanics and Foundation Division,* in *Proceeding of the ASCE* Vol. 95, pp. 1491–1514, November 1969.

15. *Practical Guidelines for the Selection, Design and Installation of Piles,* American Society of Civil Engineers, New York, 1984.

16. Rausche, Frank, and George G. Goble, "Performance of Pile-driving Hammers," *Journal of the Construction Division,* in *Proceeding of the ASCE* Vol. 98, pp. 201–218, September 1972.

17. Raushe, F., Fred Moses, and G. G. Goble, "Soil Resistance Predictions from Pile Dynamics," *Journal of the Soil Mechanics and Foundation Division,* in *Proceeding of the ASCE* Vol. 98, pp. 917–937, September 1972.

18. "Recommended Practice for Design, Manufacture, and Installation of Prestressed Concrete Piling," *Journal of the Prestressed Concrete Institute* Vol. 22, No. 2, March–April, 1977.

19. Rempe, D. M., and M. T. Davisson, "Performance of the Diesel Pile Hammer," *Proceedings of the Ninth International Conference of Soil Mechanics and Foundation Engineering,* ISSMFE, Tokyo, Japan, 1977.

20. Samson, C. H., T. J. Hirsch, and L. L. Lowery, "Computer Study of Dynamic Behavior of Piling," *Journal of the Structural Division,* in *Proceedings of the ASCE* Vol. 89, pp. 413–440, August 1963.

21. Sandhu, Balbir S., "Predicting Driving Stresses in Piles," *Journal of the Construction Division,* in *Proceeding of the ASCE* Vol. 108, pp. 485–503, December 1982.

22. Sullivan, Richard A., and Charles J. Ehlers, "Planning for Driving Offshore Piles," *Journal of the Construction Division,* in *Proceeding of the ASCE* Vol. 49, pp. 59–79, July 1973.

23. Thorburn, S., and J. Q. Thorburn, "Review of Problems Associated with the Construction of Cast-in-Place Concrete Piles," Report PG2, Construction Industry Research and Information Association, London, January, 1977.

24. Tidewater Construction Corporation, P.O. Box 57, Norfolk, VA 23501.

CHAPTER

20

Equipment for Pumping Water

Most projects require the use of one or more water pumps at various stages during the period of construction. Construction pumps must frequently perform under severe conditions, such as those resulting from variations in the pumping head or from handling water that is muddy, sandy, and trashy. The Contractors Pump Bureau publishes pump standards. Before selecting a pump for a given job, it is necessary to analyze all information and conditions that will affect the decision. The most satisfactory pumping equipment will be the combination of pump, hose, and pipe that will provide the required service for the least total cost.

INTRODUCTION

Pumps are used extensively on construction projects for such operations as

1. Removing water from pits, tunnels, and other excavations.
2. Dewatering cofferdams.
3. Furnishing water for jetting and sluicing.
4. Furnishing water for many types of utility services.
5. Lowering the water table for excavations.
6. Foundation grouting.

Most projects require the use of one or more water pumps at various stages during the period of construction. Construction pumps (see Fig. 20.1) must frequently perform under severe conditions, such as those resulting from variations in the pumping head or from handling water that is muddy, sandy, and trashy, or highly corrosive. The required rate of pumping may vary considerably during the duration of a construction project. The most satisfactory solution to the pumping problem may be a single all-purpose pump, or in other situations it may be better to use several types and sizes of pumps, to permit operational flexibility. The

FIGURE 20.1 | Skid-mounted centrifugal pump.

proper solution is to select the equipment that will take care of the pumping needs adequately at the lowest total cost. The cost analysis must take into account the investment in pumping equipment, the cost of operating the pumps, and any losses that will result from possible failure of the pumps to operate satisfactorily. This last fact can be critical when considering dewatering applications.

For some projects a pump may be the most critical item of construction equipment, even though it is not a direct production machine. In constructing a multimillion-dollar concrete and earth-fill dam, a contractor used a single centrifugal pump to supply water from a nearby stream. The water supplied by the pump was used to wash concrete aggregates, for mixing and curing the concrete, and for compaction moisture in the earth-filled dam. When the pump developed mechanical trouble and the rate of pumping dropped below the job requirements for several days, progress on the project suffered a loss of approximately 25%.

The factors that should be considered in selecting pumps for construction applications include

1. Dependability.
2. Availability of repair parts.
3. Simplicity to permit easy repairs.
4. Economical installation and operation.
5. Operating power requirements.

GLOSSARY OF TERMS

The following glossary defines the important terms that are used in describing pumps and pumping operations.

Capacity. The total volume of liquid a pump can move in a given amount of time. Capacity is usually expressed in gallons per minute (gpm) or gallons per hour (gph).

Discharge head. The (total) discharge head is the sum of the static discharge head plus the head losses of the discharge line.

Discharge hose. The hose used to carry the liquid from the discharge side of the pump.

Impeller. The rotating vanes within the pump housing which are driven by the drive shaft. The vanes create a partial vacuum drawing the fluid through the pump.

Self-priming. The ability of a pump to separate air from a liquid and create a partial vacuum in the pump. This causes the liquid to flow to the impeller and on through the pump.

Static discharge head. The vertical distance from the centerline of the pump impeller to the point of discharge (see Fig. 20.2).

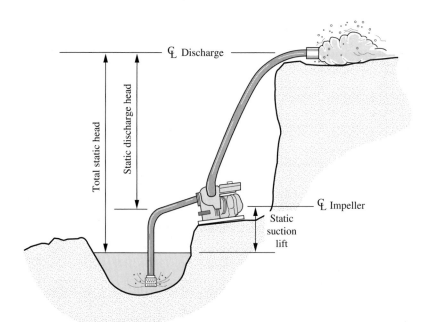

FIGURE 20.2 | Dimensional terminology for pumping operations.

Static suction lift. The vertical distance from the centerline of the pump impeller to the surface of the liquid to be pumped (see Fig. 20.2). Suction capability is limited by atmospheric pressure. Therefore, maximum practical suction lift is 25 ft. Decreasing the suction lift will increase the volume that can be pumped.

Strainer. A cover matched to the size of the pump and attached to the end of the suction hose that permits solids of only a certain size to enter the pump body.

Suction head. The (total) suction head is the sum of the static suction lift plus the suction line head losses.

Suction hose. The hose connected to the suction side of the pump. Suction hose is made of heavy rubber or plastic tubing with a reinforced wall to prevent it from collapsing.

Total head. The suction head plus the discharge head.

Volute. The housing in which the pump's impeller rotates is known as the volute. It has channels cast into the metal to direct the flow of liquid in a given direction.

CLASSIFICATION OF PUMPS

The pumps commonly used on construction projects can be classified as

1. Displacement
   a. Reciprocating (piston).
   b. Diaphragm.
2. Centrifugal
   a. Conventional.
   b. Self-priming.
   b. Self-priming trash.
   c. Submersible.
   d. Multistage.

Reciprocating Pumps

A reciprocating pump operates as the result of the movement of a piston inside a cylinder. When the piston is moved in one direction, the water ahead of the piston is forced out of the cylinder. At the same time, additional water is drawn into the cylinder behind the piston. Regardless of the direction of movement of the piston, water is forced out of one end and drawn into the other end of the cylinder. This is classified as a double-acting pump. If water is pumped during a piston movement in one direction only, the pump is classified as single acting. If a pump contains more than one cylinder, mounted side by side, it is classified as a duplex for two cylinders, triplex for three cylinders, etc. Thus, a pump can be classified as a duplex double-acting or a duplex single-acting.

The volume of water pumped in one stroke will equal the area of the cylinder times the length of the stroke, less a small deduction for slippage through the valves or past the piston, usually about 3 to 5%. If this volume is expressed in cubic inches, it may be converted to gallons by dividing by 231, the number of cubic inches in a gallon. The volume pumped in gallons per minute by a simplex double-acting pump can be expressed as

$$Q = c\,\frac{\pi d^2 ln}{4 \times 231}$$ [20.1]

where
- Q = capacity of a pump in gpm
- c = one-slip allowance; varies from 0.95 to 0.97
- d = diameter of cylinder in inches
- l = length of stroke in inches
- n = number of strokes per minute (*Note*: The movement of the piston in either direction is a stroke.)

The volume pumped per minute by a multiplex double-acting pump will be

$$Q = Nc\,\frac{\pi d^2 ln}{4 \times 231}$$ [20.2]

where N = number of cylinders in the pump.

The energy required to operate a pump will be

$$W = \frac{wQh}{e}$$

where
- W = energy in foot-pounds per min
- w = weight of 1 gal of water in pounds
- h = total pumping head, in feet, including friction loss in pipe
- e = efficiency of pump, expressed decimally

The horsepower required by the pump will be

$$P = \frac{W}{33,000} = \frac{wQh}{33,000e}$$ [20.3]

where
- P = power in horsepower
- $33,000$ = ft-lb of energy per min for 1 hp

EXAMPLE 20.1

How many gallons of fresh water will be pumped per minute by a duplex double-acting pump, size 6 × 12 in., driven by a crankshaft making 90 rpm? If the total head is 160 ft and the efficiency of the pump is 60%, what is the minimum horsepower required to operate the pump? The weight of water is 8.34 lb per gal.

Assume a water slippage of 4%. If we apply Eq. [20.2], the rate of pumping will be

$$Q = N c \frac{\pi d^2 l n}{924}$$

$$= 2 \times 0.96 \times \left(\frac{\pi \times 36 \times 12 \times 180}{924} \right) = 508 \text{ gpm}$$

Applying formula [20.3], the power required by the pump will be

$$P = \frac{wQh}{33,000e}$$

$$= \frac{8.34 \times 508 \times 160}{33,000 \times 0.60} = 34.2 \text{ hp}$$

The capacity of a reciprocating pump depends essentially on the speed at which the pump is operated and is independent of the head. The maximum head at which a reciprocating pump will deliver water depends on the strength of the component parts of the pump and the power available to operate the pump. The capacity of this type of pump may be varied considerably by varying the speed of the pump.

Because the flow of water from each cylinder of a reciprocating pump stops and starts every time the direction of piston travel is reversed, a characteristic of this type of pump is to deliver water with pulsations. The amplitude of the pulsations may be reduced by using more cylinders and by installing an air chamber on the discharge side of a pump.

The advantages of reciprocating pumps are

1. They are able to pump at a uniform rate against varying heads.
2. Increasing the speed can increase their capacity.
3. They have reasonably high efficiency regardless of the head and speed.
4. They are usually self-priming.

The disadvantages of reciprocating pumps are

1. Heavy weight and large size for given capacity.
2. Possibility of valve trouble, especially in pumping water containing abrasive solids.
3. Pulsating flow of water.
4. Danger of damaging a pump when operating against a high head.

Diaphragm Pumps

The diaphragm pump is also a positive displacement type (see Fig. 20.3). The central portion of the flexible diaphragm is alternately raised and lowered by the pump rod that is connected to a walking beam. This action draws water into and

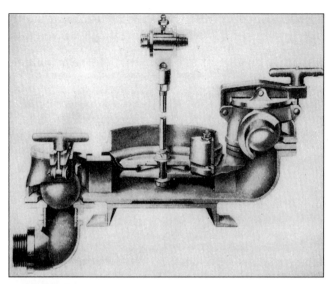

FIGURE 20.3 | Section through a diaphragm pump.

discharges it from the pump. Because this type of pump will handle clear water or water containing large quantities of mud, sand, sludge, and trash, it is popular as a construction pump. It is suitable for use on jobs where the quantity of water varies considerably, as it will diligently continue pumping air and water mixtures. The accessible diaphragm may be replaced easily.

The Contractors Pump Bureau specifies that diaphragm pumps shall be manufactured in the size and capacity ratings given in Table 20.1.

TABLE 20.1 | Minimum capacities for diaphragm pumps at 10-ft suction lifts.*

Size	Capacity (gph)
Two-in. single	2,000
Three-in. single	3,000
Four-in. single	6,000
Four-in. double	9,000

*Diaphragm pumps shall be tested with standard contractor's type suction hose 5 ft longer than the suction lift shown.

Source: Courtesy Contractors Pump Bureau.

Centrifugal Pumps

A centrifugal pump contains a rotation element, called an "impeller," that imparts to water passing through the pump a velocity sufficiently great to cause it to flow from the pump even against considerable pressure. A mass of water may possess energy due to its height above a given datum or due to its velocity. The former is potential, while the latter is kinetic energy. One type of energy can be converted into the other under favorable conditions. The kinetic energy imparted to a particle of water as it passes through the impeller is sufficient to cause the particles to rise to some determinable height.

The principle of the centrifugal pump may be illustrated by considering a drop of water at rest at a height h above a surface. If the drop of water is permitted to fall freely, it will strike the surface with a velocity given by the equation

$$V = \sqrt{2gh} \qquad\qquad \textbf{[20.4]}$$

where

  V = velocity in feet per second (fps)

  g = acceleration of gravity, equal to 32.2 fps at sea level

  h = height of fall in feet

If the drop falls 100 ft, the velocity will be 80.2 fps. If the same drop is given an upward velocity of 80.2 fps, it will rise 100 ft. These values assume no loss in energy due to friction through air. It is the function of the centrifugal pump to give the water the necessary velocity as it leaves the impeller. If the speed of the pump is doubled, the velocity of the water will be increased from 80.2 to 160.4 fps, neglecting any increase in friction losses. With this velocity, the water can be pumped to a height given by the equation

$$h = \frac{V^2}{2g} = \frac{(160.4)^2}{64.4} = 400 \text{ ft}$$

This indicates that if a centrifugal pump is pumping water against a total head of 100 ft, the same quantity of water can be pumped against a total head of 400 ft by doubling the speed of the impeller. In actual practice the maximum possible head for the increased speed will be less than 400 ft. The reduction is caused by increased losses in the pump due to friction. These results illustrate the effect that increasing the speed or the diameter of an impeller has on the performance of a centrifugal pump.

A centrifugal pump may be equipped with either an open or enclosed impeller. Although an enclosed impeller usually has higher efficiency, it will not handle water containing trash as well as an open impeller.

The power required to operate a centrifugal pump is given by Eq. [20.3]. The efficiencies of these pumps may be as high as 75%. Because of their high-efficiency factors, they will pump a greater volume with lower fuel consumption as compared to other pump types.

A centrifugal pump will pass a spherical solid one-quarter ($\frac{1}{4}$) the size of the suction opening. As an example, a 2-in. pump can pass $\frac{1}{2}$-in.-size solids. These pumps can typically pass up to 10% by volume of mud, silt, and sand.

Self-Priming Centrifugal Pumps

On construction projects, pumps frequently must be set above the surface of the water that is to be pumped. Consequently, self-priming centrifugal pumps are more suitable than the conventional types for use on construction projects. The operation of a self-priming pump is illustrated in Fig. 20.4. A check valve on the suction side of the pump permits the chamber to be filled with water prior to starting the pump. When the pump is started, the water in the chamber produces a sealed flow through channel A into the chamber, where the air escapes through the discharge, and the water flows down through channel B to the impeller. This action continues until all the air is exhausted from the suction line and water enters the pump. When a pump is stopped, it will retain its charge of priming water indefinitely. Such a pump is self-priming to heights of 25 ft when in good mechanical condition.

Effect of Altitude   At altitudes above 3,000 ft, there is a definite effect on a pump's performance. As a general rule a self-priming pump will lose one foot of priming ability for every 1,000 ft of elevation. A self-priming pump operated in Flagstaff, Arizona at an elevation of 7,000 ft will develop only 18 ft of suction lift rather than the normal 25 ft.

Effect of Temperature   As the temperature of water increases above 60°F, the maximum suction lift of the pump will decrease. A pump generates heat that is

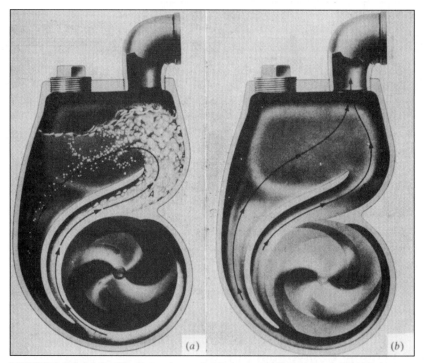

FIGURE   20.4 I Section through a self-priming centrifugal pump. (a) Priming action. (b) Pumping action.

passed to the water. Over a long duration of operation, as the heat increases, a pump located at a height that is very close to the suction maximum can lose prime.

Multistage Centrifugal Pumps

If a centrifugal pump has a single impeller, it is described as a single-stage pump, whereas if there are two or more impellers and the water discharge from one impeller flows into the suction of another, it is described as a multistage pump. Multistage pumps are especially suitable for pumping against a high head or pressure, as each stage imparts an additional pressure to the water. Pumps of this type are used frequently to supply water for jetting, where the pressure may run as high as several hundred pounds per square inch (psi).

Performance of Centrifugal Pumps

Pump manufacturers will furnish sets of curves showing the performance of their pumps under different operating conditions. A set of curves for a given pump will show the variations in capacity, efficiency, and horsepower for different pumping heads. These curves can be very helpful in selecting the pump that is most suitable for a given pumping condition. Figure 20.5 illustrates a set of performance curves for a 10-in. centrifugal pump. For a total head of 60 ft, the capacity will be 1,200 gpm, the efficiency 52%, and the required power 35 brake horsepower (bhp). If the total head is reduced to 50 ft and the dynamic suction lift does not exceed 23 ft, the capacity will be 1,930 gpm, the efficiency 55%, and the required power 44 bhp. This pump will not deliver any water against a total head in excess of 66 ft, which is called the "shutoff head."

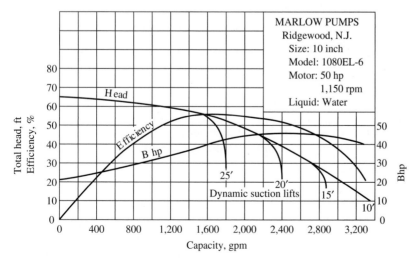

FIGURE 20.5 I Performance curves for centrifugal pump.

Since a construction pump frequently is operated under varying heads, it is desirable to select a pump with relatively flat head-capacity and horsepower curves, even though efficiency must be sacrificed to obtain these conditions. A pump with a flat horsepower demand permits the use of an engine or an electric motor that will provide adequate power over a wide pumping range, without a substantial surplus or deficiency, regardless of the head.

By studying the curves in Fig. 20.6, one can see the effect of varying the speed of a centrifugal pump. As discussed in Chapter 4 with respect to internal combustion engines in general, altitude will affect engine performance. In the case of a pump powered by a gasoline or diesel engine, there will be a 3% loss of horsepower for every 1000 ft of elevation. The loss of horsepower will reduce the obtainable operating speed and therefore also reduce maximum pumping capacity.

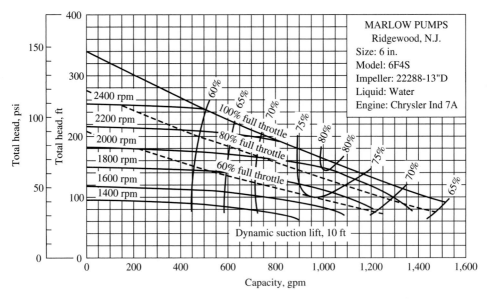

FIGURE 20.6 I The effect of varying the speed on the performance of a centrifugal pump.

The Contractors Pump Bureau publishes pump standards for several types of pumps, including self-priming centrifugal pumps; these are given in Tables 20.2 and 20.3 [1].

Submersible Pumps

Figure 20.7 illustrates an electric-motor-operated submersible pump that is useful in dewatering tunnels, foundation pits, trenches, and similar places. Figure 20.8 is a performance curve for this type of pump when operated against varying heads of water. The figure includes pertinent information related to the pump. Other types and models have different performance characteristics.

TABLE 20.2a | Minimum capacities for M rate self-priming centrifugal pumps manufactured in accordance with standards of the Contractors Pump Bureau

Model 6-M ($1\frac{1}{2}$ in.)										
Total head including friction [ft (m)]		**Height of pump above water [ft (m)]**								
		5	**(1.5)**	**10**	**(3.0)**	**15**	**(4.6)**	**20**	**(6.1)**	**25**	**(7.6)**
		Capacity [gpm (l/min)*]								
5	(1.5)	100	(379)								
10	(3.0)	96	(363)								
15	(4.6)	93	(352)	85	(322)						
20	(6.1)	89	(337)	84	(318)	68	(257)				
25	(7.6)	85	(322)	82	(310)	67	(254)				
30	(9.1)	80	(303)	79	(299)	66	(250)	49	(186)	35	(133)
40	(12.2)	71	(269)	71	(269)	60	(227)	46	(174)	33	(125)
50	(15.2)	59	(223)	59	(223)	52	(197)	41	(155)	28	(106)
60	(18.3)	42	(159)	42	(159)	40	(151)	32	(121)	22	(83)
70	(21.3)	22	(83)	22	(83)	22	(83)	20	(75)	12	(45)

Model 8-M (2 in.)										
Total head including friction [ft (m)]		**Height of pump above water [ft (m)]**								
		5	**(1.5)**	**10**	**(3.0)**	**15**	**(4.6)**	**20**	**(6.1)**	**25**	**(7.6)**
		Capacity [gpm (l/min)*]								
5	(1.5)	140	(530)								
10	(3.0)	137	(519)								
20	(6.1)	135	(511)	117	(443)						
25	(7.6)	133	(503)	116	(439)						
30	(9.1)	132	(500)	116	(439)	102	(386)	82	(310)		
40	(12.2)	123	(466)	105	(397)	100	(379)	80	(303)	58	(220)
50	(15.2)	109	(413)	92	(348)	90	(341)	76	(288)	55	(208)
60	(18.3)	90	(341)	70	(265)	70	(265)	70	(265)	55	(208)
70	(21.3)	66	(250)	40	(151)	40	(151)	40	(151)	40	(151)
80	(24.4)	40	(151)	40	(151)	40	(151)	40	(151)	40	(151)

Model 12-M (2 in.)										
Total head including friction [ft (m)]		**Height of pump above water [ft (m)]**								
		5	**(1.5)**	**10**	**(3.0)**	**15**	**(4.6)**	**20**	**(6.1)**	**25**	**(7.6)**
		Capacity [gpm (l/min)*]								
5	(1.5)	200	(757)								
10	(3.0)	196	(742)								
20	(6.1)	190	(719)	167	(632)						
25	(7.6)	185	(700)	166	(628)						
30	(9.1)	174	(659)	165	(625)	140	(530)	110	(416)		
40	(12.2)	158	(598)	158	(598)	140	(530)	110	(416)	75	(284)
50	(15.2)	145	(549)	145	(549)	130	(492)	106	(401)	70	(265)
60	(18.3)	126	(477)	126	(477)	117	(443)	97	(367)	68	(257)
70	(21.3)	102	(386)	102	(386)	100	(379)	85	(322)	60	(227)
80	(24.4)	74	(280)	74	(280)	74	(280)	68	(257)	48	(181)
90	(27.4)	40	(151)	40	(151)	40	(151)	40	(151)	32	(121)

*Liters per minute.
Source: Courtesy Contractors Pump Bureau.

TABLE 20.2b | Minimum capacities for M rate self-priming centrifugal pumps manufactured in accordance with standards of the Contractors Pump Bureau

Model 18-M (3 in.)											
Total head including friction [ft (m)]		**Height of pump above water [ft (m)]**									
		5	**(1.5)**	**10**	**(3.0)**	**15**	**(4.6)**	**20**	**(6.1)**	**25**	**(7.6)**
		Capacity [gpm (l/min)*]									
5	(1.5)	300	(1,136)								
10	(3.0)	295	(1,117)								
20	(6.1)	277	(1,048)	259	(980)						
30	(9.1)	260	(984)	250	(946)	210	(795)	200	(757)		
40	(12.2)	241	(912)	241	(912)	207	(784)	177	(670)	160	(606)
50	(15.2)	225	(852)	225	(852)	202	(765)	172	(651)	140	(530)
60	(18.3)	197	(746)	197	(746)	197	(746)	169	(640)	140	(530)
70	(21.3)	160	(606)	160	(606)	160	(606)	160	(606)	138	(522)
80	(24.4)	125	(473)	125	(473)	125	(473)	125	(473)	125	(473)
90	(27.4)	96	(363)	96	(363)	96	(363)	96	(363)	96	(363)

Model 20-M (3 in.)									
Total head including friction [ft (m)]		**Height of pump above water [ft (m)]**							
		10	**(3.0)**	**15**	**(4.6)**	**20**	**(6.1)**	**25**	**(7.6)**
		Capacity [gpm (l/min)*]							
30	(9.1)	333	(1,260)	280	(1,060)	235	(890)	165	(625)
40	(12.2)	315	(1,192)	270	(1,022)	230	(871)	162	(613)
50	(15.2)	290	(1,098)	255	(965)	220	(833)	154	(583)
60	(18.3)	255	(965)	235	(890)	205	(776)	143	(541)
70	(21.3)	212	(802)	209	(791)	184	(696)	130	(492)
80	(24.4)	165	(625)	165	(625)	157	(594)	114	(432)
90	(27.4)	116	(439)	116	(439)	116	(439)	94	(356)
100	(30.5)	60	(227)	60	(227)	60	(227)	60	(227)

Model 40-M (4 in.)									
Total head including friction [ft (m)]		**Height of pump above water [ft (m)]**							
		10	**(3.0)**	**15**	**(4.6)**	**20**	**(6.1)**	**25**	**(7.6)**
		Capacity [gpm (l/min)*]							
25	(7.6)	667	(2,525)						
30	(9.1)	660	(2,498)	575	(2,176)	475	(1,798)	355	(1,344)
40	(12.2)	645	(2,441)	565	(2,139)	465	(1,760)	350	(1,325)
50	(15.2)	620	(2,347)	545	(2,063)	455	(1,722)	345	(1,306)
60	(18.3)	585	(2,214)	510	(1,930)	435	(1,647)	335	(1,268)
70	(21.3)	535	(2,025)	475	(1,798)	410	(1,552)	315	(1,192)
80	(24.4)	465	(1,760)	410	(1,551)	365	(1,382)	280	(976)
90	(27.4)	375	(1,419)	325	(1,230)	300	(1,136)	220	(833)
100	(30.5)	250	(946)	215	(815)	195	(738)	145	(549)
110	(33.5)	65	(246)	60	(227)	50	(189)	40	(151)

*Liters per minute.

Source: Courtesy Contractors Pump Bureau.

TABLE 20.2c | Minimum capacities for M rate self-priming centrifugal pumps manufactured in accordance with standards of the Contractors Pump Bureau

				Model 90-M (6 in.)					
Total head including friction [ft (m)]		**Height of pump above water [ft (m)]**							
		10	**(3.0)**	**15**	**(4.6)**	**20**	**(6.1)**	**25**	**(7.6)**
		Capacity [gpm (l/min)*]							
25	(7.6)	1,500	(5,678)						
30	(9.1)	1,480	(5,602)	1,280	(4,845)	1,050	(3,974)	790	(2,990)
40	(12.2)	1,430	(5,413)	1,230	(4,656)	1,020	(3,861)	780	(2,952)
50	(15.2)	1,350	(5,110)	1,160	(4,391)	970	(3,672)	735	(2,782)
60	(18.3)	1,225	(4,637)	1,050	(3,974)	900	(3,407)	690	(2,612)
70	(21.3)	1,050	(3,974)	900	(3,407)	775	(2,933)	610	(2,309)
80	(24.4)	800	(3,028)	680	(2,574)	600	(2,271)	490	(1,855)
90	(27.4)	450	(1,703)	400	(1,514)	365	(1,382)	300	(1,136)
100	(30.5)	100	(379)	100	(379)	100	(379)	100	(379)

				Model 125-M (8 in.)					
Total head including friction [ft (m)]		**Height of pump above water [ft (m)]**							
		10	**(3.0)**	**15**	**(4.6)**	**20**	**(6.1)**	**25**	**(7.6)**
		Capacity [gpm (l/min)*]							
25	(7.6)	2,100	(7,949)	1,850	(7,002)	1,570	(5,943)		
30	(9.1)	2,060	(7,797)	1,820	(6,889)	1,560	(5,905)	1,200	(4,542)
40	(12.2)	1,960	(7,419)	1,740	(6,586)	1,520	(5,753)	1,170	(4,429)
50	(15.2)	1,800	(6,813)	1,620	(6,132)	1,450	(5,488)	1,140	(4,315)
60	(18.3)	1,640	(6,207)	1,500	(5,678)	1,360	(5,148)	1,090	(4,126)
70	(21.3)	1,460	(5,526)	1,340	(5,072)	1,250	(4,731)	1,015	(3,841)
80	(24.4)	1,250	(4,731)	1,170	(4,429)	1,110	(4,201)	950	(3,596)
90	(27.4)	1,020	(3,861)	980	(3,709)	940	(3,558)	840	(3,179)
100	(30.5)	800	(3,028)	760	(2,877)	710	(2,687)	680	(2,574)
110	(33.5)	570	(2,158)	540	(2,044)	500	(1,893)	470	(1,779)
120	(36.6)	275	(1,041)	245	(927)	240	(908)	240	(908)

				Model 200-M (10 in.)					
Total head including friction [ft (m)]		**Height of pump above water [ft (m)]**							
		10	**(3.0)**	**15**	**(4.6)**	**20**	**(6.1)**	**25**	**(7.6)**
		Capacity [gpm (l/min)*]							
20	(6.1)	3,350	(12,680)	3,000	(11,355)				
30	(9.1)	3,000	(11,355)	2,800	(10,598)	2,500	(9,463)	1,550	(5,867)
40	(12.2)	2,500	(9,463)	2,500	(9,463)	2,250	(8,516)	1,500	(5,678)
50	(15.2)	2,000	(7,570)	2,000	(7,570)	2,000	(7,570)	1,350	(5,110)
60	(18.3)	1,300	(4,921)	1,300	(4,921)	1,300	(4,921)	1,150	(4,353)
70	(21.3)	500	(1,893)	500	(1,893)	500	(1,893)	500	(1,893)

*Liters per minute.

Source: Courtesy Contractors Pump Bureau.

TABLE 20.3a | Minimum capacities for MT rated, solids-handling, self-priming centrifugal pumps manufactured in accordance with the standards of the Contractors Pump Bureau

Model 6-MT ($1\frac{1}{2}$ in.)										
Total head including friction [ft (m)]	Height of pump above water [ft (m)]									
	5	(1.5)	10	(3.0)	15	(4.6)	20	(6.1)	25	(7.6)
	Capacity [gpm (l/min)*]									
5 (1.5)	100	(379)								
10 (3.0)	96	(363)								
20 (6.1)	89	(337)	84	(318)	68	(257)				
30 (9.1)	80	(303)	79	(299)	66	(250)	49	(186)	35	(133)
40 (12.2)	71	(269)	71	(269)	60	(227)	46	(174)	33	(125)
50 (15.2)	59	(223)	59	(223)	52	(197)	41	(155)	28	(106)
60 (18.3)	42	(159)	42	(159)	40	(151)	32	(121)	22	(83)
70 (21.3)	22	(83)	22	(83)	22	(83)	20	(75)	12	(45)

Model 11-MT (2 in.)										
Total head including friction [ft (m)]	Height of pump above water [ft (m)]									
	5	(1.5)	10	(3.0)	15	(4.6)	20	(6.1)	25	(7.6)
	Capacity [gpm (l/min)*]									
5 (1.5)	185	(700)								
10 (3.0)	183	(693)								
20 (6.1)	178	(674)	164	(621)	132	(500)				
30 (9.1)	169	(640)	164	(621)	132	(500)	105	(397)	75	(284)
40 (12.2)	164	(621)	164	(621)	132	(500)	105	(397)	75	(284)
50 (15.2)	150	(568)	150	(568)	132	(500)	105	(397)	75	(284)
60 (18.3)	135	(511)	135	(511)	132	(500)	105	(397)	75	(284)
70 (21.3)	88	(333)	88	(333)	88	(333)	88	(333)	68	(257)
80 (24.4)	40	(151)	40	(151)	40	(151)	40	(151)	40	(151)

Model 18-MT (3 in.)								
Total head including friction [ft (m)]	Height of pump above water [ft (m)]							
	10	(3.0)	15	(4.6)	20	(6.1)	25	(7.6)
	Capacity [gpm (l/min)*]							
20 (6.1)	310	(1,173)	265	(1003)				
30 (9.1)	305	(1,154)	265	(1003)	200	(757)	115	(435)
40 (12.2)	300	(1,136)	265	(1003)	200	(757)	110	(416)
50 (15.2)	275	(1,041)	260	(984)	200	(757)	105	(397)
60 (18.3)	215	(814)	215	(814)	200	(757)	100	(379)
70 (21.3)	170	(644)	170	(644)	170	(644)	100	(379)
80 (24.4)	87	(329)	87	(329)	87	(329)	87	(329)
90 (27.4)	25	(95)	25	(95)	25	(95)	25	(95)

*Liters per minute.

Source: Courtesy Contractors Pump Bureau.

TABLE 20.3b | Minimum capacities for MT rated, solids-handling, self-priming centrifugal pumps manufactured in accordance with the standards of the Contractors Pump Bureau

Model 33-MT (4 in.)								
Total head including friction [ft (m)]		**Height of pump above water [ft (m)]**						
		10	**(3.0)**	**15**	**(4.6)**	**20**	**(6.1)**	**25**	**(7.6)**
		Capacity [gpm (l/min)*]						
30	(9.1)	550	(2,082)	460	(1,741)	350	(1,325)	240	(908)
40	(12.2)	540	(2,044)	455	(1,722)	350	(1,325)	240	(908)
50	(15.2)	500	(1,893)	430	(1,628)	340	(1,287)	230	(871)
60	(18.3)	450	(1,703)	395	(1,495)	320	(1,211)	220	(833)
70	(21.3)	370	(1,401)	360	(1,363)	300	(1,136)	210	(795)
80	(24.4)	275	(1,041)	275	(1,041)	260	(984)	180	(681)
90	(27.4)	190	(719)	190	(719)	190	(719)	150	(568)
100	(30.5)	100	(379)	100	(379)	100	(379)	100	(379)

Model 35-MT (4 in.)								
Total head including friction [ft (m)]		**Height of pump above water [ft (m)]**						
		10	**(3.0)**	**15**	**(4.6)**	**20**	**(6.1)**	**25**	**(7.6)**
		Capacity [gpm (l/min)*]						
30	(9.1)	585	(2,214)	500	(1,893)	350	(1,325)	240	(908)
40	(12.2)	585	(2,214)	500	(1,893)	350	(1,325)	240	(908)
50	(15.2)	585	(2,214)	500	(1,893)	350	(1,325)	240	(908)
60	(18.3)	545	(2,063)	500	(1,893)	350	(1,325)	240	(908)
70	(21.3)	495	(1,874)	480	(1,817)	350	(1,325)	240	(908)
80	(24.4)	430	(1,628)	420	(1,590)	340	(1,287)	240	(908)
90	(27.4)	320	(1,211)	320	(1,211)	260	(984)	220	(833)
100	(30.5)	100	(379)	100	(379)	100	(379)	100	(379)

Model 70-MT (6 in.)								
Total head including friction [ft (m)]		**Height of pump above water [ft (m)]**						
		10	**(3.0)**	**15**	**(4.6)**	**20**	**(6.1)**	**25**	**(7.6)**
		Capacity [gpm (l/min)*]						
30	(9.1)	1,180	(4,466)	975	(3,690)	715	(2,706)	350	(1,325)
40	(12.2)	1,175	(4,447)	950	(3,596)	715	(2,706)	350	(1,325)
50	(15.2)	1,160	(4,391)	935	(3,539)	715	(2,706)	350	(1,325)
60	(18.3)	1,150	(4,353)	925	(3,501)	715	(2,706)	350	(1,325)
70	(21.3)	1,120	(4,239)	900	(3,407)	715	(2,706)	350	(1,325)
80	(24.4)	950	(3,596)	875	(3,312)	700	(2,650)	350	(1,325)
90	(27.4)	700	(2,650)	700	(2,650)	600	(2,271)	350	(1,325)
100	(30.5)	450	(1,703)	450	(1,703)	450	(1,703)	300	(1,136)
110	(33.5)	200	(757)	200	(757)	200	(757)	200	(757)

*Liters per minute.

Source: Courtesy Contractors Pump Bureau.

FIGURE 20.7 | Section through an electric-motor-operated submersible pump.

With a submersible pump, there is no suction lift limitation, and of course, no need for a suction hose. Another advantage is that there are no noise problems. For construction applications a pump made of iron or aluminum is best, as other materials are much more prone to damage when the pump is dropped. The power cord for a submersible pump should have a strain relief protector, as it is rare that someone does not inadvertently lift the pump by the power cord.

There are basically two size categories for submersible pumps, small fractional horsepower size pumps and larger pumps having one horsepower and large power units. The small pumps, typically, $\frac{1}{4}$, $\frac{1}{3}$, and $\frac{1}{2}$ hp units, are for minor nuisance dewatering applications. The one horsepower and larger pumps are for moving large volumes and/or high head conditions.

LOSS OF HEAD DUE TO FRICTION IN PIPE

Table 20.4 gives the nominal loss of head due to water flowing through new steel pipe. The actual losses may differ from the values given in the table because of variations in the diameter of a pipe and in the condition of the pipe's inside surface.

Model S3B1 .... Centrifugal, single stage

Discharge .... 3 in.

Solids handled .... 3/8 in.

Horsepower 6

Hertz 60

RPM 3,450

Voltage .... 230 volt, 1 phase, 7.2 kW
230/460 dual voltage, 3 phase, 6.8 kW
or 575 volt, 3 phase, 6.8 kW

Cable .... #10 gauge, 50-ft length

Weight (pump and cable) .... 125 lb (approx.)

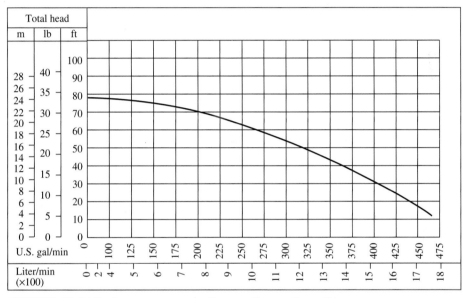

FIGURE 20.8 | Performance curve for Gorman-Rupp submersible pump.

Source: The Gorman-Rupp Co.

The relationship between the head of fresh water in feet and pressure in psi is given by the equation

$$h = 2.31\,p \qquad\qquad \text{[20.5]}$$

or

$$p = 0.434\,h \qquad\qquad \text{[20.6]}$$

where

h = depth of water or head in feet

p = pressure at depth h in psi

Table 20.5 gives the equivalent length of straight steel pipe having the same loss in head due to water friction as fittings and valves.

TABLE 20.4 | Friction loss for water, in feet per 100 ft of clean wrought-iron or steel pipe*

Flow in U.S. (gpm)	Nominal diameter of pipe (in.)													
	$\frac{1}{2}$	$\frac{3}{4}$	1	$1\frac{1}{4}$	$1\frac{1}{2}$	2	$2\frac{1}{2}$	3	4	5	6	8	10	12
5	26.5	6.8	2.11	0.55										
10	95.8	24.7	7.61	1.98	0.93	0.31	0.11							
15		52.0	16.3	4.22	1.95	0.70	0.23							
20		88.0	27.3	7.21	3.38	1.18	0.40							
25			41.6	10.8	5.07	1.75	0.60	0.25						
30			57.8	15.3	7.15	2.45	0.84	0.35						
40				26.0	12.2	4.29	1.4	0.59						
50				39.0	18.5	6.43	2.2	0.9	0.22					
75					39.0	13.6	4.6	2.0	0.48	0.16				
100					66.3	23.3	7.8	3.2	0.79	0.27	0.09			
125						35.1	11.8	4.9	1.2	0.42	0.18			
150						49.4	16.6	6.8	1.7	0.57	0.21			
175						66.3	22.0	9.1	2.2	0.77	0.31			
200							28.0	11.6	2.9	0.96	0.40			
225							35.3	14.5	3.5	1.2	0.48			
250							43.0	17.7	4.4	1.5	0.60	0.15		
275								21.2	5.2	1.8	0.75	0.18		
300								24.7	6.1	2.0	0.84	0.21		
350								33.8	8.0	2.7	0.91	0.27		
400									10.4	3.5	1.4	0.35		
500									15.6	5.3	2.2	0.53	0.18	0.08
600									22.4	6.2	3.1	0.74	0.25	0.10
700									30.4	9.9	4.1	1.0	0.34	0.14
800											5.2	1.3	0.44	0.18
900											6.6	1.6	0.54	0.22
1,000											7.8	2.0	0.65	0.27
1,100											9.3	2.3	0.78	0.32
1,200											10.8	2.7	0.95	0.37
1,300											12.7	3.1	1.1	0.42
1,400											14.7	3.6	1.2	0.48
1,500											16.8	4.1	1.4	0.55
2,000												7.0	2.4	0.93
3,000													5.1	2.1
4,000														3.5
5,000														5.5

*For old or rough pipes, add 50% to friction values.

Source: Courtesy Contractors Pump Bureau.

RUBBER HOSE

The flexibility of rubber hose makes it a desirable substitute for use with pumps in place of pipe on many jobs. Such hose may be used on the suction side of a pump if it is constructed with a wire insert to prevent collapse under partial vacuum. Rubber hose is available with end fittings corresponding with those for

TABLE 20.5 | Length of steel pipe, in feet, equivalent to fittings and valves

Item	Nominal size (in.)											
	1	$1\frac{1}{4}$	$1\frac{1}{2}$	2	$2\frac{1}{2}$	3	4	5	6	8	10	12
90° elbow	2.8	3.7	4.3	5.5	6.4	8.2	11.0	13.5	16.0	21.0	26.0	32.0
45° elbow	1.3	1.7	2.0	2.6	3.0	3.8	5.0	6.2	7.5	10.0	13.0	15.0
Tee, side outlet	5.6	7.5	9.1	12.0	13.5	17.0	22.0	27.5	33.0	43.5	55.0	66.0
Close return bend	6.3	8.4	10.2	13.0	15.0	18.5	24.0	31.0	37.0	49.0	62.0	73.0
Gate valve	0.6	0.8	0.9	1.2	1.4	1.7	2.5	3.0	3.5	4.5	5.7	6.8
Globe valve	27.0	37.0	43.0	55.0	66.0	82.0	115.0	135.0	165.0	215.0	280.0	335.0
Check valve	10.5	13.2	15.8	21.1	26.4	31.7	42.3	52.8	63.0	81.0	105.0	125.0
Foot valve	24.0	33.0	38.0	46.0	55.0	64.0	75.0	76.0	76.0	76.0	76.0	76.0

Source: Courtesy The Gorman-Rupp Company.

iron or steel pipe. As a rule of thumb, total length of hose on a centrifugal pump should be less than 500 ft and less than 50 ft for a diaphragm pump.

Hose size should match the pump size. Using a larger suction hose will increase the pumping capacity, e.g., 4-in. hose on a 3-in. pump, but it can also cause overload of the pump motor. A suction hose smaller in size than the pump will starve the pump and can cause cavitation. This in turn will increase the wear of the impeller and volute and lead to early pump failure.

A discharge hose sized larger than the pump will simply reduce friction loss and can increase volume if long discharge distances are involved. If a small discharge hose is used, friction loss is increased and therefore pumping volume reduced.

Table 20.6 gives the loss in head in feet per 100 ft due to friction caused by water flowing through hose. The values in the table apply for rubber substitutes.

SELECTING A PUMP

Before a pump for a given job is selected, it is necessary to analyze all information and conditions that will affect the selection. The most satisfactory pumping equipment will be the combination of pump and pipe that will provide the required service for the least total cost. The total cost includes the installed and operating cost of the pump and pipe for the period that it will be used, with an appropriate allowance for salvage value at the completion of the project. In order to analyze the cost of pumping water, it is necessary to have certain information, such as

1. The rate at which the water is pumped.
2. The height of lift from the existing water surface to the point of discharge.
3. The pressure head at discharge, if any.
4. The variations in water level at suction or discharge.
5. The altitude of the project.
6. The height of the pump above the surface of water to be pumped.

TABLE 20.6 | Water friction loss, in feet per 100 ft of smooth bore hose

Flow in U.S. (gpm)	Actual inside diameter of hose (in.)											
	$\frac{5}{8}$	$\frac{3}{4}$	1	$1\frac{1}{4}$	$1\frac{1}{2}$	2	$2\frac{1}{2}$	3	4	5	6	8
5	21.4	8.9	2.2	0.74	0.3							
10	76.8	31.8	7.8	2.64	1.0	0.2						
15		68.5	16.8	5.7	2.3	0.5						
20			28.7	9.6	3.9	0.9	0.32					
25			43.2	14.7	6.0	1.4	0.51					
30			61.2	20.7	8.5	2.0	0.70	0.3				
35			80.5	27.6	11.2	2.7	0.93	0.4				
40				35.0	14.3	3.5	1.2	0.5				
50				52.7	21.8	5.2	1.8	0.7				
60				73.5	30.2	7.3	2.5	1.0				
70					40.4	9.8	3.3	1.3				
80					52.0	12.6	4.3	1.7				
90					64.2	15.7	5.3	2.1	0.5			
100					77.4	18.9	6.5	2.6	0.6			
125						28.6	9.8	4.0	0.9			
150						40.7	13.8	5.6	1.3			
175						53.4	18.1	7.4	1.8			
200						68.5	23.4	9.6	2.3	0.8	0.32	
250							35.0	14.8	3.5	1.2	0.49	
300							49.0	20.3	4.9	1.7	0.69	
350								27.0	6.6	2.3	0.90	
400									8.4	2.9	1.1	0.28
450									10.5	3.6	1.4	0.35
500									12.7	4.3	1.7	0.43
1,000										15.6	6.4	1.6

Source: Courtesy Contractors Pump Bureau.

7. The size of pipe to be used, if already determined.
8. The number, sizes, and types of fittings and valves in the pipeline.

Examples 20.2 and 20.3 illustrate methods of selecting pumps and pumping systems.

EXAMPLE 20.2

Select a self-priming centrifugal pump, with a capacity of 600 gpm, for the project illustrated in Fig. 20.9. All pipe, fittings, and valves will be 6 in. with threaded connections. Use the information in Table 20.5 to convert the fittings and valves into equivalent lengths of pipe.

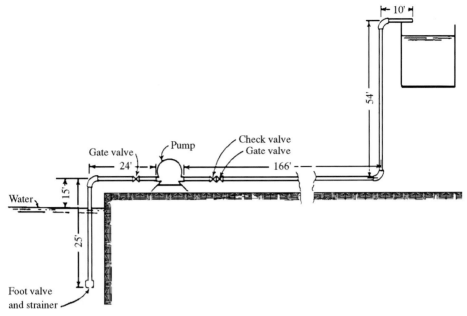

FIGURE 20.9 I Pump and pipe installation for Example 20.2.

Item	Equivalent length of pipe (ft)
One foot valve and strainer	76
Three elbows at 16 ft	48
Two gate valves at 3.5 ft	7
One check valve	63
Total	194
Add length of pipe (25 + 24 + 166 +54 + 10)	279
Total equivalent length of 6-in pipe	473

From Table 20.4, the friction loss per 100 ft of 6-in. pipe will be 3.10 ft. The total head, including lift plus head lost in friction, will be

Lift, $\qquad$ 15 + 54 = 69.0 ft

Head lost in friction, 473 ft at 3.10 ft per 100 ft = 14.7 ft

Total head = 83.7 ft

Table 20.2d indicates that a model 90-M pump will deliver the required quantity of water.

Sometimes the problem is to select the pump and pipeline that will permit water to be pumped at the lowest total cost. Example 20.3 illustrates a method that can be used to select the most economical pumping system.

EXAMPLE 20.3

In operating a rock quarry it is necessary to pump 400 gpm of clear water. The pump and pipeline selected will be installed as illustrated in Fig. 20.10. It is estimated that the pump will be operated a total of 1,200 hour per year. Compare the economy of using 4- and 6-in. steel pipe for the water line. Assume that the pump will have an economic life of 5 years and that the pipeline and fittings will have a life of 10 years. Also, assume that the cost of installing the pipeline will be the same regardless of the size, so that this cost may be disregarded.

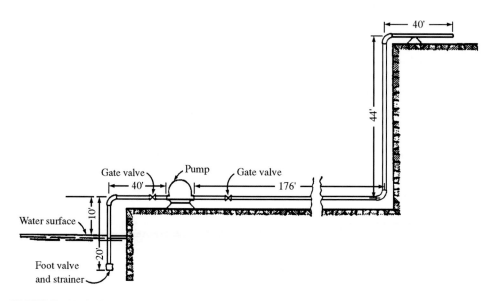

FIGURE 20.10 | Pump and pipe installation for Example 20.3.

Consider the use of 4-in. pipe. The total length of pipe will be

Item	Equivalent length of pipe (ft)
One foot valve and strainer	75
Three elbows at 11 ft	33
Two gate valves at 2.5 ft	5
Pipe (20 + 40 +176 + 44 + 40)	320
Total equivalent length	433

The total head, including lift and head lost in friction, will be

Lift, $\qquad\qquad$ 10 + 44 = 54.0 ft

Head lost in friction, 433 ft at 10.4 ft per 100 ft = 45.0 ft

$\qquad\qquad\qquad\qquad$ Total head = 99.0 ft

A model 125-M self-priming pump, with a capacity of approximately 800 gpm, will be required for this installation.

Consider the use of 6-in. pipe. The total equivalent length of pipe will be

Item	Equivalent length of pipe (ft)
One foot valve and strainer	76
Three elbows at 16 ft	48
Two gate valves at 3.5 ft	7
Pipe (20 + 40 + 176 + 44 +40)	320
Total equivalent length	451

The total head, including lift and head lost in friction, will be

Lift, $\qquad\qquad\qquad$ 10 + 44 = 54.0 ft

Head lost in friction, 451 ft at 1.4 ft per 100 ft $\quad$ = $\quad$ 6.3 ft

$\qquad\qquad\qquad\qquad$ Total head = 60.3 ft

A model 40-M self-priming pump, with a capacity of approximately 585 gpm, will be satisfactory for this installation. The excess capacity of this pumping system is an advantage in favor of using 6-in. pipe.

The cost of each size pipeline, fittings, and valves will be:

Item	Size pipe	
	4 in.	6 in.
A 320-ft pipeline	$766.00	$1,340.00
One foot valve and strainer	24.00	36.00
Three elbows	12.00	24.00
Two gate valves	168.00	216.00
Total cost	$980.00	$1,616.00
Depreciation cost per year, based on 10-yr life	98.00	161.60
Depreciation cost per hr, based on 1,200 hr/yr	0.08	0.14

The combined cost per hour for each size pump and pipeline system will be

Item	Cost per hr	
	4-in. pipe	6-in. pipe
Pump	$2.62	$1.74
Pipe, fittings, and valves	0.08	0.14
Total cost per hr	$2.70	$1.88

This analysis shows that the additional cost of the 6-in. pipe is more than offset by the reduction in the cost of the smaller pump.

WELLPOINT SYSTEMS

In excavating below the surface of the ground, the constructor may encounter groundwater prior to reaching the bottom of an excavation. In the case of an excavation into sand and gravel, the flow of water will be large if some method is not adopted to intercept and remove the water. Dewatering, temporarily lowering the piezometric level of groundwater, is then necessary. After the construction operations are completed the dewatering actions can be discontinued

Header pipe   Riser pipe   Well spacing 2 to 5 ft

FIGURE 20.11 | Parts of a wellpoint system.

and the groundwater will return to its normal level. When planning a dewatering activity, it should be understood that groundwater levels change from season to season and as a result of many factors [4].

Ditches located within the limits of the excavation may be used to collect and divert the flow of groundwater into sumps from which it can be removed by pumping. However, the presence of collector ditches within the excavation usually creates a nuisance and interferes with the construction operations. A common method for controlling groundwater is the installation of a wellpoint system along or around the excavation to lower the water table below the excavation bottom thus permitting the work to be done under relatively dry conditions.

A *wellpoint* is a perforated tube enclosed in a screen that is installed below the surface of the ground to collect water in order that the water may be removed from the ground. The essential parts of a wellpoint are illustrated in Fig. 20.11. The top of a wellpoint is attached to a vertical riser pipe. The riser extends a short distance above the ground surface, at which point it is connected to a larger pipe called a "header." The header pipe lies on the ground surface and serves as the feeder connecting multiple risers to the suction of a centrifugal pump. A wellpoint system may include a few or several hundred wellpoints, all connected to one or more headers and pumps.

The principle by which a wellpoint system operates is illustrated in Fig. 20.12. Figure 20.12a shows how a single point will lower the surface of the water table in the soil adjacent to the point. Figure 20.12b shows how several points, installed reasonably close together, lower the water table over an extended area. A group of wellpoints properly installed along a trench or around a foundation pit will lower the water table below the depth of excavation.

Wellpoints will operate satisfactorily if they are installed in a permeable soil such as sand or gravel. If they are installed in a less permeable soil, such as silt,

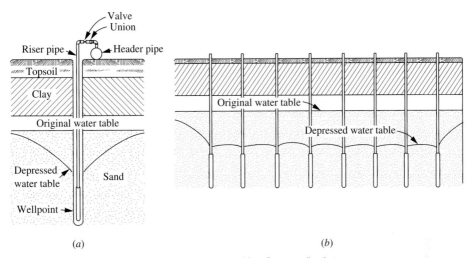

(a) (b)

FIGURE 20.12 | Water table drawdown resulting from wellpoints.

FIGURE 20.13 | Single-stage wellpoint installation.

it may be necessary first create a permeable well. A permeable well can be constructed by sinking, for each point, a large pipe, say 6 to 10 in. in diameter, removing the soil from inside the pipe, installing the wellpoint, filling the space inside the pipe with sand or fine gravel, and then withdrawing the large pipe. This leaves a volume of sand around each wellpoint to act as a water collector sump and a filter to increase the rate of flow for each point.

Wellpoints may be installed at any spacing, but usually the spacing varies from 2 to 5 ft (see Fig. 20.13) along the header. The maximum height that water can be lifted is about 20 ft. If it is necessary to lower the water table to a greater depth, one or more additional stages of wellpoints should be installed, each stage at a lower depth within the excavation.

Installing a Wellpoint System

If the soil conditions are suitable, a wellpoint is jetted into position by forcing water through an opening at the bottom of the point. After each point is jetted into position, it is connected through a pipe or a rubber hose to a header pipe. Header pipes are usually 6 to 10 in. in diameter. A valve is installed between each wellpoint and the header to regulate the flow of water. The header is connected to a self-priming centrifugal pump that is equipped with an auxiliary air pump to remove any air from the water before it enters the pump proper.

Capacity of a Wellpoint System

The capacity of a wellpoint system depends on the number of points installed, the permeability of the soil, and the amount of water present. An engineer who is experienced in this kind of work can perform tests that will provide data to make a reasonably accurate estimate concerning the capacity necessary to lower the water to the desired depth. The flow per wellpoint may vary from 3 or 4 gpm, in the case of fine to medium sands, to as much as 30 or more gpm for coarse sand. Figure 20.14 presents approximate flow rates to wellpoints in various soil formations.

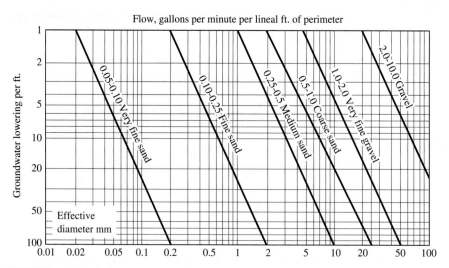

FIGURE 20.14 | Approximate flow through various soil formations to a line of wellpoints.

Source: Moretrench America Corporation.

Using the chart in Fig. 20.14 will aid in the selection of the size of pumps that should be used with a wellpoint system. As an example, consider that it is necessary to dewater a pit that is 15 ft deep and that the water table is 5 ft below the surface of the ground. It is known that the soils to be encountered are fine sands. Therefore, starting with a water lowering requirement of 10 ft (15, 5) on the left side of the chart proceed horizontally to the Fine Sand diagonal. From the intersection of the horizontal projection and the Fine Sand diagonal, project a vertical line down to the flow rate numbers on the bottom of the chart. Consequently a flow of 0.5 gpm per foot of header pipe could be expected for these conditions.

When excavating 45 to 50 ft below the surface of the water in the Colorado River for the cutoff wall for the Morelos Dam, the contractor installed three main stages, with a supplemental fourth stage of wellpoints to enclose an area of 15 acres. A total of 49 pumps serviced 2,750 wellpoints. The maximum pumping rate was 17,400 gpm, with 2,150 wellpoints in operation. This gave an average yield of 8.1 gpm per point and 528 gpm per pump.

DEEP WELLS

Another method for dewatering an excavation is the use of deep wells. Large-diameter deep wells are suitable for lowering the groundwater table at sites where

- The soil formation becomes more pervious with depth.
- The excavation penetrates or is underlain by sand or coarse granular soils.

In addition, there is a requirement that there be sufficient depth of pervious materials below the level to which the water table is to be lowered for adequate submergence of well screens and pump. The advantage of deep wells is that they can be installed outside the zone of construction operations as illustrated in Fig. 20.15.

FIGURE 20.15 I Deep wells used to dewater an excavation.

SUMMARY

Construction pumps must frequently perform under severe conditions, such as those resulting from variations in the pumping head or from handling water that is muddy, sandy, or trashy. The required rate of pumping may vary considerably during the duration of a construction project. The most satisfactory solution to the pumping problem may be a single all-purpose pump, or in other situations it may be better to use several types and sizes of pumps, to permit operational flexibility. The proper solution is to select the equipment that will take care of the pumping needs adequately at the lowest total cost. The total cost includes the installed and operating cost of the pump and the required hose or pipe for the period that they will be used.

In excavating below the surface of the ground, the constructor may encounter groundwater prior to reaching the bottom of an excavation. A common method for controlling groundwater is the installation of a wellpoint system along or around the excavation to lower the water table below the excavation bottom thus permitting the work to be done under relatively dry conditions. Another method for dewatering an excavation is the use of deep wells. Critical learning objectives would include:

- An understanding of the types of pumps available.
- An understanding of the effect altitude and temperature have on a pump's performance.
- An ability to determine pump performance from appropriate charts.
- An ability to calculate friction losses.
- An ability to select a suitable pump based upon project conditions and the quantity of water to be moved.

These objectives are the basis for the problems that follow.

PROBLEMS

20.1 A diaphragm pump is best suited for which type of application?
   a. Muddy water
   b. Slow seepage
   c. High-volume pumping

20.2 The maximum practical suction lift for a self-priming centrifugal pump is
   a. 35 ft
   b. 50 ft
   c. 25 ft

20.3 A suction hose sized larger than the pump size will:
   a. Increase capacity
   b. Decrease capacity
   c. Leak

20.4 A diaphragm pump is classified and operates as which type pump?
   a. Rotary
   b. Centrifugal
   c. Positive displacement

20.5 Pump engine speed affects:
   a. Pressure
   b. Volume
   c. Both

20.6 At higher elevations, above 3,000 ft, a pump

    a.   Engine runs better

    b.   Has reduced suction ability

    c.   Pressure is reduced

20.7 A two-cylinder duplex double-acting pump, size 6 × 12 in., is driven by a crankshaft, which makes 120 rpm. If the water slippage is 7%, how many gallons of water will the pump deliver per minute? If the total head is 100 ft and the efficiency of the pump is 60%, what is the minimum horsepower required to operate the pump? The weight of water is 8.33 lb per gal.

20.8 The centrifugal pump whose performance curves are given in Fig. 20.6 will be used to pump water against a total head of 50 ft. The dynamic suction lift will be 10 ft. Determine the capacity and efficiency of the pump and the horsepower required to operate the pump.

20.9 A centrifugal pump is to be used to pump all the water from a cofferdam whose dimensions are 70 ft long, 50 ft wide, and 12 ft deep. The water must be pumped against an average total head of 45 ft. The average height of the pump above the water will be 12 ft. If the cofferdam must be emptied in 15 hr, determine the minimum model self-priming pump class M to be used based on the ratings of the Contractors Pump Bureau. (349 gpm, Model 40-M, approximately 600 gpm)

20.10 Use Table 20.2 to select a centrifugal pump to handle 300 gpm of water. The water will be pumped from a pond through 460 ft of 6-in. pipe to a point 30 ft above the level of the pond, where it will be discharged into the air. The pump will be set 10 ft above the surface of the water in the pond. What is the designation of the pump selected?

20.11 Select a self-priming centrifugal pump to handle 600 gpm of water for the project illustrated in Fig. 20.9. Increase the height of the vertical pipe from 54 ft to 60 ft. All other conditions will be as shown in the figure. (Model 125-M, approximately 980 gpm)

20.12 Select a self-priming centrifugal pump to handle 300 gpm of water for the project illustrated in Fig. 20.8. Change the size of the pipe, fittings, and valves to 5 in.

REFERENCES

1. The Contractors Pump Bureau (CPB), www.cimanet.com/. CPB is a bureau of Construction Industry Manufacturers Association, 111 East Wisconsin Avenue, Milwaukee, WI 53202. The Contractors Pump Bureau provides services for manufacturers of pumps designed for the construction industry, as well as pump engine manufacturers. The specific purposes of the bureau are to develop and publish standards for contractor pumps and auxiliary equipment.

2. *Foundation Engineering,* edited by G. A. Leonards, McGraw-Hill, New York, 1962.

3. Schexnayder, Francis S., and Cliff J. Schexnayder, *Understanding Project Site Conditions,* Practice Periodical on Structural Design and Construction, ASCE, May 2001. Vol. 6, No. 2.

4. *Pump Handbook,* edited by Igor J. Karassik, William C. Krutzsch, Warren H. Fraser, and Joseph P. Messina, McGraw-Hill, New York, 1976.

5. *Selection Guidebook for Portable Dewatering Pumps,* Contractors Pump Bureau, P.O. Box 5858, Rockville, MD 20855.

A P P E N D I X

A

Alphabetical List of Units with Their SI Names and Conversion Factors

To convert from	to	Symbol	Multiply by
Acre (U.S. survey)	square meter	m^2	4.047×10^3
Acre-foot	cubic meter	m^3	1.233×10^3
Atmosphere (standard)	pascal	Pa	1.013×10^5
Board foot	cubic meter	m^3	$2.359 \div 10^3$
Degree Fahrenheit	Celsius degree	°C	$t_C = (t_F - 32)/1.8$
Degree Fahrenheit	Absolute	°A	$°A = (t_F + 459.67)$
Foot	meter	m	$3.048 \div 10$
Foot, square	square meter	m^2	$9.290 \div 10^2$
Foot, cubic	cubic meter	m^3	$2.831 \div 10^2$
Feet, cubic, per minute	cubic meters per second	m^3/s	$4.917 \div 10^4$
Feet per second	meters per second	m/s	$3.048 \div 10$
Foot-pound force	joule	J	1.355×1
Foot-pounds per minute	watt	W	$2.259 \div 10^2$
Foot-pounds per second	watt	W	1.355×1
Gallon (U.S. Liquid)	cubic meter	m^3	$3.785 \div 10^3$
Gallons per minute	cubic meters per second	m^3/s	$6.309 \div 10^5$
Horsepower (550 ft. lb/sec)	watt	W	7.457×10^2
Horsepower	kilowatt	kW	$7.457 \div 10$
Inch	meter	m	$2.540 \div 10^2$
Inch, square	square meter	m^2	$6.452 \div 10^4$
Inch, cubic	cubic meter	m^3	$1.639 \div 10^5$
Inch	millimeter	mm	2.540×10
Mile	meter	m	1.609×10^3
Mile	kilometer	km	1.609×1
Miles per hour	kilometers per hour	km/h	1.609×1
Miles per minute	meters per second	m/s	2.682×10
Pound	kilogram	kg	$4.534 \div 10$
Pounds per cubic yard	kilograms per cubic meter	kg/m^3	$5.933 \div 10$

To convert from	to	Symbol	Multiply by
Pounds per cubic foot	kilograms per cubic meter	kg/m³	1.602×10
Pounds per gallon (U.S.)	kilograms per cubic meter	kg/m³	1.198×10^2
Pounds per square foot	kilograms per square meter	kg/m²	4.882×1
Pounds per square inch (psi)	pascal	Pa	6.895×10^3
Ton (2,000 lb)	kilogram	kg	9.072×10^2
Ton (2,240 lb)	kilogram	kg	1.016×10^3
Ton (metric)	kilogram	kg	1.000×10^3
Tons (2,000 lb) per hour	kilograms per second	kg/s	$2.520 \div 10$
Yard, cubic	cubic meter	m³	$7.646 \div 10$
Yards, cubic, per hour	cubic meter per hour	m³/h	$7.646 \div 10$

Note: All SI symbols are expressed in lowercase letters except those that are used to designate a person, which are capitalized.

Sources: *Standard for Metric Practice,* ASTM E 380-76, IEEE 268-1976, American Society for Testing and Materials, 1916 Race Street, Philadelphia, PA 19103.

National Standard of Canada Metric Practice Guide, CAN-3-001-02-73/CSA Z 234.1-1973, Canadian Standards Association, 178 Rexdale Boulevard, Rexdale, Ontario, Canada M94 IRS.

B

Selected English-to-SI Conversion Factors

In general, the units appearing in this list do not appear in the list of SI units but they are used frequently, and it is probable that they will continue to be used by the construction industry. The units meter and liter may be spelled metre and litre. Both spellings are acceptable.

Multiply USC (English) unit	by	To obtain metric unit
Acre	0.4047	Hectare
Cubic foot	0.0283	Cubic meter
Foot-pound	0.1383	Kilogram-meter
Gallon (U.S.)	0.833	Imperial gallon
Gallon (U.S.)	3.785	Liters
Horsepower	1.014	Metric horsepower
Cubic inch	0.016	Liter
Square inch	6.452	Square centimeter
Miles per hour	1.610	Kilometers per hour
Ounce	28.350	Grams
Pounds per square inch	0.0689	Bars
Pounds per square inch	0.0703	Kilograms per square centimeter

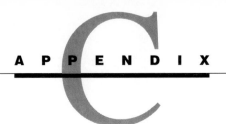

Selected U.S. Customary (English) Unit Equivalents

Unit	Equivalent
1 acre	43,560 square feet
1 atmosphere	14.7 lb per square inch
1 Btu	788 foot-pounds
1 Btu	0.000393 horsepower-hour
1 foot	12 inches
1 cubic foot	7.48 gallons liquid
1 square foot	144 square inches
1 gallon	231 cubic inches
1 gallon	4 quarts liquid
1 horsepower	550 foot-pounds per second
1 mile	5,280 feet
1 mile	1,760 yards
1 square mile	640 acres
1 pound	16 ounces avoirdupois
1 quart	32 fluid ounces
1 long ton	2,240 pounds
1 short ton	2,000 pounds

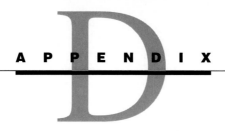

Selected Metric Unit Equivalents

Unit	Equivalent
1 centimeter	10 millimeters
1 square centimeter	100 square millimeters
1 hectare	10,000 square meters
1 kilogram	1,000 grams
1 liter	1,000 cubic centimeters
1 meter	100 centimeters
1 kilometer	1,000 meters
1 cubic meter	1,000 liters
1 square meter	10,000 square centimeters
1 square kilometer	100 hectares
1 kilogram per square meter	0.97 atmosphere
1 metric ton	1,000 kilograms

INDEX